Global City Typologies

Global City Typologies explores the historical, cultural and socio-economic transactional forces in the development of existing cities through to newly planned and emerging cities. Individual chapters address different sets and typologies of global cities to analyse their comparative evolution and standing today. The separate parts and individual chapters have been grouped around 125 different established, planned and emerging cities and analysed according to different typologies and thematic categories that comprise historic cities, early trading cities, planned cities, emerging global cities, mega cities and megalopolitan agglomerations. These span five continents – including the industrial cities of Chicago and Manchester, new capitals such as Brasília and New Delhi, innovative cities such as Singapore and Tel Aviv and mega cities such as Mexico City. The book is fully illustrated throughout with modern and historical maps, which enables visualisation of the forces that have shaped ongoing development of these major global cities. This is an essential book for students and professionals in urban design and planning, administrators, economists, designers and developers.

Nigel C. Lewis is a consulting engineer, planner and designer who has practised in the UK, and across western and central Europe, the mid-East, Africa, Asia and the USA. As part of multi-disciplinary design and building teams, he has been responsible for the planning, design and realisation of both private and publicly financed facilities and urban infrastructure.

Global City Typologies

Transactional Forces in Urbanised Development

Nigel C. Lewis

Routledge
Taylor & Francis Group
NEW YORK AND LONDON

Designed cover image: Nigel C. Lewis and S. Hare. Photos from iStock.

Row 1 – Vancouver, New York, London, Moscow, Tokyo; Row 2 – San Francisco, Miami, Algiers, Cairo, Beijing; Row 3 – Mexico, São Paulo, Nairobi, Delhi, Shanghai; Row 4 – Santiago, Buenos Aires, Cape Town, Jakarta, Sydney.

First published 2025
by Routledge
605 Third Avenue, New York, NY 10158

and by Routledge
4 Park Square, Milton Park, Abingdon, Oxon, OX14 4RN

Routledge is an imprint of the Taylor & Francis Group, an informa business

Library of Congress Cataloging-in-Publication Data
Names: Lewis, Nigel C., author.
Title: Global city typologies : transactional forces in urbanised development / Nigel C. Lewis.
Description: New York, NY : Routledge, 2025. | Includes bibliographical references and index.
Identifiers: LCCN 2024061061 (print) | LCCN 2024061062 (ebook) | ISBN 9781032404950 (hardback) | ISBN 9781032404967 (paperback) | ISBN 9781003353348 (ebook)
Subjects: LCSH: City planning—Cross cultural studies. | Cities and towns—Cross cultural studies.
Classification: LCC HT166 .L477 2025 (print) | LCC HT166 (ebook) | DDC 307.1/4—dc23/eng/20250212
LC record available at https://lccn.loc.gov/2024061061
LC ebook record available at https://lccn.loc.gov/2024061062

ISBN: 978-1-032-40495-0 (hbk)
ISBN: 978-1-032-40496-7 (pbk)
ISBN: 978-1-003-35334-8 (ebk)

DOI: 10.4324/9781003353348

Typeset in Berling
by Apex CoVantage, LLC

Access the Support Material: www.routledge.com/9781032404967

To my family – for Ada & Frankie, Sky & Levi

Contents

About the Author

Nigel C. Lewis, as engineer, planner and designer, initially trained at Arup, has been responsible for and has been practising internationally in the planning, design, development and project management of major urban infrastructure and transport projects worldwide.

He holds both a Master of City Planning (MCP) from the University of Pennsylvania and Master of Arts (MA Oxon) in Engineering Science & Economics, and also is a Chartered Engineer and Member of the Institution of Civil Engineers, Member of the Royal Town Planning Institute, and Member of the American Institute of Certified Planners.

He has practised in the UK and across western and central Europe, the mid-East, Africa, Asia and the USA. As part of multi-disciplinary design and building teams, he has been responsible for the planning, design and realisation of both private and publicly financed facilities and urban infrastructure.

He has also published *Design and Order: Perceptual experience of built form, Principles in the planning and making of Place* (Wiley Blackwell, 2020) and *Road Pricing: Theory and Practice* (Thomas Telford, 1992).

Figures

Note: All modern configuration maps are from SnazzyMaps unless otherwise indicated.

PART II

PART III

PART IV

PART V

PART VI

Tables

Preface

After graduation and professional qualification, I have had the privilege to have been employed in practice working on the realisation of numerous infrastructure and urban projects internationally in many countries. Having lived and worked in both Europe, the UK and the USA, I have undertaken assignments across Africa, Asia and the Americas, enabling me to interact with many different societies in different regional environments.

With increasing urbanisation and current climate change, the challenges are becoming significantly more urgent in order to provide adequately for larger populations in less well-resourced regions. Guided by the ***'C-40 Cities'*** group, a network of nearly 100 mayors of the world's leading cities, these cities are seeking united action to confront the climate crisis. Their mission is to undertake accelerating actions to overcome climate change whilst building equitable and thriving communities globally.

From my work experience, I have selected sets of around 125 cities for analysis (80% of which I have had the opportunity to work in or visit) and being able to experience the achievements and challenges of these evolving cities which I have described under different typologies. In particular, besides different continents, I have aimed to focus on the Global South, not just 'mega-cities' but also 'megalopolitan agglomerations' that are emerging as peripheral areas are being extended yet further outwards.

I should like to thank my work colleagues and mentors, friends, family and all of those people whom I have had the benefit to interact with, who have specially inspired me. This has enabled me to better understand the different transactional forces and challenges in cities and countries globally where increasingly over 50% of the population live in urbanised environments, many now being affected negatively by climate change.

Part I

Introduction and Urban Theory

1.1 Introduction and Urban Theory

Settings and origins – Established, planned and emerging cities – City typologies – Urban growth theories – Location and place – Transactional forces – Ecological processes – Catalysts for change – Public infrastructure and private places – Urban expansion – Poly-nodal development

1.1.1 INTRODUCTION – CONTEXT AND APPROACH

In this millennium we are in uncharted territory, with high rates of population growth and urbanisation leading to increased concentration and densification of people in expanding centres. With the centripetal pull of these centres, there are challenging demands for both physical and social infrastructure to support the growing needs of each generation of urban inhabitants.

Consequently, these challenges impact the natural environment and settings in supporting the human population and their transactional requirements for habitation, employment, welfare and recreation. Such 'affordances' differ in different regions of the world.

With their given topography, urban footprints are determined essentially by both the settled built form and the inter-weaving arteries that serve to support and move population and services within the urban fabric. With the pressures of space, this forces development either longitudinally or vertically depending on the available resources, which generates the corresponding urban morphology.

The resultant built 'place' has an ambience that may be liveable and workable for inhabitants, or else may suffer from congestion, contagion or pollution. The resultant 'infrastructure' networks may be performant and afford mobility, or be inadequate, lack capacity and become gridlocked. Such levels of quality of life are determined by these factors.

Given this, the goal is to address different sets of typologies of ***established, planned and emerging global cities*** and to analyse their comparative evolution and current standing today.

This includes assessment of their:

- *Origins* and *natural settings* and topography,
- Early *human settlements* and economic activities, and the historical and political forces that may have impacted their development,
- Basic *built form*, footprint, landmarks and places that have taken root,
- Communication and *transport networks* that have determined the resultant urban morphology and configurations that currently exist today, and
- *Human population*, mobility, density and distribution patterns, and their settlement and administration.

This approach includes exploring the ***transactional forces*** that have led to the evolution and morphology of each city and its urbanised area, whether it is contained and preserved, as say Venice, extended and expanded as Los Angeles, managed and organised as Singapore, pressurised and sprawled as Cairo and Kolkata, Mexico and Mumbai.

Globalisation, urbanisation and agglomeration are all dominant forces – yet these are pushing against the constraints of available natural resources, air and water quality, contagion, climate change and species extinction. A better understanding and awareness of these forces and 'place-boundedness' is needed to lead to future improved policy formulation and governance at all levels.

Human settlements have evolved over time, buffeted by natural, socio-political and technological changes. Their layout and organisation have shaped the morphology of cities and their evolution through the ages. Such places act not only as local activity or behaviour settings, but also as links or paths of communication in larger inter-connected networks. These are the two distinctive traits that serve to

DOI: 10.4324/9781003353348-2

underline the approach undertaken in analysing both the individual *built form* of cities and also their key *arteries* that form routes of communication within and between other settled areas.

1.1.2 SUSTAINABILITY

In global terms, we are now in the Anthropocene Age succeeding the Holocene Age, where the rate of change is rapidly exceeding that which we have experienced previously. Since the Industrial Revolution and advancement of production using new technologies, in the 21st century, we have witnessed the 'Great Acceleration', with unprecedented growth in human population and urbanisation, most of which has been concentrated in areas of the 'Global South'.

Accompanying this has been a significant increase in consumption that has included limited elements of natural resources such as forests and fish. With the extraction and burning of fossil fuels for energy utilisation, negative outcomes have resulted with an increase in carbon dioxide emissions that has impacted the biosphere and triggered tipping points that are leading to global warming and climate change.

This is the current reality whereby unbalanced growth and over-consumption has led to excess emissions and climate tipping events that are destabilising different regions around the world. From the poles to the equator, for oceans and islands, forests and deserts, the landscape is suffering a distinctive climate and ecological crisis that is now impacting the sustainability of entire rural, built and urbanised areas.

1.1.3 CITY TYPOLOGIES AND HUMAN ACTIVITIES

Cities of today and *city typologies* can be addressed under different categories (see Box 1.1 – City Typologies), that is:

- Established natural cities,
- Planned cities, and
- Emerging global cities and agglomerations.

City form determinants are generated from different causations (see Box 1.2 – City Form Determinants) due to their:

- Natural setting and topography,
- Human settlements, and
- Socio-economic activities.

From their *natural settings* emanate the resulting *built form* of cities and their respective patterns of development and settlement.

Human activity generates a series of *transactional forces* of development (see Box 1.3 – Transactional Forces) which are derived from the essential basic needs of survival and sustenance and through activities of:

- Agriculture and fishing, animal husbandry, or forestry and mining,
- Crafts and weaving, manufacturing and building,
- Protection and defence, and
- Worship and religion, organisation and administration.

These in turn lead to collective transactions in terms of *trading* with the movement of people, goods and services between different groups and communities and different specialisations. From established settlements, the need for access and communication between communities is generated either via land or sea, in order to facilitate such exchanges.

Hence emphasis is not only on the development of built forms themselves, but also on the network of byways and arteries in order to facilitate both travel and trade. With the advancing stages of development and increasing population and densities, we have reached the current situation whereby often up to half of a modern large city's footprint may be dedicated to streets and highways for movement and storage of vehicles for individuals' mobility.

The method of enquiry adopted is to seek to determine the key determinants of urban form and city typologies from original driving forces to their subsequent arrangement of having generally *privately* financed built form and *publicly* financed urban infrastructure and networks. Both of these, although at times contradictory, aim to secure a reasonable sense of place with efficient mobility for the widest mix of peoples.

1.1.4 URBAN GROWTH THEORIES

In terms of spatial structure, there are a number of different theories concerning the development and *configuration* of urban form. These include:

- *Radial axes*, along transportation spokes and corridors, as outlined by Richard Hurd (1903) in addressing city land values.
- *Concentric zones*, where separate zones of social groups radiate out from a central commercial district, as posited by Ernest Burgess (1923) in Chicago.

BOX 1.1 – CITY TYPOLOGIES

The typology of cities is addressed in accordance with their given settings, form determinants and forces that have shaped the development of their built form.

City typologies can be classified under three main categories:

- Established natural cities,
- Planned cities, and
- Emerging global cities, as well as urban agglomerations.

City form determinants result from their:

- Natural setting and topography,
- Region and climate,
- Human settlements, and
- Socio-economic activities.

Natural settings are typically:

- Coastal, estuary or river,
- Lake, valley or island,
- Hill or mountain,
- Plain, forest or savanna, etc.

Resultant *built form* of cities may be:

- Organic,
- Irregular,
- Centralised,
- Polycentric.

With types of *settlement* that are:

- Nuclear or concentric,
- Sectoral or axial,
- Grid or orthogonal,
- Core-satellite,
- Cluster or hybrid,
- Edge or peripheral.

And *Arteries* and *networks* which are:

- Axial,
- Orthogonal,
- Tree and branch, etc.

BOX 1.2 – CITY FORM DETERMINANTS

The method of enquiry into 'city typologies' has been influenced by the prescient bibliography outlined by **Louis Wirth** in Park and Burgess's urban compilation of *The City* as published at the time of the industrialising cauldron of Chicago in the 1920s, as a tentative scheme of classification for the study of cities. This classifies city form and type principally through consideration of the attributes of:

- City definition – geographic, historic, economic, sociological, cultural, ecological.
- City history – ancient, mediaeval, modern.
- City type – topographical, functional, transactional, structural (planned/unplanned).
- City hinterland – trading (specialisation), commuting, administrative, satellite, peripheral/edge.
- City organisation – natural, ecological, neighbourhood, community, zoning.
- City networks and facilities – utilities, communications, streets, welfare, social, recreational, cultural, civic.
- City growth – expansion, distribution, migration, mobility.
- City demographics – age, racial cohort, density.
- City life – labour and economic activity, social organisation, communities.
- City places – urban, suburban, rural.

In addition, different types of layout are summarised, having increasing levels of complexity, for example:

- Linear – longitudinal or with extended ribbon development.
- Recti-linear.
- Square or quadrilateral and internally subdivided orthogonally with a grid.
- Circular, with a central symbolic feature.
- Curvi-linear, with curved or serpentine lines.
- Radio-centric with radial and circular paths.
- Irregular or star shaped.
- Hybrid.

A.E.J. Morris, in *History of Urban Form before the Industrial Revolution*, defines the set of essential 'urban form determinants' as:

- Natural location – topography, climate and natural resources.

Man-made:

- Primary forces – religion, defense, trade and administration,
- Secondary forces – manufacturing and production, consumption,
- Tertiary forces – distribution and servicing, and
- Digitalisation and information processing.

BOX 1.3 – TRANSACTIONAL FORCES

The development and globalisation of cities has resulted from a number of transactional forces, as a result of their given settings and paths of communication.

Settings **–** From their natural settings and given surrounding geographical environment, humans made settlements in the most advantageous locations. Whether coastal or estuarine, island or hilltop, groups have settled collectively to benefit from protected, strategic or advantageous locations. With early fishing, livestock or agricultural activities for subsistence, any surplus was available for exchange or barter.

Trading **–** Markets developed and trading occurred, which set the scene for the beginnings of wider trading patterns between settled places in a region and then developed further beyond. This led to routes of communication across land and trading routes along rivers and across seas, which led to the cross-fertilisation of goods and peoples. Even long maritime routes were developed for valued commodities as consumables, crafted products or minerals that were shaped resulting in added value to the items.

Siting **–** Such movements resulted in the siting of built form, the accumulation of buildings in centralised areas and the gravitation of people around marketplaces. As these economic activities increased, the densification of settlements occurred, with the need for provision of basic services, water and waste, shelter and accommodation.

Transport **–** These settlements led to the development of and need for supporting networks of paths and roads in order to transport people and goods within and between centres. The development of technology assisted with this, from the transformation of the marine sector from sail to steam ships, and the railway sector and highway sector to the prevalence of urban rail systems and automobiles. From walking to horse drawn and powered transport, the range and also the speed of transport has increased enormously, creating the need for enhanced networks, hubs and transfer stations between modes. This enhanced ground mobility has created its own challenges of congestion, pollution and parking of individual vehicles when not in motion. To this has been added the challenges of transport by air and aviation.

Migration **–** Such networks and facilities have created both push and pull forces of attractiveness, of settlements yet more populated as new livelihoods are sought, opportunities created and that have a comparative advantage over less well-endowed areas. With shortages of resources, patterns of migration develop, usually of the most able or those with least dependencies, and those that have benefitted from education or training to secure formal jobs. Those less fortunate can only join the growing informal sector in seeking a basic means of living and ending in a path of poverty. Unable to afford decent housing, their mode of dwelling is insecure and impermanent. These forces often result in a vicious circle of despondency with no or only limited access to basic services.

Real estate **–** In contrast, the formal sector has developed markets in the acquisition and trading of real estate, and in concert with the public sector in the purchase and development of land so as to increase its purchase or rental value through creative financing and borrowing mechanisms. As a result of favourable zoning and regulation, corporate bodies have been able to leverage their financing so as to create valuable property portfolios that are also traded in the market. Hence, these forces result in the realisation and shaping of basic built form to be used for residential, commercial or industrial purposes.

Information services **–** Further transformations have occurred with the development of the digital economy, innovation of information services and new technologies that have enabled companies to operate transnationally and globally. With immediate electronic transactions, these forces have resulted in the 'death of distance' and the ability of corporations to be more nimble in creating and responding to different markets, so impacting the location and employment of different sources of labour.

- *Sectors*, where different sectoral zones result due to topography or access routes, having different rent patterns, as outlined by Homer Hoyt (1936) in America.
- *Multiple nuclei*, where development tends to occur and cluster around several discrete nuclei having similar or distinct activities, as described by Chauncy Harris and Edward Ullman (1945).
- *Inverse concentric zones*, where the central zone may be purely symbolic or administrative, as in some traditional cities, with outlying districts of development.

Traditionally towns and cities tended to originate along river valleys and places with natural harbours for trading. With agricultural surplus, mercantilism and trade developed with the exchange of goods. As transportation routes improved, further clustering occurred and concentration of peoples resulted. With specialisation, and availability of natural resources, new economic activities developed, leading to the making and manufacturing of local goods.

Another way of looking at settlement is in terms of *place*:

- *Central places*, that facilitate the exchange of goods and services, as service centres for adjoining areas as posited by Walter Christaller (1933) in Southern Germany, leading to a series of tiered cities within a net of inter-connected hexagonal market areas.
- *Primate cities*, defined by geographer Mark Jefferson (1939), that are at least twice as large in size as the next largest national city and more than twice as significant.
- *Edge cities*, the new urban frontier as defined by Joel Garreau (1991), urban centres that contain all of the functions of a city but coalesced and agglomerated at peripheral locations having extensive communications.

In terms of *industrialisation*, economies are generally classified as tending to migrate from primary agricultural activities, through secondary manufacturing to tertiary service activities. Each of these activities has a spatial dimension, having impacts of either territorial concentration or dispersal.

Having a set of basic functions supported by non-basic functions, the transitional economic base of a city can grow, leading to the development of a mix of residential, industrial and commercial areas. The transport networks then have strategic importance in being developed so as to act as arteries to service both internal urban mobility and external movement between regions.

1.1.5 LOCATION, PLACE-BOUNDEDNESS AND CHARACTERISTICS

With the successive transformation of the service sector through to the information economy, this has had an impact on businesses, that on the one hand requires a 'central place' for financial and business services, and yet on the other hand more dispersed distribution of productive places to cities or regions that have a comparative economic advantage.

This leads via various path dependencies to a new form of configuration of businesses, with some functions 'place-bound' and other transactions for which distance is neutralised through the introduction of modern information technology.

Such locational patterns develop, dependent on the availability of resources and labour, and with the creation of *transnational* companies. This in turn leads to *globalisation*, where due to the importance of place, clustered centres or regions of agglomeration occur.

With increased *urbanisation* from a human perspective, a sense or 'spirit of place' – 'platzgeist' – however is also most important, such that inhabitants can grow through a process of self-actualisation in a city that provides an appropriate balance of work, living and dwelling for all stages of the life cycle.

This leads to the requirement for more fundamental principles of building, of sustainability, re-useability and 'well-building' that takes into account resource preservation, energy conservation, air, noise and water quality management and well-being and mindfulness.

Dependent on a place's attractiveness, push-pull factors develop that lead to further migration of population, with increasing industrialisation and urbanisation and migration to peri-urban 'edge cities'. In seeking to benefit from relative economic comparative advantage, the industrial manufacture and distribution of services becomes transnational and leads to further increased urbanisation.

1.1.6 TRANSACTIONAL FORCES AND ECOLOGICAL PROCESSES

The main driver of development is the *transactional forces* that determine the functional outcome and morphology of urbanised areas. Such typological and thematic categories result, for example, in:

- Market and trading centres, industrial and manufacturing centres.
- Ports and terminals, service centres, information and IT centres and innovation hubs.
- Regional clusters, etc.

This is the situation that we have reached today, where with a world population of around 8 billion people, over one half of us live in densely urbanised areas. Our challenge is to manage this going forward in a sustainable, supportable and smarter manner for future generations across the globe, failing which communities will be alienated and left to struggle in the informal sector.

R.D. McKenzie, in 'The Scope of Human Ecology', outlined the evolution of urban areas essentially in terms of a sequence of *ecological processes*, for example:

- *Concentration*, which leads to different distribution of population in residential areas,
- *Centralisation*, which leads to the focusing of functions around pivotal points of activity,
- *Decentralisation*, the converse, which as a result of competition and mobility, leads to settlement away from the centre,
- *Specialisation*, leading to the separation of functions and uses in urban areas into more homogeneous areas,
- *Segregation*, by contrast, results in the distinguishing or separation of uses and functions and
- *Succession* as the process whereby land uses or settlements evolve, either by natural evolution or forced invasion.

In consequence with migration from rural to urban areas, increasing industrialisation has led to high rates of urbanisation. But as eventually transformations of patterns of economic activity have developed, *de-industrialisation* and aging has also taken place, leading in some areas to the marginalisation of communities and the creation of extensive peripheral urban areas and informal activities. This has created its own sub-class of groups of people.

1.1.7 CATALYSTS FOR CHANGE

Accordingly, cities have grown around the world as a result of a multiplicity of different factors that are determined by their natural physical setting, patterns of human activity and settlement, and transactional and economic forces. One of the prevailing forces is that power of 'proximity', the latent strength of cities as described by Edward Glaeser that enables cities to build human networks, facilitate the communication of ideas, innovate and lead to a reduction in transaction costs.

Whilst individual cities, national capitals or cities evolved historically due to their location on, say, the Silk Route, or within the Hanseatic League or the Holy Roman Empire, other trading and political forces have led to significant changes in their development.

Kent Calder has outlined how, in the political arena, transformations have occurred due to recent changes in the triad of information and communications technology, financial transformations and geo-political changes. With increased connectivity, and the rising power of markets and increased civic autonomy, these have each contributed to institutional changes in the 'penumbra of power' of government institutions, with the rising prominence of political forums and intensification of grassroots activism under local leadership, which have all been pivotal in these changes.

All of these *institutional forces* have led to a deepened inter-dependence between regions and cities with multiple series of interconnected networks.

1.1.8 PUBLIC INFRASTRUCTURE AND PRIVATE PLACES

As a result of the different transactional forces, we have seen urban areas grow and be transformed into cities with dense cores and suburban and peripheral areas. These areas have further coalesced variously into 'conurbations', as labelled by Patrick Geddes. Further aggregation has led to 'agglomerations' but also 'edge cities', and marginal peripheral areas have developed.

In many areas a concomitant loss of a 'sense of place' has developed, resulting in a 'non-place urban realm', as described by Melvin Webber. Christopher Alexander also posits that 'a City is not a Tree' but rather a semi-lattice of linkages.

Notwithstanding this, the *public* sector generally has provided the essential infrastructure, street and utilities to make cities work and to provide services in a sanitary manner. The actual built form and morphology, beyond monuments and collective social facilities, by contrast, has been predominantly the domain of the *private* sector in designing, financing and realising residential, commercial and industrial facilities for its own pecuniary benefit.

Whilst regulated by public administrations, this has generally been the pattern of development under differing typologies, and it is these factors which will be explored in different cities and regions around the world in the following chapters.

1.1.9 URBAN EXPANSION PATTERNS

With increasing rates of growth of population and of urbanisation, there are different patterns of urban development around the world, typified by distinctive strategies in different regions.

In the *Global South*, urbanised areas are generally expanding at new locations outside and beyond existing settlements, with informal settlements extenuating at the periphery (see Box 1.4 – Global South). But whilst urban areas are growing and extending, their overall density is not increasing.

In older established cities having already dense urbanised cores, residences and workplaces are becoming decentralised, with new growth centres also developing at outer areas that serve as individual centres of attraction.

As a result, both of these trends are leading to the settlement and development of resources at the periphery, resulting in the development of 'poly-nodal centres' rather than concentrated mono-centric structures as formerly.

The net effect of these patterns, however, is that whilst the size of urbanised built-up areas is increasing with additional population, the result is that average urban densities are not increasing as measured by the number of persons per square kilometre.

The forces that are driving urban expansion of population growth and urbanisation rates, of improved transport, cheap land and income inequality, are all leading to extending built-up areas of urbanised land at lower densities, which is resulting in more scattered and less compact development of land resources. Hence, this dispersal of land occupation is leading to increased urbanised land cover and more fragmented landscapes, requiring continuing investment in additional infrastructure services (see Box 1.5 – Determining Forces in Population Settlements).

1.1.10 DIGITAL ECONOMY AND POLY-NODAL DEVELOPMENT

As transactional forces have evolved from former trade and trading relationships, and urbanisation has developed with the growth of population and migration from outer areas, de-industrialisation and globalisation have impacted the pattern of urban settlements. As their morphology has evolved, the impact of the digital economy is breaking down the spatial and temporal barriers of work, residence and leisure.

This is emphasising the breakdown into poly-nodal centres, where the benefits of proximity, density and diversity can thrive. However, the forces of agglomeration remain and with a focus on the knowledge economy and non-trade services such that a new type and structure of communities is emerging. These require supporting transport networks and utilities to enable them to flourish in a sustainable manner under an increasingly challenged climatic environment.

Accordingly, the following chapters will demonstrate the evolution of established historic and early trading cities through planned and capital cities and their typologies, to the emergent global, innovative and transnational cities of today and the resultant mega-cities and megalopolitan agglomerations on each continent, especially Asia.

BOX 1.4 – GLOBAL SOUTH

In addressing different city typologies around the world, it is useful to distinguish between 'natural' and 'planned' cities, due to their origins as being either more natural and organic or artificial and man-made.

In a similar vein, a distinction is being made between cities in the most 'developed' parts of the world and those that are in 'emerging' regions that are generally considered as being part of the 'Global South'. This nomenclature that is now quite well founded encompasses most of the cities in sub-Saharan Africa, South Asia and South America, as compared with the more northern, traditional and industrialised cities in Europe and North America.

The term came into common parlance subsequent to the UN Commission *Brandt Report – 'North South: A Programme for Survival'* that was published in 1980, concerning international development. Due to their particular population challenges and of urbanisation and globalisation, these countries have targeted more recently a set of Sustainable Development Goals, as formulated by the UN General Assembly Resolution in 2015*.

In consequence, under these circumstances, it is practical to address countries and cities having these common developing challenges together in the emerging 'mega-cities' under the 'Global South'. It should also be recognised that approximately one half of those people who live in urbanised areas actually settle in 'informal peri-urban suburbs'. These are heterogeneous areas with many different types of land ownership, settlement and governance. In consequence, such emerging Asian cities typically show characteristics of binary opposites, that is, having dense central areas but also extensive informal peripheral areas, leading to poly-nodal peripheral development.

'Desa-kota'

A further phenomenon in the Global South that has been developing, especially in the urbanising areas of South-East Asia, of this intensive sprawl in peri-urban areas, is a high density of population that undertakes essentially intensive agricultural cultivation in restricted pockets of urban areas.

Coined by Terry McGee using the Indonesian terms 'desa' (village) and 'kota' (city), this co-existence of settlements and different forms of land use is typically found in the agglomerations around the Jakarta Capital Region, or 'Jabodetabek' (Jakarta – Dapok – Tangerang – Bekasi), an area of approximately 30 million people.

Typically sprawling along road corridors, unregulated, there is an intensive mix of occupations, erratic and regularly changing, that is a source of income for many urban poor. Such spontaneous areas evolve continuously as the participation in this informal sector unfolds and activities spin off each other, seeking earnings either off the land or from remnants of other processing activities, often having high rates of female participation.

* UN Sustainable Development Goals (17) – No poverty. No hunger. Good health. Quality education. Gender equality. Clean water. Affordable energy. Decent work. Sustainable industry. Reduced inequality. Sustainable communities. Responsible consumption and production. Climate action. Marine life. Biodiversity. Peace and justice. Partnerships.

BOX 1.5 – DETERMINING FORCES IN POPULATION SETTLEMENTS

Michael Batty, in *Inventing Future Cities*, emphasises that city form and structure are a result of multiple interactions of forces and decision-making processes by both public and private actors. Neighbourhoods and networks evolve, with 'wholes' developing from the series of 'parts' that emerge and evolve in different patterns over time. This is as a result of a set of principles:

Firstly, due to the law of size distributions – G. Zipf's power law, whereby objects are ordered regularly from the one largest to many smaller objects. That is, where cities are ranked in decreasing population size, then the rank of a city will be inversely proportional to its population. Cities that exist as part of a hierarchy of settlements accordingly tend to merge into one another as clusters are bound together through networks into a predominantly whole urban realm.

Secondly, under the paradox of the modern metropolis cited by E. Glaeser, that 'proximity' is pivotal to the success of modern cities in order for interactions to be maximised between people. Most significantly, the deterrent effect of physical distance becomes less important as increasingly the digital economy is energised by the use of instantaneous information via the computing 'cloud' in the ether.

Then thirdly, according to the basic economic model of J.H. von Thunen and as elaborated further by W Alonso in his general theory of land rent, that cities evolve around their central business districts. As settlements develop around the original market, rents increase concentrically as they approach the centre. In addition, this is supported by the assumption as outlined by H.G. Wells that the distribution of population is wholly dependent on the means of transportation that is available.

Finally, under the law of geography as cited by W. Tobler that nearer locations insert a greater pull than further locations (viz. the city and its hinterland). This has been recognised and summarised by F. Cairncross in the 'Death of Distance' in that it is a result of the modern revolution in information technology, as already cited.

In consequence, there are three measurable criteria that are considered pivotal in assessing city form, scale and mobility:

- Interactions – in terms of daily traffic flows (volume and speed of individual and commodity movements),
- Density – in terms of average persons inhabiting each square kilometre and built floor space (square metres) per land area, and
- Proximity – siting or contiguity of individuals, facilities and networks.

As clusters of development occur, *interactions* are observable between nodes and links by volume and/or vector. They are also visible in the third dimension, vertically, insofar as built form is developed in denser blocks with higher stories. Similarly, *density* of form and occupation is measurable. These determine as an outcome the effective levels of *proximity* between places and persons.

Finally, it is noted that with respect to the growth of businesses, the selected location of a business can lead to both economies of scale and also agglomeration economies. More specifically, *localisation economies* may result from clustering and increased concentration, and *urbanisation economies* may result from increased specialisation and diversification.

1 Zipf, George – *Human Behaviour and the Principles of Least Effort*, Cambridge, MA, 1949.
2 Glaeser, Edward – *Triumph of the City*, New York, Penguin, 2012.
3 von Thunen, Johann – *Isolated State*, 1825 (Trans. 1966).
Wells, H.G. – *The Probable Diffusion of Great Cities*, 1902.
Alonso, William – *Location and Land Use, Toward a General Theory of Land Rent*, Harvard University Press, 1964.
4 Tobler, Waldo – Computer Simulating Urban Growth in the Detroit Region, *Economic Geography* 46, 1970.
5 Cairncross, Francis – *The Death of Distance*, Harvard Business Review, 2001.

Further Reading

Alexander, Christopher – *A City Is Not a Tree*, Design, Feb. 1966.

Angel, Shlomo – *Planet of Cities*, Lincoln Institute of Land Policy, 2012.

Batty, Michael – *Inventing Future Cities*, MIT Press, 2018.

Burgess, Ernest – The Growth of a City, *Proceedings of the American Sociological Society*, 18, 1923.

Burgess, Ernest – *The Urban Community*, University of Chicago Press, 1926.

Calder, Kent – *Global Political Institutions*, The Brookings Institution, 2021.

Christaller, Walter – *Central Places in Southern Germany*, Prentice Hall, 1933.

Garreau, Joel – *Edge City*, Doubleday, 1991.

Geddes, Patrick – *Cities in Evolution*, Williams and Norgate, 1915.

Glaeser, Edward – *Triumph of the City*, Penguin Press, 2011.

Harris, Chauncy & Ullman, Edward – The Nature of Cities, *Annals of the American Academy of Political and Social Science*, 242, Nov. 1945.

Hoyt, Homer – *The Structure and Growth of Residential Neighborhoods in American Cities*, Federal Housing Administration, 1939.

Hurd, Richard – *The Principles of City Land Values*, Record and Guide, 1903.

Jefferson, Mark – The Law of Primate Cities, *Geographical Review*, 29, Apr. 1939.

McKenzie, RD – Human Ecology, *Encyclopaedia of the Social Sciences*, 314, 1931.

Morris, AEJ – *History of Urban Form before the Industrial Revolutions*, Pearson Education, 1994.

Park, Robert & Burgess, Ernest – *The City*, University of Chicago Press, 1925. (Bibliography – L Wirth).

Seabrook, Jeremy – *Cities*, Pluto Press, 2007.

Webber, Melvin – *The Urban Place and the Non-Urban Place Realm, in Explorations into Urban Structure*, University of Pennsylvania, 1964.

Part II

Established Historic Cities

2.0 Introduction – Historic Cities

Through their fortunes of location and geo-physical forces, certain settings have found themselves to be in a pivotal position geographically, which with their natural resources have served to become the point of attraction for the gathering of local peoples. By having either a commanding position by a natural water inlet or an elevated hillside site, such natural features afford protection that is suited for human settlement. With such inherent advantages, these areas tended to develop into places where people could assemble and then trade with each other. Gradually their networks extended into the hinterland until they formed an established presence in the region.

This is how physical development fundamentally occurred, albeit with a greater complexity of inputs and outputs of both human and natural resources. With development of a core, as its attractiveness grows and as a result of the push-pull of different forces, built forms and arteries of movement are created. This is the making of a new 'centre', a man-made place in a natural setting.

Earlier civilisations developed in such a manner, benefitting from the found natural resources in the vicinity or from land that was readily able to be used for arable purposes or for grazing for pastural purposes. Such was the beginning of primary agricultural activity for food and subsistence.

The gathering and coalescing of tribes, based on their natural instinct for survival, brought their own tensions for families, clans and their leaders, who were driven by their particular needs and aspirations. Dependent on their resources and ingenuity, larger associations developed in the region until they were able to organise themselves under defined rules, pagan or religious, that afforded some stability to their ways of life and livelihood.

In such a manner, the groupings of peoples developed until their critical mass enabled them to fully inhabit, overcome and populate adjoining regions. Such outreach was to lead to the creation of leaders who, with the assembly of supporting forces, were able to develop the makings of empires and their subsequent dynasties.

SETTLEMENTS AND CIVILISATIONS

In the Mesolithic period humans started to transition to a more settled way of life, and farming began in earnest during the Neolithic period. This included in the fertile crescent of Mesopotamia, in China and in Central America, and in Europe along the coastlines. In Africa pastoralism prevailed; hunter-gatherers gravitated around estuaries and inland rivers, and animal husbandry expanded. But sedentism was to lead to concepts of territorialism and land ownership, as farming was extended on fertile areas, which was the catalyst for the construction of shelter and permanent buildings using local materials.

Communities began to settle along the major rivers, particularly around the River Nile and Nubia in Egypt – also the Yangtze and Yellow rivers in China under the Shang dynasty, and the Tigris and Euphrates rivers in Mesopotamia, Sumer and Assyria, and the River Indus and Gujarat.

Similarly in Europe, settlement developed along such major rivers as the River Danube. In Meso-America, settlements were in different geographies in more isolated mountainous areas. Earlier settlements developed along the Pacific coast of the Mexican trans-isthmian region, and highlands as the Peruvian Andes Cordillera.

As people started to become settled around protected settings, their activities developed from purely agricultural to more creative activities. So too their settlements became more concentrated with social groupings and

DOI: 10.4324/9781003353348-4

family occupational specialisations, such as artisans, farming, trading, etc. As this led to increased productivity, with the manufacture of tools and domestic craft production, any surplus was available for trading.

With further skills and development of scripts and writing, new methods of recording activities facilitated trade and correspondence and more specialised functions around which groups of people gravitated and settled. This clustering was the beginning of the emergence of large towns and cities, with larger areas having common goals that attracted further migration of peoples (see Support Material: Annexe 2.0A – Early Urban Civilisations).

HISTORIC CITY TYPES AND SIZES

The earliest groups of recognisable towns or cities that have been found in different regions are under different civilisations: Babylonia and Assyria in Mesopotamia, Thebes and Nubia in the Egyptian Nile Valley, Harappa and Mohenjo-daro in the Indus Valley, Zhengzhou and Anyang in China, Minoan Knossos and Mycenae in Greece, and Monte Alban in the Oaxaca Valley of Mexico and Chavin de Huantar in Peru.

Each of these has been endowed with extraordinary archaeology and history, each dependent on its own particular natural setting during different civilisations. As either a walled city (Zhengzhou), mountain site (Monte Alban), fortified citadel (Mycenae), on a flood plain or riverbank (Thebes), their configuration and form was determined by their particular geographical location and natural setting.

Accordingly, this section addresses those ancient largest established cities that have evolved to the present day from favourable natural settings whose beginnings were either imperial, theistic or ecclesiastic in nature, and that have significant historic importance. Historically cities have been realised of different types, essentially for defence or protection, religious or political purposes, agricultural trade or mercantile purposes, manufacturing and services and distribution.

In terms of size, the earliest places that could be categorised as proto-cities were initially over 10,000, as Uruk in Mesopotamia, then areas that had populations of up to 100,000, as Thebes and Babylon. In the next stage period BCE these increased to 200,000 at Memphis and Alexandria. They then rose significantly to 500,000 at Chang'an and Rome, followed by Constantinople, Carthage and Cordoba. The only cities that reached 1 million were Luoyang, Alexandria, Rome and Baghdad in the period from 0 to 1000 CE (see Support Material: Annexe 2.0B & C – Historic Largest Cities).

Accordingly, this **Part II** addresses the evolution of established historic cities, their origins, setting and purposes, as outlined (see Table 2.0.1 – Established Historic City Types).

Table 2.0.1
Established Historic City Types

		City	Topography	Plan	Trans. Forces	Type
2.01	**ESTABLISHED CITIES**					
2.1	***Imperial***					
		Alexandria	Coastal	Linear	Maritime	Egyptian
		Athens	Hills	Organic	Imp. Capital	Classical
		Rome	Riverine/hill	Axial	Imp. Capital	Classical
		Constantinople	Coastal	Dual	Imp. Capital	Euro-Asian
		Isfahan	Riverine	Axial	Prov. Capital	Persian
2.2	***Theistic***					
		Jerusalem	Hill	Divided	Religious	Cultural
		Makkah	Desert	Core	Religious	Cultural
		Amritsar	Plain	Radial	Religious	Cultural
		Benares/Varanasi	Riverine	Organic	Religious	Indian
2.3	***Former Capital***					
		Chang'an/Luoyang	Riverine	Recti-linear	Capital	Cultural
		Nanking	Riverine	Sectoral	Capital	Cultural
		Peking	Plain	Radial-grid	Imp. Capital	Global
		Kyoto	River valley	Recti-linear	Capital	Cultural
		Samarkand	River valley	Organic	Capital	C Asian
2.4	***Historic European Capital***					
		London	Riverine	Radio-centric	Capital-services	Transnational
		Paris	Riverine	Radio-centric	Imp. Capital	Cultural
		Lisbon	Coastal/hills	Axial/organic	Capital	Cultural
		Madrid	Plain	Zonal	Imp. Capital	Cultural
		Milan	Plain	Radio-centric	Reg. Capital	Cultural
		Munich	Riverine	Sectoral	Sta. Capital	Cultural
		Vienna	Riverine	Radio-centric	Imp. Capital	Neo-classical
		Budapest	Riverine/hill	Radio-centric	Capital	Twinned
2.5	***Academic and Ecclesiastic***					
		Amsterdam	Riverine	Canal Ring	Maritime	Cultural
		Prague	Riverine	Core & Sectors	Capital	Neo-classical
		Florence	Riverine	Core & Sectors	Reg. Capital	Cultural
		Barcelona	Coastal	Grid & Axes	Sta. Capital	Cultural
		Dublin	Estuarine	Radio-centric	Capital	Cultural
		Edinburgh	Coastal	Sectoral	Capital	Cultural

Table 2.0.2
Established Historic Cities Population Growth (thousands)

CITY		1950	2000	2020	*2035**	Area
Athens	Greece	1,347	3,179	3,153	*3,169*	Urban agglomeration
Rome	Italy	1,884	3,708	4,257	*4,471*	Metropolitan area
Alexandria	Egypt	1,037	3,546	5,281	*7,169*	City Proper
Istanbul	Turkey	967	8,744	15,190	*17,986*	Urban agglomeration
Isfahan	Iran	184	1,382	2,132	*2,563*	City Proper
Jerusalem	Israel	121	664	932	*1,144*	City Proper
Makkah	Saudi Arabia	148	1,168	2,042	*2,521*	City Proper
Amritsar	India	340	981	1,377	*1,875*	City Proper
Varanasi (Benares)	India	363	1,204	1,665	*2,266*	Urban agglomeration
Luoyang	China	145	1,213	2,387	*3,078*	Urban agglomeration
Nanjing	China	1,037	4,279	8,847	*11,506*	Urban agglomeration
Beijing (Peking)	China	1,671	10,285	20,463	*25,366*	Urban agglomeration
Kyoto	Japan	1,100	1,470	1,470	-	City Proper
Samarkand	Uzbekistan	168	361	579	*727*	City Proper
London	UK	8,361	7,273	9,304	*10,556*	Urban agglomeration
Paris	France	6,283	9,737	11,017	*12,065*	Urban agglomeration
Lisbon	Portugal	1,304	2,672	2,957	*3,125*	Metropolitan area
Madrid	Spain	1,700	5,014	6,618	*6,975*	City Proper
Milan	Italy	1,883	2,985	3,140	*3,251*	Metropolitan area
Munich	Germany	831	1,202	1,538	*1,625*	City Proper
Vienna	Austria	1,615	1,549	1,930	*2,154*	City Proper
Budapest	Hungary	1,618	1,787	1,768	*1,784*	City Proper
Amsterdam	Netherlands	851	1,005	1,149	*1,237*	Urban agglomeration
Prague	Czechia	935	1,172	1,306	*1,352*	City Proper
Florence	Italy	565	683	708	*732*	Metropolitan area
Barcelona	Spain	1,809	4,355	5,586	*5,869*	City Proper
Dublin	Ireland	626	989	1,228	*1,455*	Urban agglomeration
Edinburgh	UK	453	451	537	*604*	Urban agglomeration

Note: **Forecast*
Source: Global Database of Metropolises 2020 – UN-Habitat (2020), UN DESA Population Division.

2.1 Imperial Cities

Greek and Roman Capitals – Upper Nile Egypt – Byzantine and Ottoman Settlements at the Bosphorus – Safavid Isfahan

Alexandria, Athens, Rome, Constantinople and Isfahan

2.1.0 INTRODUCTION – ANCIENT EMPIRES

The makings of grand Imperial Cities which were to evolve under those Empires that developed across the Eurasian continent over many centuries. In the 5th century BCE in southern Europe, under Greek dominance, the principal cities of Athens, Corinth and Sparta vied for competition in the Peloponnese.

The Hellenistic empire was to develop thereafter in southern Europe beyond the creation of local naval bases around the Aegean. Its Imperial reach was extended under Alexander of Macedonia, who having seized Egypt from the Persians, established a port as a new capital, to be named Alexandria, at the head of the River Nile, one of the first city states. As Egypt's principal waterway, the Nile served as the main axis around which the First Kingdom had been founded. As New Egyptian Kingdoms evolved and dominated, there was a steady flow of peoples striving to settle in the Nile Delta.

The Roman Empire was developed later under Augustus, at the end of the 1st century BCE as it expanded its overseas territories from Italy into Egypt and across North Africa. Its pre-eminence began to wane when Constantine took over Byzantium, renaming the city Constantinople in the 3rd century CE which was to become the centre of a great Byzantine Empire that was to be followed much later by the Ottoman Empire across Asia Minor.

Other land-based Eurasian empires were more focused on trading routes where centres developed at key locations along these routes. Such imperial states evolved under the Mongols, covering enormous sprawling rural areas beyond Mongolia across the Central Asian steppes with localised centres at Bukhara and Samarkand.

Further east, the Ming dynasty subsequently overthrew the Mongols who had taken over the Khanates in China. In securing defensive protection from construction of sections of the Great Wall of China in the north, the early Ming capital was created in Nanking. More defensive walls were built and with the excavation of the Grand Canal between the Yangtze and Yellow Rivers, and construction of the Forbidden City, the capital was able to be moved to Peking.

Other imperial centres were to develop also under the later Persian Empire at Tabriz, Qazvin and Isfahan which was to become the new Safavid capital with its magnificent 'maidan' and mosques. Beyond in turn in South Asia, the Mughals controlled northern India for a long period with a centre in Delhi before subsequently moving it south to the Deccan.

The last Asian Empire was essentially the Meiji who overthrew the Tokugawa Shogunate in Japan. With a coastal area extending from Manchuria in the north to Indo-China in the south, this region contained several significant trading posts that were to develop into strategic urban centres. Most of these former imperial capitals which are illustrated have grown and developed to become significant densely urbanised centres in the 21st century today.

2.1.1 ALEXANDRIA, EGYPT

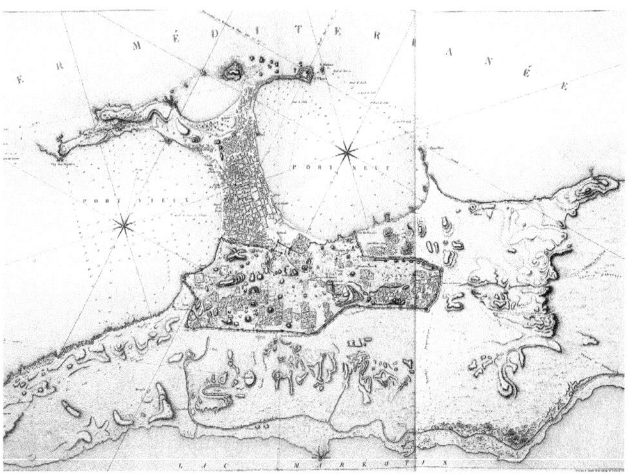

FIGURE 2.1.1 Alexandria – early configuration, 1798.
Source: British Library.

DOI: 10.4324/9781003353348-5

Background and Origins

Alexandria (Al-Iskandariyah) is located to the west of the wide Nile delta in the Mediterranean Sea, on the northern coast of Egypt. Founded by Alexander the Great in 330 BCE under his imperial exploits, Alexandria played a pivotal role as a port in the eastern Mediterranean and became an important part of the following Hellenistic civilisation and growing to a population of around 0.5 million. In the reign of the Pharaohs under Ptolemy I, it was capital and remained the seat of the country's administration under the Roman and Byzantine Empires for nearly a millennium until the Muslim conquest in 640 CE.

One of the main ingredients of the local economy was grain which was shipped and exported to the Roman capital. On a narrow strip of land between the Mediterranean and the inland Lake Mareotis, the city was designed on a gridiron plan, with different neighbourhoods, having an Egyptian quarter (Rhakotis), Greek quarter and Jewish quarter. The city's heart was the Royal quarter with its fine buildings, as the Serapeum Temple in the Egyptian quarter.

Extending from the city centre, an artificial causeway, the Heptastadion, was built on the island of Antirrhodos where there were Royal palaces, including that of Cleopatra. With its 'T' shaped peninsula, this created two natural harbours, the Great Harbour to the east and Eunostos to the west. The most distinctive structure standing on the reef was the great Pharos Lighthouse with its three tapering stages, one of the 'seven ancient wonders of the world'. This dominant feature assisted navigation of the large naval fleet that was based in the harbour.

The sacred royal Necropolis where Alexander was eventually buried became a site of special significance at the intersection of the two main boulevards, the Canopic Way and the Street of Soma, with the Gates of the Sun and the Moon in the city walls. Key buildings here were the great Library and Museum with the library with its papyrus rolls being the centre of learning in Hellenistic times. Under the Ptolemaic Dynasty, polymaths, astronomers and philosophers studied here, as it became a major cultural and scientific centre. Sadly, with climatic changes and earthquakes, in the 4th century CE, all of this land subsided, the coast eroded, and most of these classic edifices were lost.

Under subsequent Roman occupation and development around 30 BCE, the walled city became one of the centres of early Christianity under the Patriarchate of Alexandria, despite ensuing schisms as with the Arians and the eventual Coptic Orthodox church. This evolved further under the subsequent Byzantine empire which was overrun with the eventual Muslim conquest.

The city remained a significant port in mediaeval times and during the Ottoman period. Sultan Qait Bey built a fortress on the island of Pharos in 1480. During the Napoleonic period a number of forts were built by the French to the south of the city in order to protect it. Around 1820 a new waterway, the Mahmodia Canal was built to connect the River Nile with Alexandria under Ali Pasha, but it became impassable at times and un-navigable. The area evolved then until the Khedivate of Egypt was established in 1867 with the first railway line having been built to connect Alexandria with Cairo on the Nile in 1856.

Further to the east beyond Damietta and Lake Manzala, during the 1850s particular interest arose in developing a new canal from Port Said to connect the Mediterranean Sea with the Red Sea, east of Cairo as an improved route for shipping. Ferdinand de Lesseps secured a concession from Sa'id Pasha, the Khedive of Egypt for piercing the isthmus and to construct a canal to connect Ismailia in the south with Port Said then passing through the Small and Great Bitter Lakes to Suez (Canal des deux Mers). After both considerable financing and construction difficulties, the canal was eventually completed and opened to shipping in 1869. This 'Suez Canal' was a significant project that was to transform the entire pattern of world trade going forward with a direct sea connection from the Mediterranean to the Indian Ocean, so avoiding the long transit around Africa and the Cape of Good Hope for a large volume of shipping and maritime trade.

Modern City – Buildings and Arteries

Within Alexandria, trading internationally has continued and modern port terminals are active at the Western Port, Dekhela and Abu Qir. Highways connect with the International Coast Road to Port Said and the Cairo to Alexandria desert road and also rail connections via the Misr and Sidi Gaber rail stations and tramway network. A new Metro for Alexandria is being planned for construction paralleling the Corniche, and has an existing tramway system of two routes across the city.

As a linear city, hemmed in by a long ridge, a new eight lane Highway, by the Corniche was built along the waterfront in 2006, cutting off the beach from the city. Ancient marshlands were lost and the sea floor is starting to erode, being susceptible to storm surges. Today the city of Alexandria is quite different with its main commercial port and Arab fort of Qait Bey and the long corniche lined with modern buildings. The back streets are assorted and congested, home to a large resident urban population. Once very cosmopolitan, it is a traffic filled place and cramped, that has lost much of its former character. The affluence of the Cotton Exchange has now gone and the labyrinthian Arab quarter remains as a maze with bazaars, mosques and their meuzzins and localised nodes of activity.

One of the striking new buildings that echoes the former great library is the new 'Bibliotheca Alexandrina' library (designed by Snohetta) with its wedge-shaped glass roof sloping down to the Mediterranean. Once known as the 'Pearl of the Mediterranean', the city is striving to cope with its expanding population growth and burgeoning oil and gas industry from neighbouring Port Said and Suez. The population of the city is approx. 6 million, a significant place in the region.

Transactional Forces and Typology

With its ancient history and due to its strategic location, Alexandria developed and served as a significant port and trading centre in the eastern Mediterranean. With the long River Nile nearby, it has facilitated the trade of grain and Egyptian cotton in early eras around its great natural harbour. Located by the wide river delta, since the realisation of the Suez Canal trading between the Mediterranean and the Red Sea has been channelised through Port Said via Ismailia and the Great Bitter Lake to Suez. With its widening and deepening, this now carries enormous maritime freight volumes between Europe and the Indian Ocean. As a coastal city developed around its original grid plan, its Great Harbour was enclosed originally by the Hepta Stadion Causeway linking with Pharos Island. Modern development has now extended linearly along the coast to a new Western Port around which the metropolis is extending along the Corniche and also inland at the new nodes of Al Amaria and Borg El Arab. New development is continuing to be extended aggressively in a linear manner westwards to El Alamein along the coast, the site of World War II tank battles, with new high rise developments and nodal centres. The Nile Delta now has an increased propensity to flooding.

2.1.2 ATHENS, GREECE

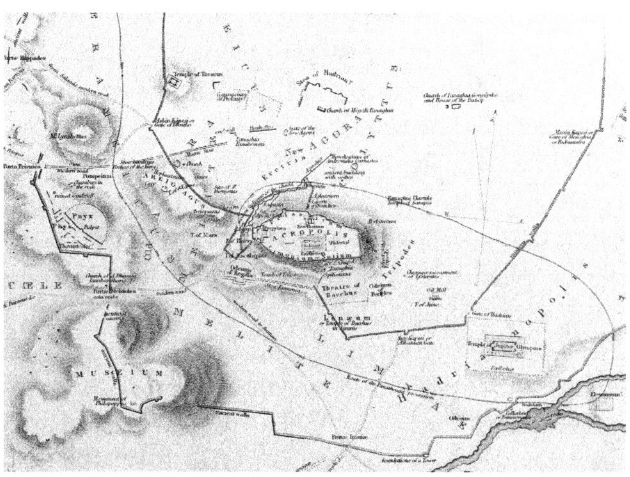

FIGURE 2.1.2 Athens – early configuration, 1844.
Source: British Library.

Background and Origins – Ancient City

Athens (Athina) is located around a group of hills that include the Hymettus Mountain range in the Attica basin in central Greece. It is several miles inland from the Mediterranean port of Piraeus.

The city of Athens has two striking natural geographic features, the Aeropagus Hill of polished limestone and Lykavittos Hill. Next to the former is the Acropolis rock, fortified with Cyclopean walls above which the Parthenon temple holds a commanding position, built in celebration of the goddess Athena.

Ascending from the Panathenaic Way, the entrance to the Acropolis is via the main gateway that passes the Propylaea and the ionic 'Temple of Athena Nike' up towards the main Temple with the original tall statue of Athena within its inner cella. With its external peripheral Doric columns, numbering 8 by 16 each side, all finely described with the effect of entasis, the original pediment and frieze, metopes and triglyphs of the building, would have been decorated and multi-coloured, with the marble sourced from nearby Mount Pentelikon.

The separate Ionic Erechtheion and its porch has four Caryatids, and around the Acropolis there is the Theatre of Dionysos and Roman Theatre of Herodes Atticus. At the foot of the Acropolis lies the significant Ancient Agora, an area that was originally the heart of daily activity with its temples, stoas, schools, commercial shops and markets.

Other classic buildings include the 'Temple of Hephaisteion', the 'Temple of Apollo' and the 'Tower of the

Winds', a marble octagonal structure that served both as weather-vane and water clock. With each side facing the principal directions of the compass, the walls are elegantly sculpted with friezes that personify each of the eight winds, that is: Boreas, Kaikias, Apeliotes, Euros, Notos, Lips, Zephyros, and Skiron.

The classical city of Athens in ancient times was the home and the centre of traders and adventurers, a bustling city. It was also the centre for philosophers where the concept of 'democracy' came to be debated and practiced. Its dominant buildings reflected a new order and style that imposed itself upon the inhabitants, reflecting the strength and aspirations of the rulers and their people.

The heart of the city was the 'polis' that developed from the political, religious and military organisation of its citizens by its leaders. This was recognised in its special annual festival, the Panathenaia, in honour of the city's patron goddess, Athena.

The city was the centre of imperial power during this Hellenic period and Age of Pericles. Its period of particular influence was essentially during the century following 480 BCE, the 'Golden Age' of Athenian democracy. The Parthenon was built, philosophers, playwrights and historians flourished and the city state of Athens was dominant over rival cities such as Thebes and Sparta.

A pair of long external walls was built under Pericles to connect the city with the port of Piraeus. However, wars with other regions as Persia and Macedonia developed and with disease and epidemics, its power waned and under the Roman Empire it became a free city. Significant changes were made in the built fabric during the Roman period especially with the development of the Agora and Roman Stoa.

Subsequently Athens was to come under the influence of the succeeding Ottoman Empire and the Parthenon itself was converted into a mosque with an external minaret. With occupation by the Turks, the Parthenon was badly destroyed during a subsequent explosion in 1687 CE, which caused considerable damage.

Modern City – Buildings and Arteries

Today Athens is a bustling European city on the edge of the Mediterranean region of the Middle East, a rather polluted and stifling place, especially in summer with excessive traffic around Syntagma Square and the centre. The older districts remain, such as Monastiraki, the heart of Ottoman Athens with its mosques, and the dense interwoven houses and tavernas beneath the Acropolis, the ancient Stadium and with numerous small squares, housing and apartments.

The municipality of Athens comprises several different districts, and contains the oldest squares of Omonia and Syntagma, adjacent to the Greek Parliament building and with the market and commercial areas at Plaka, just beneath the Acropolis, and Kolonaki at the base of Lykavittos Hill. The city centre is congested, and commercial and residential development has extended north to new areas of Marousi and Kifissia, and south to Glyfada.

The new ring road managed by Attiki Odos has been built to connect with the new Athens International Airport Eleftherios Venizelos (ATH) whilst the old Hellenikon Airport has been closed and is being redeveloped as an urban development project. In 2004 the Olympic Park was developed for the Summer Olympics and much of the urban infrastructure was improved at this time. The original Suburban Railway line to Kifissia was upgraded and the Metro expanded with two new lines built to connect the centre of the city with the port of Piraeus, now expanded under Chinese financing.

Since joining the EU in 1981, the rate of urbanisation and modernisation of the region of Attica has continued to develop as the population has grown with in-migration. The population is approx. 3.75 million in the metropolitan area of Athens, despite a slow down with the economic crisis in 2008. The Greater Athens Area now extends to include 40 municipalities.

Most proud of its ancient heritage and with its imposing and impressive classical sites and monuments, these assets remain to remind us of its former dominant Hellenistic age and influence around much of the Mediterranean and surrounding islands.

Transactional Forces and Typology

With its development around a number of hills, Athens grew from its early civilisation into a classical 'city-state' and then on to its current sprawling metropolis. With its commanding position in Attica and with its nearby port, it became a strategic location in the Saronic Gulf and the Aegean Sea. As the 'polis' developed, this comprised not just physical features such as the Acropolis and Agora, but also a system of governance and administrative procedures around which public life was organised, with its 'virtues', the forerunner of democratic principles. This extended to 'synoecism' or the incorporation of adjoining communities, as a concentration of power, was a driving factor in facilitating foreign trade and communications in the region.

From this, Athens benefitted from the Mycenean culture that developed in Greece in a similar manner as from the Minoan culture in Crete. In the subsequent Hellenic period, the primacy of Athens was challenged by other ascendant city-states before succumbing to Roman rule, being recognised as a 'free' city.

In late antiquity, Athens had lost its leadership but regained this in Byzantine times with an increase in trade. After independence from the Ottomans, as capital of an independent Greek state, Athens began to modernise and grew and extended northwards, beyond central Plaka and Monastiraki and Kolonaki with residential areas towards Marousi and Kifissia, and south beyond Glyfada and towards Piraeus with new industrial areas. This metropolitan area has now expanded extensively with inward migration and new commercial activities resulting in significant densification and congestion.

Athens as a hilly city has grown organically around its geographical natural constraints. With the dominant Acropolis and Lykavittos Hill, its residential communities nestle under the hillsides. Newer planned settlements and commercial areas have been developed beyond in suburban areas to the north along road corridors, and with the city now bounded by the peripheral Ring Road 'Attiki Odos'. Industrial activity is concentrated around the separate marine harbour of Piraeus where ferries connect to the outer Greek islands. Accordingly, the city has features of a hillside city type along with new poly-nodal centres that are developing in outer areas.

2.1.3 ROME, ITALY

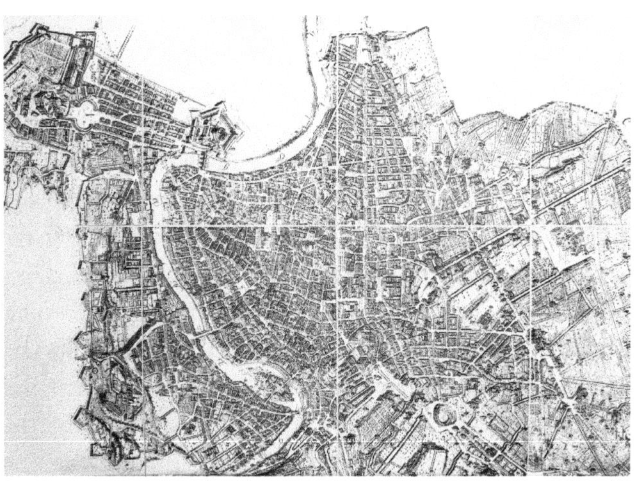

FIGURE 2.1.3 Rome – early configuration 1756 (Falda).
Source: British Library.

Background and Origins – Ancient City

Rome (Roma) is located in the Lazio region of central Italy. According to legend, Rome was founded around 750 BCE with the finding of the twins, Romulus and Remus, on the banks of the River Tiber. From its early beginnings and development of the Italian peninsula by the Etruscans, the Roman people founded the Republic of Rome around 500 BCE. With its earliest Republican walls and fortifications around the seven hills, it became a formidable site. The buildings on the Palatine Hill and the realisation of the Forum with dense buildings for the residents, with the use of local tufa stone, assisted in the creation of more permanent buildings within the boundary of the city (pomerium).

Under its first Emperor Augustus, in 25 BCE the urban fabric of the city was developed with basilicas, temples, forum, circus and theatres around Campus Martius. With realisation of the aqueducts, Aqua Appia and Marcia, clean water was channelled to the city and was available to citizens for baths. Drainage was installed like the 'Cloaca Maxima', so enhancing municipal services for the inhabitants.

In terms of trading, with the limited size of the River Tiber, access from the sea was not easy, so sea-faring ships importing grain would need to anchor off the estuary at nearby Ostia and have goods trans-shipped to smaller vessels in order to be transported into the city of Rome. During the subsequent period the Roman Empire flourished with overseas imperial adventures and extended its reach all around the Mediterranean from Gaul in the west, to Cappadocia and Cyrenaica in the east. Many straight single-track roads were built that radiated out to connect with new centres.

Under its prevailing rule of law, its organisational skills were applied to development of new sites around a standard planning system, having regular gridded walled towns with perpendicular 'cardo' and 'decumanus' layout, central forum, market and urban facilities. This model was replicated across all countries of the empire with roads, buildings and aqueducts forming significant elements of the built environment.

Its reach was extensive and had a significant impact on such different towns across Europe and North Africa as Pompei, Syracuse, Carthage, Caesarea, Lutetia, Massalia, Eboracum and Aquincum. Under the Roman Empire a series of Imperial towns and cities were developed that comprised coloniae, municipia, civitates and important tribal centres, market and administrative centres. The population of Rome itself is estimated to have risen to between 0.5 and 1 million at this time.

In 64 CE a great fire occurred throughout the city of Rome which led to the need for complete rebuilding and overhaul of the city under Emperor Nero. After this, new buildings were constructed as the elliptical Colosseum and huge Circus Maximus and the palatial baths of Caracalla, open to all Roman citizens. The predominant technique used was slim kiln fired bricks, laid in typical bond, 'opus latericium' that prevailed throughout the city.

Despite suffering from periodic bouts of Antonine plague, especially around 175 and 250 CE, there followed a prosperous period before the 3rd century CE, when Christianity was on the rise. However, the influence of Rome started to decline with the transfer of the centre of the Roman Empire from Rome to Byzantium. In the 6th century CE the city was overrun by Goths and Vandals, and Christian churches built that were linked with the burial of Saints, had to be located outside the city walls. A new set of Aurelian Walls were also built.

Whilst the Roman Empire started to decline, the remaining power in Rome arose through the Papacy and in 800 CE Charlemagne was crowned Holy Roman Emperor in Rome. But conflicts between the Pope and Holy Roman Emperor weakened the papacy, and with the schism in 1309 CE, the papacy was eventually forced to move to Avignon.

Renaissance City

It was not until the 15th century that the next period of ascendancy in Rome occurred. During this period of the Renaissance, the Popes aimed to restore Rome as the undisputed centre of Catholicism. As rulers of the papal states in Italy, the papal base was moved from the Lateran Palace where its main residence was developed, to the Palace of the **Vatican**. The Vatican Library, the Stanze Apartments and the newly built Sistine Chapel, subsequently decorated by Michelangelo and St Peter's Basilica which was rebuilt by Bramante, became the main centre of worship. In both the fields of art and architecture, significant new churches and palaces were built under the Medici and artistic achievements flourished.

However, with the unpopular actions of simony, that is the selling of ecclesiastical offices for cash, political disunity in the states developed and with disputes with the Protestants, in 1527 CE the city of Rome was sacked by the Holy Roman Emperor Charles V. Its resilience however led the Eternal City into a period of Counter-Reformation. Michelangelo was commissioned to redevelop the Capitoline Hill and villas such as Villa d'Este at Tivoli were designed for eminent cardinals.

Under one of the most significant periods of development in Rome during the Renaissance at the time of Pope Sixtus V, the street network was extended and integrated with overall visual effect, unifying areas between buildings and public spaces. This involved the construction of connections between pilgrimage churches and attained cohesion of the road system in the centre of the city. It was achieved through the realignment of streets and installation of landmark obelisks at key junctions in creating 'reciprocal vistas' in either direction. In particular, this included along Strada Felice, from Santa Croce to Piazza del Popolo via Santa Maria Maggiore. This is intersected at right angles by Strada Pia, which links Porta Pia with Piazza Quirinal.

The other significant redevelopment that was undertaken during the period was at the Capitoline Hill, the former seat of the Senate where Michelangelo was commissioned to redevelop the Campidoglio (Capitol Piazza) with new facades to the three grand Palaces. In so doing, he also created a magnificent space in the monumental square centred on the statue of Marcus Aurelius, trapezoidal in shape. Open on one side it leads down a long flight of steps from the patterned oval marble flooring.

In terms of city-scape, other monumental squares are the later Piazza Navona by Bernini (c 1650) and Piazza di Spagna that leads up to Santa Trinita dei Monti (c 1725) and Piazza del Popolo (c 1820). At the Piazza del Popolo, after the twin domed Santa Maria churches were built straddling the Strada del Corso on the southern side, two new streets, Via del Babiuno and Via di Ripetta were built that intersect at equal angles on entering the square at the placement of the square's central obelisk. Although only completed after many years in the 19th century by G Valadier, this is another striking monumental oval place in the centre of Rome. These works that were undertaken across the city in realigning squares and avenues, taking their axial nature into consideration, with strengthening of bridges, led to the now familiar layout of much of the city with its elegant squares. Around the centre these landmarks besides the Capitol and the Campidoglio, complement the old Forum and Palatine Hill, with Campo De' Fiori, Piazza Navona, Piazza della Rotunda and the Pantheon.

Geographic Areas

In terms of topography, each of the areas of the seven hills of Rome have distinctive geography.

Capitoline Hill – At the summit of the Capitoline Hill, at the time of the Roman Empire, reached by a zig-zag path from the Forum, was the Temple of Jupiter, at one of the most symbolic locations. Redesigned in the 16th century CE by Michelangelo around an elegant square, actually a trapezium, and accessed by a flight of stairs, the Cordonata, it has striking palaces and facades on three sides of the square, the Piazza del Campidoglio, with its geometrically patterned marble flooring. These are the Palazzo Nuovo, Palazzo dei Conservatori and Palazzo Senatorio, the seat of the City Government (SPQR – Senatus Populusque Romanus) with its statutes, including the 'She-wolf' that became the symbol of Rome, are located.

Palatine Hill – Adjacent to the Capitol is the Palatine Hill, a defensive settlement once the residence of emperors and aristocrats in the Roma Quadrata, with its mix of palaces, villas and gardens. This included that the Domitian Palace and Domus Augustana, the home of the 'August' emperors. Now mostly ruins, they adjoin the Farnese gardens and extensive Roman Forum. Nearby the Forum was a centre for social and commercial activity at the time, with its meeting halls and basilicas, shops and food stores and Trajan's Market, and temples and monuments. It was also a place for the Senate (Curia) and political meetings, legal administration and tax collection, and religious assemblies at several temples, as Saturnalia at the Temple of Saturn. At the end of the Forum is the Colosseum, the great elliptical amphitheatre with arched entrances and tiered seating built by Emperor Vespasian for gladiatorial combats.

Quirinal Hill – Another of the seven hills, the Quirinal Hill which was home to the Baths of Diocletian and Piazza Esedra, now Republica, was mainly residential in imperial times. The city was fed by water from the Acqua Vergine aqueduct, which was also used to feed the Rococo Trevi Fountain with its travertine sculptures that was built in 1762. Nearby is the Via Veneto. The Palazzo del Quirinale, built as the summer residence of the Pope in 1573, that became the residence of the King, is now the official residence of the President of the Republic with its secluded gardens. This neighbourhood with its several palaces has now been transformed with monumental buildings along the Via Nazionale and Via XX Settembre that now houses various Ministries of State.

Viminal Hill – The Viminal Hill is the smallest of the hills between the Quirinal and Esquiline Hills where the current Ministry of the Interior 'Il Viminale' is located on the edge of the Monti District. Here also is the Palazzo Massimo alle Terme and the Teatro dell' Opera and now Rome's modern 'Termini' central railway station.

Esquiline Hill – Adjoining the Quirinal and Monti District is the Esquiline Hill which is one of the largest of the Seven Hills. Here was the site of Domus Aurea, Nero's large 'golden' palace that covered a large part of this area. Nearby were the Baths of Trajan. More recently the area has been built over but features several churches that include the large Santa Maria Maggiore, commenced in the 4th century with many mosaics.

Caelion Hill – The Caelion Hill which overlooks the Colosseum contains the vast ruins of the Baths of Emperor Caracalla that were fed by aqueducts. The Via Appia crosses the area and leads to the still existing Porta San Sebastiano Gate in the old city Aurelian wall in the south-east of the city. This is now quite an open and green area beside the archaeological ruins and has several fine churches and the 16th century Villa Celimontana and its wooded park.

Aventine Hill – The Aventine Hill to the south alongside the River Tiber has fine views across the river of St Peter's. The top of the hill is crowned by the early Christian Basilica of Santa Sabina. Nearby is the park of Circus Maximus, the largest stadium in ancient Rome, the home of chariot races. By the Aurelian wall at Porta San Paolo is the imposing Pyramid of Caius Cestius. To the north towards the Capitol is the Arch of Janus and the Temples of the Forum Boarium, the rectangular Temple of Fortuna Virilis (Portunus) and the elegant circular Temple of Hercules with its Corinthian columns built in 120 BCE.

Vatican City – The other district on the opposite side of the River Tiber across the bridge with Castel Sant' Angelo, and its fortifications, is the Vatican City, the residence of the Popes, within its own walled confines. The first Basilica of St Peter's was built in 349 CE by Emperor Constantine. During the Middle Ages before the move of the Popes to Avignon, the Lateran Palace was the residence of the Popes with the Basilica of San Giovanni in Laterano, where the Popes were crowned.

At the beginning of the 16th century CE, Pope Julius II commissioned Bramante to build a new enlarged basilica. With a huge dome designed by Michelangelo and facade, the new Basilica was completed in 1626 and in the 1660s Bernini embellished the external piazza with a sequence of spaces with a set of semi-circular colonnades that connected the square and its central obelisk with the façade and dome of St Peter's Basilica.

In the Palaces behind the Basilica are the Vatican Museum, with the Belvedere Palace and spiral staircases and Pinacoteca gallery, and the famous Sistine Chapel and the Raphael Rooms. The ceiling and spandrels of the Sistine Chapel were painted by Michelangelo, and the altar frescoes of the 'Last Judgement'. Following plans by Alberti for a new district Borgo Leonino, a new connecting road was built from St Peter's along the Via della Conciliazione to Castel Sant' Angelo just south of the Borgo. Under the Lateran Treaty, the position of the Vatican State 'within a state' was protected in 1929 within the Republic of Italy. It has continued to retain its own independence and as centre of the Roman Catholic Church although the current population is less than 1,000 people.

There are two other further distinct Roman riverside areas, *Trastevere* and *Campo de' Fiori*.

Trastevere – On the west bank south of the Vatican is the district of Trastevere. An area of old quarters with the Aurelian walls and mediaeval towers, narrow cobbled streets and churches, its character was typified by the Piazza di Santa Maria and adjoining church. The neighbouring district is Janiculum along Lungo-tevere with the Villa Farnesina an elegant Renaissance villa designed by Peruzzi and subsequently owned by Cardinal Farnese, with Palazzo Corsini behind and the Botanical Gardens.

Campo de' Fiori – On the east side of the River Tiber is Campo de' Fiori which overlooks Tiber Island and Ponte Fabricio. With a series of fine palaces, Campo de' Fiori is a bustling market square dominated by the Palazzo Farnese. Adjoining are the Palazzo Spada with its colonnaded gallery and Palazzo Ricci with its frescoed facade and Palazzo Cenci. Opposite the island is the former Jewish Ghetto and the neighbouring Theatre of Marcellus with its curved outer wall.

Finally, there are a number of other Districts in the city. The centre contains the *Pigna District* and the Piazza della Rotunda with the inspirational Pantheon Temple and its coffered hemi-spherical dome and central oculus. Built by Emperor Hadrian around 120 CE, it contains tombs and shrines. In the Piazza del Gèsu is the first Jesuit church designed by Vignola under the Counter-Reformation. The Palazzo di Montecitorio initially by Bernini has been enlarged and now serves as the Parliamentary Chamber of Deputies. The Palazzo Borghese has an inner courtyard with other fine palaces in the neighbourhood and the column of Marcus Aurelius commands the centre of Piazza Colonna.

In the adjoining *Ponti District* is the long oval Piazza Navona, the site of a former Roman stadium. The centrepiece in front of the Sant' Agnese in Agone Church is the 'Fontana dei Quattro Fiumi' representing the four great continental rivers of the Ganges, Nile, Danube and Plata, with its Egyptian obelisk and Fontana del Nettuno at the northern end. With other palaces, the most distinctive one is Palazzo Massimo alle Colonne, with its convex colonnaded facade, following the line of the street. Further west is the Chiesa Nuova, with its broad Baroque facade.

To the north beyond Campo Marzio, is the city's north gate, Porta del Popolo by the Piazza del Popolo in which three radiating roads were built, the main Via del Corso, Via di Ripietta and Via del Babuino. The latter leads into the Piazza di Spagna with the Fontana della Barcaccia from which leads a series of stairs, the 'Spanish Steps', up to the Trinita dei Monti church above. Beyond the city walls is the Villa Borghese and its gardens. The baroque park was laid out by Cardinal Scipione Borghese with sculptures and fountains. Nearby is the Renaissance Villa Giulia designed by Vignola, now an Etruscan Museum.

Modern City – Buildings and Arteries

The Italian peninsula historically was a collection of City States and it was not until after Napoleon's time that in 1870 under Garibaldi's efforts, Rome was declared the capital of a new unified Italy under King Vittorio Emanuele II. New streets were cut through the capital and the massive white marble monument to the King was imposed and inaugurated in the city centre in 1911 overlooking the Piazza Venezia.

During the subsequent 1930s the 'Esposizione Universale Roma' (E U R district) was developed with the Palazzo dei Congressi for the planned Universal Exposition, cancelled in 1942. During the beginning of the 20th century, the country had had to endure a period of fascist dictatorship and the Republic was only established after the end of World War II.

With the layout of the city of Rome determined by its hills and rivers, and its surrounding natural topography, it has been shaped by historical developments over the years. Many of the ancient sites exist as remains and several of the key Renaissance sites still dominate the city, as the Piazza del Campidoglio, the Piazza Navona and Piazza del Popolo as described.

The urban fabric is embellished by its fine streetscape and with a number of landmarks, including obelisks, columns, statues and fountains. Several bridges cross the River Tiber which flows down to the port of Ostia. New urban roads and infrastructure have been built including an orbital ring road, 'Grande Raccordo Annulare' (GRA).

The commuter rail network connects with the main Termini central station and High-speed TAV rail lines connect from Roma Tibertina station with the north and Milan and south to Naples. The Metropolitana metro line with its three lines is being extended and there is also a dense tram and bus network. Two international airports are located at Leonardo da Vinci Rome Fiumicino (FMO) and Ciampino (CIA), and the port of Civitavecchia serves the city.

Since the end of World War II, a number of international agencies have located to the city, namely UN FAO and IFAD and World Food Programme. This has encouraged supporting services, research and IT companies to become established, with the manufacturing and commercial sector developed, and several major Italian companies are headquartered here as ENEL and ENI. It also has become a major fashion capital.

The 'Metropolitan City of Capital Rome' comprises a large number of separate municipalities. The population of the city is approx. 2.8 million in the 'Comune' and 4.4 million in the metropolitan area which also includes the separate Vatican City. With its long history and heritage, and one of the oldest Universities, La Sapienza, the City of Rome or the 'Eternal City' continues to be one of the most desired places for international tourists to visit.

Transactional Forces and Typology

Located by the River Tiber, the ancient city of Rome is situated around seven hills not far from the Tyrrhenian Coast. With its historical fortifications of the Servian and then Aurelian Walls, the city developed in distinctive quarters around the hills, especially the Palatine Hill. With dependence on the importation of grain, this was fulfilled by trade from around the Roman Empire that extended across the Mediterranean. With its central Forum, temples and basilicas, this was the centre of activity and administration of Ancient Rome at its peak.

With the fall of the Western Roman Empire, its situation, power and population declined, which was not restored until the Renaissance. With the rising dominance of the Holy Roman Empire, the original Saint Peter's Basilica had been built on Vatican Hill, on the western side of the River Tiber, eventually to be replaced by the newer St. Peter's Basilica consecrated in 1626.

Rome became the capital of the Papal States and under the Renaissance expanded its influence. But it was not until the 19th century that Italy was unified, with Florence as its temporary capital, with Rome reverting to being the capital in 1871. Buffeted by later fascist forces and two World Wars, it was not until the period of post war reconstruction and since the formalisation of the European Union Treaty of Rome in 1957, that Rome re-established its stability and re-emerged as prime European capital city.

As a riverine and hill city, the city of Rome has developed organically around its seven individual hills with separate Vatican City. With its historic core, parts of the metropolitan area have been industrialised at its fringes and with a modern exhibition area, EUR, in the south. With its peripheral ring road, the GRA and new Metro lines and industrial port at Civitavecchia, the urban infrastructure has been developed and upgraded and some green areas remain. Whilst preserving its unique civic architecture and antiquities, modern traffic pollution and congestion remain challenging in the centre. Typified by its hilly characteristics, however its distinctive quarters with planned piazzas and streetscapes from Renaissance times, these still shape the historic centre.

2.1.4 CONSTANTINOPLE – ISTANBUL, TURKIYE

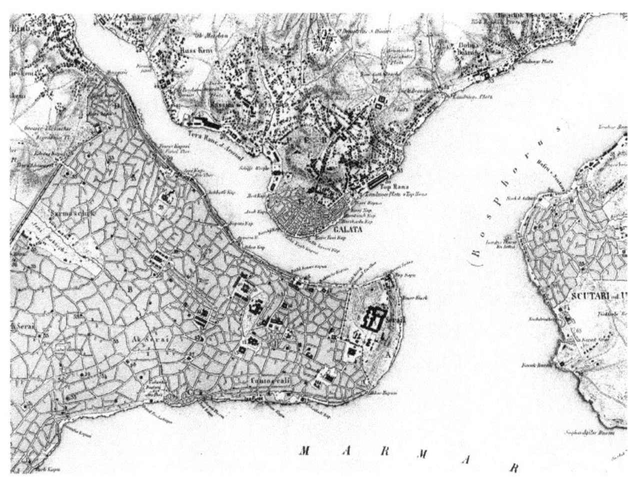

FIGURE 2.1.4 Constantinople – early configuration, 1851.
Source: iStock.

Background and Origins

Settlements and Eras

From the Mediterranean and the Aegean Sea, passing the Dardanelles through the Sea of Marmara, the narrow Bosphorus waterway is reached which interconnects with the Black Sea. Here Seraglio Point is a promontory on the western side that overlooks the Bosphorus which signifies the eastern end of the European continent, facing the Asian continent on the opposite side.

Adjacent is the Golden Horn where a natural harbour exists that could be used to protect the land from invasion by ships by sea, an ideal location for a settlement. On the landward side, where high walls could be built, this was the making of the first protected fortress on this imposing site.

Due to its strategic location, this site was developed to become a leading imperial city and port that commanded several different empires, firstly within the Roman Empire named Byzantium, then in the Byzantine Empire as Constantinople, and in the Ottoman Empire as Istanbul.

As a gathering point of different tribes and peoples, its geophysical location on the Bosphorus Strait between the Sea of Marmara and the Black Sea, with a natural harbour in the Golden Horn, afforded a perfect maritime environment for shipping. During the Hellenistic period it was on the east-west silk and spice routes and also along Via Egnatia across northern Greece to China.

The original Byzantion settlement around 667 BCE on the promontory of Seraglio Point on the European side was protected on the western end by a series of defensive walls. Rather like Rome, there are seven hills in this area and during the Hellenistic period a palace, temple and agora were built. A corresponding settlement was established across the Marmara Sea on the opposite Asian side at Chalcedon, now known as Kadikoy.

With constant threats from various groups of Goths, Huns and Vandals, over the subsequent centuries, new lines of fortification were built beyond the original walls. The city endured continuous periods of political upheaval with other invading groups of Bulgars, Slavs and Rus, and was finally overcome by the Ottomans. Here the Byzantine fleet famously chained the entrance across the Golden Horn but the Ottomans made a deal with the resident Genoans who enabled them to transfer their boats overland into the harbour.

With its significant location on an important trade route, from the earliest age under Hellenistic influence, it was subsequently subsumed under the Roman Empire to become **Byzantium** in 64 BCE. Despite localised conflicts, Constantine became ruler and succeeded in moving the capital of the Roman Empire in 324 CE to found a 'new Rome' whose capital came to be known here as **Constantinople.** With a place of learning, the 'Pandidakterion', this was in effect an early university of the classical world.

With its strategic position on the coastal waterway, it also had its own water supply delivered via Roman aqueducts and with a grand underground water cistern, to service the collective population. It was developed further during this era to have a Great Palace, forum, hippodrome, public buildings and basilica. As part of this ongoing development, the first Hagia Sophia Church was also built.

During the subsequent Byzantine period there were continuous conflicts between different regional tribes and peoples, but the city of Constantinople was protected by its walls and sea defences. During the 11th–12th centuries CE there were a series of Christian Crusades that eventually succeeded in invading, which was followed by a period of influence from Genoa with the settling of a Genoese colony at Pera opposite the Golden Horn.

Subsequent incursions came from the Ottomans under their Muslim faith, who crossed the Hellespont and besieged the city and established a Sultanate under Sultan Mehmet II in 1453. The Sultans brought considerable wealth and style to the development of the city. The Ottoman Sultans established their own Empire and caliphate that extended out and beyond from the 15th century and ruled from the renamed **Konstantiniyye** or **Istanbul**, in their Topkapi Palace with its array of fine courtyards and individual pavilions.

Historic Places and Districts

The Topkapi Palace is entered by passing through the Imperial Gate and the 'Gate of Salutations' with the Divan complex and then through other courts and the 'Gate of Felicity'. The next courtyard contains the Throne Room and central Library, each beautifully decorated stone buildings. The 'Pavilion of the Holy Mantle' contains sacred Islamic relics and collections of miniatures and manuscript texts. Besides the Treasury, the Divan was the meeting place of the Imperial Council and the Grand Viziers.

Beyond in the next courtyard is a series of gardens populated with smaller pavilions and the exquisitely tiled Baghdad Pavilion, and the Konyali Restaurant. The innermost sanctum is the Harem, a labyrinth of exotically decorated rooms, baths and inglenooks where the Sultans' wives and concubines inhabited the Seraglio.

Adjoining the Palace is Sultanahmet, where two of the city's most significant monuments face each other across Sultanahmet Square, site of the former Roman hippodrome, containing the two great religious buildings of the Blue Mosque and Hagia Sophia.

The Hagia Sophia 'Church of Holy Wisdom', rebuilt under Emperor Justinian, with its huge central dome, nave, apses, upper gallery and byzantine mosaics, acted as the centre of the Orthodox Church. Its exterior with its reused stones and 'spolia' has been buttressed on several occasions over the years in order to afford protection against the earthquakes that occurred in the years of 896, 1317 and 1346 CE.

After 1453 it was converted into the 'Ayasofya', a Mosque when the symbolic mihrab and minber were built and eight large circular wooden plaques with calligraphic inscriptions were hung at the corners of the gallery. In 1572 four tall minarets were completed externally to complement the mosque.

In the 20th century, under Mustapha Kemal, President Ataturk's secular regime in Turkiye, it was converted into a Museum in 1935 and has remained as one until President Erdogan reconverted it back into use as a Mosque in 2021.

The Blue Mosque was built in Istanbul in the early 17th century covered in Iznik blue tiles. With its principal, secondary domes and semi-domes and six slender minarets, the courtyard and its arcades balance the whole building. With beautifully decorated blue tiles, the light enhances the inner size and scale of this building.

Next to Sultanahmet is the Bazaar Quarter, characterised by the narrow streets around the Spice Bazaar, and the New Mosque at Eminönü leading down to Galata Bridge. In addition, there is the covered Grand Bazaar, with a meandering realm of painted vaults and stores selling rugs and kilims, textiles and jewellery, and mixed oriental products, with a labyrinthal layout.

The zenith of the Ottoman Empire was under Süleyman the Magnificent during the 16th century. During this time in the middle of this district, the grand Süleymaniye Mosque, designed by the architect Mimar Sinan, was built and opened in 1557, directly aligned with Mecca. With its surrounding courtyards, this 'Külliye' complex was a charitable organisation with a school, hospital and caravanserai lodgings. The interior of the mosque is a soaring space with a height double the diameter of the dome. The tomb of Süleyman is located within, along with the tomb of its architect Sinan, subsequently located on the periphery.

The city has grown apace with different neighbourhoods and Districts. North of the Golden Horn and opposite the district of Eminönü is Beyoglu, on steeply rising ground which is overlooked by the Galata Tower with its distinctive conical roof, originally built in the 13th century by the Genoese inhabitants.

The main pedestrianised street in Beyoglu is Istiklal Caddesi, once known as the 'Grande Rue de Pera', where the more modern Pera Palace Hotel was located for rail travellers across Europe on the Orient Express during the 19th century. The grand Haydarpasa Station was built in 1908 designed by German architects, as the main rail terminal for the Ottoman railway line to the east to Damascus and Baghdad. A historic funicular railway the 'Tunel' also connects the district in Beyoglu with Karakoy and its ferries.

In the north is the huge open modern Taksim Square at the other end of Beyoglu, that is often the scene of large assemblies and demonstrations. Ataturk had requested the French planner Henri Prost to assist in westernising the city, who proposed the creation of Gezi Park in the 1930s.

Other landmarks in the city of Istanbul, are the grand Dolmabahce Palace on the waterside of the Bosphorus built in 1856 for the Ottoman Sultans with its ceremonial features and elegant fountains, and Yildiz Palace in Besiktas, now the residence of the President.

Modern City – Buildings and Arteries

There are other Districts on the European side which have their own distinctive characteristics such as Sisli, Besiktas and Ortakoy. Straddling the Bosphorus now is the striking set of modern highway suspension bridges that span between the European and the Asian side.

Lit at night by coloured lights, an alternative attractive journey is the bustling ferry below across the Bosphorus to either side, an evocative although short 'sea' trip. The regular ferries make frequent crossings from the western side across to Kadikoy and Usküdar districts on the eastern Asian side. Further travel up the Bosphorus affords views of the villas or individually designed wooden 'yalis' along the waterside.

The city has a network of new expressways, that carry large levels of goods and passenger traffic. The O-1 Highway which forms the inner ring road and crosses the First Bosphorus (15th July Martyrs) Bridge between Ortakoy and Beylerbey was opened in 1973. The O-2 Highway acts as the outer ring which traverses the Second Bosphorus (Fatih Sultan Mehmet) Bridge that opened in 1988 between Rumeli Hasari and Anadolu Hasari, the old castles of Europe and Asia. These are very heavily trafficked tolled routes across these bridges. The Third Bosphorus (Yavez Sultan Selim) Bridge further north was opened in 2016 which carries both road and rail traffic.

In addition, since 2016 the new 'EurAsia' double deck bored road tunnel for cars has been in operation to the south near the new Marmary rail tunnel. Each of these provide significant carrying capacity across the Bosphorus Straits.

In terms of public transport, from its early dependence on trams, the city now has a network of Metro Lines, M1 and M2 that crosses the Golden Horn, M3 and M6 on the European side, and M4 and M5 on the Asian side, with three more lines planned. In addition to this modern metro system, there are light rail and bus rapid transit routes. There is a major airport on each European and Asian side of the city – the smaller being Sabiha Gokcen (SAW) on the Asian side.

The city has a commuter suburban railway system to the older main terminals, of Sirkeci on the European side and Haydarpasa on the Asian side. These have been connected by the Marmaray immersed rail tunnel that crosses under the Bosphorus and connects between Yenikapi and Haydarpasa Stations. This provides an integral rail route now from Halkali on the western side to Gebze on the eastern side, 76 km. in length that was opened in 2019. With these improvements and increase in capacity, they alleviate congestion at peak times of the day.

During the latter half of the 20th century, the population of Istanbul grew rapidly, along with the expansion of the city limits. With an increase in both human and financial capital, these inflows have created substantial growth on both the European and Asian sides. Formal real estate construction, manufacturing, financial and information services have all contributed to growth, especially within the Levent business district. Due to its geopolitical location, on the confluence of both the east and the west continental areas, the city plays a pivotal role in the region.

The current population of the administrative area is approx. 15 million in the urbanised agglomeration. The population density of the whole City Region is approx. 25,200 pers./sq. km. It is still expanding and has a high rate of urbanisation with migration from both eastern and western sides. There is an imbalance with only 35% of the residents being on the Asian side, and 65% of the residents on the European side, with potential scope for further expansion on the Asian side. There remains a considerable volume of informal settlements and 'gecekondu'. With the ongoing refugee crises in the region, and political uncertainties, this risks unsettling planned future developments.

Major Projects

Istanbul has a Metropolitan Mayor who has authority over 39 municipal Districts, an area which is also coterminous with the Istanbul Parliamentary Province. The most recent Istanbul Master Plan defined an agreed set of high-level municipal objectives in terms of globalisation, ecological balance, culture, economics and quality of life.

However, in parallel the President of the Republic launched a directive for a set of mega-projects, without consultation in the Metropolitan area, that would have a significant impact on development, namely for the realisation of the Third Bosphorus Bridge and Ring Road, construction of a Third International Airport and also a new Istanbul Canal as a parallel waterway to the Bosphorus Straits between the Marmara and the Black Sea.

Each of these projects has a potential significant impact on economic activity, transport and mobility. In recent years there has been a great amount of construction of high-rise real estate, commercial areas and condominiums on both sides of the city, which has acted as a strong force of attraction to the city with continuing high rates of migration especially on the Asian side.

Whilst the new Third Bridge and new Istanbul Grand Airport (IST) have both been realised and opened, the full impact of these projects has yet to be fully assessed. In addition, the environmental impact and affordability of the planned 45 km. canal in a zone having tectonic faults and with the natural Lake Terkos, is highly sensitive. These will have both local and regional impacts, and mitigating measures remain to be fully addressed.

In effect, the most significant material development has been the realisation during the last 50 years of five fixed crossings across the challenging Bosphorus waterway. More than anything else, these infrastructure links have served both to integrate and facilitate movement each way across different European and Asian neighbourhoods acting as one enlarged metropolitan area and having strategic significance in the region.

With Istanbul comprising two rather distinct areas either side of the Bosphorus, the Istanbul Development Agency (ISTKA) has responsibility for overseeing the future development of the region as outlined in the Regional Plan for 2023. Whilst guided by the Tenth National Development Plan and the National Strategy for

Regional Development, it has defined the vision, strategy and goals for the immediate future of Istanbul.

In summary, these are encapsulated as aspiring to being an 'Unique Istanbul, a city of innovation and culture with creative and free citizens' with a 'globally decisive, high value-added innovative and creative economy' and a 'fair sharing, inclusive and learning society' with 'joyful authorities, urban spaces and sustainable environment'.

Whilst these objectives are well expressed, however, with the politico-economic and population challenges at the geo-physical location in the Near East where Europe meets Asia at the Bosphorus, it is not just building new infrastructure that is required, but also managing population pressures, and building institutional capacity in the metropolitan region in this new digital age. These are the challenges for the 2050 Metropolitan vision of Istanbul in modern Turkiye.

Overall, Istanbul, former Constantinople, has a wealth of history, charm and character. It has been described by local author Orhan Pamuk as having a prevailing atmosphere of 'melancholy', especially evident with the humidity and in the winter season and during different seasonal winds and storms. Its eclectic bringing together of peoples from both the east and west, and with its dramatic features along the Bosphorus, despite its density and challenges of everyday congestion, make it a unique and memorable urbanised setting.

Transactional Forces and Typology

With its unique geographical setting at the junction of the two major land masses, European and Asian, and on a dividing waterway that connects the Mediterranean with the Black Sea, Istanbul has arisen and been settled as a result of the movement and trade of peoples crossing the Bosphorus. With a natural harbour and man-made connecting roads, it has evolved as a pivotal location on the land and maritime Silk Routes for exchange of textiles, spices and oriental crafts.

From classical and Byzantine times, groups of people have travelled and traded across the region via the city. At its zenith under Ottoman times, people settled around the Golden Horn and its adjoining area. Whilst formerly a barrier, the challenging Bosphorus crossing has been connected in modern times with two tunnels and three bridges, which has integrated the city by bringing the two continents together, contained by its enveloping ring road and enhanced infrastructure connections. With improved access, settlements are extending on the Asian side which retains the pull of attraction with scope for future development and with increased migration from Asia and Arab countries.

From its original confined development around fortifications on the Golden Horn, the city is now characterised as a much larger expanding bi-polar city with its twinned developments on either side of the Bosphorus. The realisation of the new crossings has now integrated the city into more of a dual city with twinned poly-nodal centres. With its growth and expansion, this is developing well beyond a Metropolitan into a Megalopolitan region with developments both east and west, and north and south to the Black Sea and Sea of Marmara respectively.

2.1.5 ISFAHAN, PERSIA (IRAN)

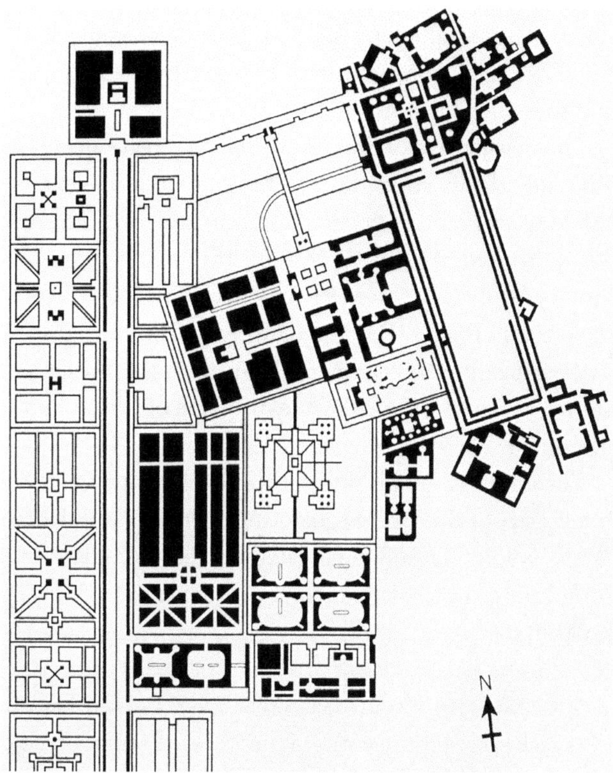

FIGURE 2.1.5 Isfahan – early configuration, 1900.
Source: AP.

Background and Origins

The city of **Isfahan** (Esfahan) in central Persia is located by the Zayandah River, with its ancient bridges of Shahrestan and the long 'Si-o-Se-Pol' (Bridge of 33 Arches), and by the Zagros foothills not far from the ancient Achaeminid capital of Cyrus the Great during the 6th century BCE at Persepolis. From its early beginnings, the city is understood to have developed from its Sassanid area of Jay that merged with a Jewish settlement at Yahudia.

During the 10th and 11th centuries CE, Isfahan was a prosperous walled city with craftsmen in tiles and textiles, with its prominent Friday Mosque.

Whilst Persia has had numerous capitals over the ages, Isfahan was made capital under Shah Abbas I in 1598, which corresponds to 1000 AH in the Islamic calendar. Shi'ite Islam was promoted and prevailed during the subsequent Safavid Empire which dominated the region particularly in the 16th and 17th centuries when the Mughals dominated in India and the Ottomans in Anatolia.

The old city with its religious dominance was connected by a built line of domed mosques and covered suq that wound its way across the city to the Friday Mosque, Masjid-i-Jami. The Sassanid city had had its ancient origins in the Zoroastrian religion, which was followed by the Islamic era and the Seljug Dynasty.

The master plan of Shah Abbas was to create a new city alongside the existing old city that in turn leads down to the long stone bridge over the Zayandah river. This city was organised around three axes, the main artery Chahar Bagh, former vineyard now tree-lined, the axis of the Maidan-i-Shah (Royal Square) lying about 15- to this axis, that knits the commercial centre to the Palace, and another axis determined by the direction of Mecca for the mosques.

The city was rebuilt round the old piazza to create the fine 'Image of the World Square' (Maidan-i-Naqsh-i-Jahan). With its four monumental gateways that dominate each side of the arcaded Maidan, these provide access to four fine buildings:

- The Ali Qapr (High) Gateway with its raised balcony which serves as the entry to the Imperial Palace of 'forty pillars', Audience Hall and Eight Paradises Mansion and gardens, on the western side.
- The opposite Gateway on the eastern side, that leads to the Sheikh Lutf-Allah Mosque, the Shah's private mosque.
- The Qaisariya (Imperial) decorated Gateway, which leads through to the Royal Bazaar, where a long arcade and suq connects a new piazza to the old one to the north and onwards through to the Friday Mosque.
- The southern Gateway leads to the recessed Masjid-i-Shah (Imperial) congregational Mosque that is turned 45°– and is flanked by two minarets and recessed with a small pool.

The doorways are all richly decorated with half-dome 'iwans' and 'muqarnas' in rich colours with tiles and Islamic calligraphy that make these most distinctive.

This ensemble created a magnificent capital city for the Safavid Empire which became an active trading centre, located on crossing trade routes of central Asia until being eventually overthrown in 1722 and the capital subsequently moved to Mashhad and then to Tehran in 1786.

Modern City – Buildings and Arteries

From its ancient origins, the city began to be modernised during the 1930s and infrastructure upgraded. With its expanding population it developed post war but was to change significantly since overthrow of the last Shah of Persia in 1979 when it had just started to industrialise as the 'White Revolution'. Now under the current Iranian radical leadership, it is taking a different Islamic route. Besides its traditional textiles and carpet making, it remains a centre of commerce and trade.

It is now an industrial steel producer and with its new highways and Metro it is one of the largest cities in Iran, after Tehran and Mashhad, with a population of approx. 2 million people. Its character remains with its long bridges, bazaars, gardens and parks, interspersed with mosques, minarets and mausoleums. It retains the network of canals and madis (streams) for irrigation which were built during the Safavid period.

Transactional Forces and Typology

In a rich fertile region, Isfahan was sited at the interaction of the main trading routes north-south and east-west across the country in Middle Persia. Under the former empires of the Achaemenids and the Sassanids, with proximity to the routes to Persepolis and Susa, it featured on the old

Silk Route. In the Islamic era under the Seljug Dynasty, as capital the city flourished. Whilst overrun by the Mongol Timur (Tamerlane), it regained its significance under the Safavids as Shah Abbas rebuilt central Isfahan with its fine walls and gateways, axes and mosques to make it the centre of the Safavid empire.

It was during this time that a number of canals (madi) were built to provide water from the Zayandah River. With silk as an integral part of the economy, and benefiting from the trade routes, carpet weaving, ceramics, crafts and book-keeping also developed. From agriculture and aquaculture, crafts and textiles, modern manufacturing has developed subsequently with its concentration of steelmaking and scientific industries.

As a city laid out and crossed by east-west running rivers developed over many years, the historic centre is defined by the dominating geometric axes of Chahar Bagh in the Maidan and its symbolic gateways and mosques. With its long bridges and aqueducts, these form a recognised feature across the main River Zayandah. As development has taken place further out, the modern Iranian city is described by its urban core with peripheral centres that encompass the city, supported by a new connecting Metro line that traverses the city and perimeter highways. The following figures show the modern layout and configuration of the Imperial Cities addressed in this section as well as their comparative modern city populations.

FIGURE 2.1.6 Alexandria – modern configuration.

FIGURE 2.1.7 Athens – modern configuration.

FIGURE 2.1.8 Rome – modern configuration.

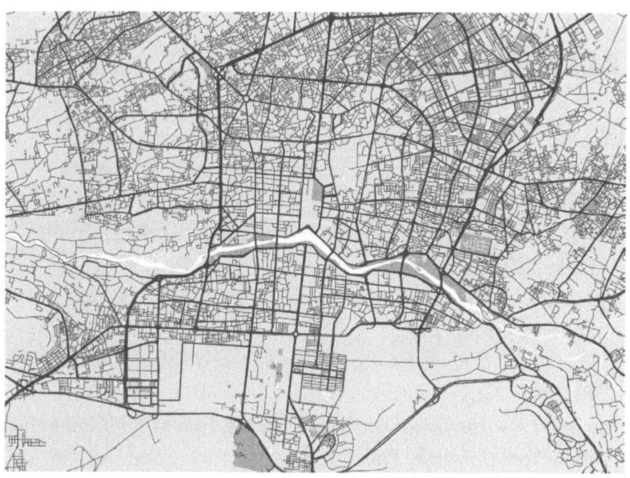

FIGURE 2.1.9 Isfahan – modern configuration.

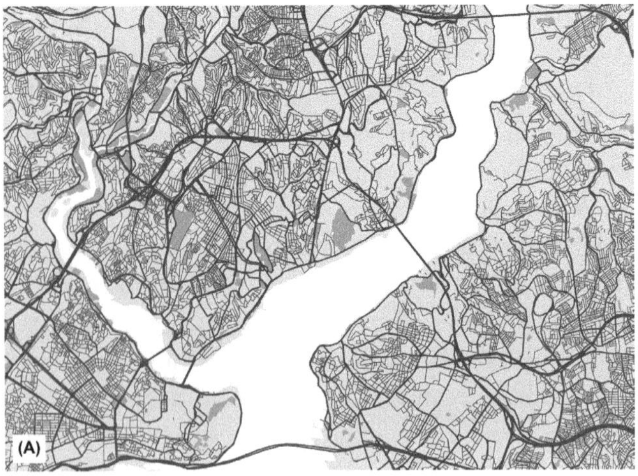

FIGURE 2.1.10A Istanbul – modern configuration.

FIGURE 2.1.10B Greater Istanbul – modern configuration.

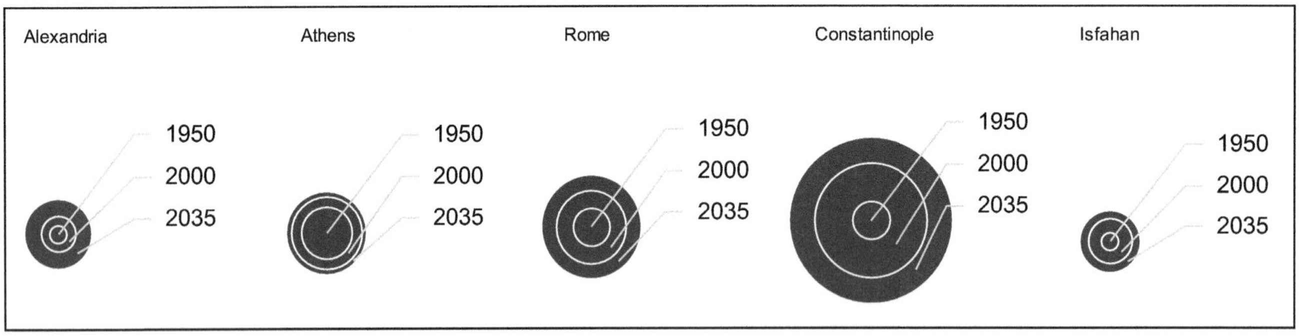

FIGURE 2.1.11 Imperial city populations.

Data source: Global Database of Metropolises 2020 – Un-Habitat (2020), Un DESA Population Division.

Further Reading

ALEXANDRIA

Duff, R E B – *100 Years of the Suez Canal*, Clifton, 1969.

Empereur, Jean-Yves – *Alexandria, Past, Present and Future*, Thames & Hudson, 2002.

Forster, E M – *Alexandria, a History and Guide* (1922), Tauris Parke, 2019.

Issa, Islan – *Alexandria, the City That Changed the World*, Pegasus Books, 2023.

Pollard, Justin & Reid, Howard – *The Rise and Fall of Alexandria*, Penguin, 2007.

Raymond A – *The Great Arab Cities of the 16th to 18th Centuries*, New York University Press, 1984.

ATHENS

Georgopoulo, M & Thanasakis, K, eds. – *Ottoman Athens, Archaeology, Topography & History, American School of Classical Studies*, Athens, 2019.

Hill, Ada – *The Ancient City of Athens*, Harvard University Press, 2014.

Niels, Jenifer & Rogers, Dylan, eds. – *The Cambridge Companion to Ancient Athens*, Cambridge University Press, 2021.

Plutarch – *The Rise and Fall of Athens*, Penguin Classics, 1973.

ROME

Addis, Ferdinand – *Rome Eternal City*, Head of Zeus, 2019.
Beard, Mary – *SPQR, A History of Ancient Rome*, Profile Books, 2016.
Fletcher, Caroline – *The Roads to Rome, a History*, Bodley Head, 2024.
Hibbert, Christopher – *Rome, the Biography of a City*, Harmondsworth, 1985.

CONSTANTINOPLE – ISTANBUL

Crowley, Roger – *Constantinople, the Last Great Siege 1453*, Faber & Faber, 2005.
Freely, John – *Istanbul, The Imperial City*, Viking, 1996.
Harris, Jonathan – *Capital of Byzantium*, Bloomsbury Academic, 2017.
Hughes, Bettany – *A Tale of Three Cities: Byzantium, Constantinople & Istanbul*, Weidenfeld & Nicholson, 2017.
Pamuk, Orhan – *Istanbul*, Faber & Faber, 2005.

ISFAHAN

Blunt, Wilfrid – *Isfahan, Pearl of Persia*, Pallas Athene, 2009.
Emami, Farshid – *Isfahan, Architecture and Urban Experience in Early Modern Iran*, Penn State University Press, 2024.
Irving, Clive – *Crossroads of Civilization – 3000 Years of Persian History*, Harper & Row, 1979.

2.2 Theistic Cities

Jerusalem – Islamic Makkah – Sikh Amritsar – Hindu Benares

Jerusalem, Makkah, Amritsar and Benares

2.2.0 INTRODUCTION – THEISTIC

From their natural geographic settings, around which settlements have grown, groupings of peoples have coalesced in urban centres with similar backgrounds and exchanged goods and artefacts. From early civilisations there was generally a common bond of belief in socio-cultural systems having transcendental or spiritual elements, divine or sacred. Particular theistic practices developed in different cultures that sought to give meaning to life, which evolved variously in different regions where populations gathered.

In early ages, settlements tended to occur at natural, sacred or pagan sites, such as those that had symbolic connotations, like particular mountains in Greece, or significant natural features like volcanoes. With assimilation with various gods of the sun or moon, astronomical features were often used to determine the optimal siting and orientation of sacred monuments, for example as in Egypt and Meso-America.

Layout and configuration were important principles also for building both in China and India. As specific religions became more formalised by region, traditions became annotated and codified, which led to established processes that were handed down to following generations, whether Taoist, Buddhist, Hindu or Islamic, each leading to a different arrangement and decoration of built forms.

These religions developed in the 'Axial Age' with texts such as the Hebrew Bible, the Analects of Confucius, the Tao Te Ching, the Bhagavad Gita and the Sutias of the Buddha. Each of their own sacred sites were upheld in individual places as having particular significance for their followers, whether Judaeo-Christian in Jerusalem, Islamic in Makkah and Medina or Sikhism in Amritsar. These theistic places have come to hold special symbolism as places of worship, as for example Varanasi in India, Angkor Wat and Angkor Thom at Siem Reap and Shinto shrines in Japan.

A significant majority of the world population is now associated with the principal religions of Islam, Judaism, Christianity, Hinduism, Buddhism, Sikhism, Jainism, Shintoism, Taoism and Zoroastrianism. There are in addition a number of indigenous folk religions that exist across Africa and Australasia and Oceania, each of which acts as a binding movement around which gatherings of people have coalesced since ancient times under organised centres of worship.

2.2.1 JERUSALEM, ISRAEL

FIGURE 2.2.1 Jerusalem – early configuration, 1877.
Source: iStock.

Background and Origins – Ancient City

Jerusalem (Yerushalayim) is at the centre of the Holy Land, an ancient walled city that stands on the Judaean hills to the west of the Dead Sea. On a hilltop with the Mount of Olives, it holds a commanding position over the valleys below. The earliest settlement of over 3,000 years ago was a ridge bounded by Siloam Valley to the east and Tyropoeon to the west. The first defensive wall was built around 1800 BCE and here the area of the ancient 'City of David' was developed during the Bronze Age.

By tradition, being the area where Solomon built the Great Temple, where Jesus was crucified and also where Prophet Muhammad visited on his 'night journey', this city represents a religious centre of great significance for

DOI: 10.4324/9781003353348-6

different groups of people following the Abrahamic religions. Since the early Canaanites, Assyrians, Babylonians and Romans, however, little evidence of their settlements remain except for the platform of the Second Temple.

As part of the extensive outer retaining walls, there is only one section of the original Western Wall remaining from the time of Herod, labelled by Jewish followers as the Wailing Wall. A second wall was later constructed to connect with the first around the temple area.

For the Islamic community, the Mosque of Al-Aqsa here remains as one of their most holy sites, after Makkah and Medina, within its walled compound. For Christians, Via Dolorosa is the route of the last steps taken by Jesus Christ before crucifixion and his tomb is in the Church of the Holy Sepulchre where he was buried.

With the exile of the Jews from Jerusalem by the Romans, and the following Byzantine period, there followed the Arab dynasties of the Omayyads, Abbasids and Fatimids. This was succeeded by the Crusaders, resulting in historical terms in a series of several different centuries of upheaval.

Under the Ottoman Empire a third set of city walls and gates in Jerusalem was constructed and strengthened. But it was not until this empire eventually collapsed that the Turkish rule ended, and in the 1920s the British mandate for the region of Palestine was instituted. At that time a plan was made for the development of the city in some outer areas to the south and west but remained unrealised. But with two world wars and after extensive periods of conflict, the independent State of Israel was eventually realised along with formalisation of the autonomous city of Jerusalem.

Quarters

Historically there were four different quarters of Jerusalem – Christian, Muslim, Jewish and Armenian, with the walled city being accessed by a number of distinctive gates.

The main feature of the *Jewish Quarter* is the Western Wall, that part of the retaining wall of the Temple Mount that was built to hold the Ark of the Covenant and recognised by the orthodox Jewish community in prayer as the Wailing Wall. The centre of the quarter, which can be accessed from either Jaffa or Dung Gate, is Hurva Square, which is largely residential, with the Roman Cardo street and remains of the Broad Wall and a number of synagogues.

The *Muslim Quarter* is the largest and most densely populated quarter. With its gates of Damascus, Herod's and St Stephens, and its busy souks and small domed buildings, it is crossed by the Via Dolorosa. The open esplanade that surrounds the octagonal Dome of the Rock with its exposed sacred bedrock is now known as Haram esh-Sharif and also contains El-Aqsa Mosque, which, although being rebuilt several times, is the main Islamic place of worship. The dominant golden Dome of the Rock built in 690 CE contains fine tilework, calligraphy and mosaics.

The *Christian Quarter* can be entered at Jaffa Gate and is a mix of stores and souks and spiritual places. The Church of the Holy Sepulchre, which contains the tomb of Christ, has been rebuilt on the site of an original basilica, approached from a small courtyard. Adjacent is the Rock of Golgotha, the place of the site of his crucifixion.

To the south is the *Armenian Quarter*, which is quieter and contains the Citadel, an ancient bastion by the city wall that dates back to the times of Herod. Armenia was the first country to recognise and make Christianity the state religion. It is an unusual place with courtyard and ramparts, Phasael's Tower and the Minaret of the Tower of David, and an area between Jaffa, Dung and St Stephens gates. Beyond the Old City to the east is the Mount of Olives with the Chapel of the Ascension, whose slopes have been used for many years as a place of burial with many tombs laid close together.

Modern City – Buildings and Arteries

Post-war after the creation of the United Nations, a Partition Plan for Palestine had been prepared by the UN in 1947 based on the original Armistice Line, which would divide the city into two separate zones of West and East Jerusalem. But wide divisions were to lead to Arab-Israeli conflicts, resulting in separation and eventually to West Jerusalem being established as the seat of government for Israel and its parliamentary institutions of the Knesset and legislative and executive administrations being established there. Further, with the Six Day War in 1967, much of the old Moroccan Quarter of Jerusalem near the Western Wall was destroyed.

Any new building that is planned in the city is to be built using the local Jerusalem stone, which has resulted generally in retaining the visual coherence of the city. Despite some of the larger governmental and institutional areas, which reflects the burgeoning administration of the state and the city, the scale of development has generally followed the organic evolution of the city.

In addition to major new buildings such as the Israel Museum, the National Library, the Bank of Israel and the campuses of the Hebrew University, there is one significant and symbolic new building, the Yad Vashem Memorial Building (designed by Safdie) on Mount Herzl with its Halls of Names and Remembrance that recognises those lost in the Holocaust.

Communications have been improved both within and outside the hilly city. In the centre a new light rail system has been integrated into the old streets of the city to supplement buses with the aim of improving public transport. With a new deep central station, a fast high-speed twin track railway line has been built that connects Jerusalem with coastal Tel Aviv and that has been constructed on long sections of viaduct and tunnel.

Controversially, however, new Israeli settlements have been built in ring neighbourhoods surrounding East Jerusalem with significantly sized new populations. Inward migration has continued into several areas of the city despite its divisions with the population of the city of approximately 1 million people.

Transactional Forces and Typology

With the city of Jerusalem located on the Judaean Hills, the Old City is sited on the hilltop by the Mount of Olives and the ancient City of David was settled in a defensive position on an elevated ridge. With the walled core of the city divided into four quarters of the Jewish, Muslim, Christian and Armenian quarters, it remains a most historic and religious centre.

With construction of successive defensive walls from Babylonian to Roman times, it was variously occupied by tribes of Israel. Jerusalem became the capital of the Kingdom of Judah and thereafter maintained its centre and in having particular religious significance. The Romans expanded the walls and built up the area around Temple Mount. Under subsequent periods of Byzantine and Islamic control, different districts were developed further. As it passed through subsequent periods of occupation from the Crusaders, Ayyubid, Mamluk and Ottoman Turks, new neighbourhoods began to be developed outside the original city walls. Now these have become yet more densely populated around rocky outcrops, extending outwards into built-up areas where the topography permits.

As a hillside city settled around its historic core, it was shaped by built fortifications around its natural contours. With the accretion of different districts and resident populations, these distinctive districts have reflected the ongoing divisions of the city. Outside the central core, the growing population has settled on the slopes beyond renamed Mount Zion, and in individual more distant nodes of settlement surrounding the city.

Contradictorily, modern-day Jerusalem is being shaped by its artificially framed borders, especially with the division of East Jerusalem, having a network of heavily fortified tall concrete walls now separating and dividing the city from the adjacent West Bank area of Palestine. With its outer areas divided between West and East, new permanent physical barriers have been constructed that weave their way circuitously around eastern Jerusalem. In certain areas this non-contiguous line has resulted in enclaves that have become physically walled off, with some of these built areas even going beyond the Green Line of the initial partition of the state.

This has had significant impacts of considerably extending the urbanised populated and built-up area around the periphery of the city and has led to the harsh division and partition between Jewish and Arab communities. In the absence of mutual agreement as to ongoing sovereignty, rights and administration, segregation remains the prevailing outcome of this divided city.

2.2.2 MAKKAH, SAUDI ARABIA

Background and Origins

Makkah, or Makkah-al-Mukarramah, is located inland from the Red Sea in the hills of western Saudi Arabia, an ancient city, the birthplace of the Prophet Muhammad. As one of the holiest cities in Islam, it is the destination for many religious visitors every year during the Haj and Umrah pilgrimages. With the Great Mosque (Masjid al-Haram) of Makkah, it has the Hira cave in Jabal an-Nour mountain nearby, the site from where it is understood that the Koran was revealed to Muhammad. Not far away is Taif, a cooler city in the mountains, and Medina farther north, and on the original caravan routes and trading centres between the Mediterranean and the Indian Ocean.

This famous shrine, the black stone Ka'bah, stands in the middle of a colonnaded sanctuary, an extensive site with its six corner minarets and mosques, where pilgrims visit on Haj every year. The Ka'bah determines the 'qibla', the orientation of the direction towards which Muslims pray, wherever they are located. Originally in a sandy wadi, built development has occurred all around this, with oratory houses overlooking the main courtyard.

Early evidence of the nomadic tribes that occupied the region has only been found from the 5th and 6th centuries CE, with some coastal settlements and other localised areas around oases on the caravan routes in what is generally a hot and inhospitable desert region. Settlements occurred in Makkah itself around the Wadi Ibrahim, a valuable source of water, where the Zamzam well is located.

With the birth of Muhammad in 570 CE, and after his divine revelations in 610 CE, he moved to Medina, returning to Makkah after the Treaty of Hudaybiyyah, promoting the Abrahamic religion of Islam since when the Islamic calendar has run from the year labelled as AH 1. With the cubic Ka'bah initially built there by Ibrahim and his son Ishmael, this was to become the most significant centre of worship for subsequent followers.

In a region that was earlier occupied by the local Quraish tribes, with incursions of the Sassanid and Byzantine empires, the centre of Makkah was developed through trade and barter from passing traders and local Bedouin. As the Islamic realm grew, Makkah was to attract pilgrims subsequently from all over the Muslim world.

However, in subsequent years the Umayyid Caliphate moved the capital to Damascus, the Abbasid Caliphate to Knifa and the Ottoman Caliphate to Constantinople. However, after the latter's modernisation under the Tanzimat, the Ottomans were challenged by the House of Saud that established the first Saudi state in the country that was to lead eventually to the creation of the current Kingdom of Saudi Arabia.

Modern City – Buildings and Arteries

The Grand Mosque area has been substantially redeveloped and expanded in recent years. With the Ka'bah and the Black Stone, covered in black brocade Kiswa, the Haram site is surrounded by an extensive wall, with tall minarets, enabling thousands of pilgrims to assemble in the area and undertake the sevenfold walk around the Ka'bah, which occurs during the annual Haj.

Surrounded by new high-rise buildings and hotels, the area has been significantly developed beyond the original confines. Beyond is the tent city of Mina, which was established in order to be able to temporarily accommodate large volumes of visitors during the pilgrimage season.

To the south of the Masjid-al-Haram, the enormous modern Abraj-al-Beit commercial complex has been built, a seven-building complex with central Royal Clock Tower and spire that required the demolition of the old Ottoman Ajyad Fortress. Whilst this has totally transformed the urban area, it was seen as being a necessary addition to the built form so as to accommodate seasonal pilgrims.

The Ottomans had originally built the Hejaz Railway, a narrow gauge system, to link Damascus with Medina. But interrupted by World War I, it was not completed to Makkah and never reached Istanbul in the north. The modern al-Haramain high-speed rail electrified double track line has now been built to connect Makkah and Medina via Jeddah and via King Abdullah Economic City (KAEC) with King Abdulaziz International airport at Jeddah (JED) that has its own dedicated Haj Terminal.

The city has four ring roads, where Makkah Gate is a landmark like the Koran Gate at the west of the city, which is accessed by the two main highways, 40 to Jeddah and 15 to Medina. There is a dedicated metro line now, the Al Masha'er Al Muqaddasseh Metro in the central area, which opened in 2010 and which connects the localities of Arafat, Muzdalifah and Mina for Haj pilgrims.

In addition, the Makkah Mass Rail Transit of four lines is planned and under construction. The population of the city is approximately 2.2 million, but this can more than double temporarily at the seasonal period of the Haj.

Transactional Forces and Typology

From its location near the south-western coast on the Red Sea of the Arabian Peninsula, its early settlement evolved from its location on the camel caravan trade routes. From its trading origins, however, its transformation was to take place after the birth of Muhammad and his divine revelations that led to the place becoming such a focal point as a place of pilgrimage. From that time onward, Makkah and its surroundings have developed to accommodate the enormous flow of annual visitors to the Holy City and the Grand Mosque. In the modern age, as transport links have improved, this has led to the increasing multitude of visitors to the city and surrounding areas such as Mount Arafat, resulting in their accommodation in areas such as Mina, with its huge but temporary 'tent city'.

The city of Makkah is centred around the al-Haram area and the Masjid al-Haram. The old city with its gates and the Ibrahim-al Khalil main street with its former traditional houses, has now been significantly developed. The modern Abraj al-Bait Complex, with its tall clock towers overlooking the square containing the Grand Mosque and the Ka'aba, has totally transformed the core of the city. Beyond, the city is now surrounded externally by a series of circumscribing ring roads that have been built to handle the seasonal trips of visitors and the new rail and metro connections have been necessitated, as in other modern cities. However, its inland hilly arid characteristics remain, with a central core and peripheral developed areas, as a unique Islamic holy city.

2.2.3 AMRITSAR, INDIA

Background and Origins

Amritsar is located in the Majha region of Punjab between the Sutlej and Ravi rivers in northern India, tributaries of the Indus River, north-west of the modern state capital of Chandigarh and close to Lahore. The holy city of Amritsar was founded in the Sikh tradition by the fourth of ten gurus, Guru Ram Das in the 16th century, known as Ramdaspur, built around a central man-made lake. As the town developed and expanded, this area was extended into a temple complex, which became the Gurdwara Harimandir Sahib, the Golden Temple, and the spiritual centre of Sikhism.

During the 18th century the Sikh gurus established military forces to resist Shah Jahan and the Mughal rulers in the region, with different clans and 'Misls'. But the Sikh confederacy was defeated in 1766 and the group disbanded. However, under Ranjit Singh, declared Maharaja of Punjab, the Sikh Empire was re-established in 1799 with the capture of Lahore and the creation of a wide empire from there that extended across the region from the Khyber Pass to Kashmir, with unification of various Sikh Misls.

During this period the Maharaja developed and refortified the city of Amritsar with a long surrounding circumferential wall and 12 gates. Whilst the Anglo-Sikh wars were resolved under the Treaty of Amritsar, further conflicts continued until the British Raj took control with their defeat in 1849, walls were rebuilt and the area was annexed to become part of the British Province of Punjab.

During World War I there was considerable unrest amongst the Sikhs and the Ghadarites seeking independence, which led to the major Jallianwala Bagh massacre in April 1919. Later, during World War II, several regiments consisting of Sikhs were engaged who fought bravely during this war on behalf of the Allies.

With the independence of India in 1947 and partition of Pakistan, further upheaval was to occur when Punjab was split, with Amritsar remaining in the Indian state of Punjab that was to have a new state capital built, Chandigarh for both Punjab and Haryana, whereas Lahore became the capital of Punjab Province in Pakistan. This was to result in much conflict between Hindus, Sikhs and Muslims during the redefinition of state borders and partial destruction of the city when the nearby Sutlej River overflowed.

Modern City – Buildings and Arteries

The city is bounded by the river from the north and the canal running diagonally to the south with built development around the cantonment in the centre. The dominant focus is the Grand Temple, besides which is the Akal Takht temple, which became the primary seat of the Sikh community. Adjacent is the Ramgarhia Bunga watch tower complex with its reddish minaret-style twin towers that overlook the temple, which are most visible as seen from the circumambulating walkway and reflected in the rectangular lake.

Roads radiate out from the centre of the city with the Circular Road around the old city. It is located on the old Grand Trunk Road (GTR) to Lahore with a new Delhi-Amritsar-Katra Expressway under construction along with a new city ring road also.

In the centre of the city Amritsar Junction Railway Station was built as part of the original Northern Railways broad gauge system in India, running east-west across the city.

The holy city of Amritsar has suffered several disastrous events over the years. In June 1984 a more recent disaster befell the city when under Operation Blue Star the Indian Army killed a large number of civilian and Sikh militants that had occupied the Golden Temple. The historic Sikh reference library was destroyed, and many lives were lost.

It is a fine city with an active textile economy, with production of woollens, pashminas and handicrafts and also the artisanal production of brass and copper utensils, which are traded.

The population of the city is approximately 1.9 million people, approximately half Hindu and half Sikh.

Transactional Forces and Typology

On a site in an area of tributaries of the Indus River, the settlement of Amritsar developed around a place where Guru Ram Das settled with its man-made lake. This was to become the centre of attraction for Sikh followers in this part of northern India around a collection of buildings and the striking Golden Temple (House of God) with its connecting causeway across the lake. It was also a prosperous centre of trading of grain in the region.

The city itself is typified essentially in having a central core, with roads radiating outward in all directions around which development has been undertaken. The focal point remains the historic core and the spiritual temple in the centre around which paths and development have gravitated.

2.2.4 BENARES, INDIA

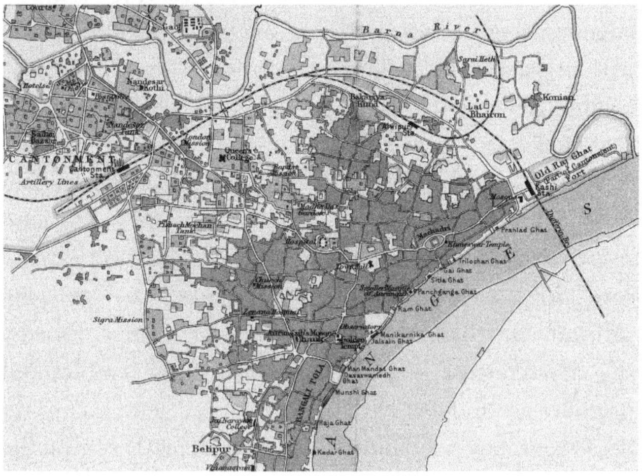

FIGURE 2.2.2 Benares – early configuration, 1930 (Murray).
Source: Maphouse.

Background and Origins

Benares, or Kashi (City of Light), now known as **Varanasi**, is located on the banks of the River Ganges in the modern state of Uttar Pradesh in northern India. A great religious hub, it is of particular significance as a holy Hindu site. Named after the two Ganges River tributaries, Varana and Asi, the old city is bounded by these two rivers in a crescent on the northern shore of the Ganges.

During the first millennia BCE this was the home of the Kashi Tribe. It was deemed to consist of four concentric circles of sacred zones, the largest Kashi that extends into the countryside, followed by Varanasi that corresponds to the equivalent of the modern city, then a smaller Avimukta zone and an Inner Sanctum of Antargriha around the central temple of Shiva Vishvanatha.

It is also believed that it is the site where Buddhism was founded, with Buddha's first sermon at Sarnath and also Parsvanatha a Jain spiritual leader is said to have been born there in the 8th century BCE. Accordingly, it has become a centre of particular religious importance and pilgrimage.

There is evidence of a very early ancient settlement that was allegedly founded by the deity Shiva, according to Hindu mythology, along with Brahma and Vishnu. During the 11th and 12th centuries CE, this was a golden age in Varanasi during the reign of the Gahadavalas and contained a large number of Hindu shrines.

This was to be followed by a long period of Muslim Sultanates when temples were destroyed and replaced by mosques. It was during the 16th century CE when the Mughals ruled that Emperor Akbar allowed the rebuilding of two large temples dedicated to Shiva and Vishnu and ghats were built along the riverside. The infrastructure of the city was developed also with construction of the long Akbari bridge.

However, in the period following Akbar's rule under Shah Jahan, the majority of the temples were destroyed and replaced with mosques. In consequence, the city had many elegant temples and mosques, and on the eastern bank of the river the sandstone Ramnagar Fort had been built in the Mughal style and with one of the earliest observatories, Jantar Mantar, built in 1737.

During the subsequent period under the Maharaja of Benares, it remained as a centre of learning, and development continued under this dynasty and under the times of the subsequent British Raj, who oversaw the development of the infrastructure, improved drainage, the filling of lakes and the channelling of waste water in Benares.

All along the riverside a large number of ghats had been built in distinctive shapes and colours. On these embankments with their stone steps, pilgrims came to the riverbank to perform ritual ablutions. Here also were a number of larger ghats where Hindus cremated their dead, as at Dashashwamedh and Manikarnika Ghat, and also with the Jain Bachraj Ghat. The River Ganges with its myths and traditions continues to serve as the focus of many religious rituals and ceremonies and although now very polluted, the waters retain their symbolic significance.

Modern City – Buildings and Arteries

Since Indian independence in 1947 Varanasi has acted as the headquarters of the Varasi District in the modern state of Uttar Pradesh, whose capital is Lucknow. Silk weaving is still a major traditional industry along with production of rugs and carpets.

In terms of transport, the main railway station Varanasi Junction (Cantt) handles large volumes of passenger traffic and from where express trains run to

Delhi, Lucknow and Ahmedabad, with other stations at Banaras and Varanasi City.

With its congested urban centre with many buses and auto-rickshaws, the ring road and several new roads are under development. The city lies on the Great Trunk Road (GTR) that runs from Kolkata to Peshawar and also on Inland Waterway NW-1. This is now being further developed as inland Varanasi Port for the transport of containers. Newer modern heavy industries have also been developed here, such as at the Benares Locomotive Works (BLW) and Bharat Heavy Electricals (BHEL).

There is a plan for a new Varanasi Metro for the city with two lines to connect BHEL and out to Sarnath. There are a number of old colleges and universities in the city, as the Sanskrit College founded in 1791 and the semicircular planned Benares Hindu University founded in 1916, with new research institutes of technology and colleges of management, which makes it a centre of education.

The population of the city is approximately 1.2 million people. However, a large informal sector remains that has poor and unsanitary accommodation across the city.

With its large volume of domestic visitors that peaks seasonally each year, these all need to be accommodated, especially during festivals, such as the Ganges Festival that occurs after Diwali. As the 'City of Light', it remains a symbolic historic place and luminous city.

Transactional Forces and Typology

Located on an old trade route that crossed the Ganges River, traditionally around this cultural centre, handicrafts and industries developed with silk and muslin, weaving and carpet making and ivory carving. With its large number of narrow streets and winding alleys lined with vernacular buildings, it comprised a dense congested trading centre with characteristic riverside ghats on the riverbanks.

With its historic urban core and alleys, Varanasi is a riverine city mainly on the west bank of the wide Ganges River, as the city continues to radiate out and expand concentrically westwards. The riverside is a deeply significant site that retains all of its spirituality. Broadly the city retains a large number of historic temples and mosques around a very tightly urbanised central core yet possesses many poor and informal residential areas on the periphery. The following figures illustrate the modern configuration of the theistic cities as described and the evolution of their modern population.

FIGURE 2.2.3 Jerusalem – modern configuration.

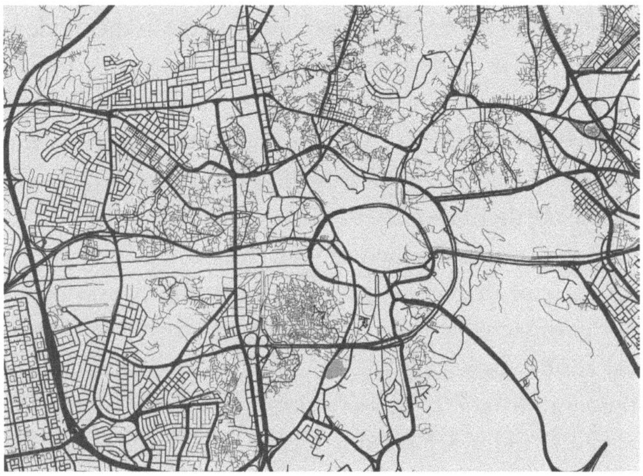

FIGURE 2.2.4 Makkah – modern configuration.

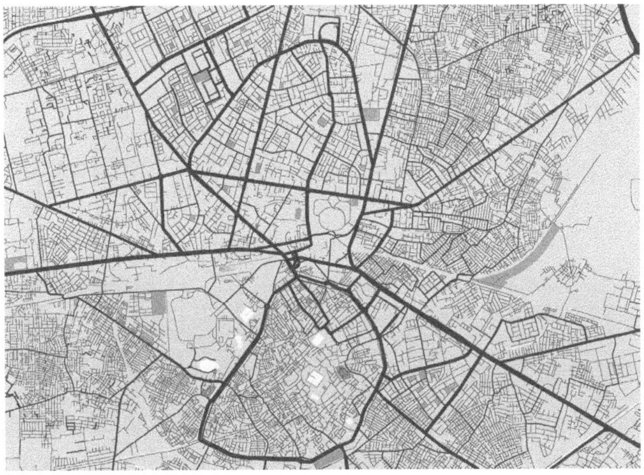

FIGURE 2.2.5 Amritsar – modern configuration.

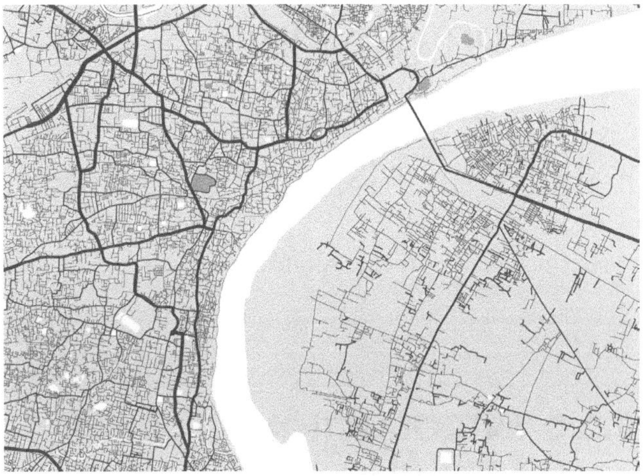

FIGURE 2.2.6 Varanasi – modern configuration.

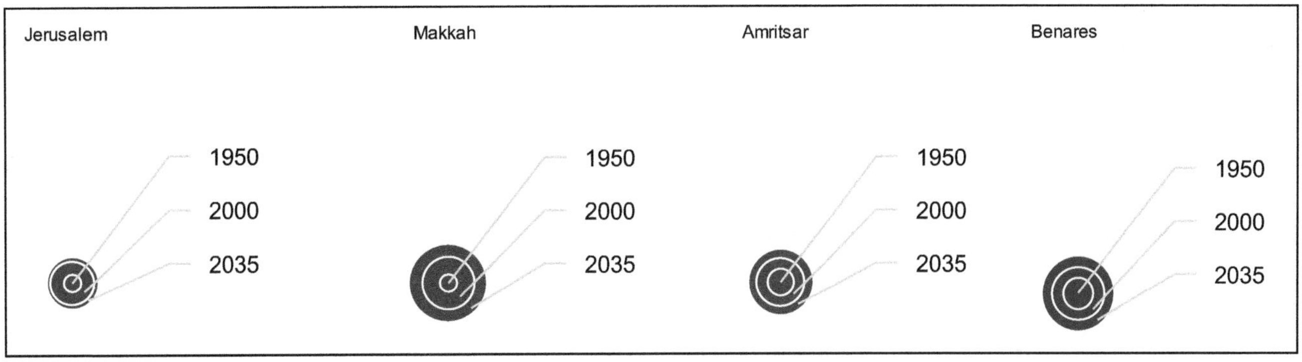

FIGURE 2.2.7 Theistic city populations.
Data source: Global Database of Metropolises 2020 – UN-Habitat (2020), UN DESA Population Division.

Further Reading

JERUSALEM

Hollis, Christopher & Brownrigg, Ronald – *Holy Places, Jewish, Christian & Muslim Monuments in the Holy Land*, Weidenfeld & Nicholson, 1969.
Lemire, V, Berthelot, K, Loiseau, J & Potin, Y – *Jerusalem, History of a Global City*, University of California Press, 2022.
Peters, FE – *The Children of Abraham – Judaism, Christianity, Islam*, Princeton University Press, 2004.
Prag, Kay – *Jerusalem* (Blue Guide) Norton, 1989.
Sebag Montefiore, Simon – *Jerusalem, the Biography*, Phoenix, 2011.
Teller, Matthew – *Nine Quarters of Jerusalem*, Profile Books, 2023.

MAKKAH (MECCA)

Peters, FE – *Mecca, a Literary History of the Muslim Holy Land*, Princeton University Press, 1994.
Sarder, Ziauddin – *Mecca, the Sacred City*, Bloomsbury, 2017.

AMRITSAR

Singh, Fanya – *The City of Amritsar*, Punjabi University, 2000.
Talbot, Ian – *Divided Cities*, Oxford University Press, 2007.
Tully, Mark – *India's Unending Journey*, Random House, 2007.

BENARES (VARANASI)

Dodson, Michael, ed. – *Banaras, Urban Forms and Cultural Histories*, Routledge, 2021.
Eck, Diana – *Banaras, City of Light*, Routledge Kegan Paul, 1983.

2.3 Former Capital Cities

Chinese Dynasties – Chang'an and Luoyang, Peking and Nanking – Japanese Kyoto – Asian Samarkand

Chang'an, Luoyang, Peking, Nanking, Kyoto and Samarkand

2.3.0 INTRODUCTION – FORMER CAPITALS

Such imperial, theistic and historic capital cities, through the passage of time, have continuously evolved in their primacy and importance. But events have occurred, whether natural or climatic, economic or political, that have substantially altered the subsequent human livelihood of several of these existing places.

Natural earthquakes, floods and typhoons have devastated cities; wars, famines and diseases have affected regions; and socio-economic forces have determined the scarcity, attractiveness and associated value of different places and areas of land. For as many causes that have led to the growth and prosperity of a city, as many other causes have arisen that have caused their decline or fall.

Such causes and events have led to the division of cities, and to changes in their administration that have resulted in cities losing their primacy or capital status. Generally, this has been in response to some exogenous shock, but gradual decay and population decline, human occupation and withdrawal have also occurred due to changing transactional, political or natural forces.

Selective cities are illustrated hereafter, outlining the life of former older or provincial capitals that, although still significant cities, due to impending circumstances have lost their role as capital of a province or state and this has been transferred to another more favoured or functional place.

Particular instances are the former capitals of Chang'an and Luoyang, Peking and Nanking in China, Kyoto in Japan and Tajik Samarkand on the old Silk Route.

2.3.1 CHINA – EVOLUTION AND EARLY DYNASTIC CENTRES

The very early origins of human occupation of the central region of China were assumed to have commenced after 'Peking Man' was found to exist in caves near Zhoukoudian in the surrounding mountains during the Stone Age. In the adjoining fertile plain of the Yellow River, which repeatedly changed its course, the tribes of Hua and Xia had inhabited the region and with the beginning of settlement, basic agricultural farming of wheat, millet and rice developed that in turn led to subsequent silk production (sericulture) and jade carving, which was to play an important role in trade and diplomacy under the early emperors.

Subsequently during the Bronze Age under the pre-imperial dynasties, of the Xia dynasty (2000–1600 BCE) and the following Shang dynasty (1600–1000 BCE) in the Axial Age, inhabitants found a way to develop more formidable tools and weapons. With the Xia capital at **Erlitou** near Luoyang and Shang capital at **Anyang**, the Shang was succeeded by the Western and Eastern Zhou dynasties with capital at **Luoyang** at the confluence of the Yellow and Luo rivers that evolved under the Warring States period (475–221 BCE) until ended by the Qin conquest.

Despite emphasis on warfare and the 'Art of War' (Senzi), the Classical Age then was also a fertile period for philosophic reflection with the Analects of Confucius, the Thoughts of Mencius, the Daoism of Laotzi, and the Book of Changes of I Ching, with its hexagrams and deliberations of yin and yang. Such major streams – Confucianism, Daoism and Legalism (of Hanfeizi) – collectively were to inspire the Chinese for many years as '100 Schools of Thought'.

Once the Qin had overpowered the Zhou and as construction of sections of the Great Walls to the north of the Yellow River were built as protection against the Hun, the central region was consolidated between the great Yellow (Huanghe) and Yangtze (Changjiang) rivers of middle states, having a consolidated capital at **Xianyang**. This region was to become the 'cradle' of the provinces of central Chinese civilisation, surrounded by an outer 'core' of other provinces.

Despite the Qin's short dynastic period (221–206 BCE), it had unified a very large region and could be considered to rule the whole Chinese ecumene. The Qin First Emperor was discovered only in recent times, buried in a fine tomb

DOI: 10.4324/9781003353348-7

surrounded by an extraordinary team of thousands of terracotta modelled warriors and horses. The general concept and moral legitimacy was that 'All under Heaven' should be ruled by a single emperor (Heaven's Son).

This was to be succeeded by one of the greatest dynasties, the Han, that ruled over a yet larger territory under the Western Han dynasty from its capital of **Chang'an** in the Wei River valley, out along the Silk and Tea Roads, west across Central Asia, to Persia and Arabia (from 206 to 0 BCE). Under the following Eastern Han dynasty, the capital was relocated again to **Luoyang** (25–220 CE). Here two palaces were built, one north and one south, separated by a long covered walkway that bisected the inner city with external tamped earth walls of 'hantu'.

After the Three Kingdoms of the Wei, Shu and Wu, there followed the many dynasties during the 'Period of Disunion', with the north and south divided (Nan and Bei), and of the six Southern and five Northern Dynasties, with centres respectively at **Pingcheng** and **Jiankang**. This was to be followed by the period of the Jin dynasty (266–420 CE) then the Sui dynasty (581–618 CE), which was finally able to reunite the Northern and Southern dynasties, with Luoyang being substantially rebuilt as the reunited capital. Symbolically this was achieved through the construction and connection with sections of the extensive man-made Grand Canal that stretched ultimately to Jianking, Sukhou and Hangzhou in connecting the great Yellow and Yangtze rivers and Huai River. A further extension was also built north-east to the coast at Tianjin with its parallel post road.

It was during the following period of the Tang dynasty (618–907 CE) that the administrative organisation was strengthened with the creation of ten, then 15 provinces, reverting to having the capital in historic **Chang'an**, and with establishment of a formal tax regime, penal code and even the levying of tolls. As a wealthy capital with its walled imperial city, palace and ministries, this was a period of affluence. However, whilst the capital had variously been moved between Luoyang and Chang'an, eventually the Tibetans, Huns and Mongols ransacked Chang'an and occupied much of Sichuan, and after subsequent rebellions, Chang'an was sacked, not to serve as dynastic capital again.

Thereafter, with an interim period of rival states of the Five Dynasties and Ten Kingdoms, the Song emperors (960–1279 CE) came to prominence. With the Northern Song capital under the Later Jin in **Kaifeng** and the Southern Song capital in **Hangzhou**, the land south of the Yangtze River now became more developed as an integrated political and cultural unit. Hangzhou with the West Lake became a prosperous capital and near Suzhou with its fine gardens. In general, borders of the country were fluid, with the country really operating as a set of different macro-regions.

At this time the Mongols were increasingly marauding in the northern region and under the leadership of Chinggis Khan, whose armies overcame the Jin capital and much of Manchuria. With most of northern China occupied, Khubilai succeeded Chinggis as Great Khan and resumed the offensive farther south in Sichuan and over the Southern Song and established the capital at **Shangdu** (which was idealised by Coleridge as Xanadu). Moving farther south, Shangdu was retained as a summer capital and a new capital was established at **Dadu**, carefully laid out, which was to form the essential basis of the future capital Beijing.

Expansion continued south to the Yangtze and beyond to Hangzhou, which was to result in the conquest of southern China becoming united under one ruler and the beginning of the Yuan dynasty (1279–1368). Further centres developed in and along coastal areas where trade prospered and populations gathered, in particular around **Canton** inland on the Pearl River. The Mongols overextended themselves by land and sea and with the creation of large provinces, not long reunited, they tended to operate as interdependent khanates. The Grand Canal was extended north to the sea, despite the meandering path of the Yellow River changing its flow into the Gulf of Bohai.

New movements developed in and around the area of the Yangtze Basin under Zhu Yuanzhang. As Hongwu emperor, he took over the Yuan capital in the north and with Nanking captured and fortified, this was to lead to the creation of the new Ming dynasty (1368–1644). With capital at Dadu, it was renamed **Bei-ping** (Northern Peace) then renamed **Bei-jing** (Northern Capital), where the new emperor resided in developing the 'Forbidden City' and instituting many reforms. Under this dynasty, sections of the Great Walls in lozenge shape with towers were built as fortifications to the north. As the population grew over the next 250 years, several significant cities came to prominence across the country. The transition from the Ming dynasty to the Qing dynasty was abrupt, with Manchu conquests from the north and taking of Nanking and Yangzhuo and the Yangtze south region. As a consequence, the middle 'Three Emperors' of the Qing dynasty (1644–1911) remained in power until the beginning of the 20th century, with its imperial capital located in Peking **(Bei-jing)**, with outlying Summer Palace, until all was to change significantly thereafter.

2.3.2 ANCIENT CHANG'AN AND LUOYANG, CHINA

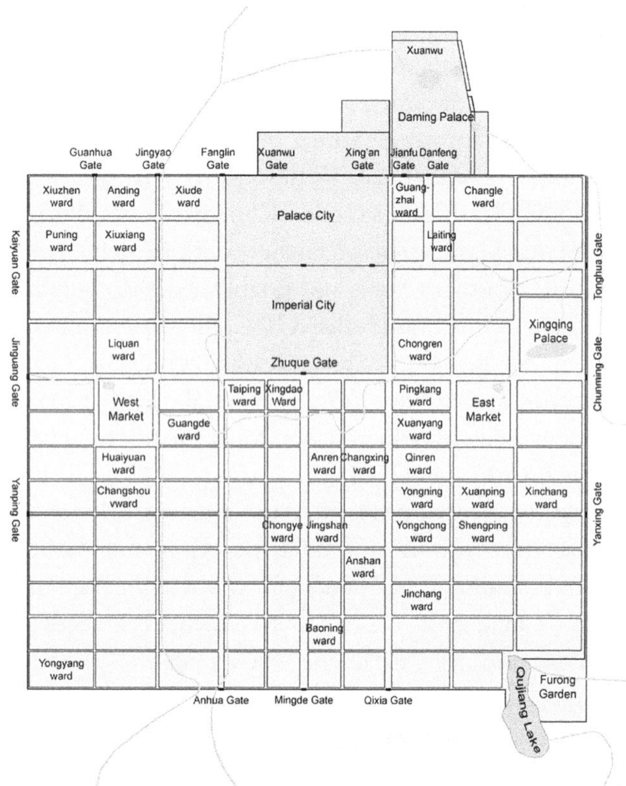

FIGURE 2.3.1 Chang'an – early configuration.

Source: WMoL.

Chang'an

Formerly **Chang'an** was one of the four great ancient capitals of China, located in the vicinity of the neighbouring modern city of **Xi'an** in central China by the Wei River. This area served as the site of the capital under several dynasties over previous centuries and in acting as a key location at the eastern end of the great Silk Road.

During the Han Dynasties, the city of **Chang'an** was established as Western Han's capital and, later, **Luoyang** became the Eastern Han capital. With the end of the Warring States period, China was, for a period, unified and the built development of Chang'an occurred around 220 BCE when the Changle Palace (East Palace) was built south of the river from the former Qin capital of **Xianyang**, with the Weiyang Palace (West Palace) added a few years later. With its long, thick perimeter walls, these encompassed an area much greater than the Forbidden City in Peking (Beijing). With the capital subsequently being moved east to Luoyang during the 'Three Kingdoms' period, during the Sui dynasty, the capital city was known as **Daxing** at an adjacent site built around Daming Palace.

Chang'an was not to revert to being capital again until the Tang dynasty when Chang'an was considerably redeveloped as a major city and laid out with a number of wards. With its palaces and perimeter walls, the city was laid out in a traditional Chinese grid pattern into nine districts with three crossing streets and avenues with connecting gates and gateways. At this time there was an inner imperial city with an outer civilian section with a series of markets. The two Great and Small Wild Goose Pagodas were built to store Buddhist sutras (scriptures) and relics. The city walls and gates were rebuilt and the Drum and Bell towers were built within the city with the West Palace and West Park.

The city declined at the end of the Tang dynasty with the Mongol invasions and eventual transfer of the capital to Luoyang, and it was not until much later modern times that the city was to flourish again. However, its influence did extend out into Japan where the cities of Nara and Kyoto came to be modelled after the layout of early Chang'an.

Luoyang

Luoyang, now in west Henan Province located in the central plains of China at the confluence of the Luo and Yellow rivers, east of **Chang'an**, was another of the four ancient capitals of China. The area's long history evolved with different individual sites acting as capital from the time of the earliest Ancient Era to the Imperial Era. Its earliest name was **Luoyi**, dating back to the Zhou dynasty. Following the basis of ancient Chinese planning rules, work commenced in 1059 BCE, laid out as a gridded city of considerable scale.

It was developed under Cheng as **Chengzhou** and subsequently as capital under the Eastern Zhou dynasty. The city was consolidated and built up around this site. It was built within a boundary of a rectangular city, with walls and regular gates having separate Northern and Southern palaces connected by a covered way, on the north bank of the River Luo.

After the Qin and Western Han periods, **Luoyang** became capital later again under the Eastern Han dynasty in the period of 25–190 CE and became a prosperous trading centre. After 190 CE, the court was moved westwards back to the more defensible **Chang'an** as capital. During subsequent periods of unrest, rebellion and disunion, China was divided during separate North-South dynasties.

Luoyang acted as the 'Middle Capital' of the Western Jin, between 265 and 313 CE. Then, under the Northern dynasty, it was capital of the Northern Wei, up to 534 CE, during which time the 'Longmen Grottoes' were carved in local rocks when Buddhism started to take hold with temples and buildings such as the Yongmin Monastery, after the earlier White Horse Monastery.

When China was reunified under the Sui Dynasty, much of Luoyang was destroyed. At this time a vast building programme was commenced to connect Luoyang with Chang'an in the west by canal, and eventually to be expanded with the 'Grand Canal' to the north, which was used as a transport route for logging for building.

During the later Tang dynasty, it became known as **Dongdu** and was dedicated as the 'Eastern Capital' between 657 and 690 and was an important place on the Silk Road. There was a significant period of building with the 'Tian Teng' Heavenly Hall under Empress Wu. Under the restored Tang in periods up to 907, much of this building, however, was demolished.

Whilst remaining as capital at the beginning of the Five Dynasties period, under the subsequent Jin and Song dynasties it was abandoned as the capital was moved to Kaifeng. Despite rebuilding around 1034, in subsequent periods under the Yuan it was destroyed and rebuilt under the Ming and the Qing dynasties. Whilst most areas of the city had been destroyed, sections of walls have survived.

Transactional Forces and Typology

Both of these capital cities were realised having distinguished patterns of development based on classic Chinese planning. With varying circumstances across middle China, with warring factions and periods of fragmentation, the physical settlement and location of the country's headquarters under different dynasties was constantly being changed. Despite that, on each occasion, the capital was established so as to reflect classic cosmic patterns of arrangement, reflecting the relationship of the earth to the 'Heavenly Kingdom' within a protected walled confine.

The guiding principles for the development of these cities were based upon the 'Rites of Zhuo', the essential principles upon which the early city of Chengzhou, the eastern capital, was built around 1036 BCE in the form of a magic square so as to maximise the flow of divine energy, 'qi', in connecting humans with heaven and earth. This cosmic pattern was derived from the position and motions of heavenly bodies, transcribed in human form, with orientation based upon the cardinal compass points and a rectilinear layout and main axis or 'ritual way'.

Early Chinese cities typically were ordered as a square and were subdivided into walled units of smaller squares of nine 'li' on each side, each side having three gates, and with streets orientated north-south and east-west. This was to mirror the geometry of heaven and earth and to channel divine energy from the centre to the periphery. Designs were built on the principles of geomancy, or 'feng-shui', in order to direct the flows of natural energy or life force and to harmonise with nature. The most important walled sites and palaces, with individual bell or drum towers were positioned accordingly at the centre where the 'qi' energy could radiate outwards. The perimeter walls surrounded planned squares leading to the cellular layout of cities with the typical ancient Chinese grid plan.

2.3.3 PEKING (BEIJING), CHINA

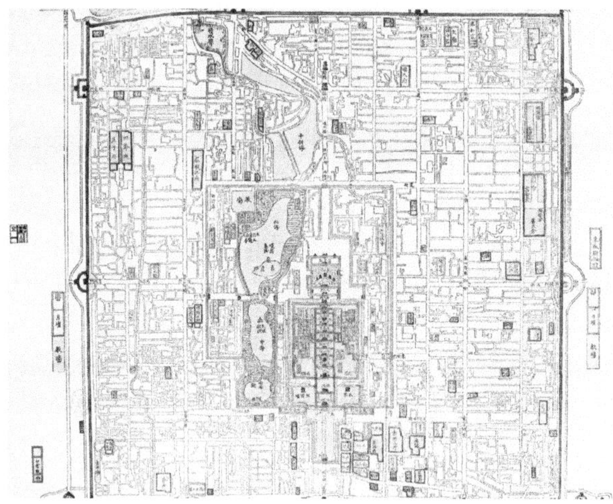

FIGURE 2.3.2 Peking – early configuration, 1900.
Source: British Library.

Background and Origins – Ancient City

In eastern Asia the great nation of China ('Zhonghua') originally centred around the middle Kingdom of the Yellow River with its Han population and has had extraordinary history over many dynasties. From the first Yellow emperors, followed by the Xia emperors, the Shang ruled the fertile River Valley with its centre at Anyang. Its northern capital **Peking**, variously named **Dadu** and **Khanbalik**, also **Beiping and** now **Beijing**, located in the north-east of the country in the plains of Manchuria in an area of dried-up lakes looking out over the Mongolian mountains, is a most enigmatic city.

All of China is monumental, in length or breadth, with its enormous land mass, the two long perambulating Yellow and Yangtze rivers, the man-made Grand Canal, the constructed Great Walls, the lengthy trade Silk Road, its surrounding mountains and Gobi Desert. Its imperial origins were based upon attaining technological developments from stirrups to wheel-carts, large ships and navigation that assisted in the development of transport, both by land and sea. Also, through its own creativity, China facilitated the development of new techniques such as paper and printing, and man-made products such as porcelain and silk.

The earliest recognition of the city as capital was during the long Zhou dynasty (1050–221 BCE), when it was named 'Ji', the 'Place of Thistles'. This was followed by the Qin dynasty, the time of the First Emperor. It was variously named – during the Jin dynasty it was known as **Zhongdu** ('Middle Capital') before subsequent dynasties when for example during the Liao dynasty it became the capital of **Yanjing** ('Southern Capital') (907–1125 CE) again, known as the 'Land of the Swallows'.

With the southern Song having its capital at Kaifeng, the tensions between the north and the south continued. In 1271 the Yuan dynasty, led by Khubilai Khan, administered the extensive Mongol Empire beyond in the north as consolidated by Chingghis Khan who had overrun the Khitans. With its upper capital at Shangdu in today's Inner Mongolia, the Yuan built the white walled capital named **Khanbalik** ('City of the Khan') or **Dadu** ('Great Capital'), with its palace on the site of modern Beijing.

In 1331, when a great bubonic plague developed, the city declined and was overrun with the last of the Yuan emperors, who were then succeeded by the Ming. The new dynasty of the Ming first established the capital in southern Nanking and went about improving the infrastructure of canals and roads. The Tonghui Canal was built from Tonzhou to Peking in 1292 to connect with the Grand Canal, which enabled grain to be transported from the south.

When old Khanbalik was rebuilt as capital, it was initially renamed **Beiping** ('Northern Peace') but was soon renamed **Beijing** ('Northern Capital'). Sections of the nearby Great Walls first made with rammed earth were repaired, strengthened and enhanced by stone and brick ramparts, as the city became the centre of a more unified country with a subsequent long prosperous dynasty with around 150 million people for the next three centuries.

The 'Purple Forbidden City', with its main axes and concentric set of walls that surround the Imperial City, was developed beginning in 1420, laid out symbolically to represent the Heavenly Kingdom in relation to earth and humankind, admired by visiting Marco Polo, who had called it 'Cambuluc'. Further change came under the overwhelming Manchu Army that led to the establishment of the subsequent Qing dynasty. With an expansionist outlook beyond Manchuria, the Qing established their zone of influence far beyond into both Xinjiang and Tibet. Under Emperor Qianlong the Imperial Gardens at Yuanmingyuan were embellished and expanded and then Beijing remained imperial capital under the later Qing dynasty right up until the end of Imperial China with the formation of the Republic of China in the early 20th century.

Forbidden City

Following Khubilai Khan having settled his capital in Peking in the 13th century, and the Ming emperors establishing their seat of power here, then the Qing emperors, the city grew in status. With the 'Purple Forbidden City' at its core, construction began in 1406 with its formal grid layout, constituting an extensive built complex deep within its perimeter of surrounding walls as a nested city, with the southern half for public audience and the northern half for private residence. With the emperor believed to be the 'Son of Heaven', he was to reside in a sacred place that should be 'forbidden' from being overlooked.

The city was approached through imposing gates along an Imperial Way at the celestial meridian via the 'Outer City' (Yongdingmen) into the 'Inner City' (Zhengyangmen) and finally to the 'Imperial City'. The 'Forbidden City' contains a series of grand raised halls hidden behind these tall walls with a number of palaces and cloisters that have been occupied by successive emperors since the 15th century.

The earth was envisaged as a cube with the heavens expanding concentrically around it, with the cosmology reflected in a series of nested rectangles orientated towards the four compass points.

Built on a defined modular system following strict codes, around a central meridian line dividing the city into eastern and western halves, the enclosed buildings were for worshipping imperial ancestors and the open-air altars for worshipping natural deities. With its sequence of spaces, squares and pavilions, from the southern Meridian Gate, 'Wumen', the central path crosses the 'River of Golden Waters' and the 'Gate of Supreme Harmony'.

The Outer Court contains sequentially the three elevated 'Halls of Supreme Harmony', 'Middle Harmony' and 'Preserving Harmony' approached by twin series of steps for public activities where the emperor's throne was located, and with the separate eastern and western 'Flourishing Gates'. These remained the exclusive domain of the Imperial Court standing out in their prevailing royal yellow colour.

The Inner Court echoes this with the smaller elevated Qiangqing Palace, 'Palace of Heavenly Purity', the 'Hall of Union' and the Kunming Palace, 'Palace of Earthly Peace' for private activities. Beyond are the Imperial Gardens with the 'Hall of the Imperial Palace', formerly a temple, with 'Prospect Hill' beyond.

All was surrounded by a continuous perimeter brick wall having four Arrow Towers at each corner. By 1553 the new walls of the Outer City supplemented those of the Inner City, built in brick with their fortified corner watch towers. To the north of the Forbidden City and today's Jingshan Park is an artificial mound colloquially called Coal Hill that was formed from excavations from the lake.

On a continuing axis beyond to the north are the Drum Tower and Bell Tower (built in 1272) that are sited amidst the old 'hutong' district of old Beijing. Here many courtyard houses were built off walled alleyways, having their own privacy, as homes and family residences. The area has continued to be subject to different waves of expansion and redevelopment of the 'Forbidden City', including during the 17th century under Emperor Kangxi, with an ambitious rebuilding programme of several palaces.

Immediately to the west of the city are the 'Lake Palaces' and the 'Summer Palace' where the Qing emperors escaped from stifling summers to Kunming Lake and the gardens on Longevity Hill. With its halls and temples, bridges and walkways, it was an attractive place to recuperate, dominated by the Buddha in the Tower of Fragrance, and the characteristic 17-arch bridge across the lake.

Farther to the south is the 'Tian Tan' Heavenly Complex laid out on a similar north-south axis from the Round Altar via the Red Step Bridge to the colourful circular 'Temple of Heaven' (Quian Dian), where heaven is circular and the earth is square, known as the 'Hall of Prayer for Good Harvests' with a series of decorated timber pillars. This was where the Winter Solstice was celebrated outside the Forbidden City and continued under the Qing dynasty.

Cultural Revolution and Modernisation

Much was to change at the turn of the 20th century, firstly as a result of the Boxer Rebellion in 1901, when the city of Beijing was occupied and divided between the 'Eight Nation Alliance' as a result of the Boxer Protocol. Yet greater changes were to evolve upon the collapse of the Qing dynasty, with the creation of the Republic of China in 1912 under Provisional President Sun Yat-sen. The 4th May Movement that emerged in 1919 at the end of the World War I led to the formation of the Communist Party of China (CCP). Although Peking lost its status as capital again, with the Second Sino-Japanese War and subsequent Japanese occupation in World War II, greater changes were still to come after the end of a gruelling civil war with the Nationalists.

With the creation of the People's Republic of China (PRC) in 1949, Beijing became the capital of new socialist China. At the northern Tiananmen Gate of the Outer City at Tiananmen, Mao Zedong proclaimed the People's Republic of China. This was in turn to lead to the beginning of the 'Cultural Revolution' with huge consequential changes. At this time the Forbidden City became dis-used and the Lake Palaces were put into use as the headquarters of the Communist Party.

From the formation of the PRC, the first policy was 'agrarian reform' across all regions, followed by a process of 'collectivisation'. With the Communist Party in power, implementation of policy was overseen by the main State Council and passed down to the next level, with each city having both a mayor and a CCP secretary. The CCP assigned a 'Class Status' to all citizens and 'Peoples Communes' were set up across the whole country.

In 1955, the Government established Mandarin Chinese based on the northern dialect around Beijing as the basis for standard spoken language – Putonghua. With development only progressing gradually, in 1958 the 'Great Leap Forward' was launched, aimed at drastically increasing production in subsequent 'Five Year Plans'. Whilst this was largely unsuccessful, in 1966, seeking to drive the revolutionary movement forward, Mao enlisted the Red Guards to force progress under his 'Little Red Book' philosophy, and aimed to throw out old customs and ways of thinking, which it pursued by ransacking old palaces, tombs and monuments with significant effect. But by 1969, the CCP had declared the 'Great Proletarian Cultural Revolution' as over, and around the same time the Sino-Soviet partnership also finally split.

Subsequently after Mao, under Deng Xiaoping, the 'Four Modernisations' were targeted to be reinforced as Agriculture, Industry and Defence, with Transport & Communications replacing the former Science & Technology. Market mechanisms began to be introduced, the Peoples Communes were disbanded and industry started to shift from the hinterland to coastal regions, with emphasis on the production of consumer goods.

With continuing emphasis on the 'Four Basic Principles' of the socialist way, dictatorship of the proletariat, leadership of the CCP and Marxist/Leninist/Maoist thought, a reforming process continued to evolve under a new constitution and General Secretary and with abandonment of the title of 'Chairman' of the Party. As Deng died in 1997, just before the hand-back of Hong Kong, and with the loosening of the central government's grip on the economy, 'individual enterprises' were promoted and the first Special Economic Zones were created at Shenzhen, Zhuhai and Shantou.

Gradually, private enterprise was allowed to take over in various industries that had been previously undertaken mainly by the State. Some controls were relaxed, which led to a surge in unemployment, and as coastal areas prospered the hinterland was left behind, with worsening inequality between urban and rural areas, and resulting in increased pollution and congestion in urbanised areas.

Modern City

In 1950, China had signed a 'Treaty of Friendship Alliance and Mutual Assistance' with the Soviet Union, who were focused on disseminating Marxist principles. In the mid 1950s, Soviet planners advised the Municipal Town Planning Commission on the development of Beijing. With an emphasis on industrial production, large work-unit components or 'danwei' were established as the basic model and a new administrative centre also was planned in the heart of the Old City.

The Japanese, as former occupiers, had had a plan to develop satellite towns, with peripheral industrial estates, and to introduce a green belt and build a new ring road on the alignment of the old outer city walls. Old courtyard houses were to be demolished and replaced with high-density apartment buildings. This densification occurred under the Soviets along with retaining major elements such as the Great Hall of the People, the Central Railway Station and the National Museum of History.

Another alternative plan was developed by two Chinese planners, Liang Sichen and Chen Zhanxiang. Guided by the Song dynasty 'Building Manual' of 1103 CE, they were interested essentially in preservation. Their approach was that any new centre should be built only outside the Old City. However, likened to the Japanese plan during the occupation that focused on the western suburbs, this 'Liang-Chen Plan' was rejected, sealing the fate for the redevelopment of the Old City.

This was accordingly undertaken individually with the widening of Chang'an Avenue, running east-west, that resulted in the destruction of historic gates and porches, in changing the axiality and leaving just part of the Great Wall and Qianmen (Outer Gate) standing behind Tiananmen Square, which was to take on the status of becoming the major public square in the city with ceremonial Tiananmen Gate to the north.

In 1959 'Ten Great Edifices' were built to celebrate the first ten years of the PRC, generally in the National Style, and with the 'Great Leap Forward' there were further plans for the extensive redevelopment of the city of Beijing. With the new east-west highway in front of the Meridian Gate in the middle of the city, Tiananmen Square was redesigned with a central obelisk 'Monument to the People' and enlarged with the 'Great Hall of the People' on the western side, and the 'China National Museum' on the other eastern side.

During the period 1969–79 and at a time of Sino-Soviet tension, a huge network of underground tunnels was built also under the city as a nuclear shelter. With the new Chang'an Boulevard running east-west right across the centre of Beijing, the 'Forbidden City' was closed for a period and extensive rebuilding undertaken. Following his death in 1976, the next year a major mausoleum for Mao was constructed, the 'Memorial Hall' in the middle of the Square where his tomb was located.

Gradually emphasis focused on returning the city to becoming the grand 'capital' of China again. With the rebuilding and expansion of Tiananmen Square, the north-south axis was restrengthened and extended beyond Prospect Hill and the Bell and Drum towers to the north. The former 'Forbidden City' now known as the 'Former Palaces' of the Imperial City, or the 'Palace Museum', this demonstrates their historical significance. The separate former 'Lake Palaces' in North Lake Park are reached across a bridge to Hortensia Island and the 'Summer Palace'.

A programme of major infrastructure development was also undertaken, with major highways built across the city. Beijing has expanded rapidly and now has a series of ten ring roads for a growing city with transport infrastructure that has been totally transformed along with new high-speed railway stations, a series of 12 underground metro lines, only commenced in 1970, and extensive public transport facilities. In 2000, a parallel northern axis to the west was created to the Millennium Monument and in 2020 the 600th anniversary of the Palaces of the Forbidden City was celebrated.

Transactional Forces and Typology

Now irrevocably anchored as the primate capital of the People's Republic of China, Beijing's consolidation as a global metropolitan centre is firmly anchored. With its long evolution through individual imperial Chinese dynasties, its history is reflected in the preserved 'Palace Museum' of the 'Forbidden City'. Now all around the central urban core, the city has grown well beyond its original walled perimeter. Its above-ground road infrastructure and generally underground rail and metro infrastructure provide arteries for both the resident and mobile working population of the metropolitan area. From this central node it is interconnected with other poly-nodal centres that have developed in Hebei Province, especially coastal Tianjin, that has led to its expansion into a truly megalopolitan agglomeration (see Chapter 6.5.1).

In 1982 urban land was transferred to full ownership of the State. As a result, this modified the forces that were instrumental in developing or remodelling urban areas. Consequently, in terms of residential areas, many of the majority of the early traditional form of buildings in Beijing, of residences or 'siheyuan', built in walled courtyards with a central gate around 'hutong', aligned in an east-west direction, with 'xiang' (lanes) running north-south, have virtually disappeared and been replaced by denser and taller modern apartment buildings.

From its early beginnings, the shape of settlement in Peking (Beijing) focused around the layout and arrangement of the Forbidden City and its central axis. With its cardinal north-south alignment and sub-divided grid arrangement and perimeter walls and gates, this was built according to the historic Rites of Zhou. Beyond this, development took place around lakes and rivers, with the traditional courtyard arrangement of 'sicheyauns' for housing extended families. With these interconnections and pathways, the lanes and alleys had created the typical built form of the original city. Only in more recent times has the form of the extended city been overtaken and shaped by the peripheral outer ring roads that have been introduced, following demolition of surrounding walls and with the city's continued growth and expansion giving modern form and shape to the entire metropolitanised area.

2.3.4 NANKING (NANJING), CHINA

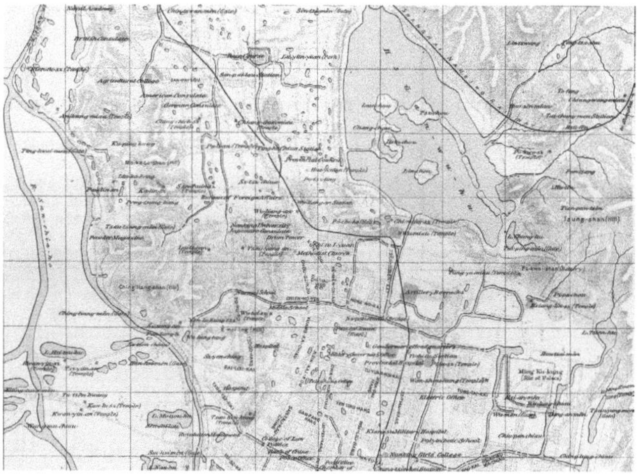

FIGURE 2.3.3 Nanking – early configuration, 1924.
Source: Maphouse.

Background and Origins

Nanking, now known as Nanjing, as the modern capital of Jiangsu Province in eastern China, is set on the banks of the lower reaches of the great Yangtze River close to the famous Purple Mountain. There is evidence of very early civilisation here during the Bronze Age when it was known as **Jinling**. With its lakes, city walls and streets shaded by plane trees, it was formerly the capital of the early 'Three Kingdoms' around 220 CE and was capital during six dynastic periods. It was one of the four Great Ancient Capitals of China, along with Chang'an, Luoyang and Peking (Beijing).

It was named **Jiankang** during the Jin dynasty, rivalling Luoyang in size. During the period of North-South division, it was the capital during the Southern Liang dynasty. But during the following Sui dynasty, it was severely razed and lost status. It was revived and named **Shengzhuo** under the Tang dynasty and then reverted to being Jiankang until the Southern Song dynasty.

During the Yuan dynasty when Khanbalik was overrun, the Mongols dominated. Subsequently the Ming took over and established the Ming dynasty capital at Nanking as Southern Capital. Under the first Ming emperor between 1368 and 1398 CE, known as **Yingtian**, there was extensive development and the great city walls were built around the natural contours of the city, bounded by the arc of the River Yangtze and surrounding hills. Having a series of 24 massive gates and inner citadels, and long ramps up to the ramparts, it was an effective fortress. With the Drum and Bell towers, the original Ming Palace was built and with the feature of Xuanwu Lake extending to the north-east of the city, with its five islands.

In 1421, the next emperor decided to rebuild Khanbalik and relocated the capital to Beijing again as Northern Capital. However, in 1644 with the Qing conquest, the Ming dynasty ended with several Ming tombs being located around the city, and Beijing was to remain as de facto capital under the long Manchu Qing dynasty. During this period the city known as **Jiangning** administered the surrounding Province of Jiangnan separately with neighbouring Suzhou.

One of the significant events that was to occur in 1842 resulted in the Treaty of Nanking being signed after the Opium Wars, which was to define the future terms of foreign trade of China. This included foreign countries gaining access to Canton, Shanghai and three other Treaty Ports. It also led to subsequent concessions being agreed with France and the USA and more Treaty Ports were created in 1858, which were considered consequently as 'unequal treaties' by the Chinese.

The city briefly became capital again, known as Tianjing, during the anti-dynastic theocratic 'Taiping Heavenly Kingdom Rebellion' period of 1851–64 but which failed with significant losses. At this time the Celestial Palace Temple was built in the city.

Republic of China Period

At the end of the 19th century, the Qing were defeated in the Sino-Japanese War and suffered several difficulties that culminated in the Boxer Rebellion. After several uprisings, and with the rise of the Revolutionary Alliance under Sun Yat-sen, this led to the collapse of the Qing dynasty after the Xinhai Revolution. Being inaugurated as Provisional President of a newly constituted 'Republic of China' (ROC) in 1912, with a new Nationalist Party, the Kuomintang (KMT).

However, with the development of the New Culture Movement, protests occurred around the country in 1919 with the creation of an urban working class seeking a way out of rural poverty. In 1921 the 'Chinese Communist Party' (CCP) was founded in Shanghai and was advised by the Soviets to form a temporary alliance with the KMT. Sun Yat-sen having promoted change in his writings under the 'Theories of Sun Wen' also developed a plan for the development of national railway trade routes to connect with the main ports, but he died in 1925. The capital was moved to Nanking, a break from dynastic tradition, in 1928.

Subsequently commandant Chiang Kai-shek led a combined northern expedition with warlords turning against the Communists in Shanghai, eventually to become supreme head of a separate Nationalist Army. During the next decade, with the Nationalists taking hold of Nanking, they prepared a Capital Plan for the city to be upgraded into a modern capital, the 'Great Plan of the Capital' (1929). Significant rebuilding of both buildings and infrastructure began to be undertaken, but the grand plan was unrealised.

Whilst the KMT continued under the leadership of Chiang Kai-shek, the Japanese had invaded Manchuria in 1931 and occupied renamed 'Manchukuo' with a capital at Xinjing. As a result of a full-scale invasion and the brutal massacre known as the 'Rape of Nanking' by the Japanese in 1937, the Nationalist government had to abandon Nanking to the Japanese and retired to Chongqing.

As this period of Japanese domination prevailed, the KMT focused on aiming to oust the Communists. But although strengthened under the new leadership of Mao Zedong, the Communists were forced to retreat on the 'Long March' from Jiangxi to Zunyi and via Sichuan to Yan'an and north of Xi'an, which was a challenging time for the Communists under the direction, leadership and 'thoughts' of Mao. Strengthened and reorganised, and gathering wider support, the Red Army pursued the civil war and with growing ranks the 'People's Liberation Army' (PLA) overcame resistance and the Communists finally prevailed. They were able to celebrate the formation of the 'People's Republic of China' (PRC) in 1949, with the national capital being moved back to Beijing.

Modern City – Nanjing

Now named Nanjing, the Yangtze River flows from the west and north around the city out to the Yellow Sea. This major river system is linked with the Qinhuai River that runs through the city and was also connected to the man-made 'Grand Canal' that was built northwards. Overlooking the city is the Purple Mountain, mainly wooded but often coloured with purple mist that gives it its name. Here the Linggu Temple with the tomb of the first Ming emperor is set and also the Mausoleum of Sun Yat-sen, who also resided in the Provisional Presidential Palace, is located just outside the city.

The central part of the city is still defined by the extensive city walls that encompass the city with its many gates and fortifications. The city is also defined by water with the Xuanwu Lake, Mouchou Lake and its circuitous and circumscribing waterways that give character to the city.

The local economy was based on a significant primary agricultural region with forestry, animal husbandry and fisheries. Now with a significant newer industrial sector based on iron and steel, chemicals, manufacturing and electronics, there are major multinational companies based here. In addition, in the tertiary sector numerous research and development centres have been established and it has become a major IT, biotech and pharmaceutical centre with research institutes and new universities, as well as being developed as a major financial centre.

As a transport hub, Nanjing is well connected through the Beijing to Shanghai north-south high-speed railway line, having the new major Nanjing South Railway Station and with a new airport link to Nanjing Lukou International Airport (NKG), and the Shanghai–Wuhai–Chengdu east-west high-speed line. In terms of public transport, it has 12 new subway metro lines and a dense net of urban busways.

Modern **Nanjing** has a series of major bridges and tunnels that cross the wide Yangtze River:

- First twin deck truss Nanjing Yangtze River Bridge – road & rail (1970), Dashengguan arch rail bridge (2010), Second cable road bridge (2001), Third cable road bridge (2005), Fourth suspension bridge (2012) and Fifth river bridge.
- Nanjing Yangtze River road tunnel (2010), Metro Line 10 tunnel, Metro Line 4 and Yanziji River tunnel.

The city also contains an important and active inland port with its deep river and Longtan Container Port on the River Yangtze. It is well connected with adjoining major expressways, inter-connected with the G25 Ring expressway in addition to a dense network of national highways.

There are a number of dominant new buildings in the centre that include the Golden Eagle Tiandi Towers and the tall stepped Zifeng Tower (designed by SOM). With its range of universities and research centres, new museum and opera house, it is a vibrant modern city that has grown comprehensively in recent times. The population of the city of Nanjing and its six urban districts is approximately 9.3 million people. There are five further outlying suburban districts that are growing, which include Luhe and Pukou north of the river, and Jiangning, Lishui and Gaoshu.

Transactional Forces and Typology

Nanjing has been as instrumental as Southern Capital in China as Beijing has been as Northern Capital over many periods and dynasties. As the influence of different emperors ebbed and flowed over the centuries, Nanjing has been shaped under these dynasties. Located in the middle part of the country, it has been defined by the great Yangtze River that arcs around the city, acting on the one hand as a port enabling trade, and on the other hand as a barrier to crossing. However, located in a rich agricultural plain, this has benefited the resourcing of the city, with its extensive city walls, which acted as formidable fortifications. In more modern times, the city became the centre of the Nationalist movement and the Republican government for a short period in the early 20th century.

Accordingly, the city of Nanjing is defined firstly by its relationship with the Yangtze River. Secondly, it is defined by the extensive length of built city walls that surround the original rectangular centre and that were extended and modified over successive dynasties. With its predominant north-south axis, the city has a gridded plan with areas that have been developed organically outwards. From its fine historic urban core, it is surrounded by perimeter river tributaries also that define the centre. Its older distinctive Drum and Bell towers also dominate the centre as in other similar older Chinese cities. Now growing rapidly outwards, newly developed areas connected by modern road and rail infrastructure are acting as poly-nodal growth centres to the south and east of the city.

2.3.5 KYOTO, JAPAN

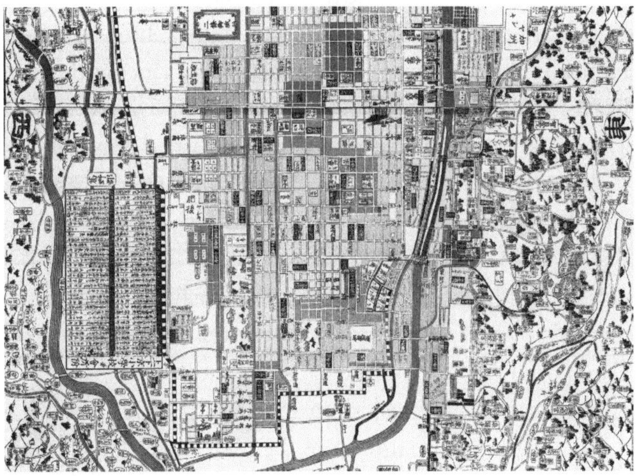

FIGURE 2.3.4 Kyoto – early configuration, 1890.

Source: Maphouse.

Background and Origins

The old city of **Kyoto** was founded in 794 CE as 'Heian-kyo' (capital of peace and tranquillity), in the Kansai highlands region of western Honshu Island of Japan by the Kano River. It became the capital of Japan with the seat of the

Imperial Court and was modelled on the ancient Chinese city of Chang'an, with its layout designed on 'geomancy' principles, palaces and parks.

The city was laid out on a grid pattern surrounded by walls and a moat. The main axial street ran north-south, Suzaku Avenue, to connect the southern Rajomon Gate with the Imperial Palace aligned with point of contact with assumed 'Heaven'. Here the main group of buildings were built and rebuilt several times as the result of fires. The palace was settled on its current site in 1331 with the dominant main Shishinden Hall of State surrounded by a gravel ceremonial courtyard and other residential buildings. Adjoining this were the landscaped Imperial Gardens.

The city benefits from a supply of fresh water from Lake Biwa to the east from where aqueducts were built. In the 16th century, daimyo Toyotomi Hideyoshi built new streets to double the number of streets running north-south. In the centre also 'Nijo-jo' castle compound was built in the 16th century with its boundary of encircling walls. In 1788 a great fire destroyed many of the city's temples.

The balance of power eventually shifted during the Tokugawa Shogunate period from the city of Kyoto to Edo. Kyoto remained the capital until the end of the Shogunate and with the transfer of the Imperial Court to Edo, was renamed To-kyo (eastern capital). In 1869 at the time of the Restoration under Emperor Meiji, Kyoto was renamed Sai-kyo (western capital).

Besides the Imperial and Sento palaces, with a number of significant temples as the old To-ji Temple, Chion-in, Nishi Hongan-ji and Nanzen-ji, there are many fine gardens in Kyoto. The Ginkaku-ji, Silver Pavilion, is in the east within a grand landscaped setting, and distinguished Kingkaku-ji, Golden Pavilion, is clad in gold leaf silhouetted around a lake. In addition, the Ryoan-ji Temple with its famous Zen dry rock garden, is located to the west.

Outside the city to the south is the 'Fushimi-Inari-Taisha Shrine', an extraordinary trail with the winding Senbon Torii path having thousands of consecutive red 'torii' gates erected that have created an extended corridor along a 4 km. length to the summit of the hill. The Katsura Imperial Villa and Gardens also lie to the south-west of the city along with Shugaku-in Gardens.

Modern City – Buildings and Arteries

Modern Kyoto comprises 11 wards (ku) and has several different attractive outlying districts, such as Arashiyama and Takao. Beyond the Kano River to the east is the Higashiyama district (eastern mountains) with its secluded gardens and temples, and beyond from Ginkaku-ji to the Kumano Nyakuo-ji shrine is the 'Philosopher's Walk' by the canal, particularly scenic during the cherry blossom season. It has numerous museums and cultural institutions that reflect the deep history of the city, its shrines and temples.

Overlaid around its historic core, it now has the modern JR railways and high-speed lines, and two new subway lines, Karasuma (K) and Tozai (T) lines, and tramways and a network of streets and highways.

Key modern landmarks include the enormous and futuristic main Railway Station (designed by Hiroshi Hara) which was opened in 1997 to serve both JR and Shinkansen services and the imposing Kyoto Tower. With its University and Institute of Technology, it retains a strong design tradition and is a vibrant modern city. The population of the city is approximately 1.5 million and double in the metropolitan area.

Transactional Forces and Typology

Located in the highlands of Japan and with its long imperial history, the local economy developed based on traditional Japanese crafts, porcelain and sake brewing. As a major centre, with its ancient temples, palaces and gardens, Kyoto was an historically significant centre in Honshu.

Geographically, the main feature is the river valley, which flows southwards from the hillside to the ocean, where rivers have taken different paths over the years. Whilst channelled in modern times, the water volumes change enormously across the seasons and are contained by a number of built-up levees.

Located in the middle of Honshu Island and surrounded by nestling hills, the city developed between the junction of the Kamogawa and Katsuraga rivers where separate sections of river were given different names. In the natural floodplain of these rivers, they have been channelled as built development has steadily increased. With the distinctive walled rectangular area of the Imperial Palace and adjacent Nijo-jo Castle, and its numerous Buddhist temples and Shinto shrines, the city is shaped by its historic temple enclaves around and across the city.

The city is also sited just to the north of the former capital Nara, and on the developed Tokaido Corridor between Tokyo, Nagoya and Osaka. With its role formerly as capital, this established itself as a strategic centre in the region and a pivotal location on the island.

Now framed by new infrastructure with its distinctive railway lines and the elevated Shinkansen High Speed Line, its modern communication facilities are superimposed on the original street grid. An inland city, orientated with its rivers, essentially aligned north-south, it is now crossed by modern east-west road and railway infrastructure. A city of classic and historic built form, and as a former Imperial capital, it has managed to retain its secluded religious sites and shrines for contemplation in alluringly private landscaped spaces.

2.3.6 SAMARKAND, UZBEKISTAN

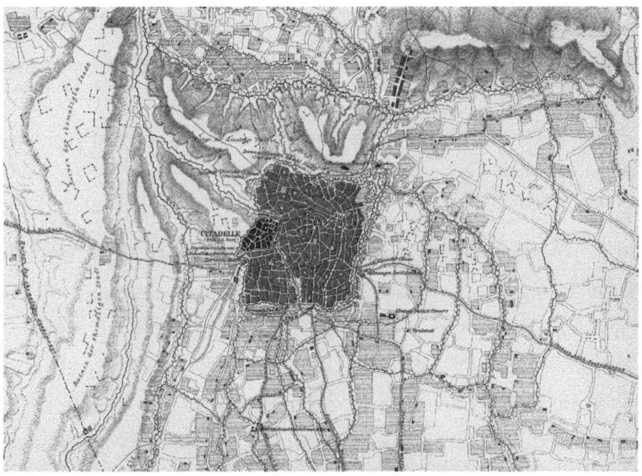

FIGURE 2.3.5 Samarkand – early configuration, 1865 (Perthes).
Source: Maphouse.

Background and Origins

Samarkand is one of the world's oldest cities located in the Zarafshan River Valley, being established by an oasis in the 7th century BCE as a trading point on the Silk Route between China, East Asia and southern Europe. Under the Achaeminid Persian Empire of Cyrus the Great, Samarkand was the capital of the Satrap of Sogdia, a central province in the region that benefitted extensively from the silk trade.

It was also known as Afrasiyab, or Maracanda, named after the first King of Sogdiana, the land between the Oxus and the Jaxartes, as described in the Shahnama, and became a major settlement surrounded by a lengthy perimeter wall. It was conquered by Alexander the Great in 329 BCE after his crossing of the Hindu Kush and subsequently came under Iranian and Turkish rule under the Umayyad Caliphate.

During the next century, a network of trade routes began to be established to China, with the Northern Silk Road running to the north of the Tian Shan mountain ranges beyond the Taklamakan Desert, enabling connection with Tashkent in the north, Kashgar in the east and Bukhara in the west.

During this period there were diverse religious communities that included Zoroastrianism. During the Abbasid period, the earliest beginning of papermaking developed with establishment of the first paper mill in Samarkand. Agriculture was also developed around the city, fed by canals and lakes to irrigate vegetation and orchards.

It was later conquered by Mongols in the 13th century, whose empire under Chinggis Khan with its military decimal organisation grew extensively across much of the Eurasian continent. It also became home to the fierce Timurid Renaissance in the 14th–15th centuries under Timur (Tamerlaine), whose mausoleum (Guri Amir), with its fluted dome, was erected in the middle of the city and in the centre of the region known as 'Transoxiana'.

He created the Registan grand square in the centre of the ancient city with its three monumental buildings on each side of the courtyard. These included the 'Ulugh Beg Madrasah' built in 1420 and the 'Sher-dor' and 'Tilya-kori' Madrasahs built in the 17th century. These were schools of learning, flanked by tall minarets, decorated by glazed blue and golden mosaics, with Tilya-kori as the Grand Mosque. Astronomy also began to be developed then at Ulugh Beg's Observatory in Samarkand with its unusual large-scale half submerged mural quadrant sextant.

The original city was very attractive and had two main gardens, the 'New Garden' and the 'Garden of Heart's Delight'. Most of the mud-brick houses in the city also had courtyards and gardens and geometric ponds – water fountains and decorative trees were common. Colour was very important with the blue domes of the mausoleums, a colour that could 'ward off the evil eye', which was prevalent as well as gold vaulting and fabrics.

In the late 16th century, the Shaybanids took control of Samarkand and the capital was moved to Bukhara, whereafter the city's influence declined.

In the 19th century, the city came under Imperial Russian rule and became capital of the newly formed Samarkand Oblast of Russian Turkestan, during which time the Trans-Caspian Russian Railway (5′ foot gauge) was built on the path of the old Silk Route from the Caspian Sea, through Samarkand and on to Tashkent, crossing the Oxus River (Amu Darya) that flows into the Aral Sea.

Modern City – Buildings and Arteries

After the Russian Revolution, in 1924 the Soviet Socialist Republic of Uzbekistan was formed with Samarkand as the capital. New commercial and industrial districts were built by the Russians to the west of the old city.

In 1930, the capital was transferred to Tashkent, where a major earthquake occurred later in 1966. It subsequently has been redeveloped as a modern Uzbek Islamic capital with its tall TV tower and former Soviet-style apartment blocks. As an independent landlocked country since August 1991, the Republic of Uzbekistan is a major cotton and also oil and gas producer.

The former major monumental Bibi-Khanym Mosque in the city has now been rebuilt and the new city has a number of new larger administrative and commercial buildings.

In terms of infrastructure, internally the city centre has a dense network of urban streets. Externally, modern high-speed powered trains (TALGO) travel between Bukhara and Samarkand and Tashkent, as new Central Asian railways.

The population of the city of Samarkand is approximately 0.5 million and double in the metropolitan area. The city acts as the capital of the Samarqand Region.

Transactional Forces and Typology

Besides its success as a major cotton-producing region, the city of Samarkand grew as an important stopping point on the old Silk Trade Routes with China. Whilst the Silk Routes were not so named until mapped late in the 19th century by the German Baron von Richthofen in seeking a suitable rail alignment through the mountains, they really consisted of a network of individual trade roads that were used for the transport of commodities such as silk and textiles, grain, jade, glass and silver. Nestling besides the northern trade route, which was divided on either side of the Tian Shan mountain ranges, Samarkand provided support to travelling caravans. With its cultivation of paper and artistic disciplines using glass and ceramic mosaics, it developed its own distinctive styles of decoration that were most evident in its local architecture. Despite occupation by different invaders, it established itself as a significant place in this more remote region of Central Asia.

As an inland city bounded by mountains, its centre is dominated by the grand Registan Square and its skilfully orientated monumental buildings, once described by Lord Curzon as being 'the noblest public square in the world'. With its prevailing Islamic architecture, its style and ornamentation are highly decorative and much has been preserved. Despite the development of more recent Soviet-style built-up areas on the perimeter, the historic core of the early Islamic City remains intact, with radial streets running outwards to the south-west. The following figures illustrate the modern configuration of these former capital cities as described and the evolution of their modern population.

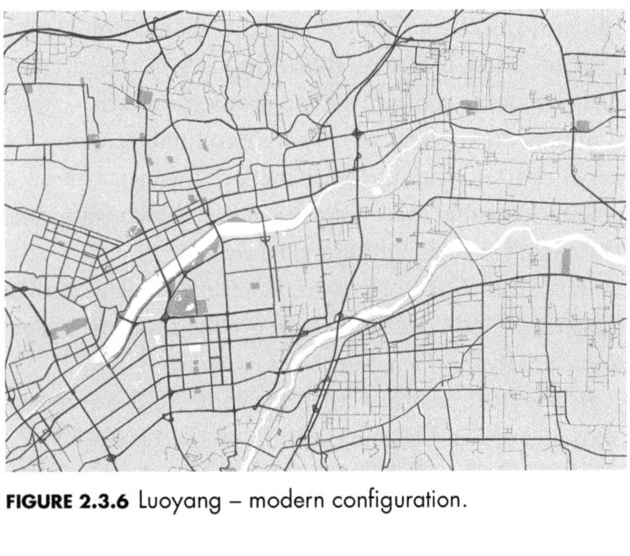

FIGURE 2.3.6 Luoyang – modern configuration.

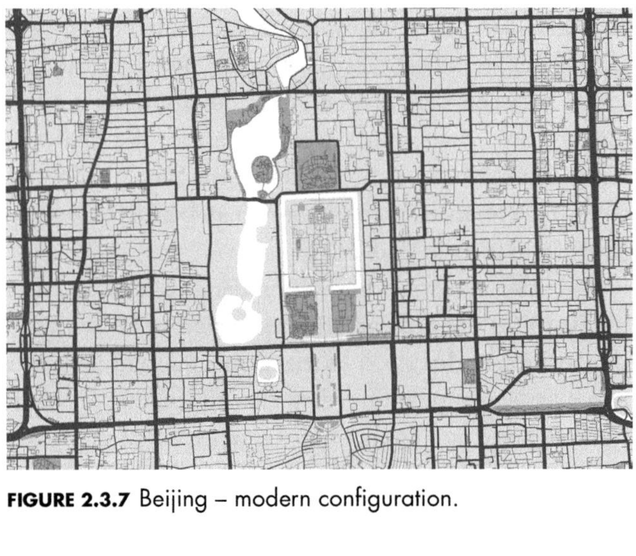

FIGURE 2.3.7 Beijing – modern configuration.

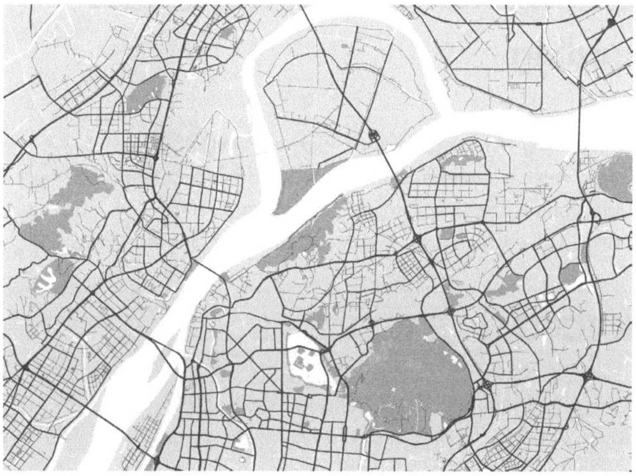

FIGURE 2.3.8 Nanking – modern configuration.

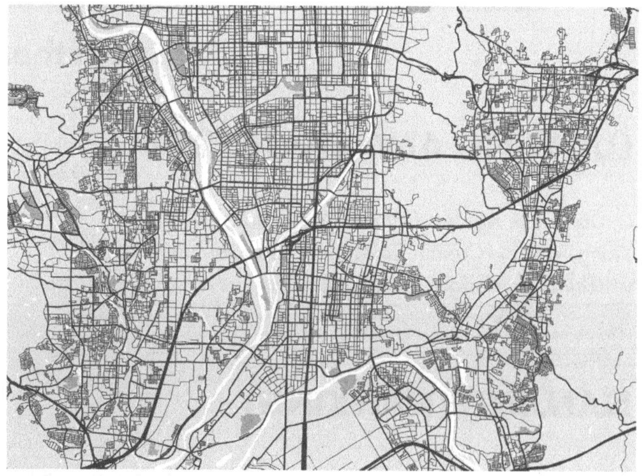

FIGURE 2.3.9 Kyoto – modern configuration.

FIGURE 2.3.10 Samarkand – modern configuration.

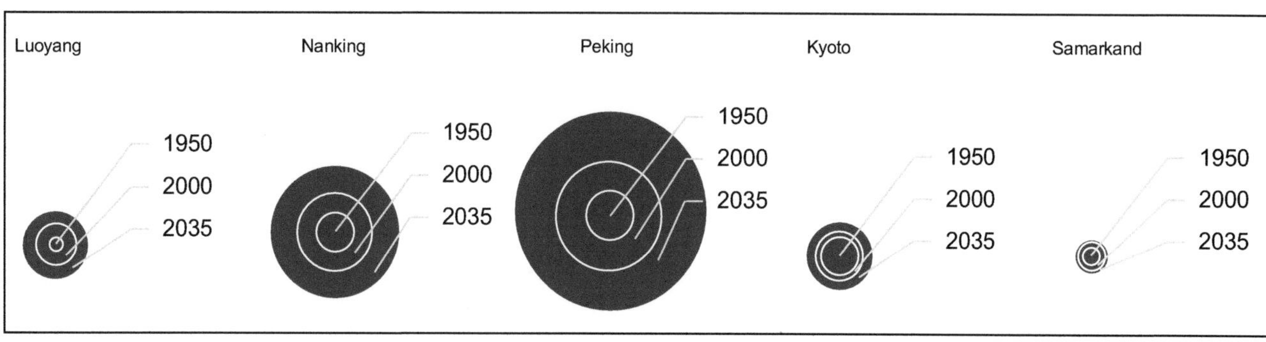

FIGURE 2.3.11 Former capital city populations.

Data source: Global Database of Metropolises 2020 – UN-Habitat (2020), UN DESA Population Division.

Further Reading

LUOYANG AND CHANG'AN

Li, Dun – *The Ageless Chinese*, Macmillan, 1978.
Xiong, Victor – *Capital Cities and Urban Form in Pre-Modern China*, Routledge, 2018.

NANJING (NANKING)

Fei, Si-yen – *Negotiating Urban Space, Urbanization and Late Ming Nanjing*, Harvard University Asia Center, 2010.
Musgrove, Charles – *China's Contested Capital, Architecture, Ritual and Response in Nanjing*, University of Hawaii Press, 2012.
Till, Barry – *In Search of Old Nanking*, Hong Kong Joint Publication, 1982.

BEIJING (PEKING)

Barmé, Geremie – *The Forbidden City*, Profile Books, 2008.
Clements, Jonathan – *Beijing, a Short History*, Haus Publishing, 2022.
Friedmann, John – *China's Urban Transition*, University of Minnesota Press, 2005.
Keay, John – *China, a History*, Harper Press, 2008.
Lorge, Peter – *Imperial China*, One World Publications, 2021.
Man, John – *The Great Wall*, Bantam Press, 2008.
Yanxin, Cai – *Chinese Architecture*, Cambridge University Press, 2011.

KYOTO

Kerr, Alex – *Another Kyoto*, Penguin Books, 2018.
Paine, Robert – *The Art and Architecture of Japan*, Yale University Press, 1992.
Plutschow, Herbert – *Historical Kyoto*, Nippon Shuppan Hanbai, 1994.
Shonagon, Sei – *The Pillow Book*, Penguin, 2006.
Young, David – *The Art of Japanese Architecture*, Tuttle Publishing, 2019.
Young, David – *The Art of the Japanese Garden*, Tuttle Publishing, 2019.

SAMARKAND

Hansen, Valerie – *The Silk Road*, Oxford University Press, 2015.
Manz, Beatrice – *The Rise and Fall of Tamerlane*, Cambridge University Press, 1989.
River, Charles, ed. – *Samarkand, the History and Legacy of One of Asia's Oldest Cities*, Charles River Publishers, 2017.

2.4 Historic European Capital Cities

Commercial and Imperial Capitals – Historic Trading Cities – Austro-Hungarian Capitals

London, Paris, Lisbon, Madrid, Milan, Munich, Vienna and Budapest

2.4.0 INTRODUCTION – HISTORIC CAPITALS

Peoples have tended to coalesce and associate together at locations having an inherent geophysical advantage in their natural setting, for example due to commanding height or adjacency to a natural waterway. Man-made development of appropriate built forms was used to protect and nurture communities responsive to their functions and according to their level of sophistication.

Beyond human protection and sustenance, trade became the predominant activity as a mode of exchange of goods and services between different groups of peoples. Such groups coalesced and individual centres of learning developed in order to provide education in religion, philosophy, language, history and science. Such centres evolved for both natural and human sciences, art and liberal arts, around which teachers and specialists gravitated in central places.

As peoples became more closely grouped, different forms of association and organisation developed. Leaders emerged and encouraged the formation of groups for protection (armies) for production and development (workers), for welfare (doctors and nurses), for administration (clerks and lawyers) and for other tasks. With these groupings, different specialised types of accommodation were required in order to facilitate these tasks, in consequence leading to the creation of barracks, housing, assembly halls, dwellings, schools, etc. in urban areas.

As the exchange of goods increased, markets were developed, that is, physical marketplaces for different raw commodities (e.g. fruits and vegetables, fish and meat), crops (e.g. corn, wheat, tea, coffee, tobacco, spices), manufactured commodities (e.g. textiles, linen and silk, wool) and building materials (e.g. brick, timber, rope).

This led to transactions in goods and to the provision of services, whether physical, administrative or financial. As trade in goods in the primary sector expanded, so secondary services developed at an increasing level of complexity in different neighbourhoods of cities. As these monetary transactions increased, there was an increasing propensity for trade and further exchange of services in these countries.

In Europe, historically centres developed at river estuaries and along the principal rivers. Either at coastal locations, like London and Lisbon, or on major rivers, as the Seine (Paris) and Danube (Vienna and Budapest), these centres were also crossing points for trade and commerce. As settlements expanded, they were the natural locations for development as capital cities. Other more land-locked locations, such as Milan and Munich, which are more centrally located on trading routes, also developed as pre-eminent cities. With many centuries of consolidation, they have evolved historically into primary centres and capitals of administration, trade, learning and culture.

2.4.1 LONDON, ENGLAND

FIGURE 2.4.1 London – early configuration, 1676 (Oliver).
Source: British Library.

DOI: 10.4324/9781003353348-8

Background and Origins – Early City

London is the capital city of the United Kingdom, located on the Thames estuary in the south-east corner of the island of Great Britain. The earliest inhabitants and tribes such as the Iceni created settlements on either side of the river where local trading began, at the lowest point where, subject to tidal flows, river crossings were possible before the broadening estuary.

Around 55 BCE the area was invaded by the **Romans**, who bridged the river near gravel mounds by the Walbrook, built roads and city walls and gates, started to fortify the north bank of the city and built fora and basilica, naming it **Londinium**. They also introduced some form of administration to this area, the first 'City of London'. Due to its favourable geographical characteristics centred around the River Thames as a natural port and bridgehead, coastal shipping trade was developed along with travel farther inland up the river.

The 'City of London' developed distinctively as a business district with its 25 ancient wards, headed eventually by Aldermen, equivalent to the Hundreds in the Shires, subdivided into precincts and with its own guildhall. Its extent was defined by the boundaries created by the original Roman walls on the north shore of the River Thames. Redeveloped after the Romans by King Alfred around 890 CE, with a renewed street plan, it was known as **Lundenburg**, a fortified town or borough.

After the Roman occupation, there followed a period of conquest by Germanic tribes, essentially Angles, Saxons and Jutes, who began to expand northwards into the regions beyond the south-east of England, along with Danish and Viking invasions. It was not until 927 CE that the regions of the country, the heptarchy of East Anglia, Mercia, Essex, Wessex, Sussex, Kent and Northumbria were unified into a Kingdom of England. The **Anglo-Saxon** centre prospered here under the name **Lundenwic** just to the west of the former fortified Roman area.

The next significant period was after the **Norman** conquest in 1066, in which England was invaded by the French Normans under William the Conqueror, who became king after the Battle of Hastings in Sussex. London became the capital and there followed a period of consolidation and introduction of a detailed administrative survey that secured the current knowledge through inventory of the entire country, of both land and resources, within the so-named Domesday Book that was completed in 1085, although excluding the City of London.

Mediaeval City

At this time the White Tower and Castle was constructed as a separate centre of power in the City of London, and a monastic abbey was commenced in the City of Westminster, as St Peters Abbey, where a royal palace was also constructed as primary residence for the royals. The city and street pattern developed in the mediaeval period with the first multi-span stone bridge built across the River Thames in 1209, to connect the City of Southwark in the south, where the archbishop had the Palace of Lambeth, with the City of London.

This became a centre of commerce, with houses and shops built on the bridge itself and applying tolls for crossing traffic at each side. It was granted its own powers as a 'commonality of citizens' able to elect its own mayor as from 1215, a real 'merchant's city'. It had its own church, Old St Paul's Cathedral, that was developed on a potential former Roman temple site on Ludgate Hill.

During the **Plantagenet** rule, in 1215 the great charter Magna Carta was created by King John to introduce the protection of rights and individual freedoms of the English people. Under the later King Edward, construction of the abbey and palace continued, which were to become the centre of power around Westminster.

During the 14th century, despite the period in 1349 of the Black Death when the population declined 40%, as foreign trading developed, a special walled free trade zone was created in the City of London, Stalhof or steelyard, whereby commodities could be directly loaded onto German ships from warehouses in Upper Thames Street. Here was a significant dedicated trading enclave that was not ended until the Anglo-Hanseatic Wars in 1475.

The region of London has several attributes essentially with its natural setting nestling along a river estuary with settlements since Roman times on each bank and with bridge crossings. It also features the seat of royalty with its palaces, court and related institutions developed over many centuries. Both of these assets aided its development, firstly as an attractive port with its mercantile pattern of trading stretching out across Europe and colonies beyond. Secondly these combined, brought together considerable political and commercial power that was centred around the royal capital and executive in the City of Westminster and the mercantile and financial square mile of the separate City of London.

At the time of King Henry VIII, commerce extended with the formation of the Navy and the cloth trade developed with European merchants. During the period of

Tudor rule, the Benedictine monastery adjoining Westminster Abbey was dissolved to become a 'Royal Peculiar', a church that was directly responsible just to the sovereign. The Palace of Westminster with Westminster Hall was established as the seat of the Houses of Parliament for the hereditary Lords and elected Commons to form a government, and the Palace of Whitehall was realised as the residence for the English monarchs. Accordingly, the 'City of Westminster' was developed as a district distinct from the City of London, having its own statutes and 'Letters Patent'. (Quaintly the king still has to ask specific permission to enter the City of London.)

In the early 17th century, there were some noticeable urban developments. Covent Garden, an open area of land, was developed into a residential area around a central piazza designed by Inigo Jones, in the style of the Palais Royal in Paris. Slightly farther east, Lincoln's Inn Fields was also developed, with terraces built around a tree-lined square not far from Temple Gardens where future Inns of Court were to be located.

City Rebuild

The severest tragedy was to occur during the time of the later **Stuart** rule, with the Great Plague that occurred in 1665, followed by the Great Fire of London in the following year. This devastation led to a significant transformation and rebuilding of the City of London that had been totally destroyed but which did not spread to the City of Westminster. Many houses, churches and Old St Paul's Cathedral were destroyed due to having closely located wooden and thatched buildings with jettied floors and inadequate firebreaks.

As a result, new plans were created for rebuilding the whole area, including proposals by John Evelyn with plazas, by Robert Hooke with square blocks, and by Christopher Wren with a grid and diagonals. It was rebuilt essentially following much of the former mediaeval street pattern but with new building codes introduced such that wider streets and standardised house types of brick and stone rather than of wooden construction were built under the Rebuilding Act of 1667. It was also the occasion to rebuild St Paul's Cathedral with its great dome that was commissioned by Wren and was finally completed in 1711. A further London Building Act of 1774 consolidated previous legislation and standardised four types or 'rates' of buildings in terms of size and materials. This led to a certain uniformity in terms of design and fenestration.

Turning the misfortune of the Great Fire to its benefit, the city was able to be rebuilt in a more robust manner with stricter building codes and with revised public street plans that soon became infilled with private developments. The River Fleet was canalised and new riverfronts and quays were developed. New districts were developed with different markets and trading specialties of textiles and clothing, food, fruit and fish, manufacture or assembling.

During the years between 1600 and 1800, the death rate in London generally exceeded the birth rate, hence the population only increased as a result of in-migration from rural areas. Diseases and plagues were prevalent and historically had been a recurrent problem, especially in built-up areas.

Under the next Georgian or Regency period of **Hanoverian** rule, London as a whole grew into a significant commercial and financial centre. With increasing population and following aristocratic wealth, new building and development turned to the centre of London and the West End, with a series of planned residential estates, squares and fine buildings and mansions which commenced in Russell Square and St James in the late 17th century. These 'Great Estates' were developed in the hands of just a few rich families.

In the West End and in Mayfair, residential squares were especially fashionable, with Berkeley, Grosvenor and Hanover Squares, around Oxford and Regent streets, and Cavendish Square, Portman and Manchester Square were also developed. The other significant development was the evolution of Regent Street during the Georgian period, the most elegant north-south street running between Piccadilly and Regents Park. With its main shopping districts, theatres and museums, the West End of London became a most cosmopolitan and commercial area.

At the end of the 18th century the Duke of Bedford parcelled out estates and Bloomsbury was developed around Bedford Square. The other major landowner was the Duke of Westminster, who developed a number of large estates. Other builder-developers also entered the market, creating areas such as Hans Town in Knightsbridge by Henry Holland and Thomas Cubitt who developed Belgravia. Subsequent developments were in Portland Place connecting with Nash's more speculative development of Regents Park Terraces, with Crescent and Carlton House.

Common land and open space were always an aspiration for inhabitants of London. During the mid-17th century Moor Fields was protected as a public open space to the north of the financial City of London and with Lincoln's Inn Fields and the legal Inns of Court and Chancery and Gray's Inn Walks.

Royal parks were carefully preserved, that is, central Hyde Park and St James Park, the Regents Park, Greenwich Park and Richmond Park on the peripheries. But also, open ground such as Hampstead Heath and Highbury Fields in the north, Battersea Park, Wandsworth and Clapham Commons south of the river, all afforded opportunities for outdoor recreation and access for centre city inhabitants to open air, and views of St Paul's Cathedral were protected.

Victorian City

London was to change further, significantly during the subsequent **Victorian** period, as new infrastructure was developed to improve sanitation and new industries were started that were facilitated by the introduction of coal and steel, which was to significantly change patterns of building and transport across the country. From wool and textiles to canals, steam power and railways, new methods of industrial production and distribution led to an increase in consumption of many different commodities.

Sanitary conditions remained important, with several tributaries flowing into the River Thames, such as the Rivers Fleet and Walbrook, which needed to be managed. Real change only resulted with the works of the engineer Joseph Bazalgette in creating mains interceptors, which is only just being properly finalised with realisation of the new 'Tideway Tunnel' built down the centre of the River Thames today. With improved sanitation, public health improved, more social facilities were built and new by-laws and regulations were introduced to control development. The streets and highways were improved with dedicated sanitation and water supplies, and a network of radiating highways and railways was built. The built-up residential areas began to expand out to the suburbs in all directions and dense ribbon development of terraced housing along these transport axes.

During the 19th century, industrialisation developed extensively, leading to subsequent urbanisation of most of London. With the increase in population, this led to its own problems in the provision of adequate housing and new methods. It also led to various philanthropic institutions being established, for example the Peabody Trust in the creation of model dwellings. This resulted in a tradition of social rental housing at affordable rates that was also similarly developed in the industrialised areas of Manchester, Sheffield and Birmingham. In 1844 the Metropolitan Buildings Act established the Buildings Office to regulate the layout and form of different classes of buildings.

The tradition continued in London with, for example, new estates such as the Boundary Street Estate at Shoreditch, planned by Owen Fleming in 1893. Other 'cottage estates' were also developed at more peripheral sites, such as at Hammersmith and Acton, at locations determined by new tramway lines. In 1894 the London Building Act was enacted to control the light and fire safety of buildings.

Transport – Marine and Metropolitan

With the growth and expansion of the British Empire, the River Thames had become the main channel for trade and a series of new docks and warehouses was developed downstream before the estuary, in particular for the East and West Indies Trading Companies, with specialised handling facilities for different commodities arriving and departing by maritime transport.

The Square Mile of the City of London became the centre of the world's commodity markets, such as for wool and cotton, tea and coffee, sugar and wheat, minerals and ore, with the stock exchange, corn exchange and metal exchange established for trading with separate markets at Billingsgate (fish), Smithfield (meat), Spitalfields (poultry) and Covent Garden (fruits and vegetables). Supporting finance and insurance and brokering services were also developed.

The volume of goods passing through the Pool of London increased significantly. Specific docks were developed and for particular commodities such as East India docks – import and export, West India docks – import and export, Victoria and Albert Royal docks, Millwall docks, Limehouse basin, London docks – tobacco, St Katherine docks – eastern and western, Surrey commercial docks – timber. As a tidal river estuary, dock basins were built with secure warehousing for goods to be unloaded at the quays, but a large volume was also transferred into lighters and discharged up-river. This led to a large demand for dock workers and labour.

A significant benefit for trade also was the use of common Sterling Bills extensively around the world in trading commodities. Trade, however, was interrupted with different periods of war, in particular the Boer War in South Africa, the Great War in Europe and the Great Depression during the 1930s.

During the Victorian period there had been extensive expansion of new railways with private development of routes across the country. Particular lines radiated out from London, like the Great Western Railway from London, Paddington to Bristol, Didcot and Swindon designed with a wider seven feet (7′) gauge. Whilst operational from 1840, this was eventually converted to standard gauge for convenience in 1892. Other lines were developed for steam railways from the great London Termini of Euston, St Pancras and Kings Cross running north on the northern side of the Thames.

With the development of the railways and the main rail terminals in London, there was growing congestion in the urban streets. As a solution, the strategy was to go below ground. An extended network of lines was developed by 'Underground Electric Railways of London' (UERL) promoted by Charles Yerkes from Chicago, fed by a network of tramways and feeder bus lines. This contributed to extending development and urban sprawl especially in north and west London. Supported by Frank Pick, this was a busy rail-building period with considerable growth.

A series of tunnels was built using a 'cut and cover' method to link the city with new stations and running east-west underground to connect the mainline railway stations of Paddington, Euston and Farringdon, known as the Metropolitan Line. Another of the earliest lines was the District Line, which was built near the embankment to connect Cannon Street, Charing Cross, Victoria and Earls Court Stations. Subsequent lines were built by bored tunnel, of a smaller diameter, and because of the tightness of the train in the tunnel itself, the underground system became known as 'the Tube'.

Administration

In terms of administration, a 'Metropolitan Board of Works' had been established in 1855 to oversee the realisation of necessary sewers, sanitation and infrastructure. As population increased, housing conditions deteriorated and in the late 1880s there was severe deprivation and poverty in many areas of London. This was highlighted by Charles Booth's Survey that plotted the location of the 'undeserving poor', which led to the subsequent 'Housing of the Working Classes Acts' that endeavoured to address some of these extreme slum conditions.

In 1889 the 'London County Council' (LCC) was established as the world's first metropolitan government for the whole urban region of London. The administration of London has evolved gradually from its Councillors and Alderman initially representing the 28 metropolitan boroughs. Its powers have been extended from urban planning to council housing, education, health, transport and tramways and a new county hall was commissioned and opened in 1922, evolving under varying political control over the years of the Progressive, Municipal Reform and Labour Parties. In 1938 the London Greenbelt Act was passed that has effectively acted as the barrier for containment in the development of the city beyond which open land was to be retained for green space and recreation.

As the capital and metropolitan centre of the whole of England, London is a very large and extended city with multiple neighbourhoods from the centre to the suburbs that have quite different characteristics. Divided between north and south by the River Thames flowing eastwards towards the estuary, the West End, East End, **City of Westminster** and the **City of London** each have different features, facilities and resources.

The financial might of the City of London grew with its markets, exchanges and trading houses, all centred around its original 'coffee house' meeting place culture and is administered by a separate City of London Corporation, distinct from Westminster.

Symbolic buildings that are key landmarks such as the Houses of Parliament, Westminster Abbey, St Paul's Cathedral, Buckingham Palace and St James Palace, each play a vital role for the Crown, clergy, laity and the people of England. With its universities, colleges and seats of learning, the law courts, teaching hospitals, professional and institutional headquarters, all contribute to the depth of metropolitan activities.

Stretching from Richmond in the west to Greenwich in the east, and from Hampstead in the north to Wimbledon in the south, London contains also a series of significant communities in separate boroughs and developments in the Thames Basin that retain an individual sense of 'village-ness' gravitating around and outside the centre.

Modern City

Despite the period of trauma during World War II, when the whole city came under attack by air from Nazi Germany, known as the Blitz, this was only overcome in the Battle of Britain by securing air superiority. With a long period of rehabilitation and recovery post-war required, with evolution and emergence from its past colonial trading ties, the city needed to be rebuilt and modernised to find and secure new emerging markets.

Housing was also a priority, and many areas needed replacement and renewal in addition to planned constructions on the outer perimeter of the city in order to provide new socially affordable homes.

With mixed commercial and residential areas and cultural institutions, it is both a work and playground for residents and visitors. As an administrative and business centre, educational or residential place, with its markets and commerce, it is a highly dynamic transactional world city.

Surrounded by a green belt where development has been restricted and the peripheral M25 orbital highway, the metropolitan area is a mix of industrial, commercial and residential development for around 10 million people. Its transport axes of both road and rail have played a key part in the shaping and configuration and growth of the city, its immediate surroundings and hinterland extending across Greater London.

Transactional Forces and Typology

Since the earliest times, the River Thames and its estuary has been the source of movement up- and downstream in accordance with its diurnal tidal variations. This also provided the challenge of crossing the river, which was secured by individual bridge crossings built at various locations downstream that afforded opportunities for settlements to develop on either side of the river. As a suitable location for trade with inhabitants of neighbouring regions and countries, maritime trade was the source of its historical development.

Through different eras, the Port of London developed and was to become pivotal in the expanding reach of the country and its empire. From the trade in raw materials and commodities, then as new technologies were developed, the exchange of manufactured goods increased, leading to London becoming the centre of trade and services. So it evolved as a centre of business and commerce and with the growth of financing institutions, trading markets and services around which migrating labour gravitated seeking gainful employment. This multiplicity of resources served as the ongoing basis for the expanding urban economy.

With its riverine and estuarine location around the London basin, from the earliest settlements around river crossings, London has expanded in a radial manner. With natural zones that developed on either side between the tributaries of the River Thames, the city has developed into a series of distinctive neighbourhoods. These now form localised sectors in the concentric and radial development of the city. Whilst separate poles of development such as the 'West End' and 'East End' still remain, there are now distinctive administrations for the Cities of London, Westminster and Southwark and Greater London.

Farther out, satellite towns also developed after the city grew through ribbon development on new railway corridors as additional housing was required to accommodate the rising urban population. With the constraints of the surrounding protected 'green belt', growth has been contained and enclosed now by bands of both inner and outer circular roads. This is now a major metropolitan conurbation but has been constrained from attaining real megalopolitan proportions.

2.4.2 PARIS, FRANCE

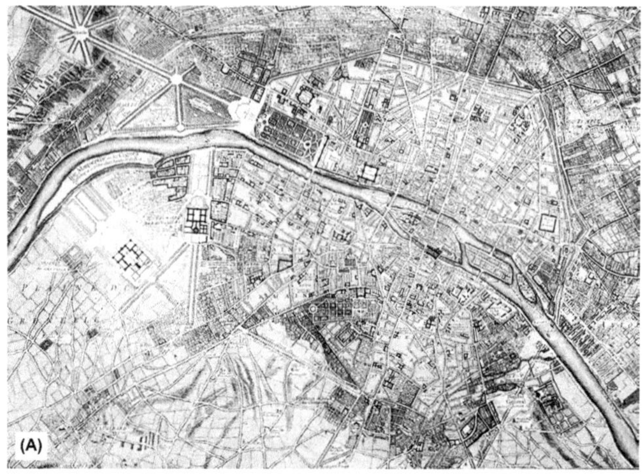

FIGURE 2.4.2A Paris – early configuration, 1754 Rocque.
Source: British Library.

Background and Origins

Paris was first occupied by a Celtic tribe, 'the Parisii', in an area known as Lutetia in northern Gaul. The Romans developed a fortification on the left bank at a crossing of the River Seine at L'Ile de la Cité, on the axis of the present Rue St Jacques and where the Temple of Jupiter was built. Later, as the marshes around the riverbanks were drained, settlements were able to be expanded further after the Gallo-Roman period and the population began to grow subsequently in mediaeval times.

Around 1250 CE, walls were built by Philippe Auguste on each side of the river on surrounding land adjoining this island and L'Ile St-Louis, and further fortifications were built in a semi-circular form on the right bank by Charles V in 1370. These were further extended under Louis XIII and created an arc around which subsequent development occurred.

Across France Christianity was promoted under Charlemagne. The Church played a key role in daily life, and the Capetian dynasty followed the Frank and Merovingian rulers. The city grew in importance and significant religious buildings were commenced, such as Notre Dame Cathedral in 1163 on L'Ile de la Cité and Sainte-Chapelle and the Basilica of Saint-Denis. Interspersed with periods of religious wars, developments evolved on the right bank also, with a marketplace and trading places and residential areas.

In mediaeval times the growing city was encircled by the defensive walls with gates on both the north and the south banks and the central bridge crossing was rebuilt, lined with buildings on top of its structure. It was not until the 16th century that permanent development was undertaken with the new stone bridge 'Pont Neuf' at the tip of the island that was completed in 1604, the previous timber one having collapsed.

After the Renaissance, urban design was initiated under Henri IV when a set of geometric royal squares with statues in the centre and buildings were laid out commencing with the triangular Place Dauphine at the apex of the island and the Place Royale, now named Place des Vosges, in the Marais. This elegant square became the prototypical residential square that was to define civic space in Paris, which with its ordered facades and arcades, and absence of through traffic, created a distinct sense of place around its central statue.

The celebratory Place des Victoires, in contrast, was circular around an equestrian statue, having through streets that acted as a public thoroughfare. Another even grander square was also planned, Place de France, on a semi-circular layout with eight radiating axes, but this was never built. Adjacent to the Tuileries Palace and Gardens, a further scheme was developed by Mansart of the octagonal Place Vendome, in honour of Louis XIV, which took many years to be realised with its original central equestrian statue.

In terms of streetscape, a scheme was developed by Le Notre for extending the axis beyond the Tuileries westwards on a new tree-planted avenue, les Champs Elysées, that was realised by 1740, rising towards the Butte de Chaillot and then falling gently downwards towards Neuilly. This dominant avenue across central Paris was to define the main east-west axis complemented by the eventual realisation of streets radiating out around the star shaped Place de l'Etoile.

The final design task was the connection of the Champs Elysées with the Tuileries and Cours de la Reine. As a result of a competition, J-A Gabriel designed Place Louis XV, which by 1790 would integrate the intersecting axes with l'Eglise de la Madeleine and future Pont de la Concorde across the River Seine, now known as Place de la Concorde around its centrally relocated Egyptian obelisk.

During the 17th century under Louis XIII, Cardinal Richelieu had orchestrated further development of new river bridges, the building of Sorbonne University and Palais-Cardinal, subsequently the Palais Royal, that would face the grand Louvre building. Although Louis XIV, the 'Sun King', had transferred his court to Versailles, the city continued to grow, and the older city walls were demolished and replaced with Grands Boulevards. There followed the Age of Enlightenment, a period of new scientific and philosophical ideas when the aristocracy prospered, and new town mansions (hôtels particuliers) were built in the centre of the city.

However, much was to change with the succeeding French Revolution, with the storming of the Bastille and overthrow of the monarchy in 1789 that led to the founding of a new République. Between 1785 and 1791 the Walls of the Férmiers Généraux were built under Louis XVI to act as a barrier around the city and for collecting customs and tolls from people and goods entering the city. Later a further ring of fortified walls was also built, although these were removed eventually.

Following a period of upheaval, Napoléon was to seize power in France and a 20-year period of war and external conquest with aspirations seeking to expand a European empire was followed until his eventual fall and exile, and subsequent reburial in l'Hotel des Invalides. His legacy was to be the establishment of a centralised administration along with codification of laws and reform of the national educational system.

Urban Transformation

With the restoration of the monarchy for a subsequent period, Louis Napoléon declared the Second Empire in 1850, and the next phase of development of the city under Napoleon III was entrusted to Baron Haussmann as Préfet de la Seine to undertake a programme of public works. This was intended to clear unsanitary streets and to create a better ordered capital with a geometric grid of main avenues and boulevards, like that emanating from the Place de l'Etoile. Here 12 new boulevards were created to radiate out from the renamed Place de la République, a star of avenues from the dominant Arc de Triomphe erected to celebrate Napoleon's former victories.

Haussmann's strategy for the redevelopment of Paris, the 'grand transformation', was centred around la Grande Croissée de Paris, the central intersection of the main north-south axis of two new boulevards, Strasbourg and Sébastopol, with the east-west axis of Rue de Rivoli and Rue St Antoine around L'Ile de la Cité. New streets were to connect with newly developed railways and the six individual mainline railway station termini built at the Gare du Nord and Gare de l'Est, Gare Saint-Lazare, Gare Montparnasse,

Gare Austerlitz and Gare de Lyon in different quarters of the city.

These main axes, along with the other long, straight boulevards constructed in the 19th century by Haussmann, have essentially determined the current structure of the layout of Paris with its Grands Boulevards radiating out from landmarks in individual squares, such as the Place de l'Opéra with its fine Opera House and the Place de la République.

The pattern of adjoining buildings along these boulevards was laid out in accordance with setbacks and height restrictions dependent on the given street width under the redefined Paris Building Code so as to retain uniform street facades.

New parks and squares were also built across the city, with the Parc Monceau and Parc de Montsouris, and the former quarry at Buttes-Chaumont was landscaped using vantage points sited at the end of long avenues, and all of this was to enhance the vistas whilst promenading around Parisian streets. As overhauled by Haussmann, its boulevards became conducive to walking and perambulating between different commercial neighbourhoods and arrondissements, and in and around covered passages.

New infrastructure was also installed with aqueducts for water supply, sanitation with main underground sewers and gas lighting along the main streets. This transformed many areas of the city and new residential and commercial buildings were erected along boulevards having a consistent civic scale and order, maintaining the typical building height of 5–6-storey facades. Outer areas of the city were also annexed, and new large parks laid out at each extremity of the city, such as the Bois de Boulogne and the Bois de Vincennes. Despite these improvements, the population density was very high, double that of similar areas of London, and there were many areas of poor housing.

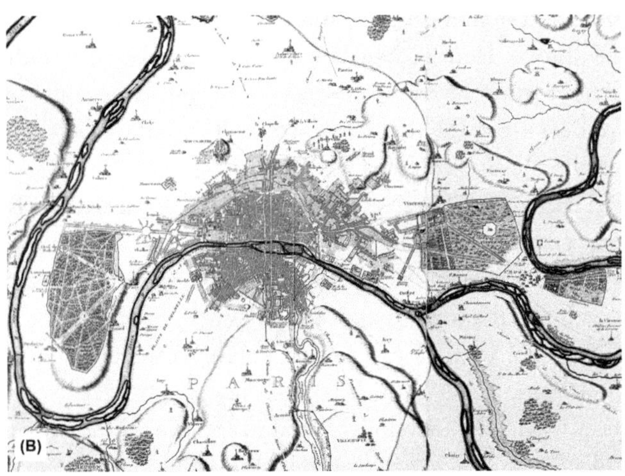

FIGURE 2.4.2B Greater Paris – early configuration, 1844.
Source: British Library.

The next significant event for France was defeat by the Prussians in 1870 and the loss of the region of Alsace-Lorraine. With the departure of the Prussians from Paris, the city was left in a state of anarchy from which the Paris Commune was to arise, and with many monuments torn down, was not to be overcome until 1875 when the new constitution of the Third Republic was formalised.

At the end of the 19th century, Viollet-le-Duc set about restoring Notre Dame Cathedral, and two major exhibitions were also held in the city. In 1889 the Universal Exposition was to celebrate the centenary of the French Revolution when the Tour d'Eiffel was constructed at the Champ de Mars, and then in 1900 another Universal Exposition was held when the Grand Palais and Petit Palais were built. The first line of the underground Paris Métro was also opened at that time.

Subsequently the two world wars were each to have a significant effect on the entire population and the city of Paris with its occupation in 1940 by the Germans until its liberation by the Allied Forces in August 1944. Fortunately spared from mass destruction, the city managed to retain the majority of its former, older and classical buildings.

Individual Quartiers

As a compact yet dense city, Paris has a number of different districts, or 'quartiers', that have quite distinctive character on both sides of the River Seine.

The *Marais* to the east has a number of grand mansions (hôtels particuliers) around the elegant brick and stone Place des Vosges with its arcades, mansard slate roofs and dormer windows. Nearby is the old Jewish quarter and the site of the Bastille. The adjoining neighbourhood of *Châtelet les Halles* is the site of the former fruit and vegetable market that was demolished and has been redeveloped.

The central *Jardin des Tuileries* runs along the quai of the right bank of the River Seine extending from the galleries of the Louvre museum with its recent modern glazed pyramid entrance (designed by Pei) and parallel Rue de Rivoli. The Palais Royal, with its colonnaded courtyard, is nearby, as well as the octagonal Place Vendôme, Bourse du Commerce, Madeleine and the Rue Royale that leads to the Place de la Concorde and bridge leading across to the Assemblée Nationale. Around the Opéra by Garnier are the Grands Boulevards, with several large department stores and the commercial district with the Bourse, Banque de France and Bibliothéque Nationale.

On the left bank, the *Quartier Latin* is the bohemian district that surrounds Sorbonne University. With its ancient

Cluny Monastery, Boulevard St Germain and perpendicular Boulevard St Michel, this was the intellectuals' hangout, with its cafés and bookstores. Here also is the Panthéon, where great men and women of France have been honoured and buried, and the nearby Place St Sulpice and Church.

Just to the east is the *Jardin des Plantes*, and on the western side is the *Luxembourg Gardens* with its palace, both green places of relaxation. Southwards these lead to the commercial area of *Montparnasse*, another legendary quarter, with the sprawling cemetery of Montparnasse.

Towards the western end of the city are *les Invalides* and hotel with its golden dome that houses the tomb of Napoléon, and the park of the Champ de Mars that is dominated by the Eiffel Tower. On the opposite bank by the *Trocadéro* is the site of the former Universal Exposition and gentle incline to the Palais de Chaillot.

Farther north the grand Avenue des Champs Elysées, with its breadth and long axis gradually rises from the Place de la Concorde westwards up to the Arc de Triomphe that celebrated Napoleon's victory at Austerlitz, the pinnacle monument used as a focus at times of national celebration.

Beyond, farther north, is *Montmartre*, a charming village on the hill with the white Sacré Coeur church and its ovoid dome and basilica that commands the hill (Butte) with winding steps up to the open 'parvis'. With a cemetery and its artist colony, the windmill Moulin de la Galette still remains as a charming neighbourhood.

On the other side of the River Seine is the outer *XVI Arrondissement* residential quarter that leads to Auteuil with its Art Deco buildings and smart villas. Other outer neighbourhoods now modernised are the former slaughterhouse district round Canal St Martin, with the *Parc de la Villette*, transformed with a science museum, Cité des Sciences, an exhibition hall and concert hall.

Modern City – Buildings and Arteries

With the end of the war in Europe, a period of analysis and planning commenced. During the 1950s a Plan d'Urbanisme Directeur (PUD) was established for the city of Paris and a new land-use zoning policy was established. A debate on the development of the city had been held under the auspices of the organisation DATAR (Délégation de l'Aménagement du Territoire et a l'Action Regionale).

During the 1960s and 1970s, a programme of major projects, 'Grands Projets', was proposed and developed aimed at modernising the city, but which did not fit comfortably within the existing urban fabric. Some neighbourhoods had been carefully renovated, such as the Marais, but others were torn apart, like les Halles.

One of the iconic development projects was the Centre Pompidou next to les Halles area (designed by Piano & Rogers and Arup). The daring design for the proposed structure, with iron 'gerberettes' and glass facade with external escalators diagonally towards the roof was a shock, occupying an entire block, sited amidst the traditional old Paris quartier, with a scheme completed in 1977 that challenged the existing urban fabric as a new 'Grand Projet'.

An entirely new area was also developed by EPAD (l'Etablissement Public pour l'Aménagement de la Region de la Défense) as 'La Défense', a region to the west of l'Etoile as an intensive commercial area with a number of tall towers, such as Tour First. This was aimed at preventing overdevelopment in the fabric of the old city, as had occurred with the earlier implantation of the Tour Montparnasse by focusing on one targeted development area.

The first generation of skyscrapers was built here in the 1960s with its own Boulevard Circulaire highway, but the next higher generation of towers was not to be built until 50 years later. It has now consolidated to become the commercial business centre of the city, as a separate development pole having corporate headquarters located there. The Grande Arche was located symbolically at la Tete Défense along with the exhibition centre, CNIT.

Farther out, a programme of New Town developments was also pursued, with whole new communities being established at Créteil, Marne la Vallée, Evry and St Quentin Yvelines. These included social housing quarters, Habitations à Loyer Modéré (HLM), that needed to be connected with new public transport infrastructure to the centre. Such Grands Ensembles as built, however, became social problems in themselves with dense poor populations at Sarcelles and at other areas such as Clichy-sous-Bois.

The 'hexagon' of France is a highly centralised country and is renowned for its 'dirigisme' and has evolved differently from its Anglo-Saxon neighbours. With its Napoleonic Code and administration, its 'state-centric' approach has differed from more liberal market attitudes. In recent times it has been able to direct resources to major urban developments in the capital, such as these Grands Projets, and new developments as at La Défense.

Administratively, the city of Paris is administered by the Conseil de Paris with a mayor from the Hôtel de Ville for the 20 arrondissements. A regional government for the region of the Ile-de-France is also elected. Numbered from 1 to 20, the arrondissements are numbered in a clockwise manner expanding spirally from the centre (Ile de la Cité) that reflects the manner of growth and expansion of the city towards the perimeter, although arrondissements 1–3 have now been combined to form Paris Central.

In 2016, a plan for the Métropole du Grand Paris was agreed as an administrative structure that combines Greater Paris and its suburbs and Départements that extend over 800 km^2. The population of the city is approximately 2.1 million and is of a high density of approximately 21,000 per km^2, whereas the larger metropolitan area has a population of approximately 12 million. Post-war there has been significant in-migration from Maghreb and sub-Saharan African countries, which has put particular pressure on the housing sector.

The goal is to manage the future development of the whole region so as to balance both its infrastructure and its transport needs. As a result, extensive new rail lines have been planned and built to better connect the outer regions of the area that include the other neighbouring Départements of Hauts-de-Seine, Seine Saint Denis and Val de Marne, Argenteuil and Val d'Oise.

With its protected historic centre, it has encouraged modern commercial development to be developed at the periphery, as at la Défense and farther out with a series of new satellite towns and housing developments. With dense road traffic both within and along the riverbanks and around the city on the Boulevard Periphérique and the Francilienne that connects with the national Autoroute highway network, there is pressure on the transport sector.

In terms of infrastructure, two axes of heavy rail rapid transit, Réseau Express Régional (RER), were built underground, initially north-south and east-west. These have been extended to complement the existing métro system, which itself has a very dense urban coverage in the central area with over 300 stations. Further the new Grand Paris Express set of rapid transit lines in the Ile-de-France region was also realised to complement this network for the 2024 Summer Olympics.

Benefitting from its own cultural and historical legacy through the ages, with iconic landmarks such as Notre Dame Cathedral, the Louvre, and the Champs-Elysées, with districts such as the Quartier Latin and Montmartre and fine squares like Place des Vosges, the Tuileries and its gardens, Paris is a most delightfully laid out city. With its history, style and architecture, despite its frequent congestion from its attraction for visiting tourists, Paris retains its inner charm as the City of Light.

Transactional Forces and Typology

From its island in the middle of the River Seine, the location of Paris served historically as a natural place of settlement around a potential crossing point of the river. As a fortification, settlements grew on each riverbank and were gradually expanded to encompass the islands and its Roman temple. As a strategic location in northern France, it grew as the capital and the seat of government, and its ascendancy rose in the 16th century, with the new stone-built Pont Neuf and adjoining formal buildings. With centralisation of the State, grander buildings were built to symbolise the role of the kings and the influence of the monasteries, and hence the city evolved in a manner reflecting the changing role of the Crown, church and the aristocracy.

Whilst the French Revolution was subsequently to reshape relationships between the State and the people, it was followed by the Napoleonic Empire, during which period expansionary ideals were pursued across Europe that were celebrated by grander building in the capital. Under the Second Empire during the 19th century, a period of significant rebuilding was also entered into that was to align the city with a series of Grands Boulevards as designed by Haussmann.

This reshaped the layout of the centre of Paris with its new long boulevards and street frontages with uniform facades, which are still in evidence despite some modernisation. Typical building designs comprised commercial shops at street level, with residential apartment houses above and smaller accommodations at roof level, with mansards or dormers in accordance with defined ordinances.

Industrialisation was to reshape outer areas and districts by the canal, and steel and ironwork was introduced into building. This was to influence new structures, such as the main railway stations, exhibition halls and the symbolic Tour d'Eiffel built for the Universal Exposition. The need for additional housing was fulfilled post-war with dense modern HLM developments built in outer areas to replace ageing housing stock, and new satellite towns were planned also on virgin sites.

With the introduction of the railways and station termini built on either side of the river, the rail networks followed by new roads as the encircling Boulevards des Maréchaux, have shaped the city. Most significantly, the construction of the underground métro system has served beneficially to support the movement of people around the densely populated city.

The overriding order of the city with its uniform streetscapes has been retained in many neighbourhoods with street frontages that have a sense of scale, proportion and continuity throughout the city. This architectural balance and layout, stemming from classical principles and supported by strict building codes, has enabled the charm of the city to be retained.

As a riverine city with two central islands, the city has grown in stages, along both the left and the right banks. With more recent river crossings, the city has radiated out from the centre, subsequently reshaped by the overlain axes of boulevards that were rebuilt under Haussmann. Typically, it operates as a highly centralised radial city. Post-war the city has continued to grow, and the metropolitan area of Greater Paris now extends well into outer areas around the meandering Seine. With a series of new satellite towns and residential 'grands ensembles' with denser housing, the suburban pattern of the outer city development has changed, with poles of development interconnected by the new RER and Grand Paris rail network that extends out and beyond to the peripheral areas.

2.4.3 LISBON, PORTUGAL

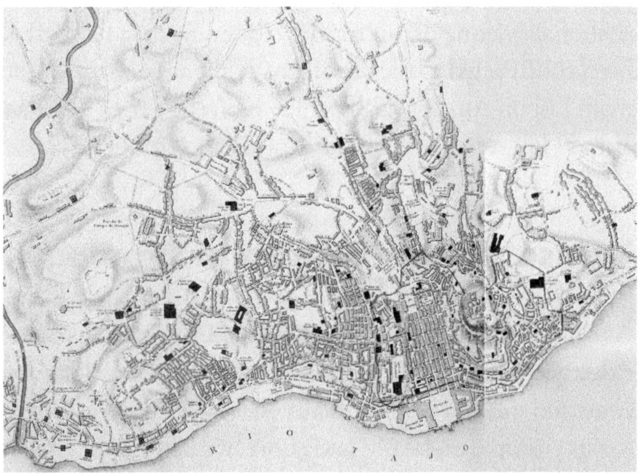

FIGURE 2.4.3 Lisbon – early configuration, 1844 (SDUK).
Source: British Library.

Background and Origins

The city of **Lisbon** (Lisboa), known as Olissippo in Roman times, is located on the Iberian Peninsula on the Atlantic coast, alongside the banks of the River Tagus (Tejo), steeped in history as it has expanded from its original marine settlement.

In former times Lisbon was within the Roman province of Lusitania. Thereafter it was invaded by different groups of barbarian tribes, including Berbers and Arabs from North Africa, and came under rule of the Taifa. During the Reconquista period, it was besieged by the Knight Crusaders and Christianity was re-established, and Lisbon became the capital of the new Portuguese territory in 1255.

During the 15th century, in the Age of Discovery, Portuguese explorers reached out beyond the Azores to Africa and India under Vasco da Gama and the Far East, and subsequently westwards to Brazil, where it shared the divided South American continent with Spain under the Law of the Indies.

As a result of its trading with newly discovered worlds in Asia and South America, under King Manuel a new riverside palace, Paco de Ribeira was constructed on the banks of the Tagus in the city along with an arsenal, warehouses and dockyards, along with the individual Casa da India, da Guiné, da Flandres, and dos Escravos (slaves) to handle its new global interests. New areas were also planned as Barrio Alto with its maritime craftsmen located there.

In 1755 the city of Lisbon was devastated by a major earthquake that destroyed much of the city. Under the direction of the Marques de Pombal it was rebuilt with new districts, of the central commercial district and lower town (Baixa Pombelina). Other urban enhancements were also undertaken, such as the Passeio Publico since revamped into the Avenida da Liberdade.

With the end of the Portuguese monarchy, in 1910 the country fell under several autocratic regimes until after World War II and the awaited end of authoritarianism and post-colonialisation with the establishment of the third Democratic Republic in 1974.

The city has several districts that have quite different distinctive characteristics that include industrial Alcantara, fado Alfama, Moorish Mouraria, Pombaline Baixa, upper Barrio Alto, monastic Belem, cultural Chiado, waterside Parque dos Nacoes and riverside Beato, each interwoven within and around the seven hills of the city.

With the fortified mediaeval castle São Jorge elevated on the hill, its cathedral and its colourful buildings and houses, development stretches in undulating corridors out of the city towards the adjoining hills. In the lower town of Baixa with its walls of individual buildings covered in blue ceramic tiles, azulejos, there is the Chiado district. Here the pavements are correspondingly built with small pieces of limestone and basalt stones laid in colourful mosaic patterns.

With its seven hills and generally white building facades and tiled roofs, its strategic harbour faces out to the Atlantic where during the Age of Discovery expeditions were sent as far afield as the Indies, Brazil and Africa. From this a strong trading tradition developed as a maritime port linking different continents.

After the city was rebuilt, with the commercial centre and lower town, delightful squares were constructed, such as Praca dos Restauradores, Praca do Rossio, da Figueira and the Praca do Comercio with their patterned pavements that overlook the coastal area leading down to the river. With a number of elegant gardens within the city, such as Eduardo VII Park and individual gardens such as the Museum Gulbenkian, the city has an attractive townscape.

Modern City – Buildings and Arteries

One movement that significantly affected modern Lisbon was a result of the decolonisation of the former Portuguese colonies in Africa. Since the 1960s and 1970s after independence and post-colonial conflicts, there has been significant immigration into the city of nationals from Cape Verde, Guinea Bissau, Angola and Mozambique.

This had two effects, firstly of a significant increase in informal housing or 'barrios informais' on vacant land plots, and provisional neighbourhoods, 'barrios provisorios', where migrants settled without formal title. With many areas being developed autonomously or on a self-build basis, this had a significant impact especially on the peripheral areas of the city. Secondly, in other areas, significant formal high-rise mass housing blocks have been built throughout the city in an endeavour to address the chronic housing deficit.

In terms of transport besides the dense urban grain of streets, the most striking modern structure across the river is the double-deck suspension bridge that dominates the strait, inaugurated in 1966 and named Ponte 25 de Abril after the 1974 revolution, that spans from Amadora to the southern part of the city Almada Barreiro. Here also is the striking statue of Christ the Redeemer with outstretched arms, built in 1949, similar to that in the city of Rio de Janeiro, Brazil.

Since Portugal joined the European Union in 1986 with a modernising economy, central areas have been renovated. In addition, the old port area was redeveloped for Expo '98 with new pavilions and joined the same year with a long new multi-span bridge, Ponte Vasco da Gama, that connects the other side of Lisbon with the area of Montijo across the bay.

Some of the historic urban trams still run in the streets of Lisbon. Four major new metro lines have been built across the city, some of which are very deep under the hills. The main railway stations at Santa Apolonia, Rossio and Cais do Sodré serve commuter lines to the suburbs, and ferries ply regularly across the River Tagus.

Farther west along the coast beyond the former Alcantara docks is Belem where the Portuguese World Exhibition was held in 1940. With its more recent commanding explorers' statue, Padras dos Descobrimados, Monument to the Discoveries, built in the 1960s under Salazar, and the Torre de Belem and the old Monastery of Jeronimo, this is a fine example of Manueline architecture. The area behind Belem has become re-animated beyond the monastery and museums with the new Museum of Art, Architecture and Technology (MAAT) (designed by Amanda Levete) on the Tagus riverfront.

Whilst industrialisation around the port wharves has declined, the service sector in the city has been growing. The city has a population of approximately 0.5 million in the centre and approximately 3 million in the metropolitan area. The city has retained much of its local culture and remains a highly attractive and liveable city with a fine ocean outlook dominated by the tall suspension bridge across the Tagus River.

Surrounded by a ring road, the city is now criss-crossed by a large number of new highways that have been built in every direction across the city, often half underground with the goal of alleviating urban traffic. However, these have often been constructed to the detriment of the urban fabric. The main Lisbon Airport (LIS), which is on the site of a former airbase, lies very close to the central districts of the city.

Transactional Forces and Typology

As a coastal city Lisbon has developed from its maritime existence on the Iberian Peninsula with trading along the coast and inland from the Tagus estuary. With its outward-looking people and early navigators, its sailors and explorers had reached new lands in the Caribbean and the West Indies, as well as beyond Africa to India and the East Indies.

Upon these adventures it built substantial trading arrangements that led to the development of wharves and docks along the banks of the Tagus in Lisbon. As a hilly city, shaped by its natural topography, different neighbourhoods developed around the centre, from Baixa Chiado to Barrio Alto. As a result of a major earthquake in 1755, the whole city had to be rebuilt, that was undertaken under the plans of the Marques de Pombal, which was an opportunity to realign the major axis of the city, along Avenida da Liberdade to Parque Eduardo VII.

In more recent years across the bay in the less developed southern area is Almada, now connected by the major road-rail suspension bridge to the south and Montijo where industrial and shipbuilding facilities are located. Montijo has also been connected by a new long viaduct, Ponte Vasco da Gama, from the main city where now there is more space for further development. The older dock area to the east by Lisboa Oriente has been substantially rebuilt and regenerated since Expo '98, not far from the city airport that nestles very closely to the centre of the city.

As a hilly coastal city, the city has been developed on buildable slopes rising up and out from the centre. On the right bank of the River Tagus, it extends radially northwards and westwards towards Belem and Cascais. It is a city that reflects typical Portuguese vernacular building that has developed organically with its white houses and blue tiles and clay roofs.

Heavily criss-crossed now by new roads and viaducts that disrupt the urban fabric, the city is also suffering from significant sprawl on marginal slopes as new multi-storey apartment buildings are built to house migrant families, particularly from former Portuguese African colonies. Here the forces of informal development are putting pressure on the limited available land space. Outside the immediate urban enclave, new poly-nodal settlements are beginning to be established as overspill, such as at Pontinha, Amadora and Odivelas, now reached by new metro lines.

2.4.4 MADRID, SPAIN

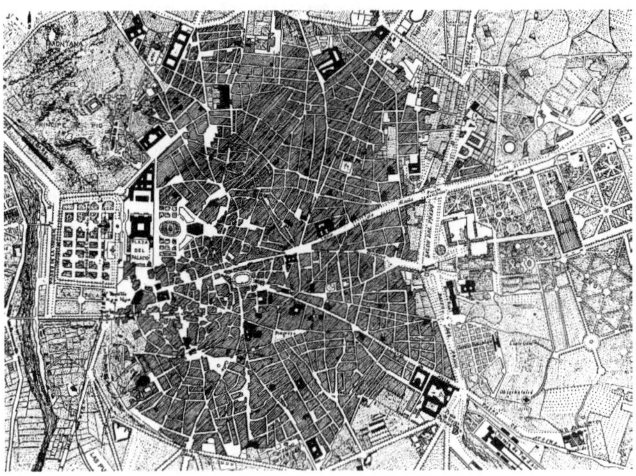

FIGURE 2.4.4 Madrid – early configuration, 1899.

Source: iStock.

Background and Origins

Settled as an early castle town by Moorish invaders and established around 852 CE, it was named Mayrit in the centre of the Iberian Peninsula, partially surrounded by the Manzanares river. In the 11th century, under the Christian Reconquest, encompassed within the Kingdom of Castile, the city was declared the capital of the Kingdom of Castile as **Madrid** by Felipe II in 1561.

Located at the geographical centroid of Spain, with the unification of Castile and Aragon, under the Spanish Habsburgs the city grew around arcaded Plaza Mayor with the Catedral de San Isidro and Palacio de Santa Cruz, known as the Madrid de los Austrias. It was to be followed by the long War of the Spanish Succession, which resulted in the crowning of Felipe V as the first Bourbon King.

Under the Bourbon Dynasty, the city was further shaped with the centre moving eastwards towards the Paseo del Prado with its ornate and baroque buildings, such as the Palacio Real on the site of the original alcazar, and the Royal Armoury. With the grand Museo del Prado and the Museo Thyssen-Bornemisza, and the later corner Edificio Metropolis and other neo-classical buildings, this gave the district its distinctive character with the adjoining Parque del Retiro.

Next came the following Napoleonic period with the Peninsula War and occupation before defeat of the French and the subsequent drafting of Spain's first constitution.

Farther north in La Castellana, grand boulevards such as the Paseo de la Castellana and fashionable Paseo de Recoletos were built to meet at the Plaza de Colon, giving a sense of grandeur to the capital. The more modern Art Deco Palacio de la Musica and Plaza de Toros de Las Ventas bull ring brought some variety to the city.

During the 19th century the city was being modernised with the Ensanche (enlargement) of streets and the development of new residential neighbourhoods replacing formerly overcrowded districts, and the city was expanded under the Plan Castro. New neighbourhoods were developed and axes such as the Gran Via extended and railways and terminal stations such as Atocha began to be built. At the turn of the 20th century the first 5 km. section of a planned long linear city as envisaged by Soria y Mata to the east of Madrid was developed on either side of a tramway, although never attaining its full length of 50 km.

In more modern times, the capital of Madrid was heavily affected by the Spanish Civil War that broke out in 1936 with General Franco coming to power. From being mainly an agricultural country, post-war there were a number of significant challenges. With the political differences between the left and the right, the country had been held back under the Franco regime. After ostracism for many years, it was not until after the end of this era that the economy was revitalised. With his death in 1975 and a new constitution, and subsequent accession to the European Community in 1986, the economic situation improved. There was a high level of migration to the city from rural areas and especially from South America and the modernisation and transformation of several sectors.

Modern City – Buildings and Arteries

Madrid covers an extensive area which includes several parks, such as El Retiro and the largest forested area of Monte de El Pardo, much of these being royal assets with urban parks and gardens that adjoin royal palaces, for example Sabatini Gardens. Madrid is home to several universities, including one of Europe's oldest, the Complutense University of Madrid, founded in 1293. With the Ciudad Universitaria, the Technical University of Madrid, UAM and modern business schools it has a strong educational sector.

In terms of the economy, Madrid has a strong financial centre and the service sector predominates. The modern style Edificio Espana and Torre de Madrid (design Otamendi) were built in the 1950s and the Torrespana TV Tower was built in 1982. To the north of the city a new commercial district of AZCA has been built with its towers such as Torre Europa. Substantial commercial and real estate developments have been realised in the centre with the modern Puerta de Europa, a striking inclined pair of towers at the Paseo de la Castellana. Also towers such as Torre Picasso and the Cuatro Torres Business Area with the tall Torre de Cristal, Cespa, Espacio and PWC have been realised.

Besides its role as national capital hosting the Palacio de las Cortes and Palacio del Senado, Madrid is also capital of the Commune de Madrid and the City has a mayor responsible to the Pleno who presides over 21 districts from the Ayuntamiento de Madrid. In recent years several additional districts have been annexed around the city.

In terms of transport, the city has a series of major highways and boulevards with intervening structures such as the Segovia Viaduct. Madrid has an inner ring road, the M-30, and additional ring roads, the M-40, M-45 and M-50, and several underground sections of tunnel. The main radial Autovias are the A-1 to A-6 inclusively that lead respectively from Madrid to Irun, Barcelona, Valencia, Sevilla, Badajoz and Coruna.

It has an extensive rail network and new high-speed inter-city rail lines (AVE) to other major cities and dense commuter rail services and bus services. The main Adolfo Suarez Madrid-Barajas International Airport (MAD) has been developed with four main terminals.

Significantly, as a result of a change in land-use law in 1998, a period of speculation in the real estate sector occurred. Building permits were secured and many new peripheral housing developments were built around the city. With finance generously available from regional savings banks (Caja), further areas were annexed and urbanised, new infrastructure was built and there was a significant change from renters to owner-occupiers, with increased urbanisation leading to large increases in the development of built-up land areas on the periphery of the city. However, this housing bubble burst ten years later, with elevated land values having been driven by speculation. Regeneration projects are being undertaken in selective areas such as Madrid Nuevo Norte and Campamento.

The population of the central city is approximately 3.3 million and the growing metropolitan area has double this population. It has incurred a high rate of migration in recent years, in particular from Latin American countries. With its increasingly hot summers and heat island effects, it has become a stifling city for Madrilenos with limited shade for shelter as a result of the effects of climate change, and water supply remains critical.

Transactional Forces and Typology

Rather than having a coastal or riverine location as many primate capitals do, Madrid is well located centrally at the centroid of the country in the Iberian Peninsula. During the early 18th century, a concerted effort had begun to create a series of royal highways to connect central Madrid with other regional cities. Whilst of strategic importance, after the Napoleonic wars a series of canals was also planned to complement this, but the grandiose plan was never fully realised, with only 300 kms. achieved due to the challenging changes in elevation, such as on the Ebro Canal. This lack of an effective transport system held back regional development, which was only assisted subsequently with the introduction of the railway system and eventually modern highways.

From its more elevated location between Castile la Mancha and Castile and León, and despite its early unification before Habsburg and Bourbon influences, it was not until the 18th century that its destiny was fully within its own control. The path was varied, with periods of dictatorship and democracy, which was to shape the evolution of the modern capital.

As an elevated inland city in an undulating plateau, the Meseta, Madrid is sited with radial routes running northwards and southwards beyond the city. From its original Moorish fortress, its Old City evolved organically around the Plaza Mayor. With its Royal Palace subsequently developed on the bluff overlooking the Rio Manzanares, Bourbon Madrid was then extended to the east around Paseo del Prado with its squares and palaces. The area of La Castellana was next developed around Paseo de la Castellana and Paseo de Ricoletos with the Plaza de Colon.

From its smaller gridded neighbourhoods, the city has grown radially bounded by the Manzanares gully to the west and ring road to the east. With a newly developed commercial area of AZCA to the north, the city has both a dense historic urban core and areas of recent oversized housing developments that were realised on the periphery by speculators at the end of the 20th century.

2.4.5 MILAN, ITALY

FIGURE 2.4.5 Milan – early configuration, 1572.
Source: Historic Urban Plans.

Background and Origins

Milan (Milano) is situated in Lombardy on a flat plain in the Po Valley in the north of Italy between the River Po to the south, and the Alps and the Great Lakes of Maggiore, Como, Iseo and Garda to the north. With a concentric layout that reflects the system of man-made interconnected canals, the city was walled initially at the time of the Roman Empire and then during mediaeval times.

First settled around 600 BCE and named Mediolanum by the Celtic Insubri tribe, it came under the control of the Romans who created the Province of Cis-Alpine Gaul. As the capital of the Western Roman Empire, under Emperor Maximilian, considerable building was undertaken with a new city wall, palaces, baths, circus and houses. In 313 CE under Emperor Constantine, the Edict of Milan was declared to treat Christians benevolently across the Roman Empire as it became the prevailing religion.

During the Middle Ages, there were incursions by the Huns, Ostrogoths and Visigoths, and disputes between the northern Lombard city states. In 1183 Milan became a Duchy, which it remained as until the 19th century. With its series of Dukes in the Visconti and Sforza families in the Renaissance, during this period the city flourished. The Naviglio Grande Canal was also built to connect the Vicino River to run into Lake Maggiore at this time.

Thereafter coming under the rule of the French, then subsequently the Spanish and Austrian Hapsburgs, it was not until Napoleon's defeat in 1815 that Lombardy reverted to Austrian control under the Congress of Vienna. Finally, with Garibaldi's uprising and the Risorgimento, there was political unification with the creation of the Kingdom of Italy in 1861.

Besides the early Roman buildings, there are several fine basilicas, the Church of Sant Ambrogio, Sant Eustorgio, San Lorenzo Maggiore and San Simpliciano e Santa Maria della Passione. The most imposing edifice is the gothic cathedral, built over several centuries, that dominates the central Piazza del Duomo. Santa Maria della Grazie in the west is also well known for its housing of Leonardo da Vinci's mural *The Last Supper*.

In the historic centre of the city are the Palazzo della Regione and the Palazzo Reale, homes of the Visconti and Sforza families. The vast Castello Sforzesco with its tower and gardens, Ospedale Maggiore and Biblioteca Ambrosiana are imposing buildings along with the significant Pinacoteca di Brera Gallery. After the Renaissance, further buildings were developed under Austrian influence, such as the famous Teatro alla Scala, the Royal Palace and the Royal Villa.

With its circumferential walls, the city was reshaped in Napoleonic times with new boulevards, squares such as Porta Ticinese in the south and the Arch of Peace at Porta Sempione in the north of the city. With the fine Galleria Vittorio Emanuele II completed in 1868, with its iron and glass arcaded roof leading off the Piazza del Duomo towards La Scala, it was the perfect location for a 'passegiata'. Both Via Monte Napoleone and Corso Venezia remain prestigious modern shopping streets.

At the end of the 20th century the area of Bizocca was industrialised with the establishment of the factories of Pirelli and Breda that would be the beginning of the automotive and railway industries.

After World War I, the new Central Railway Station was developed, with its grand frame and classical fascist design influences. World War II was another upheaval, and a subsequent period of reconstruction was required. Some of the large businesses, such as Breda, were consolidated by the State Instituto per la Ricostruzione Industriale (IRI), for example, under the umbrella of 'Finmeccanica'.

Modern City – Buildings and Arteries

During the 1950s and 1960s, new groups of buildings were realised in the city, such as the brick-clad BBPR-designed Torre Velasca, Ponti's slender yet tall concrete Pirelli Tower, and much new public housing. More recently a new commercial business district Porto Nuova has been developed to the north of the city with the UniCredit Tower, the Diamond Tower and Bosco Verticale, a pair of green vertical 'forested' towers covered in vegetation (designed by Boeri Studios), completed in 2014. Another area under re-development is a neighbourhood of the old fair, Fiera-Milano, now known as City Life. It has three skyscrapers making a design statement, one straight, one curved and one twisted. With further residences designed by Libeskind and Hadid, this has been an area of urban regeneration. The regeneration area Milano Innovation District (MIND) is located to the north-west of the city.

Milan is a highly urbanised area and has had separate waves of immigration post-war from southern Italy, Eastern Europe and Africa. As a previously significant manufacturing region, more recently it has become a financial and media centre. As a fashion hub the Quadrilatero della Moda is an elegant district with fashion houses in the centre of the city.

In terms of rail, with its location in northern Italy, Milan is well connected by TAV high-speed inter-city trains with both Turin and Rome and has two international airports at Malpensa (MXP) and Linate (LIN). There is good access to the adjacent Italian lakes and Switzerland and a new major trans-alpine rail line connection to Austria via the Brenner Base Tunnel (BBT) is under construction. It also has four modern metro lines now extended to Porta Nuova and a system of urban tramways. For highways, there is a network of modern Autostrada, including the BreBeMi (A35) corridor between Brescia, Bergamo and Milan, and Autostrada degli Laghi (A8) and the Tangenziale highway around the city (A58). The population of the Comune is approximately 1.4 million and 4 million in the wider metropolitan area.

Transactional Forces and Typology

From its early beginnings with its strategic location in a river plain to the south of the Alps and the Great Lakes, Milan was chosen as the capital of the Western Roman Empire and under the Duchy of Milan. As an early city-state, at the crossroads in Northern Europe, travel was facilitated east-west but constrained north-south by the adjacent Alpine mountain range. With its agriculture and textile sector and trade, Milan had also developed a strong financial sector. Increasing urbanisation and industrialisation in Lombardy came hand in hand, with steel-making and fabrication. Based on its strength in industrial design and production, this was also to lead to its prominence in fashion.

From its origins as a walled Roman city with a central core, Milan was encircled by an external defensive wall in mediaeval times. Its concentric layout was emphasised by a network of navigable and interconnected canals, Navigli, connected by the older Fossa Interna, but have gradually been covered and used for irrigation. As the city has expanded outwards in all directions, the general radio-centric shape of the city remains. With its original nine boroughs and city mayor, the metropolitan area has now been expanded to comprise 24 municipalities as the metropolitan area for Greater Milan, with distributed growth poles extending outwards in a more suburbanised region.

2.4.6 MUNICH, GERMANY

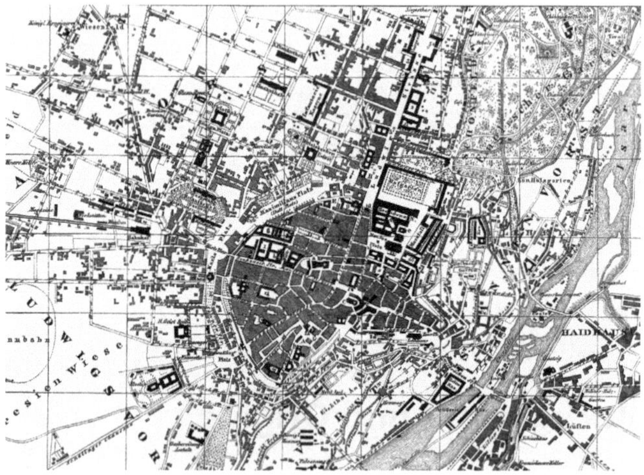

FIGURE 2.4.6 Munich – early configuration, 1854.
Source: iStock.

Background and Origins

Munich (Munchen) is located in the plains of Bavaria on the River Isar, a tributary of the River Danube, just north of the edge of the Alps. With the river running north-south, in Roman times it was crossed south of the city by the Via Julia. It was known as Munichen in high German ('to the monks') by the Benedictine monastery located in the Old Town as a market on the Salt Route by the island where a new tolled bridge was built across the river.

Settled in Roman times, in the mediaeval period it was fortified and attained city status in 1175 but continued as a key centre on trading routes. In 1255 when the Duchy of Bavaria was split in two, under the dynasty of the Duchy of Wittelsbach, Munich became the ducal residence of Upper Bavaria.

As a seat of the Holy Roman Empire, Ludwig IV was crowned emperor in 1328. With completion of the grand Frauenkirche (Munchner Dom) in 1525, Bavaria was reunited in the 16th century and became a centre of the counter-reformation. After the Thirty Years War and periods of the plague, and Habsburg occupation, under the Regency of the Bavarian electors, Munich became an important centre and subsequently was to become capital of the new Kingdom of Bavaria and archdiocese of Munich and Freising in 1806 during the Napoleonic era.

The central Old Town (Altstadt) had been a circular walled city, where three of the original five gates still stand, the Isartor (east), Sendlinger Tor (south) and Karlstor (west) around the ring, with the main square Marienplatz at the centre dominated by the Neues Rathaus and Altes Rathaus. With Peterskirche in Peters-platz and Heiliggeistkirche overlooking Viktualienmarkt, both Michaelskirche and the Frauenkirche with its twin Byzantine domes are key landmarks in the centre.

Just above, in the heart of the city, is the Residenz and its inner courtyards being the home of the Wittelsbach rulers, with the National Theatre and Residenztheater, and palaces around Max-Joseph-Platz and Maximilian Strasse. With the Renaissance Hofgarten behind, the modern Bavarian State Chancellery is also located here. Farther south is Gartnerplatz, a hexagonal square with the Staatstheater before Isarvorstadt.

To the north-west straddling Brenner Strasser and Karolinenplatz is the Museum Quarter and Konigsplatz with its Propylae, and Maxvorstadt District and the University Quarter beyond. Here are the main museums, the Glyptotek and the Neue, Alte and Moderne Pinakotek. To the north-east is Lehel and Schwabing with its Grunderzeit buildings, and the Englischer Garden beyond with the Siegestor victory arch on Ludwigstrasse north of Odeonsplatz towards Leopold Park.

Of the four grand Royal Avenues that radiate from the city centre to the suburbs, Prinzregenten-strasse runs eastwards parallel to Maximilianstrasse and bridges the River Islar just south of the Englische Garten, from Lehel to the wealthier Bogenhausen and Heidhausen districts.

The Englische Garden stretches northwards along the river and acts as a green lung for the city of Munich today. Originally designed by Count Rumford (Benjamin Thompson), with its pagoda (Chinescher Turm), Orangery and Teahouse, it is an extensive urban park.

In the west beyond Ludwigs-Vorstadt towards Westend is Thereseinwiese and Bavaria-Park, the site of the annual Munich Oktoberfest nearby. To the north-west is Rotkreuzplatz in the district of Neuhausen and Schloss Nymphenberg beyond and Hirschgarten with its beer garden.

During the early 19th century, the fortified city walls were demolished and infrastructure and sanitation were improved. Konigsplatz was built and the first railway was built to Augsburg and Regensberg with the new central terminus station Hauptbahnof.

In 1876 Munich hosted the first German Art and Industry Exhibition that was to precede the Jugendstil movement with its progressive industrial designs, that was not un-influential in the subsequent industrialisation of the surrounding Bavarian region. But much was to change with subsequent events under the Weimar Republic across the whole of Germany followed by World War II, after which there was a long period of renewal and reconstruction, and not least subsequent re-unification of divided Germany.

Modern City – Buildings and Arteries

In 1972 the summer Olympic Games was held in Munich, to the north of the city at newly developed Olympiapark, along Olympiasee canal with the stadium and Olympiaturm tower. Here also across the peripheral ring road is the BMW Tower (designed by Schwanzer), Welt and Museum, one of Bavaria's largest auto manufacturers that emerged along with Siemens, which is also headquartered here.

The seat of the Bavarian State Parliament is located in Munich. In addition, with its academies and technical universities, Munich has become a significant centre of learning and research, not least with the centres of the Max Planck Institute and Fraunhofer Society. It has developed into not just a major industrial and commercial centre in the region, but also a centre of financial, insurance and media service companies with headquarters as Hypo-Vereinsbank Tower (HVB) (designed by Betz). Significant new commercial buildings have also been built, such as Highlight Towers (designed by Murphy/Jahn) and Hochaus Uptown Munchen.

In terms of transport, Munich has an excellent and comprehensive network of public transport systems, with the 'S' and 'U' Bahns, trams and express buses within the city. With the central Hauptbahnof and Ostbahnof across the river in the east, there are high-speed rail connections to both Berlin and Leipzig. Surrounded by an integrated motorway network of autobahns with an outer and middle ring and feeder roads, the region has a high rate of car ownership. The new modern Franz Josef Straus Munich International Airport (MUC) was opened in 1992 with two parallel runways and two independent terminals that acts as a significant transport hub in the region.

In terms of population, it is the economic centre of Bavaria and southern Germany, with a population of approximately 1.5 million in the centre and in the surrounding metropolitan area approximately 6 million people. With its breweries and beer gardens and seasonal festivals, it has an attractive work environment and is a most stimulating city.

Transactional Forces and Typology

From its earliest origins as a settlement around the islands in the Isar River and crossing point on the Salt Route, the city of Munich was developed around a circular walled fortification centred on Aldstadt. With its significance from the time of the Holy Roman Empire, and as an economic hub, it developed also as an administrative centre in the region. From this time onward it has become a vital cog in Bavaria from the Grunderzeit age, such that with subsequent intensive industrialisation it has evolved into an essential driver of activity in this area of southern Germany.

With its original circular Altstadt established on the left bank of the River Isar, the city has evolved with different districts. From the central Residenz and location of the State Parliament, the unusually large Englische Garden stretches north-east along the riverside. Outside the Museum District, the city is surrounded on the western side by the inner district of Maxvorstadt, Ludwig-Vorstadt and Isarvorstadt with the outer districts beyond Westend to the south and west. With industrialisation in separate districts, the city has had several different areas of development, now fully integrated by a most comprehensive urban transport network. Whilst a dense yet generally evenly spread urbanised area, it is interspersed with green areas, having both radial and linear characteristics and tentacles of development with an almost complete surrounding ring, Route 99, towards the Bavarian lakes beyond.

2.4.7 VIENNA, AUSTRIA

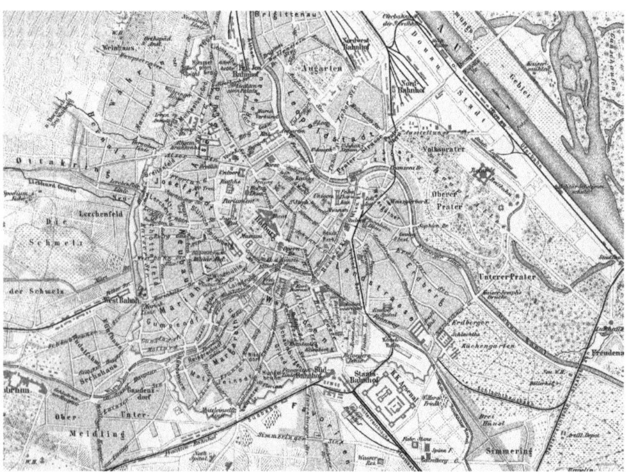

FIGURE 2.4.7 Vienna – early configuration, 1878.

Source: iStock.

Background and Origins

Vienna (Wien) is located on the River Danube in the Vienna Basin between the Alps and the western Carpathian Mountains. After early Celtic settlements, the Romans built a fortified town there called Vindobona. The compact central area of Vienna, the Altstadt, with its Baroque architecture, old churches and towers, Stephansplatz, the pedestrianised Graben, Ringstrasse, Belvedere and the Hofburg Palace are all striking elements which merge into a cohesive whole to make the city of Vienna a very distinctive place.

With its historical past under the Habsburgs, it was the de facto capital of the Holy Roman Empire for several centuries. Having to resist various Turkish sieges, after the wars and captured by Napoleon in 1809, it then became the capital of the Austro-Hungarian Empire.

Historically the centre of the old mediaeval city was the Innere Stadt, which was essentially the original fortress. In 1850, further areas called Vorstadt were annexed, to be incorporated into the city, and the old city walls were razed. After the design competition for a new 'Grundplan', a circular boulevard was constructed, the Ringstrasse, alongside which imposing public buildings were built that include the Rathaus, Burgtheater, university, Parliament, several museums and Staatsoper.

The Wien River flows through the centre of the city. The Donaukanal, which is a natural branch of the River Danube that bifurcates from the main river, was regulated in the 17th century to form a straight direct by-pass, effectively creating a long separating island, Donauinsel.

Bounded by the Danube Canal, as a result of the competition in 1858 won by L von Forster, the Ringstrasse was constructed between the bastions and the glacis of the former fortifications. The outer city limits followed the Linienwall that had been built in 1704, which acted as a cordon at which an urban toll was charged. Farther beyond a significant ring road, the Gurtel was built, which now carries urban trams. As more suburbs were subsequently annexed and integrated as districts into the administration of the city, the Linienwall was eventually demolished at the end of the 19th century. In 1905, a Forest and Meadow Belt was established around the city, which was to serve in terms of crisis as a source of fuel wood.

There are still green spaces such as the Stadt and Rathaus Parks and around the Burgtor, with its many museums, churches with spires, bell towers and steeples, which are very visible. Within the Ring, St Stephans Cathedral is at the centre with its tall spire and decorated tiled roof that dominates the main square and shopping area in and around Karntnerstrasse.

The Hofburg Palace is an intrigue of courtyards along with the Spanish Riding School at Josefsplatz, and the City Hall, Parliament Building, Burgtheater and museums dominate the western section of the Ring. Just beyond and ascending the hill is the Belvedere Upper Palace, which commands a fine view over its gardens and the Lower Palace.

Besides the Baroque and Rococo buildings, Karlsplatz Station and the Secessionist Pavilion in Friedrichstrasse display extensive Jugendstil architectural designs, and the fine Karlskirche dominates with its twin columns.

With established art museums, discrete coffee houses, theatres and the famous Musikverein, the former residence of Mozart, and musical traditions such as the Hofmusikkapelle, there is an abundance of cultural history and entertainment in the city.

With the Biedermeier era that seemed to look back to former eras, as modern components have been blended into the city, the city retains all of its former charm. At its zenith at the turn of the century with its population peaking in 1914, and with its musical, philosophical and psycho-analysis schools, Vienna became known as the City of Music and of Dreams. Additional features such as the imposing Karl-Marx-Hof residence apartment block was built in the 1920s, along with the famous Prater wheel.

But the city suffered significantly in the early 1930s under the Anschluss until the defeat of Nazi Germany. After the end of World War II Vienna came under the control of the four-power Allies with the central international zone policed by one power on a rolling monthly basis, rather than being subdivided into separately controlled sectors, as occurred in Berlin.

Modern City – Buildings and Arteries

In 1955, Austria finally regained its sovereignty as a neutral state, but it was not to join the European Union until much later, in 1995. Additional built development of the city took place on the right bank of the River Danube. Here alongside the straight Donau Kanal, now controlled by regulating locks, in the 1970s developments were extended around the UN City complex and Vienna International Centre and to create the Austria Centre Vienna (ACV). This is now the home to a number of UN and international institutions, such as UNIDO, IAEA and the OSCE. In the distance can be seen the tall Donauturm Telecom tower.

Farther beyond is Donauinsel, a narrow island that lies between the Danube River and the excavated Neue Donau channel, which was created specifically to provide flood protection for the city. Outside the city is Schonbrunn with its elegant palace and Kaiserlinden, the fine rows of lime trees.

After a number of other central European States (viz. Hungary and Poland, Czechia, Slovakia and Slovenia) joined the enlarged EU in 2004, who came to be named the Visegrad group, the city of Vienna found itself nearer the centre of gravity of these countries. New businesses became headquartered in Vienna and with an emphasis on science, research and development, tech and start-up companies have developed in the region.

In terms of transport, the centre has a highly integrated and efficient bus, tram and metro system especially round the Ring. With its 'U' and 'S' Bahns, connections with suburbs farther east are also in operation managed by OBB. Vienna International Airport (VIE) is located in Schwechat to the south-east of the city.

Architecturally, the modern Hundertwasser Museum, and new modern art museums such as MUMOK have been realised. The number of other new buildings is extensive, but in each individual era, significant contributions have been added to the city that have further embellished its architectural standing.

The population of the city is approximately 2 million, with approximately 3 million in the metropolitan area. There was much migration across Europe as a result of World War II and in recent years, particularly from the countries of ex-Yugoslavia and central Europe after the collapse of the Soviet Empire. Poor areas of housing deprivation do still exist, such as at Simmering and Meidlung with tenement housing.

Transactional Forces and Typology

As a landlocked area in an Alpine region settled originally by Celts and Romans, the principal axis of communication in Vienna is the River Danube in its long perambulation through and across Central Europe. As it by-passes the inner city, it acts as the main transit pathway for carrying freight and bulk materials. The mediaeval city grew around the Wien River, with the Altstadt as the centre where the cathedral and the palace predominate. With its role as the capital of the Holy Roman Empire, its administration has grown since the Habsburgs, with the administrative governmental buildings concentrated around the rebuilt Ringstrasse that encircles the inner city.

More recently, post–World War II, as new institutions such as the United Nations were created, a new growth pole has developed, the home of the UN City and Convention Centre on the expanded Donauinsel. With the commercial and business centre remaining in and around the old city, this modern administrative enclave provides a countervailing pole on the other north-eastern bank of the River Danube of an enlarged city, with the industrial sector mainly in the south-east as a new expanding R&D research hub.

As a riverine city astride a major European river, the city is tightly centred around its original mediaeval core. Its formerly fortified walls have now been replaced by the distinctive Ringstrasse, with its array of formal buildings aligned alongside this wide boulevard. With its good integrated public transport systems, underground and street trams, the city has developed in concentric rings with radiating axes southwards and westwards. Its building fabric has been carefully preserved in a manner appropriate for such an historic capital, and it remains as one of the most iconic central European capital cities.

2.4.8 BUDAPEST, HUNGARY

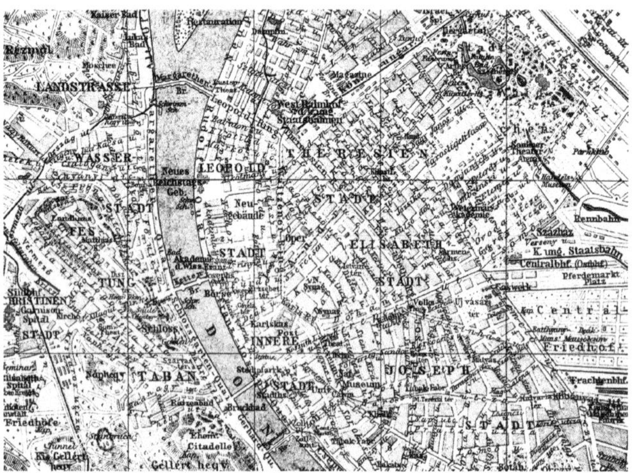

FIGURE 2.4.8 Budapest – early configuration, 1898.
Source: iStock.

Background and Origins

The city of **Budapest** has ancient origins with its Celtic settlement and original Roman development alongside the River Danube as Aquincum in Hungary. After a period of Ottoman rule in the 16th and 17th centuries, with O'Buda behind the hill and the town of Pest on the other side of the river, these towns combined with Buda to become the city of Budapest. There followed a period of Austro-Hungarian rule with its monarchy when capitals were both jointly in Vienna and Budapest.

With its landlocked location at the centre of Europe, the broad Danube River flows down from the mountains and through several other Balkan countries on to the Mediterranean. With its natural configuration with Margaret Island in mid-river, there are deep aquifers which have enabled warm thermal waters to rise and channelled, used as Turkish baths such as at the Kiraly and Gellert hotel.

On the **Buda** side stands the old castle on Castle Hill and Matyas Church, besides which the elevated Fisherman's Bastion was built later in the 19th century. Here with the National Museum is a fine vantage point overlooking the River Danube to Pest. The whole of this fortified castle district was built in and around the existing natural rock. The city also had a line of outer defences, but they were demolished in 1808.

Given the twin nature of the city on either side of the river, a set of major bridges was built that span the wide river. After Margaret Bridge, the elegant iron chain bridge Szechenyi Lanchid was designed by William Clark with a 200 m. central span, with the Elizabeth Bridge and the Freedom Bridge both farther downstream. All of these bridges were destroyed by the German forces during their retreat in World War II and had to be rebuilt.

Across the river is **Pest** with the main commercial street Vaci Utca, now a bustling pedestrianised street. With its characteristic restaurants and coffee houses, several are in the Art Deco style, such as Café Gerbeaud. Alongside the river on the left bank also is the striking large gothic Parliament building located between the Margaret and Chain bridges.

The city has one of the original steam-powered underground rail systems, the Millenium Railway, which opened in 1896 from the centre to the Millennium Exhibition in City Park. This has been complemented by a more extensive underground metro system which has also been recently extended with a new Line 4 that runs under the river between Keleti and Kelensfold. The main railway has two termini, one named Budapest Deli is located in Buda and the other of Nyugati Palyudvar is in Pest. The main Budapest International Airport Ferenc Liszt (BUD) is located at Ferihegy.

Modern City – Buildings and Arteries

With a number of large squares in the centre and a series of grand boulevards, Budapest is a fine city that retains much of its older historic architecture. As outlined, the city has very different characteristics on either side of the river.

Pest is very different from rocky Buda, being built on an alluvial plain and has been developed organically around a number of squares with its markets and mix of commercial and residential buildings and has many active retail stores.

With its 23 districts, the city is surrounded by peripheral ring roads to the east with inner Erzsebet Korut then the outer ring that also contains the right of way for the urban tram towards Arpad Bridge at Margaret Island. With its parks, churches, museums and university, Budapest is filled with cultural institutions in addition to being home to the national Parliament. The Museum District, in particular of Varosliget, has been developed and expanded with a number of new museums. The former Vituki industrial site at Ferencvaros also is being redeveloped to create a new Budapest South City Gate development with a new National Athletics Stadium beside the river Danube.

The population of the city is approximately 1.7 million and 3 million in the whole metropolitan area.

Transactional Forces and Typology

In a landlocked country, Budapest as the capital of Hungary is a twin city straddling either side of the wide River Danube. With its islands, it was found to be suitable as a crossing point and as an original Roman area of settlement around hilly Buda on the western side, where the castle was developed as a fortification, opposite flat Pest on the eastern side.

With the river able to carry heavy water-borne cargo for cross-European bulk materials, the Danube acts as a vital transport axis that runs southerly after changing its course from an easterly direction around Visegrad through to the Black Sea. Now crossed by several bridges that connect with earlier trading routes, there are radial highway routes across and around the city that connect Budapest with adjoining countries and neighbouring capital cities.

Budapest, typified by its riverine nature, comprises the three original urbanised settlements of Obuda, Buda and Pest, each now incorporated into the city. From its elevated Castle district, the city has developed radially in nature into different district sectors from the centre with the broad River Danube running as an axis north-south through it.

With each district centre linked with its river bridge crossings, the railway and underground metro now also interconnect with each of these districts. On the periphery during the Soviet era, several suburban areas were developed with a series of multi-storey brutalist concrete residential apartment blocks that still remain on the periphery.

Whilst Buda is hilly and rocky, and Pest is distinctly different and flat in nature, these two towns twinned together. The following figures illustrate the modern configuration of these historic cities and evolution of their modern population.

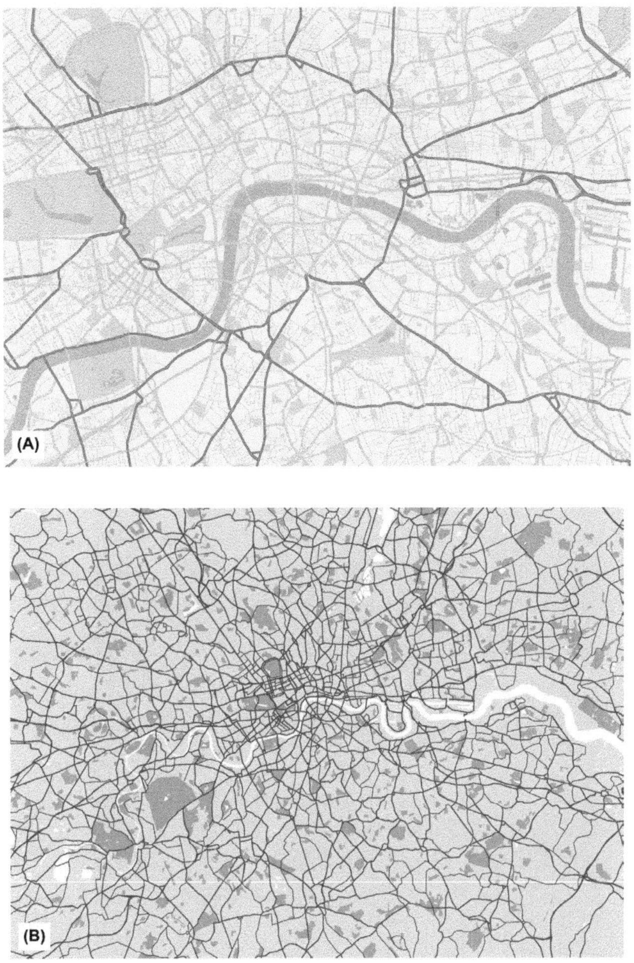

FIGURE 2.4.9A AND B London and Greater London – modern configuration.

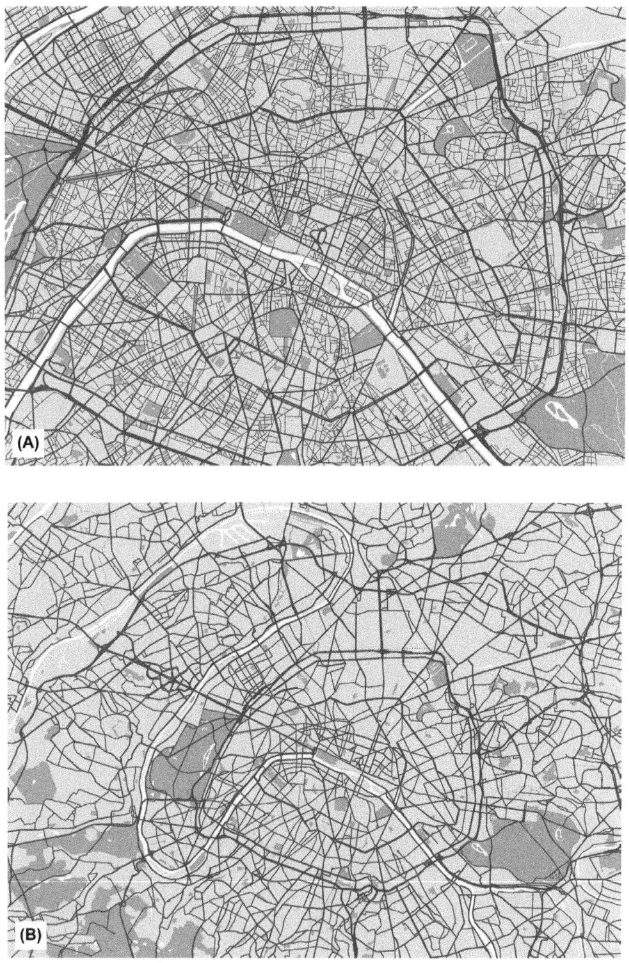

FIGURE 2.4.10A AND B Paris and Greater Paris – modern configuration.

FIGURE 2.4.11 Lisbon – modern configuration.

FIGURE 2.4.12 Madrid – modern configuration.

FIGURE 2.4.13 Milan – modern configuration.

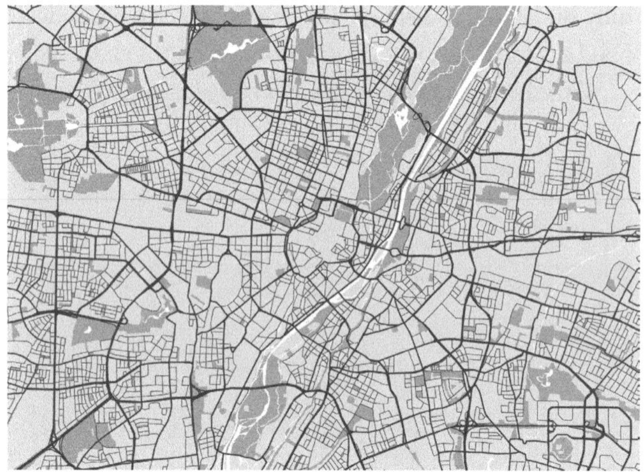

FIGURE 2.4.14 Munich – modern configuration.

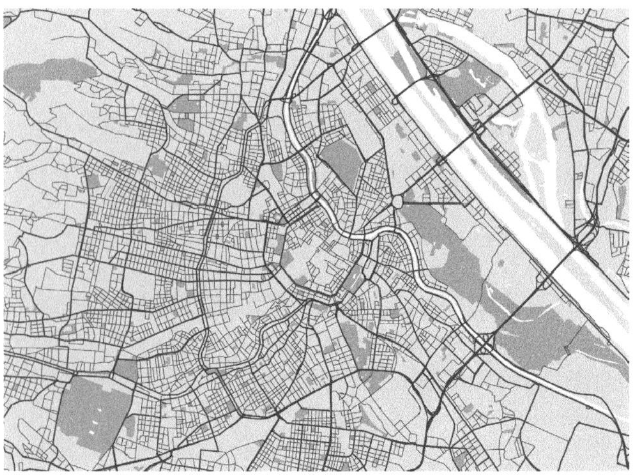

FIGURE 2.4.15 Vienna – modern configuration.

FIGURE 2.4.16 Budapest – modern configuration.

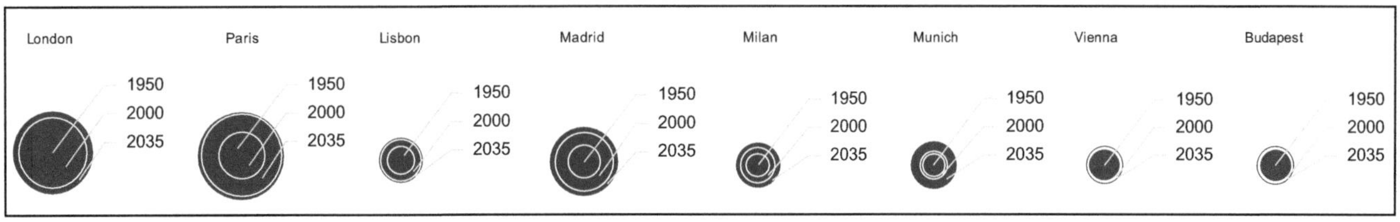

FIGURE 2.4.17 Historic European capital city populations.

Data source: Global Database of Metropolises 2020 – UN-Habitat (2020), UN DESA Population Division.

Further Reading

LONDON

Akroyd, Peter – *London, the Biography*, Chatto & Windus, 2000.
Hudson, Roger – *London, Portrait of a City*, Folio Society, 1998.
Jenkins, Simon – *A Short History of London*, Viking, 2019.
Rasmussen, Steen – *London, the Unique City*, MIT Press, 1934.
Summerson, John – *Georgian London*, Pelican Books, 1962.
Sutcliffe, Anthony – *London*, Yale University Press, 2006.
Tames, Richard – *City of London Past*, Historical Publications, 1995.
Whitfield, Peter – *London, a Life in Maps*, British Library, 2017.

PARIS

Balchin, Paul – *Paris, the Shaping of the French Capital*, Routledge, 2023.
Christiansen, Rupert – *City of Light*, Head of Zeus-Apollo, 2018.
Jones, Colin – *Paris, Biography of a City*, Allen Lane, 2004.
Nairn, Ian – *Nairn's Paris*, Penguin Books, 1968.
Olsen, Donald – *The City as a Work of Art, London, Paris & Vienna*, Yale University Press, 1988.
Sutcliffe, Anthony – *Paris, an Architectural History*, Yale University Press, 1993.

LISBON

Hatton, Barry – *Queen of the Sea*, C Hurst & Co., 2018.
Jack, Malcolm – *City of the Sea*, I B Tauris Publication, 2019.
Pinheiro, Magda & Pereira, Mario – *Biography of Lisbon*, Tagus Press, 2018.

MADRID

Calvert, Albert – *Madrid, an Historical Description and Handbook of the Spanish Capital*, Arkose, 2015.
Escobar, Jesus – *The Plaza Mayor and the Shaping of Baroque Madrid*, Cambridge University Press, 2003.
Parsons, Deborah – *A Cultural History of Madrid, Modernism and the Urban Spectacle*, Berg Publication, 2003.
Stewart, Jules – *Madrid, the History*, I B Tauris, 2012.

MILAN

Bolognesi, Cecilia – *Design City Milan*, Wiley 2007.
Welch, Evelyn – *Art and Authority in Renaissance Milan*, Yale University Press, 1995.

MUNICH

Jerram, Leif & Ruger, Jan – *Germany's Other Modernity, Munich and the Making of Metropolis, 1895–1930*, Manchester University Press, 2007.
Schwandl, Robert – *U-Bahn & S-Bahn & Tram in Munchen, Urban Rail in Bavaria's Capital City*, Schwandl, 2021.

VIENNA

Brandstatter C, Metzger R & Gregori D – *Vienna 1900 Complete*, Thames & Hudson, 2018.
Olsen, Donald – *The City as a Work of Art, London, Paris & Vienna*, Yale University Press, 1988.
Robertson, Angus – *Vienna, the International Capital*, Birlinn, 2021.
Tomun, Rolf – *Vienna, Art and Architecture*, H F Ullmann Publishing, 2018.

BUDAPEST

Hajdu, Joe – *Budapest*, Austin Macauley, 2015.
Stone, Norman – *Hungary, a Short History*, Profile Books, 2019.

2.5 Academic and Ecclesiastic European Cities

Continental Europe – British Isles

Amsterdam, Prague, Florence, Barcelona, Dublin and Edinburgh

2.5.0 INTRODUCTION – ACADEMIC AND ECCLESIASTIC

As the settlement of peoples became more concentrated and centralised in Europe, this led to an increasing exchange not only in the barter of goods, but also of the exchange of information and services. With such an increase in the quantity of information, there was a corresponding interest in the extension of the type, quality and content of information.

The early number of worldwide centres of learning was small, but this developed over the ages from continent to continent and from centre to centre, as people travelled and settled in new areas. With increasing knowledge of different languages and the beginning of the ability to print and reproduce papers and books, this led to an increase in both the geographical coverage and the speed of distribution of information and new ideas.

This rate of increase in different disciplines led to the need for the establishment of specialist schools of learning. As a result, there was the propensity for groups of specialists to come together in communal bodies to form centres which became formalised as academies or departments in dedicated fields. As these became organised, they became parts of separately developing academic institutions in departments or centres of excellence in different urban locations.

Each centre tended to specialise and coalesce around those people gravitating to a particular discipline. They were able to cross-fertilise with specialists in other disciplines as academic colleges and universities were gradually developed. The earliest universities that were established in mediaeval times in Europe and the Mediterranean were in cities such as Bologna and Florence, Oxford and Heidelberg, and have flourished and developed into recognised centres of learning, contributing to the expansion of these urban centres in different countries, as listed, over many years.

Al Quaraouiyine, Fez	859 CE	Morocco
Al-Aghar, Cairo	970	Egypt
Oxford	1096	England
Cambridge	1209	
Paris	1160	France
Bologna	1098	
Padua	1222	
Naples Frederico II	1224	Italy
Siena	1240	
Pisa	1343	
Sapienza, Rome	1303	
Perugia	1308	Italy
Florence	1321	
Salamanca	1134	Spain
Complutense, Madrid	1293	
Valladoloid	1241	
Coimbra	1290	Portugal
Charles Univ, Prague	1348	Czechia
Jagiellonian, Krakow	1364	Poland
Vienna	1365	Austria
Heidelberg	1386	Germany

DOI: 10.4324/9781003353348-9

They were also supported by ecclesiastic activities in places of religious study, monasteries, churches and other significant places of learning, as for example the School of Translators at Toledo cathedral. It is not surprising that these original centres of learning have now expanded as the centres of knowledge and research within the modern information sector around which digitised data now flows so readily.

2.5.1 AMSTERDAM, NETHERLANDS

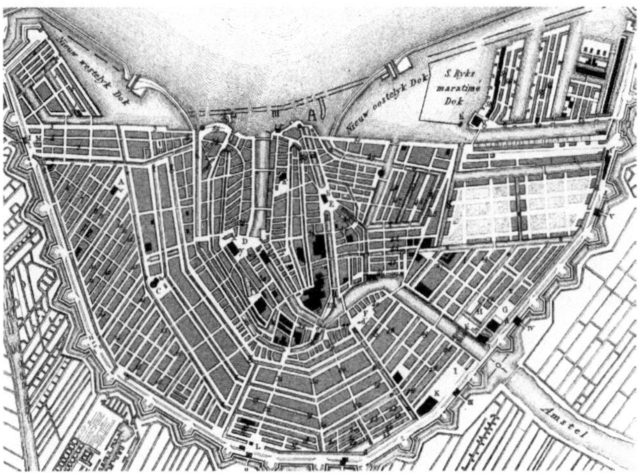

FIGURE 2.5.1 Amsterdam – early configuration, 1851.

Source: iStock.

Background and Origins

The city of **Amsterdam** is located inland from the North Sea in IJsselmeer, on low-lying ground in the province of North Holland. It was developed from the 13th century CE when the Amstel River was dammed, which led to its name of Amestelledamme. Due its natural and geographical location, the Dutch have had to find ways continuously to manage the water and high water table levels across the country. This has led to the construction of dikes and a dense network of drainage canals (grachten) and ditches (sloten) that require constant maintenance.

The whole area of Holland is very low-lying peaty marshland, with outer sandy dunes, dotted with lakes connected to the sea. In the early centuries, as settlements became established the inhabitants tried to tame the water by canalisation, but this tended only to enable sea water to penetrate further inland, creating large lakes that threatened to engulf the area. They set about building dikes with the south side of the IJ from the Gooi and the mouth of the Amstel towards Haarlem that resulted in the Zeedijk and Haarlammerdijk.

The control of the sea was vital for survival and the responsibility was given to the 'dijkgraven', or dike administrators, and 'ingelanden', or landholders. The original land area known as the Amestelle was entrusted to 'villicus', sheriffs under the jurisdiction of the Bishop of Utrecht, where a farmstead was developed around an early castle.

Early wooden dwellings were built on each side of the Amstel on infilled land by moats, the Oude Zijde and Nieuwe Zijde, old and new sides, between walls with palisades that were the beginnings of the future city. The first residents near the Amstel-dam were granted freedom from taxation and avoided paying tolls in crossing bridges and lakes across the region.

As the population grew in the 13th century, one of the most significant developments was the creation of the wooden 'cog' sailing ship that was able to carry goods on the open sea from Holland around the region of the North Sea, to the Frisian Islands and the Baltics. This provided additional communication routes besides the inland waterways passing through the Netherlands.

As trade developed, new walls were built and extended as ships were moored at the quayside. Many ships remained enclosed in the IJ in front of the city by a double row of poles that was closed off every night by wooden booms to prevent flooding. New canals also began to be built with the horseshoe shaped Singel from 1450. In its low-lying area, accordingly the city was developed around a network of canals which were man-made and which remain a key feature of the city today.

The approach to Amsterdam via the Zuiderzee and the IJ was not ideal and resulted in silting up of the harbour. This required constant dredging and became not as competitive as other harbours as Hamburg and London. From the Middle Ages onwards, areas of land and polders had continued to be reclaimed from the sea. During this period settlements developed as trade evolved with neighbouring and Hanseatic League cities in the region. Ship-building also developed with more economically designed sailing ships, 'fluits'.

Dam Square developed around the dam, becoming the centre of the city with a town hall as home of the burgo-masters with its marketplace and merchants exchange. In 1508 a long city wall was built to surround the city with three large gates. Several convents and monasteries were also built inside along with the Oude Kerk.

Whilst the first houses were made of timber, with constant risk of floods and fires, since the 15th century stone and brick materials were introduced for construction of houses with stepped gables and facades with tall chimneys. The facades of these houses along the canals evolved over time from stepped gables to the bell gable and horizontal gable with pointed doors and shutters and variously designed brick facades.

In 1581, Amsterdam had become the capital of the Protestant Dutch Republic and much changed under the Reformation. Subsequently new urban extensions of the city were built, known as the First and Second Extensions to be followed by a further Third Extension.

In 1607, Hendrik Staets prepared the grand 'Plan of the Three Canals' for the city based on the Gentlemen's, Emperor's and Prince's canals – Herengracht, Keizersgracht and Prinsengracht – in a half circle protected by outer walls. With residents having various trading roles, direct access by water was needed to individual merchants' warehouses. The plan was supported by legislation and codes that ensured that plots and uses were guaranteed by stipulated covenants.

With its developing ring of canals between the old side (Oude Zijde) and the new side (Nieuwe Zijde), the ring of the Singel Canal, along with Heren-, Keizers- and Prinsen-gracht, and Lijnbaansgracht in a horseshoe formation within the Singelsgracht, essentially contained the centre of the city, in the 'Grachten-gordel'. This has resulted in a distinctive sense of place, with small alleyways, bridges and houses fronting directly onto the canals, with streets lined in places with elm and lime trees. It was also protected by an outer ring of ramparts.

The city centre continued to evolve around Dam Square in the newer quarter on the western side, with the 'Stadhuis' (town hall), built in 1665 located opposite Nieuwe Kerk. South-west from the Damrak past the Rokin is the historic quarter of the Begijnhof sanctuary.

The eastern side past Nieuwmarkt contains the old quarter and Oude Kerk, with the old Jewish quarter and synagogues, and the university district. Beyond the Central Canal Ring of the three main canals continues the 'Golden Bend' between Leidsestraat and Vijzelstraat where some of the finest houses are sited with their 'well-mannered' architecture.

The Fourth Extension was to continue expansion of the city up to 1672 with consolidation in and around the Canal District. Other outer areas were also developed, for example the dockyards of Kattenburg and Ostenburg.

Trade extended in the 17th century, becoming the 'Golden Age' for the city as trade further expanded well beyond the Baltics to the Dutch East and West Indies and their colonies. In the west, the Dutch West Indies Company (West Indische Compagnie – WIC) was predominant both in the Americas and the Caribbean and had set up the colony of Newfoundland and established the settlement of 'Nieuw Amsterdam' in 1647 that was later to become New York City.

In the east the United Dutch East India Company (Vereenigde Oostindische Compagnie – VOC) was founded in 1602, listed as a public company and had a monopoly over the spice trade that extended beyond the South African Cape and the coasts of India from Mauritius to Malacca, Makassar and Batavia.

Besides trade in spices, tobacco, fabric and foodstuffs, with commodities and the produce exchange, Amsterdam became a financial centre with the world's first stock exchange established around which merchant banks gravitated. It also became a centre for diamond polishing and gold trade from South Africa.

But 1672 was to be a transformational year heralding the end of the Dutch 'Golden Age', as the economy stagnated and the country was to come under the domination of France. With the eventual separation of the Netherlands between North and South, Amsterdam became the capital of the Kingdom, with Den Haag as the seat of government.

It was not, however, until quite later in the 19th century that one of the most significant developments of construction of the Noordzeekanaal, North Sea Canal, opened in 1876, provided a direct connection west to the North Sea from Amsterdam. Suitable for larger ships, the city's harbour was enlarged and new quays were built. In addition, with the first railway built from Amsterdam to Haarlem, the Central Station was opened in 1889 right in front of the city centre, cutting off direct access to the IJ.

With the realisation of this new canal link, new dockyards and quays were built on the IJ to accommodate larger vessels, especially freight, with the new dockyards of Oostelijk Havangebied with the quays of Sporenburg, Borneo-, Java- and KNSM-eiland.

In 1892, a new connection south with the River Rhine was also built, the Merwede Channel, that was to form the Amsterdam-Rhine Canal towards Utrecht. This vital direct connection built on the shortest route to the North Sea through the locks at IJmuiden enabled seafaring vessels to reach the port of Amsterdam directly, further enhancing trade and commerce.

In 1876, a further new Extension Plan for Amsterdam was prepared by Kalff. This involved the annexation of additional land areas and housing developments beyond the outer ramparts, and a further annexation in 1896 of the area known as Amstelveen. The opportunity was taken for the creation of a number of parks, Oosterpark, Westerpark and Vondelpark and remaining area that led to the eventual creation of Museumplein.

Due to its low-lying precariousness and with fear of invasion, during the period between 1880 and 1920, a line of defensive fortifications was built around the outskirts of the city known as 'Stelling van Amsterdam'. This was a line of forts that could provide protection by flooding the area in time of war.

With increasing expansion there were pollution and environmental problems and a decision which was to become quite transformational in the city centre was for the backfilling of certain canals in the Jordaan neighbourhood, the Elandsgracht and Rozengracht for health reasons.

A new phase in the development of the city was to evolve with the annexation of new districts. In 1894 Plan Zuid was prepared by Berlage for the realisation of a new district centred around a new station and Minervalaan. In 1922 an even larger annexation of the former rural municipalities such as Nieuwendam, Ransdorp, Sloten and Buiksloot, other territories were to be included under a new Plan West, again by Berlage around Mercatorplein. All of this was to significantly extend the perimeter of the city.

Another significant event was the publication of the General Extension Plan in 1935 (Algemeen Uitbreidingsplan van Amsterdam – UAP), a visionary plan for the entire municipal territory. Based on functional separation plans of industry and housing, it concentrated industrial zones at Amstel I and II around the harbour with satellite garden cities for housing interspersed with green wedges.

Whilst interrupted by World War II, with occupation and destruction, it was later pursued under modernist designs, with dense new neighbourhoods of working-class housing near industrial districts of Amsterdam West and garden cities around Sloterplas and Buitenveldert in the south. Progressively these areas became overdeveloped and of poor standard; they started to decay, requiring replacement.

One special area was Bijlmermeer in Zuidoost, a major housing estate planned in 1965 on a honeycomb grid with a series of ten-storey blocks, with separate areas of accommodation and recreation laid out on modernist planning principles. However, with its scale, density, layout and access routes, social problems persisted and in 1992 a major programme of redevelopment had to be undertaken with several high-rise blocks demolished.

During the 1960s and 1970s, with increasing population, government had introduced new spatial planning policies to assist with overspill and earmarked specific localities as development poles. These included specifically around but outside Amsterdam, the places of Almere, Haarlemmermeer and Purmerend. Almere, located on the IJsselmeer in Zuidelijk Flevoland, has grown rapidly as a multi-nuclear city with individual characteristics, with approximately 0.2 million residents. Haarlem-mermeer contains Nieuwe-Vennep and Hoofddorp near Schiphol airport and houses approximately 0.15 million. With Lelystad beyond, these areas are fulfilling their function as overspill for the city.

Subsequently, in the 1980s policy was to be refocused on urban densification and renewal. With former flight to the suburbs, the city started to concentrate on renovation of the existing fabric, reconstruction of the housing stock and retention of more affordable housing nearer the centre.

In addition, the benefits of a 'compact city' were recognised rather than one of functional separation. Accordingly, the focus turned to areas along the riverbanks and waterfront to be redeveloped and rebuilding of the former dockyards. Also, the Masterplan Zuidas was developed for the area around Zuid Station with a mix of commercial and residential areas. This contributed to urban development moving out in the direction south-west towards Amsterdam Airport at Schiphol (AMS).

Further, as a result of 'Omgevingsvisie Amsterdam 2050', the future vision is for continued urban densification with a focus on the Zuidas Area, Westelijk Havengebied and areas North of the IJ. With growing areas such as Amstelkwartier, Sluisbuurt and IJburg, an entirely new area of 'HavenStad' is also eventually envisaged.

Modern City – Buildings and Arteries

Post-war, a major new building programme was undertaken as standardised blocks of apartments were developed under the 'New Pragmatism' in the areas of Slotervaat, Geuzenveld and Osdorp. Roads were improved and bicycles started to appear ubiquitously. There was a plan in 1954 for the infilling of a number of canals so as to enhance traffic circulation, but this was resisted. The construction of the IJ tunnel under the river that replaced the 'Old Tollhouse Ferry' was also realised.

Society changed in the progressive 1960s with the youth challenging with a 'White Bicycle Plan' for the city and more radical anti-authoritarianism. With the end of former Dutch colonies in the Dutch Antilles, Indonesia and South-East Asia, a new wave of migrants emerged into the city. Many buildings were torn down and with traffic redevelopment schemes around Central Station and Nieuwmarktbuurt was to be transformed, challenged by squatters although finally redeveloped under a revised scheme in 1980 after a long period of challenges.

Many Amsterdammers exited the city to new suburban communities, which was encouraged by government, with the planned development of outer new towns, for example at Almere and Lelystad. With the housing sector largely regulated in Holland, much of the housing stock was controlled by housing corporations.

Culturally, in the centre 'Museum Plein' with the imposing red brick 'Rijksmuseum' built originally in 1885, and the 'Stedelijk Museum' has been joined by the modern 'Van Gogh Museum'. Nearby is the open Vondelpark and meandering water and lakes. Another new modern building is 'NEMO Science Museum' (designed by Piano) located near the water's edge by the tunnel under the River IJ. The massive new combined Stadhuis and Opera House on Waterlooplein during construction protests came to be named 'Stopera', a name which has remained.

Further east with the Eastern Canal Ring around the Reguliersgracht is the iconic original counter-weighted double-leaf style Magere drawbridge across the River Amstel. This leads to the Plantage, once the 'plantation' area and then developed for the diamond-cutting industry.

The Kattenburg islands contained wharves for maritime shipping and the former military arsenal which is now the Scheepvaart-Museum. Across the River IJ to the north was the former shipbuilding area of NDSM, which has been redeveloped with a modern cinematic museum called the 'Eye' and A'DAM (formerly Shell) Tower nearby.

The Eastern Harbour had accommodated the deep-water steamships that travelled to and from the Dutch West and East Indies. With the loss of this traffic, the old warehouses have been converted and the Sporenberg Peninsula transformed into a residential district. The adjoining Java, Borneo and KNSM were similarly developed and now comprise riverside apartment blocks.

The municipality of Amsterdam has a population of approximately 1 million and the metropolitan area has approximately 2.5 million inhabitants, on the edge of the larger Randstad conurbation area and neighbouring Haarlem and Utrecht.

Public transport in the city is excellent, with integrated bus and tram routes and five metro lines. With its outer ring road and network of canals and bridge crossings, cars are discouraged and many areas are car-free, whilst cycling is actively encouraged. With its pedestrianised central area, it has a well-designed civic centre.

As an historical yet modern and liberal city, Amsterdam is a lively place with its characteristic recognisable traditional stepped gable facades and its old canals, parks and cultural institutions. A city of coffee houses, cosmopolitan districts and liberal values, it is a vibrant and enterprising modern European city with its former historic connections.

Transactional Forces and Typology

With its location in the low-lying environment by the mouth of the Amstel River, fishing and water-borne transport and trading were the basis of the original settlement of Amsterdam. As trade routes were extended to the adjacent Baltic Sea, with the exchange of grains and timber, it was the development of ocean-going vessels that extended the range of transactions. With both the Dutch West India and East India companies operating in locations as distant Aruba and Curacao in the Caribbean, and Batavia in Indonesia, the wider trade in spices, tobacco and foodstuffs developed.

By benefitting also through the triangular Atlantic slave trade and with gold and diamonds from South Africa, Amsterdam became a trading and financial centre with establishment of the first commodity and stock exchange in Europe. Whilst improvement of the canal and access to the North Sea has resulted in the transfer of trade, with Rotterdam's recent expansion and dominance in international shipping, especially under containerisation, Amsterdam has flourished rather as a financial and cultural centre with developing services in insurance and shipping.

Given its natural situation, the configuration of Amsterdam with its waterways is unique, with its man-made concentric rings of canals that envelope the city. Some lines of the original fortifications are still discernible, whilst several canals have been filled in and built over. The Singelsgracht typifies the horseshoe form of the central city with its pathways and canals, which are such a dominant characteristic of the city with its bridges and waterways.

With both Westpoort and the Eastern Docks, the city sits predominantly to the south of the River IJ with its circumferential roadways and Central Railway Station. As a growing radial city, new areas are being annexed, extended and built around the periphery and encompassing new poly-nodal areas of Nord, Nieuw-West, Zuid and Zuidoost and built-up areas south-west towards Schiphol Airport and south-east towards Utrecht. In addition, with reclamation of polders in the IJsselmeer, entirely new separate towns are being built besides IJburg beyond on the periphery as Almere and Lelystad but which are linked to the urban core.

2.5.2 PRAGUE, CZECH REPUBLIC

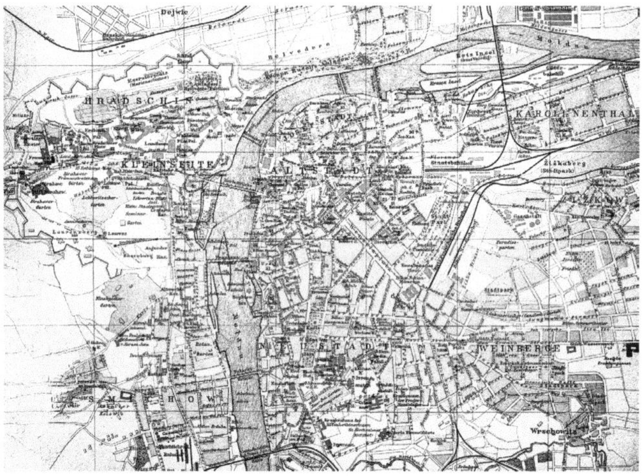

FIGURE 2.5.2 Prague – early configuration, 1897.
Source: iStock.

Background and Origins

Prague (Praha) is one of the finest old cities in central Europe, in Bohemia the capital of the Czech Republic. Whilst there is evidence of very early ancient settlement and Roman oppidum by the river, a series of different tribes – Celtic, Germanic, Slavic and Czech – had migrated around the region. The first fortified settlement was developed around the castle in the 9th century CE with the foundation of the Premystil dynasty of the Duchy of Bohemia in Great Moravia. The earliest bridge was built at the site of the ford across the Vltava River in 1170, later destroyed, but it was to form part of the foundation of the subsequent long Charles Bridge. Early developments also followed at Hradcany around the castle and the 'Mala Strana' district.

During the 14th century, the city flourished under Charles IV as king of Bohemia and also as emperor with the seat as the capital of the Holy Roman Empire. The Nove Mesto (New Town) was developed along with the new Charles Bridge, and the earliest European Charles University and St Vitus Cathedral. During the 16th century, however, the city came under the influence of the House of Habsburg and disputes between Catholics and Protestants were prevalent, followed by the 30 Years War. Much of the city was destroyed in a fire in 1689, after which it was rebuilt and renovated, and during the 18th century the city was invaded by the Prussians under Frederick the Great.

From the vantage point of Hradcany Castle and St. Vitus Cathedral overlooking the Vltava River, with its islands and bridges, from the often-flooded Mala Strana (Lesser Quarter) to the Stare Mesto (Old Town) and the Jewish Quarter, the many gothic church spires and the palaces and monasteries stand proudly where the music of Smetana, Dvorak and Janacek was composed. The route along Celetna Street from the eastern gate Powder Tower, passes the Tyn Church and Old Town Hall and Square, with the Prague astronomical clock and monument to Jan Hus, under the Old Town Bridge Tower. Then across the river on the Charles Bridge with its stone statues, and up the hill and to the Palaces and Cathedral, is a fine townscape.

Across the Maltese Square and the Lesser Quarter Square with Petrin Quarter beyond, is the church of St Nicholas by Dientzenhofer, and the monastery and church of St Thomas. Nerudova is lined with palace buildings, and the new castle steps lead up from the street to the castle quarter, through the Matthias Gate, past the chapels, gallery and gardens to the Belvedere. Predominant here is the grand gothic St Vitus Cathedral designed by master builder Peter Parler with its elegant internal fan vaulting. On the other side of the river, Nove Mesto, with its long Wenceslas Square, former mediaeval horse market, leads up to the National Museum.

The fine old separate districts of Hradcany, Mala Strana, Stare Mesto and Nove Mesto had been merged into one municipal entity to which was added Josehov, the Jewish district in 1850. During the 19th century, a period of Czech National Revival developed in response to the earlier domination by the German-speaking population. The suburbs began to be developed as a result of industrialisation and new factories resulting from the development of coal mines and steelworks in the region.

However, as a result of World War I and the defeat of the Austro-Hungarian Empire, the first Czechoslovak State was created in 1918 with Prague as its capital. Yet this was to be relatively short lived with a significant moment of upheaval in March 1939 when the German army occupied Czechoslovakia, and commencement of World War II. The city of Prague suffered considerably during this time of both material and human damage which was not to improve post-war, falling under the yoke of the following Soviet empire within the communist economic bloc of COMECON.

Subsequently, during 1968, under a period of protest of the 'Prague Spring' in seeking liberalisation, this was repressed by the Soviet Union. Although relations were temporally normalised, all was to change with the Velvet Revolution and eventual total collapse of the Soviet Bloc in 1993. With the creation of a new independent Czech Republic, having Prague as its capital, it separated from the adjacent Slovak Republic and was to become head of the Central Bohemian Region.

Modern City – Buildings and Arteries

Whilst the city fringe was shaped by the construction of Soviet-style apartment blocs post-war, the centre has been generally preserved and not radically reshaped. Since then, the city and region has been transformed to a modern economy from vehicle manufacturing to service industries and research centres and with a significant number of new buildings that have been developed.

In addition to its historic and cultural institutions, Prague has a number of universities and research institutions that complement the old and well-established Charles University. With its focus on new technologies and medicines, suburban areas of the city have been developed from Vinohrady, Pankrac Plain and beyond. Prague has become a bustling modern capital city, full of history and yet having contemporary vitality, the 'City of 100 Spires'.

Overlooked now by the tall modern concrete Zizhov Television Tower, the city has a well-integrated public transport system with three metro lines, modern tramways, buses and commuter and high-speed inter-city rail connections to neighbouring European cities. The city has two ring roads, with the long Blanke Tunnel forming part of the Inner City Ring M-0 and Outer Prague Ring D–0. Prague Vaclav Havel Airport (PRG) is located to the west of the city. From a population of 1 million in 1950, the population of the city is approximately 1.3 million and double that in the whole metropolitan area.

Transactional Forces and Typology

From its early beginnings as a crossing point on the Vltava River, tribes gravitated to this area. From its trading origins and as the centre of Bohemia, it grew into being the capital of the Roman Empire. The city, with its educational and religious institutions, was developed under the Habsburgs and with a subsequent period of Prussian intervention, and eventually industrialisation, it was overrun and shaped by critical events during the two world wars.

More recently, having come under the influence of the Soviet empire, this was to hinder its modern development until its accession to the European Community in 2004. As a primate city in the Czech Republic, it is now both an administrative centre and the centre of new high-tech businesses in the region. Nestling on the east bank by a loop in the Vltava River, Prague is characterised by the Old Town (Stare Mesto), New Town (Nove Mesto) and Lesser Quarter (Mala Strana) nestling astride the river as an historic central European city. With its organic features and the castle and palace to the west (Hradcarny), linked by its characteristic original Charles Bridge to the Old Town and Josehov, new developments have emerged farther out radially as the city has grown and industrialised.

2.5.3 FLORENCE, ITALY

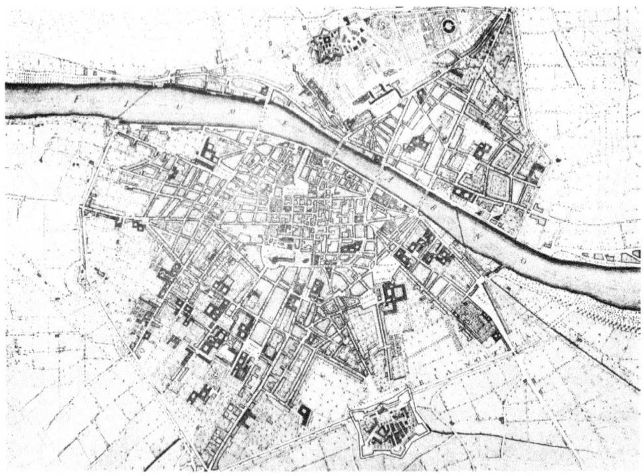

FIGURE 2.5.3 Florence – early configuration, 1755.

Source: Historic Urban Plans.

Background and Origins

Florence (Firenze), with a wealth of riches and history, is one of Italy's most elegant cities in the centre of the region of Tuscany and is known as 'La Fiorente', the flowing one. From its early Etruscan walls, mediaeval streets, Renaissance palaces and churches, fortunately enough evidence remains of these former periods of one of Italy's city-states.

The centre of the old city reflects the original Roman grid pattern north of the River Arno and the 'Piazza della Repubblica' with its arch, once the site of the main Forum. As 'Florentia', it was developed as a Roman city in a rectangular manner, with a triangular area on the other side of the river, where the Via Cassia crosses the River Arno.

Around 1175 CE the city was surrounded by peripheral walls that enclosed the city. As the city grew during the next century, a further set of external walls was built to surround the city in almost a square shape so enclosing the religious centre around the cathedral and the political centre around the Piazza della Signoria. These walls continued to be extended several times. The development of the city was facilitated partly by the patronage of one great local family, the Medici, who under their influence commissioned, developed and maintained much of the Renaissance patrimony under the Grand Duchy of the Florentine Republic.

Sited in the rolling hills of central Italy, the city of Florence straddles the River Arno. Its main landmark is the predominant cathedral in the central Piazza del Duomo. With its tall campanile and grand terracotta dome designed by Brunelleschi, with its marble facade and chapels, and separate baptistery with decorated bronze doors by Ghiberti, it presides over the whole city.

The other predominant municipal buildings are the Bargello, city hall, the original seat of government and the Palazzo Vecchio with its campanile bell tower in the Piazza della Signoria, with its fortified facade, and council chamber that serves as city hall. With its Loggia, adjoining this is the long Uffizi Gallery (designed by Vasari) where the Medici family's art treasures were kept and whose Vasari corridor leads on down to the River Arno. The University of Florence was also founded here originally in 1321 but was moved to Pisa by the Medici when they were in control of the city.

In addition, there are several very fine churches, such as Santa Croce, with its tombs of famous Florentines and cloisters, Orsanmichele and Santo Stefano al Ponte, San Lorenzo and San Marco. Around the Piazza di San Lorenzo, with its dome and unfinished facade is Palazzo Medici Riccardi, home of the Medici family, designed by Michelozzo. San Lorenzo was the parish church of the Medici family and contains the Medici tombs designed by Michelangelo and the Biblioteca Laurenziana with its fine Mannerist staircase.

Around Piazza di San Marco is the Dominican Convent of San Marco, with its cloisters and library, and nearby the fine Renaissance Piazza della Santissima Annunziata with its church and classic Spedale degli Innocenti Orphanage with arcaded Loggia by Brunelleschi. Nearby is the monumental and rusticated Palazzo Strozzi, and the Palazzo Ruccelai by Alberti and the Palazzo Corsini along the bank of the River Arno. The Via Por Santa Maria leads to the Ponte Vecchio, which is the oldest bridge in Florence, built in 1345 and containing shops and workshops straddling the three spans of the bridge, with the private Vasari corridor above and the defensive Mannelli tower at one end. The other fine church is Santa Maria Novella, which was a former Dominican monastery with its splendid nave, cloisters and Strozzi chapel, and well-proportioned main facade and volutes by Alberti.

Across the Arno River is the district of 'Oltrano' where the Palazzo Pitti features predominantly beyond the south bank, as the palace became the main residence of the Medici and subsequent base for Florentine rulers. With its royal apartments for the Grand Dukes, Palatine Gallery and inner courtyard with added side wings, this leads out to the formal areas of the Boboli Gardens and beyond with its avenue, walkways, open areas and Belvedere Fort.

Modern City – Buildings and Arteries

The active commercial centre of the city is around Via del Corso and Piazza della Repubblica, crossed by the Via dei Calzaiuoli to the north of Mercato Nuovo and Piazza della Signoria. The street for fashion is Via de' Tornabuoni. Roads run on either side along the riverbanks, Lungarno, and otherwise run radially outwards from the old city centre. The outer tangential Autostrade del Sole highway passes around the west and south of the city for inter-city traffic. The modern-tyle main train station (Stazione Centrale) built in 1935 by the Fortezza da Basso serves to connect high-speed trains (Pendolino) with the northern cities and south to Rome.

To the south of the city beyond the Boboli Gardens is Fort Belvedere and the elevated San Miniato de Monte in its picturesque rural setting. To the north is the retreat of Fiesole towards the foothills of the Apennines. Other fine Tuscan hill towns are accessible nearby, such as Siena, Arezzo and Lucca, exquisite San Gimignano with its old towers, and walled Monteriggioni in adjacent Chianti. A compact city steeped in history and culture, with its churches, museums and palaces, it is now mainly pedestrianised with limited traffic zones (ZTL) for cars in the centre and a new tram system. The population of the city is approximately 0.4 million people.

Transactional Forces and Typology

As an historic walled city, settled on the banks of the River Arno, Florence developed from its mediaeval origins to become the birthplace of the Italian Renaissance in the 15th century. With the burgeoning wool and textile industry, merchants gravitated into and around the centre. It grew as a city-state under the banking skills of parties such as the Medici family that promoted businesses and also patronage of the arts in commissioning key works by emerging Renaissance artists and architects.

With the creation of the Grand Duchy of Tuscany in 1569, the House of Medici came to rule the region for a period of over 200 years. After unification, from its early artisanal skills in leather, jewellery, clothing and furniture, the city became industrialised in the 20th century and also promoted its viticulture sector. With the restoration of its centre after World War II, the city has emphasised its cultural history, architecture and urban townscape as the 'cradle of the Renaissance'. As a mediaeval city, mainly developed within a square walled area straddling the River Arno, its several bridges connect either side of the city. At the crossroads of central Tuscany, it has radial connections beyond a dense organic urban fabric. With its covered bridge Ponte Vecchio and dominated by the cathedral with its campanile and baptistery, it retains its fine Renaissance squares and unique set of churches and has preserved most of its former precious urban features and streetscape.

2.5.4 BARCELONA, SPAIN

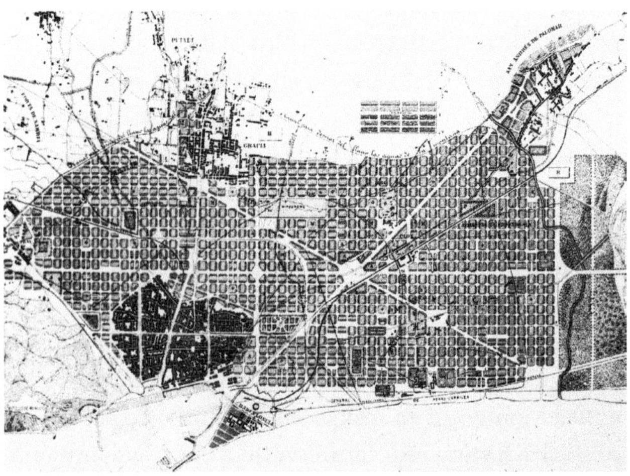

FIGURE 2.5.4 Barcelona – early configuration, 1858.
Source: Cerda.

Background and Origins

Barcelona, the capital of Catalonia, with the backdrop of the Collserola Hills and abutting the Mediterranean Sea with its rocky coast, is a most historic city in north-eastern Spain. Between the Llobregat and Besos Rivers, the three areas of the Old Town (Ciutat Vella), the Eixample and Parc de Montjuic are quite different neighbourhoods.

The splendid gothic cathedral and cloisters in the Old Town, Barri Gotic, the Montjuic Castle and Palau Nacional have become key features of the historic, the gothic and the modern blended together in a bustling Mediterranean city and port.

Populated originally by the Laietani tribe, reputedly the city known as 'Barcino' was founded in Roman times after a Carthaginian general named Barca, father of Hannibal, as a fortified military 'oppidum' with grid plan and extensive walls in a strategic location by a harbour. After later invasions by Visigoths and then Arab Moors, having their capital in Cordoba, the Hispanic Marches became a buffer zone in the Pyrenees as created by Charlemagne.

After a major plague in 1348, considerable building was undertaken with three major churches, including Santa Maria del Pi and Santa Maria del Mar, and a new district, Arrabal, was added to the existing Cuidad separated by the Ramblas boulevard, the beginning of the historic long tree-lined walkway La Rambla crossing the old town and Placa Reial.

In 1450 the University of Barcelona was founded by royal decree with the foundation stone laid for the new university building in La Rambla and the first stock exchange, Llotja, had been created and prospered from that century onwards.

The Principality of Catalonia merged after the Kingdom of Aragon and Catalunya had dominated the western Mediterranean and North Africa, as the royal lines of Castile and Aragon were united in 1479 after the marriage of Ferdinand II and Isabella. The dominant power, however, subsequently moved to Madrid in the 16th century as the maritime outlook swung rather from the Mediterranean to the Americas with the beginning of the triangular trade and slave routes from Africa.

Further confrontation arose with the 'War of the Spanish Succession' and the siege of the city in 1714 and following the Decree of the New Plan, which imposed a Cadastre tax. With the destruction of the Ribera Quarter, it was planned to be replaced by a new citadel with large, fortified bastions. Around 1750 a new area called La Barconeleta was built on an eastern triangular spit of land by the port in a grid formation where new narrow blocks of housing were built.

The realisation of Las Ramblas had also begun, in refilling the riverbed of a former stream, the Cagallel, which ran diagonally inland from the coast. As this became formalised new buildings and palaces were built along its edge, such as the Casa March de Reus. Despite incremental civic improvements, such as around Place Reial, there was no substantial change until removal of the Bourbon 'muralles' which prevented expansion of the city.

Some civic improvements came with the Jardi del General gardens by the port, then the Passeig de Gràcia with its connections to the Ramblas at the Placa de Catalunya.

In the early 19th century, with the region affected by the Peninsula War and French occupation up to 1814 and eventual end of the Bourbon regime, formalisation of a new

Spanish constitution under the Cortes of Cadiz followed, which defined Spanish liberalism and rules, rights and citizenship in its overseas possessions, such as in Spanish America.

From being a very productive agricultural region of crops, vegetables and wine, with the beginning of cotton manufacturing a large growth in population was to develop in Barcelona where people had become housed at very high densities in 'colonies industrials' with mills located by the riverside. The year 1848 was to be a year of revolutions across Europe, and in Barcelona highways and railways began to be developed as industrialisation was taking hold, with metallurgy being added to textile manufacturing. A period of 'Renaixença', renaissance, was to follow in promoting a revival of Catalanism in the region.

With pressure on accommodation, a competition was held for the design of a large empty area of urban enlargement in the centre of the city. The scheme adopted for the expansion of the Eixample district that comprised the Quadrat d'Or, centred on the Passeig de Gràcia. It was designed by Ildefons Cerdà in 1859, as a grid of streets with 550 residential blocks having internal courtyards and external chamfered corners at street intersections. With the Gran Via intersected by the Avinguda Diagonal and Avinguda Meridiana, this was developed as a major new neighbourhood with particularly distinctive characteristics.

With a cellular grid, the plan for the new area was to be a modular city interspersed with parks, with large blocks broken down into smaller barrios and housing limited to five storeys in height. However, developers took advantage and built over intended garden spaces and increased the height, resulting in overdeveloped and higher densities than originally intended, although the introduction of intermittent 'passatges' or private streets lined with houses provided some relief.

In 1869, a competition was held to design a new park to replace the Old Citadel under designs by Fontseré and Domenech, a new horseshoe-shaped Citadel Park was planned. However, this was to be transformed in 1888 when the Universal Exhibition was held in the Parc de la Ciutadella with the tall Columbus Monument from which streets branched and the Passeig de Colom, and realisation of the Palace of Industry and the Palace of Science. Under a period of evolving Catalan Modernisme, a number of new Art Nouveau buildings were also commenced, including Gaudi's Palau Guell and Casa Milà, and also eventually his extemporised church Sagrada Familia that was built over many years thereafter at the end of the Avinguda de Gaudí.

At the beginning of the 20th century Spain had avoided involvement in World War I; however, the country was to be most affected by the Spanish Civil War during the 1930s when Franco's regime took power. Although Catalonia remained staunchly Republican, most autonomous institutions of Catalonia had been abolished, not to be reinstituted until the end of a long period of dictatorship that prevailed across Spain under Franco up to his death in 1975.

Modern City – Buildings and Arteries

Barcelona has the seat of the regional government of Catalunya, the Generalitat, with its four provinces of Barcelona, Girona, Lleida and Tarragona. The mayor of the city also presides over its ten districts from the Ajuntament city hall. With Spain's accession to the European Community in 1986, the economy grew and a period of urban regeneration was undertaken, especially in the area around the harbour. Other manufacturing areas, such as the factories in Poble Nou, were upgraded and areas of urban renewal have been undertaken.

The city possesses a number of fine parks at Montjuic, de la Ciutadella, Guell, Oreneta Castle and Nou Barris, and it benefits from a number of beaches along the outlying coastline of the Costa Brava and Daurada, including Sant Sebastian, Barceloneta and Somorrostro up to the port.

Near the port, this area has been transformed since the 1992 Summer Olympics was held there around the regenerated port with its village Vila Olimpica and Olympic Harbour. The occasion of the Olympics was an opportunity to significantly redevelop several adjoining districts under a programme of urban renewal and also to upgrade many buildings, such as Palau de la Música.

The Port Authority of Barcelona is now an extensive operation with Port Vell, the old port and former shipyard, the commercial Zona Franca and Free Port nearby and is the leading container and cruise ship port in the region. The main Barcelona International Airport (BCN) is at El Prat, just the other side of the Llobregat River.

With a tradition in creative arts and crafts and textile manufacture, it holds a number of large fairs at the Fiera de Barcelona annually at its new World Trade Centre and Gran Via Centre. Barcelona is a significant manufacturing hub with several factories that include the major automobile manufacturers SEAT and Nissan. It also has a flourishing modern high-tech sector.

As a railway hub, the main railway station, Sants Estacío, serves the high-speed intercity AVE rail system and Catalonia regional lines. There is also a RENFE rail connection to the main airport. It has a comprehensive metro system operated by ATM and FGC and two modern tramways plus the heritage line and three hillside funiculars at Parc de Montjuic, Tibidabo and Vallvidrera.

There is a network of modern highways and three half ring roads, northern Ronda de Dalt B-20 (mountain side), Ronda del Litoral B-10 (coastal), and separate Ronda del Mig with intermittent tunnelled sections.

The population of the city is approximately 1.6 million, most of whom speak the local language of Catalan, and with three times that in the metropolitan area. Throughout, the city has flourished with cultural and artistic efforts as befits the home of Gaudí, Picasso, Joan Miró and other modern Catalan artists.

Transactional Forces and Typology

The city of Barcelona has developed from Roman times in comprising the mediaeval 'Old City' to the additional 19th-century expansion of Eixample and the adjoining Montjuic district. To these, newer outer industrial districts have been added at the exterior with the peripheral industrialisation continuing not only of manufacturing facilities, but also adjoining public housing and residential projects to support the growing workforce. From its primary agriculture and fishing activities, it has grown through cotton manufacturing and textiles to secondary manufacturing and automobile factories and modern service industries.

In addition, along the coastal areas a series of new housing developments has been realised yet more recently by speculators in order to accommodate seasonal visitors to the Mediterranean coast. As the centre of the Catalan region having a very strong regional identity, it has retained its language, skills and traditions that reinforce its longstanding cultural and independent artistic influences.

From its original inner Barri Gotic settlement area around a natural bay and harbour, the city of Barcelona has grown from an unplanned organic core to an enlarged city with a strong geometric pattern, reinforced by its twin diagonals of Avinguda Diagonal and Avinguda Meridian that intersect with the Gran Via de les Corts Catalanes at the renamed Placa de les Glories Catalanes.

Within this area of the planned expanded Eixample district, with its structured and chamfered corner blocks, this creates an unusual urban streetscape. Interspersed with the Parcs Montjuic and de la Ciutadella, the linear coastal strip provides a scenic edge to the city with the backdrop of the mountains to the north. Now bounded by highways and with the hills beyond, the centre is a dense urbanised area of a typical Mediterranean coastal port but that has the distinctive individual characteristics of the Catalunya region.

2.5.5 DUBLIN, EIRE

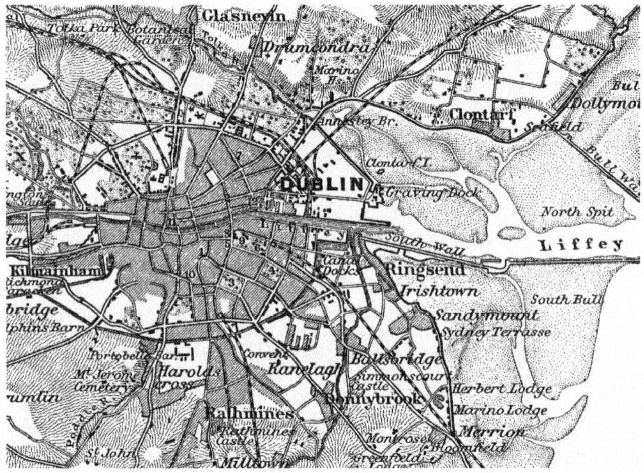

FIGURE 2.5.5 Dublin – early configuration, 1897.
Source: iStock.

Background and Origins

Dublin is the capital of Eire, located on the east coast by Dublin Bay overlooking the Irish Sea. With its Celtic origins dating back to 700 BCE, the Celtic city of Dublin straddles the River Liffey on both the north and the south banks. Originally the settlement was known as Baile Atha Cliath (the ford over the hurdles) where the Liffey converges with the River Poddle running out to the Irish Sea.

With St Patrick's mission to Ireland in 432 CE, this brought Christianity to the island and the realisation and building of several churches. But it was later, around 900 CE, that the Norse Vikings arrived and settled, then naming the place Dubh Linn (or black pool), sited near the present Wood Quay.

It was not until the 12th century that the Anglo-Norman conquest took control and developed the defences of the city with the construction of the castle in 1205 and Saint Patrick's Cathedral in 1220 in the mediaeval core of the city. Dublin and the coastal area south of the city was part of 'the Pale' that was under direct by rule by the English at that time.

Being populated with Celts, Vikings and Anglo Normans, when Henry VIII of England passed the Act of Supremacy in 1541, this made him king of Ireland also. Monasteries were dissolved and Protestantism introduced.

As the island was colonised, plantations were set up throughout the land during the following reign of Elizabeth I. In 1585, Ireland was mapped and divided into 32 counties, and Trinity College was established in 1592 on the site of an Augustine monastery as a significant Protestant centre of learning, as the sole college of the University of Dublin.

But an event of most significance was the defeat of the English Catholic King James II, who was defeated by the Dutch Protestant William of Orange at the Battle of the Boyne north of Dublin in 1690. This led to a period of Protestant dominance and Catholic repression, when physically, under the Protestant Ascendancy, further squares were developed, such as St Stephen's Green and subsequent town houses in creating the Georgian centre of Dublin, such as around Henrietta Street.

It was the Duke of Ormonde that established the first quays along the north bank of the River Liffey in the 1670s, with Ormond Quay Upper and Lower, and Sir Humphrey Jervis who started to develop areas previously occupied by St Mary's Abbey, that is, Jervis Street, Capel and Mary Streets. Ormonde's other legacy was the great Phoenix Park, an extensive deer park in Dublin and the Royal Hospital at Kilmainham.

Luke Gardiner MP was a great developer of several areas, such as the formal square Cavendish Row and Sackville Street to become the wide mall of O'Connell Street Upper. Other developments were the widening of Dame Street towards College Green, Lower Sackville Street, D'Olier Street and Westmoreland Street, and the relocation of the Custom House from Essex Quay to the north side of the River Liffey. Further developments under the Earl of Kildare were Merrion Square and Fitzwilliam Square and the perfectly square Mountjoy Square of 600 by 600 feet.

In 1757, an Act was passed giving the Wide Street Commissioners the ability to clear the axis leading to the castle and to form Parliament Street, which was to become of some significance and similar to Hausmann's later actions in Paris. Two long canals also were constructed, the Grand Canal to the south and the Royal Canal to the north of the Liffey that encircled the city in an oval and which terminated at Grand Canal Docks and Spencer Dock respectively.

Dublin had become a major trading port, and the sand bar in Dublin Bay had to be kept free. This was facilitated by the construction in stages of the Great South Wall, eventually all the way from Ringsend to Poolbeg Lighthouse, and subsequently of the North Bull Wall and the East and West Piers.

During the early 19th century, further civic improvements were made, which included institutional buildings, such as the King's Inns, the Bank of Ireland, Chapel Royal and Nelson's Pillar. New bridges were also built, notably, the cast iron Ha'penny Bridge in 1816.

In 1853, the Great Industrial Exhibition was held in the grounds of Leinster House, and a significant group of buildings of the National Gallery, National Museum and Natural History Museum were built. New roads of the North Circuit and South Circular Road were built and several institutional buildings, major bank theatres and museums erected.

With the imposing Custom House on the north bank, the central areas of Temple Bar and the Cathedral district (St Patrick's) and the older Viking/mediaeval area and the Liberties, each of these form quite distinctive neighbourhoods in the city.

After the Acts of Union in 1800 that had imposed Direct Rule, the Emancipation Act was passed in 1829 due to the efforts of David O'Connell, who later became Catholic mayor of Dublin in securing limited rights to vote. The subsequent period of 1845–8 was dominated by the Great Famine, with a total failure of the potato crop, which decimated much of the local population, causing real rural hardship.

Outside the city, during the mid-19th century, the suburbs began to be developed with Rathmines and Ballsbridge in the south. In the north, Phibsborough and Clantarf were developed with Victorian 'parlour' houses. In inner-city areas, many tenements were in poor condition and became unhygienic and overcrowded. By 1914, following the review of 'Housing Conditions of the Working Classes in the city of Dublin', many had been condemned.

The first suburban railway that was built from Dublin outside to Kingstown, on the south-east coast on Dublin Bay which was developed based on Stephenson's Standard Gauge rail track, opened in 1834 and was eventually extended to Bray. Other railways were built subsequently to other gauges, broad and narrow, but a new Irish gauge of 63" (inches) was invented to which this line was converted, that conflicted with the more universally accepted Standard Gauge. New lines and stations were also built as the Great Northern, Great Southern and Western besides King's Bridge.

The city of Dublin Junction Railway, the Loopline, was built elevated across the river in the central area. Then, around 1872, the horse-drawn omnibus gave way to double deck horse-drawn trams, which were electrified under the Dublin United Tramways Company by 1900.

Before World War I, pressure had mounted for Home Rule, which came to a head with the Easter Rising in 1916 and the Declaration of Independence from the General Post Office in Dublin that was severely quashed and turned the leaders into martyrs. Following this uprising and the Irish War of Independence, many of the principal buildings in the centre were badly damaged. It was not until 1922 as a result of the Anglo-Irish Treaty that the Irish Free State was realised and it was only post-World War II, in 1949, that the Republic of Ireland was created finally.

Patrick Abercrombie and the Kellys had drawn up a master plan for 'Dublin of the Future' with a 'garden city' model for some of the suburbs, such as Cabra, Crumlin and Marino, and a civic survey mandated by Patrick Geddes was undertaken in 1925. Dublin Corporation had hired Herbert Simms also as Housing Architect to increase provision of social housing and targeted replacing old tenements with medium rise, either terraced or built around central courtyards.

Modern City – Buildings and Arteries

Dublin is an elegant city having several distinctive districts. South-west Dublin was the original area that was developed around the old castle with its yards and towers by the city walls. With Christ Church Cathedral, the Royal Exchange and City Hall, it nestles against the quays of Temple Bar (the riverside walk) where tradesmen and craftsmen settled, and which has now become a bustling area of bars and restaurants around Meeting House Square.

In south-east Dublin opposite the Bank of Ireland, Trinity College is located with its green and three fine squares (Parliament, New and Fellow's Square) and its Old Library with the famous illuminated Celtic 'Book of Kells' housed in the Treasury. With famous alumni, it is neatly laid out near adjacent buildings off Grafton and Kildare Street, such as Mansion House, the National Library and Leinster House, with the national Parliament that houses the Dail and the Seanad, and other government buildings and the National Museum.

The area North of the Liffey with crossings of the river bridges, such as the pedestrian cast-iron Ha'penny Bridge with its original half-penny toll, were developed with the quays as Abbey Street and with the Abbey Theatre behind. Running north to the Georgian area is O'Connell Street with the General Post Office, and the Gate Theatre in Parnell Square. On this side the areas of North of the Liffey comprise Smithfield and the Legal Quarter with its 'Four Courts' Building (Common Pleas, Chancery, Exchequer and King's Bench) and Rotunda and the extensive Phoenix Park westwards beyond.

Dublin had developed into a compact elegant city in the Province of Leinster, encircled by the Royal and Grand Canals. However, as inner areas were becoming run-down and subject to redevelopment plans, the Liffey Quays needed to be protected. New towns to accommodate growth in population were planned on the western side of the city at Tallaght, Lucan, Clondalkin and Blanchardstown.

Subsequent to Ireland joining the European Community in 1973, and the economy improving, there was a risk of over-development with tall towers planned, especially in the Docklands Area, which started to impact the original charm of the Liffey Quays.

New traffic plans were also made that included upgrading the North and South Circular highways and a plan for a motorway box. A major development also was the staged development of the outer M50 peripheral western motorway and the Dublin Port Tunnel. With constraints at both the East Link and the West Link sections, plans were made to upgrade these, especially on the eastern side to accommodate heavy traffic generated to and from the port in the Greater Dublin Area.

In the railway sector, improvements were made on the Loopline and the introduction of the DART scheme, with electric trams and the LUAS Light Rail System. In 1994, Dublin County Council (DCC) was replaced by the three areas of Fingal, South Dublin and Dun Laoghaire-Rathdown. The Dublin Docklands Development Authority (DDDA) was established also in order to redevelop the Grand Canal Dock, and there were a number of significant and symbolic new projects as Croke Park, the tall steel Needle (designed by Ritchie) in O'Connell Street, replacing former Nelson's Pillar and a series of new bridge crossings as the new Millennium Bridge, James Joyce, Sean O'Casey and Samuel Beckett Bridges.

As the 'Celtic Tiger' economy began to take off at the beginning of the 21st century, with the presence of American tech giants, new built commercial developments and data centres grew both in the centre and on the fringes in the new millennium. A new City Development Plan was prepared in 2007 focused on improving design standards with sustainable housing guidelines and Strategic Housing Developments at appropriate densities with limits on 'Build-to-Rent' schemes in order to reduce the practice of houses becoming globally traded as financial assets.

With rising sea levels around Dublin Bay also, this is becoming a significant risk, with the need for realisation of a Liffey Barrage, which remains under assessment. With the residents' welcoming nature, the city is an attractive place for visitors with its Celtic traditions, folk music and entertainment sector. The population of the city is approximately 0.5 million and double in the metropolitan area.

Transactional Forces and Typology

The city of Dublin is located around a tribal pool at the mouth of the River Liffey and has developed since its early settlement on the south side subsequently on either side of the river. From its early Viking settlements and mediaeval castle in the English Pale, and with its early mediaeval layout, it was expanded in the 18th century with the construction of new districts and buildings, and as guided by the Wide Streets Commission, resulting in new frontages especially along the quays. The port has been an active trading centre, as linen was crafted and became a principal export. The city has acted mainly as an administrative and commercial rather than an industrial centre and in the forefront as national capital since the creation of the Irish Free State.

As its population stabilised and increased post–World War II, since the creation of the Republic of Ireland and its joining of the European Community in 1973, it has developed economically, especially in the service sector with the establishment of many modern high-tech companies in Dublin. This has led to the redevelopment and urban renewal of several new districts, especially in the regenerated Dockland Area.

As an estuarine and riverine city, the city has grown on both banks of the River Liffey. From its historical core and port, it has developed with radial characteristics, having concentric zones with sectors or fingers of development stretching both southwards and westwards. In terms of zoning, with no specific containment to the outer orbital ring road, there has been a tendency for ribbon development along these outward reaching stretches with new building on greenfield sites.

2.5.6 EDINBURGH, SCOTLAND

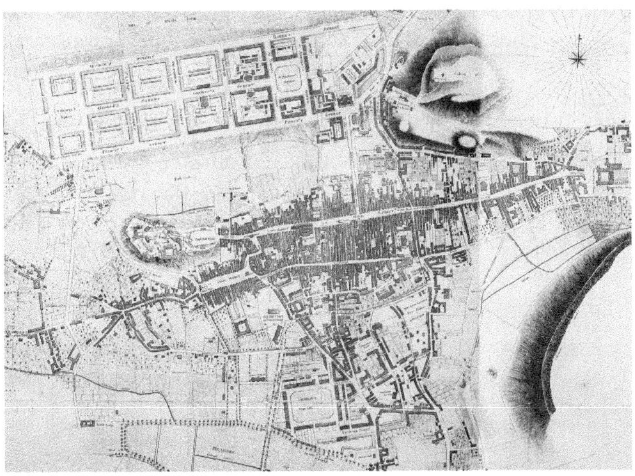

FIGURE 2.5.6 Edinburgh – early configuration, 1784 (Kincaid).
Source: British Library.

Background and Origins

The city of **Edinburgh**, from 'Eidyn', the Celtic name for the region, is located on the Firth of Forth in Scotland on the southern shore in Lothian, with the Pentland Hills to the south. With evidence of an early Bronze Age fortress on Castle Rock, as a fortified city the castle was built by King Malcolm III of Scotland on a rocky crag in the 11th century on the commanding position of Castle Hill. The Royal Burgh of Holy Rood was founded by King David I in the 12th century and it became the capital of Scotland. The city comprises several different districts nestling between the distinctive geological features of Arthur's seat, with the Radical Road running along Salisbury Crags and Calton Hill and Castle Hill.

The backbone of the city is formed around the Royal Mile, which is a linked set of streets that lead up to Edinburgh Castle. This thoroughfare runs through the heart of the Old Town with streets and alleys ('closes') that lead off from it. With its old churches, such as the High Kirk of St Giles, the National Gallery and Academy and at the eastern end the Palace of Holyroodhouse built in the 16th century, the official royal residence in Scotland next to the original Abbey of Holy Rood.

During the 18th century Edinburgh became the intellectual centre of the 'Scottish Enlightenment', with progressive thinkers such as David Hume and Adam Smith and the development of universities; the University of Edinburgh had been founded in 1583 and was located in the Old Town. With the Treaty of Union with England in 1707, spheres of influence were to change.

The **'Old Town'** as typified by a number of entrances ('pends') to alleyways ('vennels') has courtyards and narrow lanes as Old Tolbooth Wynd off the main streets Canongate and Cowgate. Nestling tightly together, it constituted individual tenements, caverns and small dwellings within the town walls. Victoria Street is a cobbled 'brae' that curves down to the lower Grassmarket from George IV Bridge and Lawnmarket above where waste-water used to be channelled down the centre, known as the 'foul burn', which was then evacuated to local farmers' fields.

New Town

Due to overcrowding in the Old Town, in the 18th century a new area was developed on a ridge just to the north, known as New Town. Laid out on a rectangular grid by James Craig after a competition, three main streets of Queen Street, George Street and Princes Street were built parallel to the

Royal Mile. With a square at each end, Charlotte Square to the west and St Andrews Square to the east, there are essentially eight areas of regular streets of terraced housing in the First New Town bordered to the north by Queen Street Gardens and to the south by West Princes Street Gardens.

These Georgian terraced granite stone houses around Young, Hill and Thistle Streets and Lanes, and Rose Streets and Lanes are elegant affluent homes. The whole of the first New Town mainly completed by 1820 is a planned, well-ordered, attractive and integral part of central Edinburgh, with an artificial causeway, the Mound, created between Princes Street and the Old Town.

To the east of New Town is Calton Hill, Broughton and Greenside. Here are the National Monument, Nelson Monument and City Observatory, quite exposed on Calton Hill. To the south is Arthur's Seat, a most unusual extinct, ragged and eroded volcano sited in an urban context where traces of early settlements have been found.

After the success of the First New Town, a Second New Town was developed north of Queen Street Gardens. This was a major development scheme undertaken by builders on either side of Great King Street towards North West Circus and crossed by Dundas Street. Just to the northwest it contains the separate geometrically designed Randolph Crescent (semi-circle), Ainslie Place (oval), Moray Place and Royal Circus (circle). A subsequent plan for the development of the Third New Town was also made to the eastern end of the city, but this was only partially completed.

Farther out, South and West Edinburgh are mainly residential suburbs and comprised tenement properties. The Union Canal was built through West Edinburgh in the 19th century to link with the Forth and Clyde Canal towards Glasgow. Stockbridge is an area to the west of the city that contains the Royal Botanic Gardens. Farther beyond is the Port of Leith and old docks around the Water of Leith and Western Harbour. Much of this has recently been redeveloped for housing.

Modern City – Buildings and Arteries

Princes Street is the main commercial and shopping street in the capital. In the lower area the railway was built running right through the centre of the city in an artificial valley of North Loch into central Waverley Station. The railway was developed around this and Haymarket Station. The city is served also by a network of streets and highways, with a recent peripheral by-pass and motorway. Both of these modes link up and connect with the three major iconic Road & Rail Bridges across the Forth Estuary to connect with Fife and the Grampians to the north. A new urban tramway has also been built which also runs out from the centre of the city to the airport.

Towards the West End of the city a new area has been developed near Haymarket as the Edinburgh International Conference Centre (designed by Farrell) that opened in 1995. This serves as a countervailing development pole to the west as a new financial district, with Fountainbridge nearby, a former industrial district that has been renewed and redeveloped. Yet further development has been undertaken farther westwards with a new Edinburgh Park (designed by Meier) besides the ring road with Edinburgh Gateway connected by the tram on to the airport.

The historic city has its special ambience of smokey 'Auld Reeky' and with its granite stone has a special baronial and Scottish feel. Its history of brewing and distilleries brings its own charm along with its annual festival, military tattoo and Hogmanay New Year celebrations. Its most distinctive new modern building is the Scottish Parliament (designed by Miralles) located near Holyroodhouse and inaugurated in 2004. The population of Edinburgh is approximately 0.5 million people, with double in the metropolitan area, out of 5.5 million in the whole of Scotland.

Transactional Forces and Typology

With its natural imposing features of Castle Rock and Arthur's Seat, Edinburgh was perceived as a stronghold around which settlements gravitated. As the Old Town grew into a burgh, it became a royal burgh in the 12th century. It became a seat of the Parliament of Scotland with its Three Estates of clerics, nobility and burgh commissioners, and developed in its role as capital and administrative centre and of the Church of Scotland.

In the 18th century international trade gravitated mainly from Glasgow towards America, whilst the tobacco re-exporting trade developed under the 'tobacco lords' generating wealth. The New Town was built in the city and trade and communications centred around the Port of Leith developed adjoining the Firth of Forth.

Traditional industries developed, from distilleries, brewing and printing to banking and newer manufacturing industries. Commercial areas were developed, and the fabric of the city was redeveloped after the construction of the neoclassical New Town with efforts to upgrade the former tenements of the Old Town. With its axial grid, main thoroughfare and squares, these developments completely transformed the centre of the city.

Whilst a riverine city of the Water of Leith, not far from the east coast, its hilly historic centre has essentially dual characteristics with the organic Old Town and later separately planned New Town, each with their distinct granite stonework. These are distinctly different neighbourhoods with a developing West End. Whilst the city extends to the coast and the Port of Leith, its more recent suburban extensions are mainly in outer areas towards the south and the west of the city, bordered by the ring road. As a bi-polar city its expansion is more radial outwards, with wedge shapes contained and limited by some form of containment from a mandated green belt. The following figures illustrate the modern configuration of these academic and ecclesiastic European cities and the evolution of their modern population.

FIGURE 2.5.7 Amsterdam – modern configuration.

FIGURE 2.5.8 Prague – modern configuration.

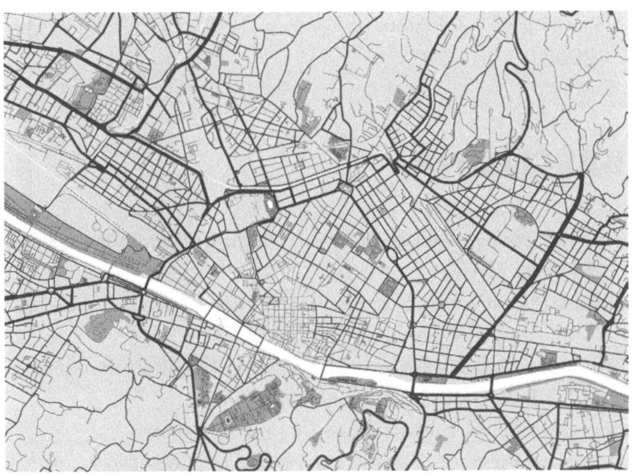

FIGURE 2.5.9 Florence – modern configuration.

FIGURE 2.5.10 Barcelona – modern configuration.

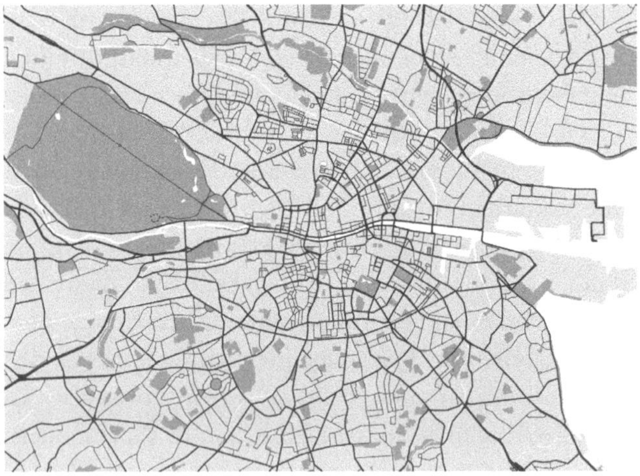

FIGURE 2.5.11 Dublin – modern configuration.

FIGURE 2.5.12 Edinburgh – modern configuration.

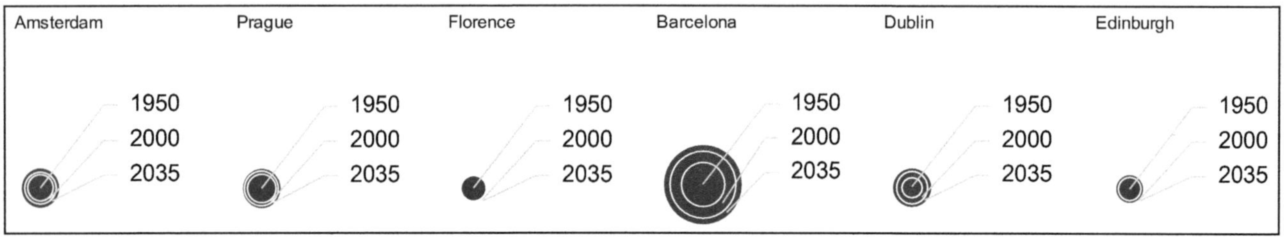

FIGURE 2.5.13 Academic and Ecclesiastic European city populations.

Data source: Global Database of Metropolises 2020 – UN-Habitat (2020), UN DESA Population Division.

Further Reading

AMSTERDAM

Abrahamse, Jaap E & Reinout, Rutte – *Historical Atlas of Amsterdam*, Thoth Publishers, 2023.

Knap, GH – *The Diligent City Amsterdam*, Junior Chamber, 1960.

Mak, Geert – *Amsterdam*, Vintage Books, 2001.

PRAGUE

Beckova, Katerina – *Prague, a City and Its River*, Karolinum Press, Charles University, 2017.

Demetz, Peter – *Prague in Black and Gold, the History of a City*, Penguin, 1998.

Sayer, Derek – *Prague Crossroads of Europe*, Reaktion Books, 2018.

FLORENCE

Crum, Roger & Paoletti, John, eds. – *Renaissance Florence, a Social History*, Cambridge University Press, 2006.

Hibbert, Christopher – *Florence, the Biography of a City*, Penguin, 1994.

Hollingsworth, Mary – *The Medici*, Apollo, 2022.

Holmes, George – *Florence, Rome and the Origins of the Renaissance*, Clarendon Press, 1986.

Kroke, Antonella, ed. – *The History of Florence in Painting*, Abbeville Press, 2013.

Najemy, John – *A History of Florence, 1200–1575*, John Wiley, 2008.

BARCELONA

Busquets, Joan – *The Urban Evolution of a Compact City*, ORO Editions, 2014.

Busquets, Joan & Perez-Ramos, P – *Manifold Grids and the Cerdà Plan*, Applied Research & Design, 2017.

Fernandez-Armesto, Felipe – *Barcelona, a Thousand Years of the City's Past*, Oxford University Press, 1992.

Hochadel, Oliver & Nieto-Galen A – *Barcelona, an Urban History of Science and Modernity 1888–1929*, Routledge, 2016.

Hughes, Robert – *Barcelona*, Harvill, 1992.

Marshall, Tim – *Transforming Barcelona, the Renewal of a European Metropolis*, Routledge, 2004.

McDonogh, Gary & Martinez-Rigol, Sergi – *Barcelona*, Polity, 2018.
Vives, Antonio – *Smart City Barcelona, the Catalan Quest to Improve Future Urban Living*, Liverpool University Press, 2017.

DUBLIN

De Courcy, JW – *The Liffey in Dublin*, Gill & Macmillan, 1996.
Dickson, David – *The Making of a Capital City*, Harvard University Press, 2014.
McDonald, Frank – *A Little History of the Future of Dublin*, Martello Publication, 2021.
Tobin, Fergal – *A City Runs Through Them*, Atlantic Books, 2024.

EDINBURGH

Carly, M, Dalziel, R, Dargen, P & Laird, S, eds. – *Edinburgh New Town, a Model City*, Amberly Publishing, 2021.
Crawford, Robert – *On Glasgow and Edinburgh*, Belknap Press, 2015.
Fry, Michael – *Edinburgh, a History of the City*, Pan Publication, 2010.
Moffat, Alistair – *Edinburgh, a New History*, Birlinn, 2024.
Youngson, AJ – *The Making of Classical Edinburgh*, Edinburgh University Press, 2019.

Part III

Early Trading Cities

3.0 Introduction – Early Trading Cities

Human settlements have tended to develop around places having suitable geophysical features in each continental land mass. These have coalesced around favourable places of communication along coasts, on rivers or at natural sites having a commanding location. Over large areas of land mass, major rivers have led to the settling of populations in gravitating together. In less-developed areas, movement of peoples has commenced from independent riverine estuaries.

With its extensive land area, the 'peninsula' of Eurasia, down to the Mediterranean Sea in the south and the North and Norwegian seas in the north, was the scene of the early movement of many marauding tribes. The major European rivers of the Volga, Dnieper, Oder, Danube, Rhine, Rhône and Loire acted as both barriers and corridors of movement across the continent.

Coastal locations or riverine estuaries were major points of settlement, for example at:

- London (Thames), Bordeaux (Garonne), Rotterdam (Nieuwe Maas), Oporto (Douro) and Nantes (Loire).

Places of settlement also developed at natural harbours, such as:

- Genoa (Polcevera), Istanbul (Bosphorus), Hamburg (Elbe), Marseilles (Rhône), St Petersburg (Neva), Copenhagen and Trieste.

Settlements also developed at the most feasible crossing points inland along rivers or where there were islands in the river where bridging was easier, or at confluences of rivers, for example at:

- Vienna and Budapest (Danube), Paris (Seine), Dnipro (Dnieper), Volgograd (Volga), Lyon (Rhône and Saône), Rome (Tiber), Berlin (Spree) and Munich (Isar).

Such early settlements were the makings of substantial sites, essentially as a result of those natural features which were typical determinants for built form. Each of these developed as significant historic cities having their own particular natural and regional characteristics, where trading and mercantile activities were able to play a key part in their development.

In a similar vein, across oceans rather than just rivers, trading around seas as the Baltic and Mediterranean or the Caribbean, as well as the South China Sea, Gulf of Guinea or the Bay of Bengal, led to the development of strategic ports. Often as a result of their natural harbours and favourable currents or trade winds these developed as strategic centres around which maritime trading developed. In particular, such trading centres grew on islands that found themselves on maritime trade routes, notably Colombo in the Indian Ocean and Makassar and others in the South China Sea.

In contrast, in landlocked countries, where topographical features often act as obstacles to communication, routes or passes are found through mountain ranges where valleys may facilitate travel. Here similarly places of settlement have arisen as way stations around which settlement has grown, for example Bogotá and La Paz, to become significant urbanised areas.

This **Part III** accordingly addresses the different types of trading cities, either mercantile or maritime, that have evolved in riverine and coastal areas as well as those at strategic locations on islands as well as those located on strategic routes on mountains in interior regions.

DOI: 10.4324/9781003353348-11

Table 3.0.1
Early Trading City Types

	City	Topography	Plan	Trans. Forces	Type
3.01 TRADING CITIES					
3.1 *Mercantile Port*					
	Genoa	Coastal	Linear	Reg. Capital	Cultural
	Bordeaux	Riverine	Radio-centric	Reg. Capital	Cultural
	Marseille	Coastal	Axial/organic	Reg. Capital	Port
	Algiers	Coastal	Organic	Capital	Port
	Liverpool	Estuarine	Radial	Industrial	De-industrial
	Glasgow	Riverine	Axial	Industrial	De-industrial
3.2 *Lowland, Hanseatic, Baltic and Scandinavian Ports*					
	Antwerp	Estuarine	Radial/organic	Maritime	Port
	Hamburg	Riverine	Radial/organic	City-state	Hanseatic
	Helsinki	Coastal	Radial	Capital	Baltic
	Stockholm	Fjord/island	Radio-centric	Capital	Marine
	Oslo	Bay	Axial	Capital	Marine
	Copenhagen	Coastal	Radio-centric	Capital	Scandinavian
3.3 *Maritime*					
	Freetown	Coastal	Organic	Capital	Trop. African
	Accra	Coastal	Radial/organic	Capital	Trop. African
	Cape Town	Coastal	Sectoral	Leg. Capital	S African
	Boston	Estuary/bay	Radial/organic	Sta. Capital	Modern
	New Orleans	River/lake	Grid and Axes	Waterway	Louisianian
	Madras/Chennai	Coastal	Organic	Prov. Capital	S Indian
3.4 *Island*					
	Havana	Coastal/bay	Grid/organic	Capital	Caribbean
	Port-au-Prince	Coastal/bay	Organic	Capital	Caribbean
	Kingston	Coastal/bay	Organic	Capital	Caribbean
	Colombo	Coastal/riverine	Organic	Maritime	Sri Lankan
	Makassar	Coastal/riverine	Organic	Maritime	Indonesian
	Auckland	Coastal	Dual/axial	Maritime	S Pacific
3.5 *Mountain*					
	Bogotá	Mountain	Axial	Capital	Lat. American
	Quito	Mountain	Organic/axial	Capital	Equat. Volcanic
	La Paz	Mountain	Sectoral	Minerals	Altiplano
	Kathmandu	Mountain	Organic	Capital	Himalayan
	Tehran	Mountain	Sectoral	Capital	Iranian
	Addis Ababa	Mountain	Organic	Capital	Ethiopian

Table 3.0.2
Early Trading Cities Population Growth (thousands)

City	Country	1950	2000	2020	*2035**	Area
Genoa	Italy	785	732	680	*690*	Metropolitan area
Bordeaux	France	430	765	969	*1,086*	Urban agglomeration
Marseille	France	756	1,476	1,608	*1,746*	Urban agglomeration
Algiers	Algeria	516	2,141	2,768	*3,496*	Urban agglomeration
Liverpool	UK	1,382	853	902	*992*	Urban agglomeration
Glasgow	UK	1,755	1,585	1,673	*1,834*	Metropolitan area
Antwerp	Belgium	759	945	1,042	*1,100*	Metropolitan area
Hamburg	Germany	1,608	1,710	1,790	*1,815*	City proper
Helsinki	Finland	366	1,019	1,305	*1,410*	Urban agglomeration
Stockholm	Sweden	741	1,206	1,633	*1,873*	Urban agglomeration
Oslo	Norway	468	774	1,041	*1,252*	Urban agglomeration
Copenhagen	Denmark	1,216	1,077	1,346	*1,477*	Metropolitan area
Freetown	Sierra Leone	92	690	1,202	*1,849*	City proper
Accra	Ghana	177	1,668	2,514	*3,632*	Metropolitan area
Cape Town	South Africa	618	2,785	4,618	*5,845*	Urban agglomeration
Boston	USA	2,551	4,036	4,309	*4,778*	Urban agglomeration
New Orleans	USA	664	1,009	992	*1,123*	Urban agglomeration
Chennai (Madras)	India	1,491	6,593	10,971	*15,376*	Urban agglomeration
Havana	Cuba	1,142	2,186	2,140	*2,194*	City proper
Port-au-Prince	Haiti	133	1,693	2,774	*3,810*	Urban agglomeration
Kingston	Jamaica	278	578	591	*645*	Metropolitan area
Colombo	Sri Lanka	400	644	613	*761*	City proper
Makassar	Indonesia	223	1,077	1,584	*2,050*	City proper
Auckland	New Zealand	319	1,110	1,607	*1,861*	Urban agglomeration
Bogotá	Colombia	630	6,329	10,978	*12,753*	Urban agglomeration
Quito	Ecuador	206	1,357	1,874	*2,353*	City proper
La Paz	Bolivia	320	1,415	1,858	*2,358*	Urban agglomeration
Kathmandu	Nepal	104	642	1,424	*2,220*	City proper
Tehran	Iran	1,041	7,128	9,135	*10,664*	City proper
Addis Ababa	Ethiopia	392	2,377	4,794	*8,939*	City proper

Note: * *Forecast.*
Source: Global Database of Metropolises 2020 – UN-Habitat (2020), UN DESA Population Division.

3.1 Mercantile Port Cities

Mediterranean- and Atlantic-Facing Ports

Genoa, Venice, Marseille, Algiers, Liverpool and Glasgow

3.1.0 INTRODUCTION – MERCANTILE CITIES

Some of the earliest forms of human settlement tended to occur on land around a riverine estuary, inlet or coastal location that had protection from nature's elements, in the form of a promontory or peninsular. With suitability as a natural harbour, such settings became places of trade, which enabled the exchange of natural resources and man-made artefacts.

By strengthening breakwaters, deepening of harbours, extending wharves and creation of canals, the marine movement of goods and products became the basis of mercantile trade. Insofar as new basic materials were imported into a region, concomitantly other 'made' goods were exported by return, leading to circuits and continuity of trade.

Such marine environments enabled a wider reach to more distant locations as ships developed increased capacity and range, eventually having mechanical rather than wind power. Designed for carrying different commodities, from natural agricultural cargoes (e.g. wheat, fruits, vegetables and spices) and manufactured bulk cargoes (e.g. timber, textiles, liquid drinks and vegetable oils), to higher-value fuel (e.g. coal and oil), different loading and unloading facilities were needed to be installed at purpose-designed ports.

Hence, ports tended to be either 'staple ports' that specialised in handling commodities from the local interior and hinterland, or 'entrepot ports' that had specific warehouses where goods could be imported and re-exported. Depending on the demand for trade and transactions, such trans-shipments led to associated transactions of finance and insurance. This in turn led to the creation of port city hubs.

Lines of communication developed to service these 'gateway ports', with transport by canal, road and rail that had increasing complexity of roads and frequency of movements from fields to factory and from supplier to producer. This is especially evident in Europe where the Atlantic-facing ports developed with the triangular trade with Africa and the Americas. In northern Europe, the western ports of Britain, Liverpool, Glasgow and Bristol were pre-eminent, and farther north Antwerp and Hamburg, and to the south Nantes and Bordeaux.

Other significant port cities developed in the Mediterranean, Barcelona and Marseille, Genoa and Venice, and Naples and Palermo in the Tyrrhenian Sea, with Algiers on the opposite north African coast. In the Mediterranean Sea there were significant trade routes within it, some circular, both east and west, north and south, linking with the Levant and northern Africa.

With increasing specialisation and the modern introduction of the use of containers, this has led to a fundamental transformation in the manner of transport and trading of goods. This has impacted not only the method of packaging and handling of commodities, but also the whole infrastructure for containerisation, with the siting and enhanced connectivity of new terminals and their handling facilities. This in turn has led to the development of larger and marginally more remote marine sites having more generous storage space away from established trading wharves in older cities.

3.1.1 Genoa, Italy

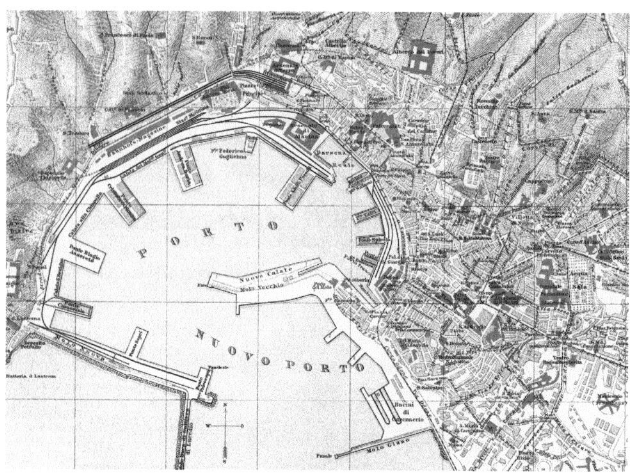

FIGURE 3.1.1 Genoa – early configuration, 1897.
Source: iStock.

DOI: 10.4324/9781003353348-12

Background and Origins

The city of **Genoa** (Genova) is located on the Gulf of Genoa in the Ligurian Sea, with the Apennine Mountains rising behind on the west coast of northern Italy. As a natural port, it was an early city-state and capital of one of the most powerful maritime republics from the 11th to the 18th century CE, during which time it was a naval power and a leader in increasing mercantile trade across the Mediterranean.

In early Roman times, it had a fortified settlement (oppidum) on the hill in the centre around which development occurred. This continued in mediaeval times when the port also developed as a centre of boat-building. The city came to be administered by the Doges and groups of wealthy merchants that retained power. They were outward looking, involved in trading, and established colonies around the Mediterranean and took part in the early Crusades to the Levante. The Genoan Republic was to rival the Venetian Republic at this time.

In the 9th century CE, the city was encircled by the first set of walls with three gates until around 1160 when these walls were further extended and able to protect local artisans and trade. The coastal strip remained as state property since the waterfront was pivotal for extra-territorial trade, which was to develop around the Mediterranean with its port and lighthouses to support the marine anchorages.

The original historic centre had been developed around a mix of squares and small alleys (caruggi) between tall buildings (creuze) from mediaeval times. However, the zenith of the city of Genoa was in the 16th century when wealth accumulated, from which time its financial centre and first banks developed, the university was founded and the arts flourished.

These developments comprised a series of fine streets (Strade Nuove) which constituted the Via Garibaldi, Via Cairoli and Via Balbi and with a series of elegant palaces that were built later that century. For these Palazzo dei Rolli and the particularly splendid Palazzo Reale that comprised a large number (163) of residences of private families, their owners were obliged according to arrangements with the State, to host notable guests during state visits to the city as public lodgings.

In the centre, there is a residential area around the Belvedere Castelleto, with Santa Maria di Castello, and the old city walls and gates and also San Lorenzo Cathedral and Piazza de Ferrari with its fountain and 'Teatro Carlo Felice' Opera House and the Doge's Palace. Right across the city there are a large number of fine churches and basilicas and distinctive villas populated around the city. There are fine parks and gardens also, such as at the Villa Durazzo Pallavicini and the extensive Monumental Cemetery of Staglieno on the hillside.

The city was to lose power, however, in the late 18th century, being annexed by Napoleon and then coming under Savoyard control until the eventual unification of Italy in 1861. During the 19th century, Genoa was to become a significant steel and shipbuilding centre, supporting its role as a major Mediterranean port. From this, major cruise ships were built and new ferry and cruising lines were established. With its growing financial resources, new financial institutions and insurance companies were also established there.

Modern City – Buildings and Arteries

Whilst the economy was initially based on agro-industry in the region, the shipbuilding in the 1960s has now declined, but it remains an active port, especially for cruise ships, and the city lies at the corner of the Milan–Turin–Genoa industrial triangle.

From the compact circular Old Harbour (Porto Antico) in the centre of the city with its distinctive Lanterna, the modern port area now stretches linearly along the Ligurian coast between Sestri Ponente and Sestri Levanti. To the west after the industrialised area, the Lungomare stretches past Pegli, Pra and Voltri in Ponente towards Arenzano and the surrounding Beigua natural park. To the east along Corso Italia, the coast stretches past the bays of Boccadasse and Sturla towards the Riviera di Levante.

The old warehouses have been restored in the centre, which has generally been preserved, but there have been a number of new landmarks, such as the Bolla (sphere – designed by Piano) and the new aquarium built for Expo '92 and the Matitone thin skyscraper (designed by SOM). A new Erzelli Great Campus science park is under development.

In terms of transport, there are three main railway stations, Genoa Brignole, Principe and Sampierdarena, which provide connections westwards to southern France, to Turin and south to Rome. There is an urban metro system, the Metropolitana, which runs across the city from Brin to Brignola. With its hilly topography, there are also two funicular systems, an inclined elevator and rack-railway line from Principe to Gianarolo.

The main Autostrada A10 runs right across the city and crosses the Polcevera and Bisagno valleys, to connect through traffic running between Nice, Genoa and La Spezia. During the 1960s a major elevated viaduct was constructed across the entire central area by the main inter-city rail line, called Ponte Morandi, an unusual pre-stressed reinforced concrete cable stayed structure to carry a tolled motorway. However, in August 2018 a major section of deck and pier collapsed killing many people. In poor condition and requiring imminent renewal, following litigation, it has been replaced by the new Genova–San Giorgio integral reinforced concrete viaduct designed by local architect Piano. Further west along the coast beyond the docks is Genoa Cristoforo Colombo Airport (GOA) at Sestri Ponente.

It is a fine ancient city that has been significantly renewed around the port area and enlarged, with a population of approximately 0.6 million that has been ageing and declining since 1970, having peaked at 0.8 million.

Transactional Forces and Typology

The Old City of Genoa was sited and originally settled on the hillside at a strategic location in north-east Italy around which trade developed, and from which it became a city-state, a pivotal maritime republic on the Ligurian Sea. Accordingly, the circular Old Port acted as a centre of trade and from there the port has been extended linearly beside a new airport up the coast where it remains very active today. With its unusual topography, the city developed in several districts around the hillsides, with its upper and lower levels and with an area of 'Strade Nuove', and now acts as an industrialised city as an expanded northern Mediterranean port.

As a coastal city with its ravines and river valleys, Genoa extends linearly east-west along the coast with perpendicular fingers into the twin Polcevera and Bisagno valleys. The distinctive semi-circular natural port acts as the focus of the city around which the adjoining original central districts were developed, with its narrow lanes and streets (caruggi). As its wealth grew from maritime trading, a series of new streets and palaces were developed in the 16th and 17th centuries, Strada Nuove and with the Palazzi dei Rolli, have significantly shaped the central part of the city, on a central axis across the city. With its maritime background, this has dictated its evolution over the centuries and with reclaimed land for both airport and port, new berths have been able to be created longitudinally along the coast, considerably extending the port's footprint and activities.

3.1.2 BORDEAUX, FRANCE

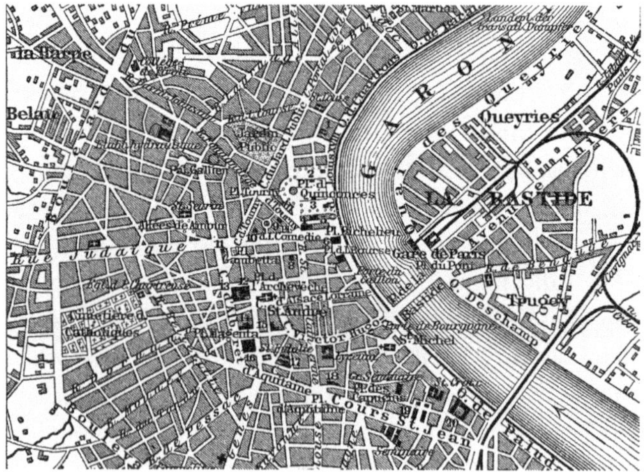

FIGURE 3.1.2 Bordeaux – early configuration, 1897.
Source: iStock.

Background and Origins

The city of **Bordeaux** is located on the Garonne River of the Gironde estuary in south-western France in the Nouvelle Aquitaine region. Formally known as Aquitaine Gaul, it was a province of the Roman Empire, laid out with cardo and decumanus, and was a key trading centre in the region. During the 12th and 13th centuries CE, it was part of the Angevin Empire of the English Plantagenet Kings having fortified ramparts.

During the following centuries it became a significant trading centre, not just regionally but with wine exports internationally across the Atlantic and with sugar and tobacco imports. With its Port de la Lune at the bend of the River Garonne, its activities grew extensively, and accordingly the quays and facilities along the riverside also grew. With its favourable climate and soil conditions, it became a centre of fine wines, with the production of five Premier Cru wines from Medoc and Graves and other clarets.

Bordeaux remained an important port for maritime trade as an Atlantic-facing port. After the Napoleonic period and the Bourbon restoration, the city continued in its development, especially from shipping traffic with French colonies in both Africa and the Caribbean. With its first bridge crossing built in 1822 and extended boulevards, new housing and hotels were built along the main streets.

The city developed essentially on the right bank with the elegant multi-span stone bridge, Pont de pierre, connecting each quarter to La Bastide. The Place Royale, now Place de la Bourse, is a focal point on the riverside with its central statue (now fountain) and main facade overlooking the Quai Richelieu and the river.

It has fine features, such as its old city wall and gates (Portes de Bourgogne, d'Aquitaine, Cailhau, Dijeaux and Grosse Cloche), the main boulevard (Cours de l'Intendance) and the triangle with Allées de Tourny and pedestrian street (Rue Sainte-Catherine) and squares (Place des Quinconces with twin rostral columns, du Parlement and Chapelet).

It also has a series of distinguished churches (Eglise Sainte-Croix, Saint-Pierre, Saint-Seurin, Notre-Dame and Saint-Michel Basilica) and palaces (Palais Rohan – Hôtel de Ville, Cour d'Honneur, Gallien and de Justice), hotels (Lalande, Poissac and Saige Boyer-Fanfrède), the fine gothic Cathédrale Saint-André and the Grand Théatre and museums (Beaux Arts and Arts Décoratifs).

In total, it is one of the finest historic French cities, full of architectural style and elegance, a UNESCO World Heritage site. Its green open areas at l'Esplanade des Quinconces and les Alliées, the Jardin Public and its green places make the ensemble an attractive urban townscape. There are a series of Art Nouveau buildings around the Parc Bordelais.

During the Prussian and recent world wars the French government was temporarily relocated to Bordeaux, with the city falling under German occupation during World War II. The Italian Navy built a massive submarine base with a dry dock (BETASOM) at this time on the Garonne River that was used by German U-boats during the Battle of the Atlantic.

Modern City – Buildings and Arteries

The centre of Vieux Bordeaux retains its elegant squares and historic buildings and facades which form an attractive townscape. With its old university, grandes écoles and modern business schools, it has a strong educational sector. With an active economy having a number of aeronautical businesses located there, it is also a centre of fine wine and gastronomy, with VinExpo held bi-annually.

Concerning contemporary architecture, there are contrasting buildings such as at the redeveloped Quartier Mériadeck, and the Tribunal d'Instance (designed by Rogers), Cité du Vin (designed by XTU) and Maison de l'Economie et de la Culture (designed by BIG), each distinctive modern buildings.

In terms of modern infrastructure, a new river crossing with a vertical lift deck, the Pont Chaban-Delmas was opened in 2013 just upstream across the River Garonne and the Pont Simone Veil. Farther downstream, as part of the city's Rocade ring road, is the major Pont d'Aquitaine suspension bridge, opened in 1967.

A network of integrated tramways was developed beginning in 2003 by the Tramway de Bordeaux across the metropolitan area. The main railway station Gare de Bordeaux Saint-Jean connects with the main modern inter-city TGV system to Paris and the south adjoining the now pedestrianised Passerelle Eiffel bridge. A major redevelopment zone, Bordeaux Euratlantique, is being built adjacently.

With its prefecture and city hall, mayor and districts, a metropolitan government for the Bordeaux Métropole was also created as the centre of the Gironde Département in 2015 for the larger surrounding urbanised Nouvelle-Aquitaine region. The population of the city is approximately 0.3 million, with 1.3 million in the metropolitan area. Bordeaux-Mérignac Airport (BOD) is located in the western outskirts.

Transactional Forces and Typology

With its location nestling on the Garonne River before it meets the Dordogne River to join the Gironde estuary, Bordeaux is located in a favourable region with hillsides well suited as vineyards. Adjacent to the Atlantic, its port was developed on the Left Bank, which grew to become one of the most active ports for commodities during the 18th century and which later benefitted from the triangular trade with Africa and the Americas. Upon this wealth and its viniculture, the city was developed around its main axes running from the 'quais' to connect with landmark features and squares, with its fine hotels and facades, and its long stone bridge crossing.

As an historic riverine city, Bordeaux has been developed on the Left Bank following its axial design, with a generous layout of boulevards, 'grands places' and palaces carefully integrated into a harmonious whole. From its initial core, the original townscape has been retained, with its historic churches, squares and cathedral, and profile and facades that benefit from their location alongside the natural curve of the riverside embankment. With one area completely redeveloped, the brutalist Quartier Mériadeck and towers, the city extends outwards in a radial direction with the surrounding circular Rocade and new bridges crossing the river Garonne and the regenerated zone of La Bastide on the opposite bank. As an economically active region it remains an important wine growing region.

3.1.3 MARSEILLE, FRANCE

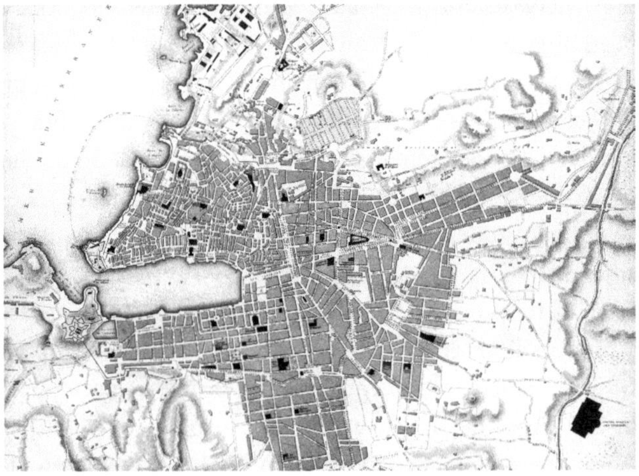

FIGURE 3.1.3 Marseille – early configuration, 1840.
Source: iStock.

Background and Origins

Marseille(s) is France's second largest city in the south of the country on the Mediterranean coast. It was named Massilia by the Romans and is located near the mouth of the River Rhône in the Mediterranean Sea in the region of Provence, now the Département Bouches-du-Rhône. It was developed as a Gallo-Roman city in the West Roman Empire as a trading hub in the Mediterranean after the earlier Phocaean traders.

Along a ragged coast on the Côte d'Azur of Provence, it developed around the old port, whose entrance is flanked by two forts, Fort Saint-Nicolas and Fort Saint-Jean. Out in the Bay of Marseille is the Frioul Archipelago of four small islands and includes the old fortress of Château d'If on the smallest of these islands.

As part of Provence, Marseille has a very different character from other areas of France. Ports have their own discrete characteristics, and Marseille noticeably has been influenced as a transit point for passengers and goods in and out of the country, especially from Corsica and with Kabyles from Algeria and other North African countries on the opposing Mediterranean coast, such as Morocco and Tunisia and former French colonies in Africa.

Declared a free port originally by Louis XIV, Marseille has since developed and now become the largest port in the western Mediterranean. The newer working harbour and extended port with its logistics and shipping facilities is very different from the old city itself with its former castle, leaving the central Vieux Port essentially for fishing and sailing activities.

Since the introduction of containerisation, whole new areas have been developed for handling containers in the industrialised area, and the port has been expanded around the Port de la Joliette. Another heavily industrialised port area nearby but to the west is Fos-sur-Mer where petro-chemicals are handled. Marseille was connected with the River Rhône by a canal, but with the collapse of the Rove Tunnel, this only enables connection from the Étang de Berre via Martigues to Fos.

As one of the main southern gateways into France, Marseille has continued to serve as a place of migration and has become a multicultural city with a large Maghrebi population. It is also a centre for the arts in attracting artists to the small neighbouring port of L'Estaque to the west and to the south are the coastal towns of Cassis and La Ciotat with its small shipyard. In the city centre and Le Quartier Panier is the Hôtel de Ville, la Vielle Charité and the Bourse in the St Ferreol commercial district along with the opera house. The Abbey of Saint-Victor is one of the oldest buildings there.

The city was occupied during World War II, and on a rocky outcrop is a symbolic place of pilgrimage, the elevated Notre Dame de la Garde Basilica, which has special significance since its exploits during the liberation of France in August 1944. Also interestingly, from the former time of the French Revolution, the song that was adopted as a national anthem and first sung by volunteers from Marseille has since become known as 'la Marseillaise'.

Modern City – Buildings and Arteries

Today the main axis of the city is the Boulevard Canebière that runs eastwards through the commercial centre with the perpendicular Cours Lieutaud at Noailles. The cathedral and the Hôtel de Ville are prominent buildings here. With its old university and grandes écoles, Marseille also houses modern business schools and new research institutions, such as CNRS and INSERM.

The economy has evolved from light manufacturing to service industries and major new buildings have been constructed that include the CMA CGM Tower (designed by Hadid) in the new Euromediterranée business district developed here during the 1990s. With its modern MuCEM Museum, opened in 2013 opposite the old forts, the area beyond the harbour has changed significantly.

The city has two main metro lines and three tramways under extension that serve the central area from the suburbs and connect with the Gare de Marseille–St-Charles terminus that has the high-speed TGV rail line that connects with other cities such as Avignon and Lyon and further metropolitan centres.

The main A7 Autoroute du Soleil and the A55 run southwards into the city and connect with the perimeter A507 to the east that connects with the A50 in the south running towards Toulon and outer A52 to Aix-en-Provence. A new cross-harbour tolled tunnel has been built under the old port, the Tunnel du Prado Carenage.

Marseille Provence Airport (MRS) is located at Marignane north-west of the city near the outer industrialised areas. Further outside the city is Le Corbusier's massive brutalist Unité d'Habitation, which opened in 1953, a seminal building with 337 apartments that in the modern international architectural field has been much emulated by designers seeking different methods of residential provision of higher-density and high-rise housing.

The city of Marseille has a mayor and 16 arrondissements in eight paired sectors with a population of approximately 0.9 million. In 2016 the larger administrative Métropole d'Aix-Marseille-Provence was created as an urbanised region with a population double that of the city.

Transactional Forces and Typology

As a city on the north coast of the Mediterranean with its original natural harbour, the role of Marseille as a port has expanded beyond its early fortifications to the extended area westwards that it occupies today. From handling commodities of bulk cargo at La Joliette, it has developed into specialised areas with containerisation as a major logistics centre with its basins and wharves that have been extended linearly to the north behind a protective breakwater and with a separate petro-chemical facility beyond at Fos. From its early fishing origins, the port is now dominated by its maritime activities all across the Mediterranean.

From its rugged coastline and with its hilly and forested areas inland in Provence, the core of the city is defined by the Vieux Port. From its early origins and dense gridded streetscape, around Rue d'Aix with its monumental Porte d'Aix, the city extends inland along valleys following areas of buildable terrain. The port has been extended linearly northwards, alongside which the main rail and road communication channels have been built. Surrounded on either side by the attractive natural zones and coast of the Côte d'Azur, it is an industrialised maritime zone, nestling within the geographic Chaine de l'Estaque and the protected communal forests to the east. As urbanisation has spread around the perimeter, this is limited within the peripheral ring highway that runs part in tunnel around the eastern edge.

3.1.4 ALGIERS, ALGERIA

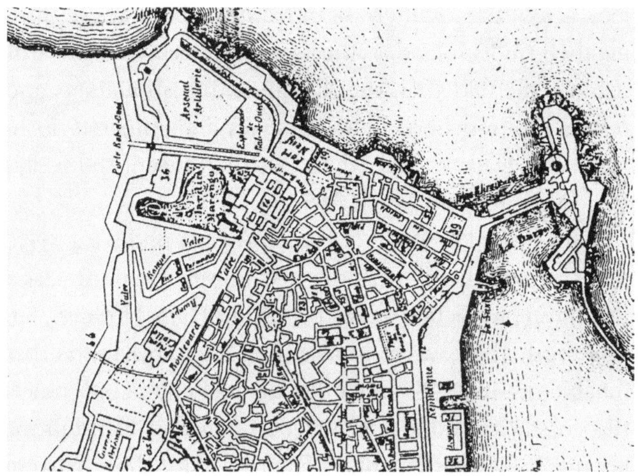

FIGURE 3.1.4 Algiers – early configuration, 1830 (Bastide).
Source: iStock.

Background and Origins

Across the Mediterranean, a rather similar counterpart port to Marseille is **Algiers** (Alger). Once a Roman port known as Icosium and later with its Arab name from the 10th century CE as **El Djazair**, it developed as a closed city in mediaeval times with walls and a citadel, becoming the capital of Algeria. During the early 14th century it acted as an autonomous actively trading city-state.

During the following period, the Spanish occupied the city from an islet, the Penon, in front of the inner harbour, but were expelled by the Ottomans in 1516. There followed a period of Barbary piracy and wars in this area of the Mediterranean, but it was protected from its geographical position with mountains and a ravine forming two sides of a triangle with the sea in between. The citadel was extended with a moated ring wall, and the upper and lower city had separate communities of Moorish, Andalusian, Jewish and Kabyle inhabitants.

With the Grand Market Street crossed by a series of alleys, the Kasbah was the heart of the city. During Ottoman times the city infrastructure was enhanced with water supply via aqueducts, oriental buildings and the Grand Mosque and an active trading city profiting from the sale of slaves.

However, in 1830 it fell under the colonial influence of the French and with its style of buildings and culture, the metropolitan capital city developed in Mediterranean style with its bustling characteristics, albeit having a different catchment area to the south of Arab peoples and being on trade routes with the Sahel countries.

The lower city was Europeanised and along the seafront with the Boulevard Front de Mer with the northern outlook across the Mediterranean from the port where ships ply their trade. Inland, as the land rises steeply southwards behind, the Old City, the Kasbah and Arab markets and restaurants remained. The mosques and minarets form particular features and whilst Mediterranean, the feel is more Maghreb and North African.

With the colonisation of French Algeria under a government general, there was a push to take over agricultural land for cotton production along the Mitidja Plain. However, farther south, there was resistance from the Kabylia and other inhabitants. In 1848, under the Second Republic of France, the coastal territories of Algiers, Oran and Constantine became full Départements of France. Whilst those Saharan territories farther to the south became more difficult to control, infrastructure started to be developed with new roads and railways leading out from Algiers and along the coast.

During World War II, Algiers was part of Vichy France and for a time was the provisional capital of France until it was invaded by the Allies in 1942 and eventually taken over. Of greatest significance was the subsequent Algerian War of Independence that was waged for many years beginning in 1954 between the French and the Front Libération Nationale (FLN). Only after considerable resistance did it lead to eventual independence of the new Republic of Algeria in 1962 that was subsequently celebrated by the Monument des Martyrs, erected in 1982.

Modern City – Buildings and Arteries

Of its different areas, the Kasbah remains a typical Arab Quarter with its homes and mosques, including the Great Mosque Djamaa el Kebir. The port has been expanded from its Inner Harbour in the north to the Southern Harbour with its ancillary facilities and protected by breakwaters. The colonial Grand Post Office building dominates the front in the centre.

In 1954 the Climat de France (designed by Fernand Pouillon) was built in the suburbs, a huge multi-storied courtyard concrete block '200 colonnes' that aimed to house the local population. It featured when the FLN sought to use it as a blockade during the Algerian War of Independence.

During the 1950s, Le Corbusier developed the plan for a dramatic modernistic scheme of a gigantic linear housing development on the hillside overlooking the Mediterranean, but this was never built. Likewise, currently there are grand plans for new modern developments in the bay proposed by EMAAR.

In terms of transport, the SNTF operates the suburban railways form the central station and to the airport. The Algiers Metro was opened in 2011 which runs along the front from Place des Martyrs to El Harrach and with the tramway between Les Fusilliées and Bab Ezzour and the eastern Cité Diplomatique. There are also a series of Télépheriques in the hilly coastal area around Bab El Oued. To the south-east of the city is located Houari Boumedienne International Airport (ALG).

The city is divided into 13 districts, the most populated being Dar El Beida, Bab El Oued, Hussein Dey and Sidi M'Hamed. The current population of the city is approximately 4 million and double that in the metropolitan area.

Transactional Forces and Typology

As a southern Mediterranean coastal port, Algiers developed as a trading centre not just in maritime terms, across and around the Mediterranean, but also inland towards the Sahel, with its trade and caravan routes into increasingly desert areas beyond the coastal plain. Despite its occupation variously by Spanish, Ottoman and French invaders, it has retained its predominant Arab culture and traditions in this part of the Maghreb region. With its agricultural and cotton production, vegetables and dates, textiles and copper, the modern economy has developed from increasing oil and gas production in the region.

From its early Roman settlement and development as a port, the city has grown around its natural harbour by the coastal plain. Expanding linearly east and west around the Bay of Algiers, the topography varies from the upper hilly Casbah to the flatter eastern area around Dar El Beida and Bordj El Bahri. As the port was developed around La Darse, the boulevard along the seafront and Hussein Dey is shared by both highway and railway across the El Harrah Oued. As built development has climbed the hillside and around the Didouche Mourad to the heights, this was a commanding position overlooking the harbour. Consequently, the city constitutes the different districts of the organic residential Arab Kasbah, the residential port area with its wharves and warehouses, a more open area along the bay and then into developed localised areas inland. As these built areas are being developed away from the centre, this is transforming the structure of the city to more poly-nodal centres in a wider urbanised area as the population continues to expand.

3.1.5 LIVERPOOL, ENGLAND

FIGURE 3.1.5 Liverpool – early configuration, 1897.

Source: iStock.

Background and Origins

Liverpool is located on the north-west coast of England on the estuary of the River Mersey and the Irish Sea. On gently rising ground, it grew on the peninsula around an inlet with a sheltered tidal dock known as the Pool.

The earliest local influence was from the Romans having a nearby fort at Chester (Deva Victrix), before coming under the Anglo-Saxon Kingdom of Northumbria and then eventually Viking settlers under Danelaw. With its name reflecting a 'creak of muddy water', Liverpool's roots can be traced back to the time of Magna Carta when the Borough of Liverpool was founded by royal charter. With its original five streets in the form of an 'H' leading up from the waterside, the settlement grew with the population trading with Ireland across the Irish Sea. A castle was built in the West Derby 'Hundred', which was to feature in the English Civil War, and the town grew as trading increased.

With expansion of trade and shipping with the Americas and the West Indies, the triangular trade grew, with cloth and coal being exchanged for sugar and tobacco and with human resources from Africa. As the trade in commodities grew, the docks were expanded such that by the end of the 18th century, over half of Britain's activity in the Atlantic slave trade was accounted for via the port of Liverpool. However, with the abolition movement, this declined and ended effectively around 1810, despite some illicit continuation.

The first wet dock the 'Old Dock' had been built in Liverpool in 1715 and then the subsequent docks of St George's, King's and Queen's Dock. The commercial heart developed around the pierhead, quayside and docks. With larger ships, the berthing requirements could only be met by holding them alongside the open water frontage. Of the surrounding ancillary buildings, each had its own specialty, either by region of origin or by commodity. These were followed by Canada and Prince's Dock, Clarence, Stanley and Albert Docks. They reflected the particular trading materials, with warehouses for imported sugarcane and molasses, cotton and tobacco, and exports of manufactured wares of steel and ceramics. The Albert Docks were of progressive design with cast iron columns, brick vaults and sited close to the dock edge so as to enable off-loading direct into the warehouse.

The first Liverpool & Manchester Railway was opened in 1830 with the maiden journey of Stephenson's steam-powered Rocket locomotive. After a long, challenging period of financing and construction, the Manchester Ship Canal also was realised with a complex series of locks and bridges to connect the city of Manchester with the Mersey estuary so as to create its own inland port for the city.

Liverpool was eventually granted its own city status in 1880. The corporation laid out the city centre with a number of squares as Great George Square and Abercromby Square. At its prime, the principal buildings in the centre of the city of Liverpool were the town hall and exchange, with classical St George's Hall and the Customs House. Whilst there were grand public buildings, there were also many areas of poor accommodation with over-crowded housing conditions during the 19th century. Some more spacious schemes were built, such as Prince's Park and Sefton Park for the better off.

Accordingly, the city grew up around its North and South Docks as a centre of trade and commerce. Around the town, new roads and turnpikes were also built to improve communications, and the Liverpool electrified Overhead Railway in the Docks was opened in 1893.

Across the River Mersey at Birkenhead in the Wirral, shipbuilding developed with the combined skills of Cammell Laird. With all of its commerce, Liverpool became a cosmopolitan city with the migration of workers and tradesmen and was a magnet for unskilled labour. With the coming of steam, passenger liners also plied across the Atlantic as new shipping lines became established in Liverpool, such as the Cunard and White Star Line, the 'steamship Kings of Water Street'.

Large businesses developed in the city around particular commodities, for example Tate & Lyle (sugar), Lever Brothers (soap) and Imperial (tobacco). Around Lever's factory in the Wirral, Port Sunlight Model Village was also developed as a residential area for its employees. A large number of commercial offices were also built in the centre.

During its time as a significant trading port, the centre of Liverpool with its waterfront Pier Head, which was dominated by the 'Three Graces' as viewed across the Mersey from Birkenhead, that is the Royal Liver Insurance Building (1908) with its cupola and twin liver birds, the Cunard Building (1914) and the Mersey Docks and Harbour headquarters. The old Albert and Canning docks warehouses, built in 1846 with their brickwork and iron frame, remained as the centre of the wool trade. Another later civic landmark was the imposing brick Anglican Cathedral designed by Gilbert Scott in 1904.

During the 1930s, the Great Depression hit the city very badly and there was extensive unemployment and poverty. With low wages, a programme of council or social housing was initiated. The city had benefitted from a comprehensive urban street tramway system that radiated out from the centre. In terms of transport, a key development was the construction of the Mersey Tunnel across to the Wirral towards Birkenhead that opened in 1934 and which greatly improved access beyond use of the regular ferries. Renamed the Queensway Tunnel, this was followed much later by another modern tolled tunnel, the Kingsway Tunnel into Wallasey, which opened in 1974.

The city suffered heavily during the Blitz. After the end of World War II rebuilding was required and with the end of the Sterling area, cross-Atlantic shipping and trade began to decline with its concomitant impact on the city. The shape of global trade was changing and with containerisation and no need for warehousing, the pattern of traffic moved to the more specialist ports in the south-east of England. A spiral of decline set in, without requirement for supporting and ancillary businesses.

Modern City – Buildings and Arteries

The post-war decline of the city continued. Much of the former 'back-to-back' housing or courtyards with narrow entries was in poor condition, condemned and torn down as a result of the economic doldrums in the 1960s; these decaying areas were replaced, and a large programme of urban renewal was initiated. With the lack of opportunities and social deprivation, a period of riots in Toxteth during the 1980s was a particular low point. It took a regeneration strategy and unusual EU funding to revitalise the city, which with additional funds from some sovereign wealth funds started to reinvigorate the real estate market in the city again.

Besides the large Anglican Cathedral, whose nave was only completed post-war, the very modern Roman Catholic Cathedral designed by Frederik Gibberd, with its symbolic crowned lantern, was completed in 1962 and other areas regenerated, especially along the waterfront.

The city comprises different quarters having distinctive characteristics, for example the Baltic Triangle, Canning Georgian Quarter, Commercial District, Cultural Quarter, new Knowledge Quarter, Islington, Ropewalks and the Waterfront.

With the closure of the South Docks, in 1989 the Merseyside Development Corporation (MDC) was founded with the goal of achieving urban regeneration in the dockland area. Much of the infrastructure was improved and areas were zoned for new development.

With the revitalised energy of the good-humoured inhabitants, proud of their diversity and with the maritime and earthy character of the city, it started its rebirth and commercial redevelopment around the docks to revert to becoming a strategic centre in the north-west of England again. The centre and the waterside have been completely redeveloped with new residential buildings, such as West Tower and Beetham Tower, the Lexington, Alexandra Tower and the Unity Buildings, that have changed the character of the area and Prince's Dock and the regenerated commercial Albert Dock.

Parts of the outer periphery fall within a protected green belt around Speke and beyond. With motorways now across the Wirral, and the latest development of the new Mersey Gateway Bridge that crosses the narrowing river farther up at Widnes, this has enhanced travel in and around the industrialised area and to Manchester.

There are significant plans for new extensive Atlantic Gateway ports in the north-west of England to be developed at Port Salford and Liverpool and Wirral Waters as an Enterprise Zone, but these remain at the planning stage. In 2014, a new surrounding urbanised city region was created as the Liverpool City Region and expanded to include the boroughs of Knowsley, St Helens, Sefton and the Wirral and Halton. The current population of the city is approximately 0.5 million, below the peak of nearly 1 million pre–World War II.

Transactional Forces and Typology

With the early mediaeval origins of Liverpool as a settlement with a harbour, maritime trade developed around the northern coastal regions and the Irish Sea. It was subsequent to the development of the Americas that the Port of Liverpool came to the fore, being located on the Western Atlantic-facing coast of the British Isles and also being close to manufacturing towns that had developed near Lancashire under the Industrial Revolution in Great Britain. This led to the development of the docks on the River Mersey and then enhanced by its canal and rail connections into the growing industrial regions.

Whilst this was to last for over a century, post–World War II the pattern of trade had significantly declined and also the type of trade with containerisation that resulted in the decline of the docks. Large areas of the city were subject to wholesale redevelopment and a new waterfront under urban renewal as a result of de-industrialisation requiring new businesses and housing also.

This transformation had been taking place before the turn of the 21st century, as the city repositioned itself under the new service and digital economy in the wider Liverpool City Region.

The city of Liverpool is well sited on the eastern side of the River Mersey facing the Wirral, a protected site suitable for harbouring marine vessels. From its early waterside wharves and wet docks were developed that were especially suited for different cargoes and commodities and specialised warehouses as appropriate. The city has grown eastwards to accommodate the growing population as new businesses gravitated to the city. From its planned civic centre, new residential neighbourhoods have developed more sporadically, radially outwards, more characteristic of ribbon development and interspersed with industrial and green areas. Whilst originally focused on its maritime Atlantic trading, the focus has now turned inwards towards its remaining manufacturing and expanding service industries.

3.1.6 GLASGOW, SCOTLAND

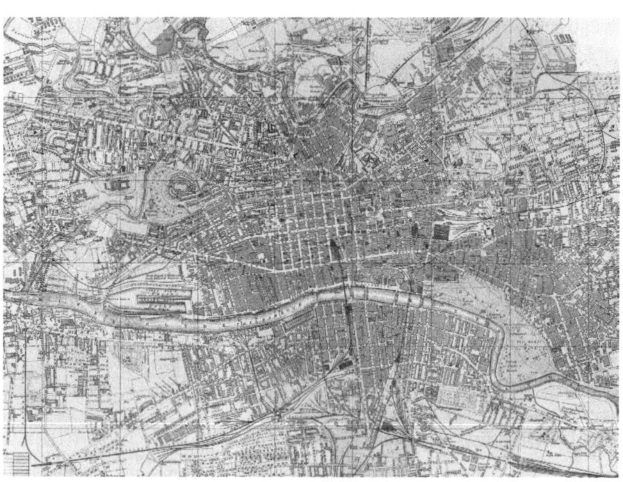

FIGURE 3.1.6 Glasgow – early configuration, 1886.

Source: Maphouse.

Background and Origins

The city of **Glasgow** is located in the West Central Lowlands of Scotland by the River Clyde, which flows westward out of the Firth of Clyde to the Irish Sea. It was settled originally by Celtic and Pict tribes and was occupied in the Roman period as part of northern Caledonia.

At a fording point, the city was understood to have been founded by the Christian missionary St Mungo in the 6th century CE and grew around Molendinar Burn with the first permanent crossing over the River Clyde recorded in 1285 at Glasgow Cross. Having a Bishopric in 1492, the area was primarily agricultural and with its own fisheries, and in 1611 became a Royal Burgh.

Other activities developed with local merchants around the riverside and as trading grew, the port was developed in deeper water towards the west. Trade developed, especially in tobacco, then in sugar and cotton as the triangular Atlantic trade developed. The new Port Glasgow had been developed as early as 1668 on the southern bank of the Firth of Clyde, near Newark Castle with a customs house as this trade expanded, promoted by the Tobacco Barons (trading tobacco, sugar, cotton, rum and indigo).

The next significant event was the development of the Forth and Clyde Canal between 1770 and 1790 to connect both the Firths of Forth and Clyde by water from Grangemouth to Bowling Basin at Port Dundas. This was a considerable undertaking of construction of over 50 km. of new canal locks and drawbridges, which serve to connect with Union Canal and Monkland Canal. Although not suitable for vessels with deep drafts, this did provide an effective transport link for freight from east-west and for the main iron and coal mines that had been developed in Lanarkshire.

Attempts at regulation of the River Clyde were made with the construction of central landbanks, narrowing and dredging in order to deepen the channel and placing markers to assist shipping. With the increase in trade, further industrialisation occurred, and with the introduction of steam power, manufacturing and shipbuilding developed outside the centre at Port Glasgow and at Clydeside. Infrastructure was developed with the construction of new railways and tramways and as the population increased, improvements in sanitation and water supply were also required. The population grew significantly at the end of the 19th century along with considerable in-migration from nearby Ireland.

The centre of the city of Glasgow is divided into different districts, principally the East End and West End, South Side and North Glasgow. Originally the mediaeval centre was characterised by streets, or 'riggs' linking the centre with pastureland for cattle to the north and west, off the High Street. Cowcaddens, where the animals were milked, Rotten Row and Candleriggs, all reflect former agricultural activities. The city centre evolved around a grid of streets on the north bank of the River Clyde.

In the 1780s, as the city was developing, a new grander plan was designed by James Barrie, with regular streets and new squares. With Queen Street and George's Square, Hanover and Frederick Streets were planned in a Regency manner rather as Edinburgh's New Town, with terraced housing and larger townhouses.

The mediaeval centre was Glasgow Cross where Trongate, Gallowgate, Salt Market and High Street intersect. With Tolbooth Cross clock tower, Castle Street and High Street, these are the oldest roads in the city. Just to the north of this is Rotten Row and the Cathedral, dedicated to St Mungo, and the old Provand's Lordship building.

Subsequently, this was complemented by the Queen's Hotel on the North Side and Glasgow City Chambers on the eastern side. The area known as the Merchant City is where the wealth of the Tobacco Barons was used to develop properties south of George Street and west to Argyle Street.

The *East End* was formerly the industrial centre, with the many different manufacturing industries that included engineering (William Arrol), shipbuilding (William Beardmore), breweries (Wellpark), and textiles (Templeton carpets) by Glasgow Green.

The *West End* is the area around Blythswood Square and Sauciehall Street intersected now by the modern motorway crossing north-south, which turns eastwards past Port Dundas, towards Kelvingrove Park, Park Circus and the River Kelvin. On Gilmore Hill is the University of Glasgow's main campus with Gilbert Scott's main building.

The *South Side* area extends into South Lanarkshire and East Renfrewshire and other residential areas, such as Cambuslang and Rutherglen that were encompassed by the city. Along the south side of the river Pacific Quay has been renewed but was formerly part of Plantation Quay and Prince's Dock Basin close to the Govan area and opposite Partick where the River Kelvin joins the River Clyde. Here, originally Robert Napier, then William Beardmore and John Elder, formed Fairfield Shipbuilders that were established in the Upper Clyde where shipbuilding flourished for around 100 years from 1880 onwards. Immediately south of the river is the Gorbals district with its dense old residential housing estates.

North Glasgow possesses some of the poorest areas of the city with Possil Park and Hamilton Hill. With many tenement housing areas, considerable urban renewal has been undertaken that includes Ruchill. One of the major industrial areas was Springburn with the Saracen Foundry of William MacFarlane, Charles Tennant Chemical Works where, with the development of a number of railway lines in the area, this led to the growth of its own heavy engineering railway works. This included North British Locomotive, Cowlairs Works, Caledonian Railways, St Rollox Works and Neilson & Co. and Hyde Park Works.

John Brown Shipbuilders was originally established at Cessnock in 1851, but moved to West Barns O'Clyde near Dalmuir, close to the Firth and Clyde Canal in 1871, opposite the confluence of the River Cart with the Clyde. The growth of this shipyard led to the development of a new town, Clydebank, in the parish of Old Kilpatrick and encompassing part of the Roman Antonine wall. Many tenement houses were built, with Thomas buildings to house the workforce, and the railway was built to provide connections with the Singer Manufacturing Company, also established there, and several warehouses for the whisky business.

The transport infrastructure across Glasgow has been developed in different stages. From early mediaeval times of 'riggs', roads and streets were built. Following intensive use of the river, a period of canal construction was then developed with the Union and Mockland canals and the major Forth and Clyde Canal built in sections to connect the waterways of both the Firths of Forth and Clyde (1770–90) from Port Dundas to Grangemouth.

Next came the railways, as individual railway lines were developed, such as Caledonian Rail to Edinburgh in 1845 and North British Railway in and around the city. With the Central Station parallel to Eglinton Street and Queen Street Station, other station terminals were built at Buchanan Street and St Enoch with other intersections at Glasgow Cross, Anderston Cross and Bridgetown Cross. The Glasgow Central Railway opened in 1896 and ran underground, east-west across the city to connect the docks and industries on the north bank for both passengers and freight. With the heavy engineering works at Springburn, with good access to coal and steel materials, this became the manufacturing centre of rail products, locomotives and related machinery.

The city had one of the largest urban tramway systems also, Glasgow's Corporation Tramways. Commencing in 1872 on a 4′7¼″ gauge from Saint George's Cross to Eglinton Toll, this was to allow railway wagons to be pulled over the same track. Originally horse drawn, it was electrified in 1898, and many new lines were built to a total coverage of approximately 100 route miles. Double-deck tram cars were used and at its peak in 1938 reached out deep into the suburbs. This was phased out, however, post-war in the 1960s and replaced by trolley buses.

Glasgow also had one of the world's oldest underground subways that opened in 1896 with a four feet (4') running gauge, configured as double track in a large circular loop, originally a cable railway extending on both sides of the River Clyde. It was converted to third rail DC electric traction in 1935 with 15 stations, but without direct interconnection between Buchanan Street and St Enoch stations. Whilst serving the central area of the city, running both clockwise and anticlockwise, it deteriorated and required significant modernisation in the 1970s with new rolling stock and systems refurbishment.

Modern City – Buildings and Arteries

Besides its heritage and Victorian architecture, the city is notable for the Glasgow School of Architecture led by Mackintosh and exemplified by the Glasgow School of Art building. It has a strong educational sector also and is home to the old University of Glasgow and the University of Strathclyde.

During World War II, the city had suffered significant bomb damage, especially around Clydebank. In 1946, a plan was prepared for regeneration and rebuilding within the city (Bruce Report), whereas Abercrombie developed the Clyde Valley Plan for the rebuilding and realisation of four new towns to accommodate growth and overspill, two of which were built at East Kilbride and Cumbernauld.

Post-war, with poor areas of housing identified, a series of Comprehensive Development Areas was defined that reflected the need to undertake extensive renewal of a number of districts of former tenement buildings. However, these were replaced by a series of 10–12-storey tower blocks often of poor quality and of prefabricated construction and with inadequate supporting social amenities, built during the 1960s. These too were found to be unsatisfactory, several of which had to be demolished, such as Hutchesontown 'C' in 1993 and the Red Road Tower Blocks in 2015. This was a most unsuccessful programme of urban renewal, and provision of residential housing had to be reassessed.

More recently, after a period of regeneration and renewal within the city centre, a number of commercial areas have been redeveloped. These include the realisation of new modern buildings, for example the Science Centre built on Pacific Quay on the site of the former Glasgow Garden Festival, Clyde Auditorium, the Riverside Museum and the SEC Centre. In addition, an International Financial Services District (IFSD) has been developed in the west of the city, where several service companies are headquartered.

Whilst major manufacturing and shipbuilding along the river has declined, most industrial activities have been replaced by more modern manufacturing and with a focus on biosciences and software IT services after an unsuccessful entry into semiconductor production. For shipping, King George V Dock remains a single-entry dock at Shieldhall (Govan) on the south bank, with most major shipping having been transferred down to the container port at Greenock and Hunterston run by Clydeport.

The modern suburban and commuter rail network is well developed with lines across Strathclyde, now run by Strathclyde Partnership for Transport (SPT). It is also connected with the mainline inter-city network West Coast Mainline to London and also to Manchester and Edinburgh. There is also a new planned BRT Clyde Fast link between the centre and the SECC.

In terms of highways, the city is encircled with motorways, with the M74 running east-west across the south of the city and connecting with the M73 running north-east into the M80, and the M8 crossing north-south in the centre at Kingston Bridge and then running into the M73 towards Cumbernauld. An earlier planned programme of extensive construction of a motorway box entirely across the city was never completed. To the west is the more recent Erskine toll bridge crossing on the M898 to Clydebank and Dunbarton.

The population of the city now is approximately 0.6 million, much less than the peak of 1.1 million in 1950, and with approximately 1.8 million in the metropolitan area.

Transactional Forces and Typology

The city of Glasgow has been shaped by its maritime activities that have developed around the River Clyde and thanks to the availability of coal and steel materials, which has afforded its industrialisation. Steam power for both fixed and mobile machinery determined its development, as after the construction of the canals and the railways, the river was channelled and shaped with its new docks. Thereafter shipbuilding became the predominant activity with its upper yards at Govan and Clydebank farther downstream. Industrialisation included both heavy engineering and railway industries that were promoted under the expanding labour force.

As labour continued to be attracted to the city, the provision of adequate housing was challenging as older tenemented housing remained in areas such as the Gorbals. As de-industrialisation continued post-war, a programme of urban renewal was required for the clearance and rebuilding of many older inner housing districts along with a policy of dispersal of the population. However, initial efforts were largely unsuccessful and the city continues to strive to this day with appropriate regeneration and renewal schemes and retraining of the labour force.

As a city, Glasgow was developed initially on the north bank of the River Clyde, upstream from the Firth of Clyde estuary beyond old Dumbarton. From a compact centre of mediaeval 'riggs' as the Merchant City, it grew outwards as new districts were developed in the East and West End. After the construction of the canals and the beginning of shipbuilding, this was located on the South Side around Prince's Dock. As a result, the city developed with industrial areas interspersed with residential areas, many with tenements suffering from poor conditions and with inadequate provision of services.

Whilst this caused further development of outer areas and following dispersal with new relief districts such as East Kilbride, the subsequent de-industrialisation of the city has led to the requirement for substantial renewal, rebuilding and regeneration of areas of the inner city. This has resulted in the extended radial nature of the city from its core, in sectors and nodal centres around the urban rim and into a wider area of the urbanised region of Strathclyde. The following figures illustrate the modern configuration of the mercantile port cities as described and the evolution of their modern population.

FIGURE 3.1.7 Genoa – modern configuration.

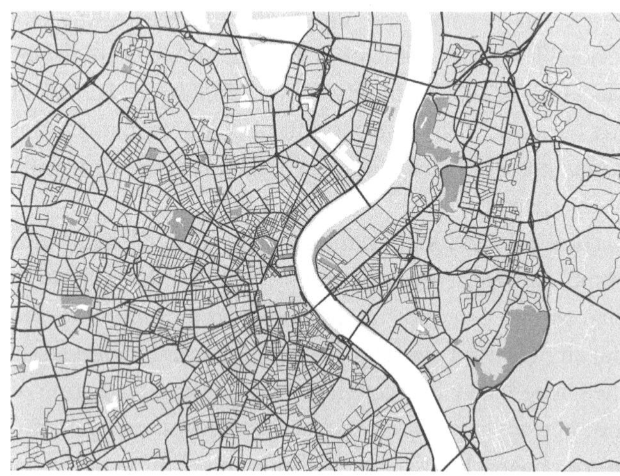

FIGURE 3.1.8 Bordeaux – modern configuration.

FIGURE 3.1.9 Marseille – modern configuration.

FIGURE 3.1.10 Algiers – modern configuration.

FIGURE 3.1.11 Liverpool – modern configuration.

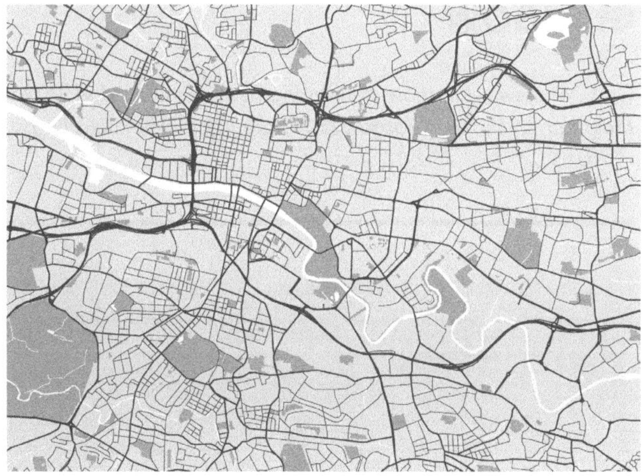

FIGURE 3.1.12 Glasgow – modern configuration.

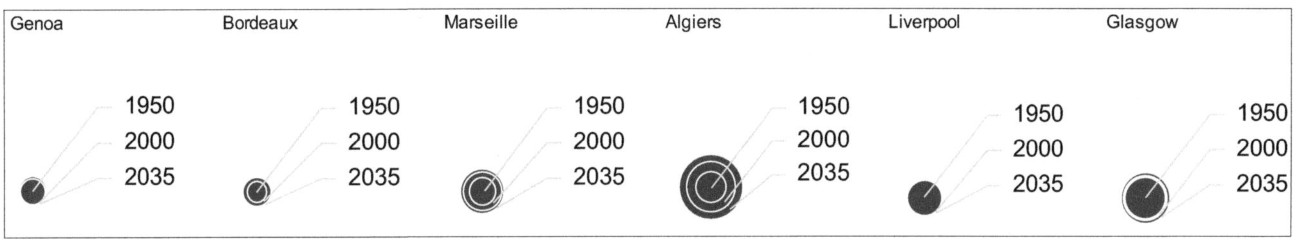

FIGURE 3.1.13 Mercantile port city populations.
Data source: Global Database of Metropolises 2020 – UN-Habitat (2020), UN DESA Population Division.

Further Reading

GENOA

River, Charles, ed. – *The Republic of Genoa, the History of the Italian City That Became Influential Across the Mediterranean During the Middle Ages*, Charles River Publisher, 2019.
River, Charles, ed. – *Republic of Venice and Republic of Genoa, the History of the Italian Rivals and Their Mediterranean Empress*, Charles River Publisher, 2019.
Walton, Nichola – *Genoa 'la Superba', the Rise and Fall of a Merchant Private Superpower*, Hurst, 2015.

BORDEAUX

Lacarce, Xavier & Philip, Marion – *Bordeaux*, Cairn, 2015.
Rosan, Xavier, ed. – *Bordeaux*, le Festin, 2016.

MARSEILLE

Crane, Sheila – *Mediterranean Crossroads, Marseille and Modern Architecture*, University of Minnesota Press, 2011.
Hewitt, Nicholas – *Wicked City, the Many Cultures of Marseille*, Hurst, 2019.
Takeda, Junko – *Between Crown and Commerce, Marseille and the Early Modern Mediterranean*, Johns Hopkins University Press, 2011.

ALGIERS

Celik Z, Clancy-Smith J & Terpak F, eds. – *Walls of Algiers*, University of Washington Press, 2009.
Duyckinck, Evert – *A Short History of Algiers*, Kessinger, 2010.

LIVERPOOL

Hollinghurst, Hugh – *Georgian Liverpool*, Amberley Publishing, 2023.
Kidd, Alan & Wyke, Terry, eds. – *Making the Modern City*, Liverpool University Press, 2016.
Sharples, Joseph – *Liverpool*, Yale University Press, 2004.
Tulloch, Alex – *The Story of Liverpool*, The History Press, 2008.

GLASGOW

Gibb, Andrew – *Glasgow, the Making of a City*, Routledge, 2021.
Meighan, Michael – *Glasgow, a History*, Amberley Publishing, 2015.
Meighan, Michael – *Glasgow, a Transport History*, Amberley Publishing, 2024.
Pacione, Michael – *Glasgow – the Socio-Spatial Development of the City*, Routledge, 2015.

3.2 Lowland, Hanseatic, Baltic and Scandinavian Port Cities

North Sea and Frisian Islands – Rhine Estuary – Hamburg and Elbe Estuary – Baltic Sea and Gulf of Finland

Antwerp, Hamburg, Helsinki, Stockholm, Oslo and Copenhagen

3.2.0 INTRODUCTION – PORT CITIES

Located in the north-western corner of Europe are the Lowlands facing the North Sea and the outward-facing marine districts. Just beyond, north of Amsterdam, are the Frisian Islands in the North Sea and coastal cities in Jutland.

Farther north is the Scandinavian peninsula, a rather geographically distinct region with its long Norwegian seaboard and inland fjords and forests that run up to the Arctic Circle. The enclosed Baltic States are rugged lands with tundra and long winters and sub-arctic climate extending to Lapland and the north.

Being relatively scarcely populated, human settlements originally gravitated around rivers and islands, with fishing being a key livelihood for the indigenous population, along with forestry. The larger settlements were coastal, with marine transport being the most common mode. The earliest tribes eventually coalesced into separate kingdoms whose national borders took shape during the 16th century.

In Scandinavia the population gravitated around the coastal centres of:

- Oslo, Stavanger, Bergen, Trondheim and Tromsø in Norway,
- Stockholm, Gothenburg and Malmö in Sweden, and
- Copenhagen and Aarhus in Denmark.

Around the Baltic Sea, settlements grew as part of the former Hanseatic trading league with freeports at:

- Kiel, Lubeck, Rostock and Stralsund in Germany and
- Szczecin, Gdynia and Gdansk in Poland.

Other port cities, which are now capitals of newly independent states since the end of the Soviet empire, included:

- Konigsberg – Kaliningrad,
- Klaipeda in Lithuania,
- Riga in Latvia, and
- Tallinn in Estonia.

Nearby are the close inland neighbouring ports of:

- Helsinki and Turku in Finland and
- St Petersburg in Russia.

The Hanseatic League developed as a confederation of specialist merchant guilds and towns in northern Germany that came to dominate trade in the North Sea and Baltic Sea in the 12th–15th centuries, stretching from Bruges to Novgorod, that with its own legal and administrative system for the exchange of various types of goods aimed to provide protection against marine piracy in the region. This led to the significant development of these Baltic and Scandinavian port cities.

3.2.1 ANTWERP, BELGIUM

FIGURE 3.2.1 Antwerp – early configuration, 1851.
Source: iStock.

DOI: 10.4324/9781003353348-13

Background and Origins

Antwerp (Antwerpen), in the Flemish region of Belgium, is located on the east bank of the wide River Scheldt and is linked to the North Sea by the Westerschelde estuary towards Walcheren. Antwerp, after the Dutch 'handwerpen' – hand-throwing – is derived from the myth of the giant Antigoon, who extracted a toll from passing boatmen, severing the hand of anyone who did not pay, as celebrated by the statue in the Grote Markt town square.

Settled originally in the 4th century CE by Germanic Franks, in the 10th century the River Scheldt became the boundary of the Holy Roman Empire and early fortifications were built. In mediaeval times the central 'burcht' area and the cathedral, the 'Church of our Lady', were enclosed by walls. Further fortifications were built around 1545 with the citadel sited at the southern end of the walled city. As the Zwin tidal inlet silted up and Bruges declined, trading through the port of Antwerp increased. It grew as a centre for merchants, firstly in commodities such as pepper and spices, in wool and cloth and manufactured items and significantly in diamonds, then in financial transactions and banking by having the first bourse (Beurs) and stock exchange and also publishing.

It became very cosmopolitan, receiving people fleeing the Spanish Inquisition and as a port for pilgrims to Jerusalem, yet with the Dutch War of Independence with Spain in 1570, the city was blockaded and sacked, and with independence of the United Provinces, it was stipulated that the River Scheldt was to be closed to navigation, so trading declined.

Subsequently Napoleon, in recognising its importance, sought to construct a new dock and deepen the Scheldt for larger ships to Antwerp, but this was not fully realised. Later, further fortresses were built by Brialmont in a double ring around the city periphery. Whereas formerly the Dutch controlled the mouth of the River Scheldt, when Belgium became an independent state in 1830, the river was freed again.

With the subsequent advance of industrialisation, trade and shipping activity increased. As the Ruhr economy took off, the railways were developed also and the Middenstatie Central Station was constructed in the centre of the city. Trade expanded with its only colony, Belgian Congo, and Antwerp developed as a commercial centre.

By 1900 the population of the city had significantly increased and in 1920 hosted the Summer Olympics. Given the necessity of keeping the River Scheldt free for river traffic, in 1933 a foot tunnel for pedestrians was opened, the Sint-Annatunnel, to connect the Linkeroever, the Left Bank, with the Right Bank.

The city suffered heavily, however, during both world wars, especially during World War II with the Battle of the Scheldt around Walcheren in 1944, with severe destruction inside and around its surrounding region.

Modern City – Buildings and Arteries

The city has several fine buildings and landmarks, such as the gothic Onze-Lieve-Vrouwekathedraal, the bourse, city hall and a number of fine churches and classic museums. The 17th century Rubenshuis, the painter Rubens' studio is still in evidence in the city centre.

With the Nieuwstad on the northern edge, the main square is the dog-legged Grote Markt with a central statue, with other squares and open spaces such as the Graanmarkt, Theaterplein and Stadspark. Zurenborg is a neighbourhood having unique Jugendstijl features. The Boerentoren (KBC Tower) is one of the oldest concrete skyscrapers in Europe, built in the 1930s. The original Palace of Justice has been supplemented recently by the modern long Lawcourts ('Butterfly Palace') with sail-type roofs (designed by Rogers) over a road tunnel and the site of the old Zuid Station. New museums include the modern contemporary art museum MUKHA in a former warehouse.

More recently the Port of Antwerp has been developed with a considerable volume of bulk cargo and with the growing petrochemical industry. The new Port Authority Building Havenhuis (designed by Hadid) is located dynamically at the harbour entrance and the MAS Museum nearby. The docks are located to the north of the city, with entrances from the Scheldt-Rijnkanaal and Kanaaldok to the inner docks of Churchilldok, Leopolddok and Albertdok.

The highway network has been developed with three tunnels under the River Scheldt. With its congested urban ring, the new Oosterweel road link has been proposed, with a long bridge or tunnel across the docks. It is surrounded by the peripheral ring road to the east of the city.

The Brabant rail network is compact and intensive, with Antwerpen-Centraal station as a terminal at the dam and is complemented with a lower-level station to facilitate through trains. It has a small metro system that radiates out from the Opera station. It is an attractive and active city with a long trading history. The population of the city is approximately 0.5 million with approximately 1.2 million in the surrounding metropolitan area.

Transactional Forces and Typology

The city of Antwerp has a long history from mediaeval times and its glory years as a maritime trading centre dealing in cloth and commodities from the East and West Indies and also in diamonds. Specialist markets developed around these commodities, which then extended to the creation of banking and financial institutions. Despite a period of closure of the river to navigation, with Belgium independence, trade developed again during the 19th century, and a network of docks was built. Despite severe destruction during both world wars, it was rebuilt post-war, although the city lost its position as trade was diverted to the expanding docks around the Nieuwe Maas in Rotterdam as containerisation took hold.

As a riverine city, Antwerp developed on the east bank of the River Sheldt. From its central urban core, the city developed organically from mediaeval times and still retains its dense urban fabric around the central places, Grotemarkt, Groenplaats and Graanmarkt. As a historic city in the Flemish region of Belgium, Antwerp retains its unusual marine identity defined rather by its series of man-made docks inter-connecting in and around with the local waterway system.

3.2.2 HAMBURG, GERMANY

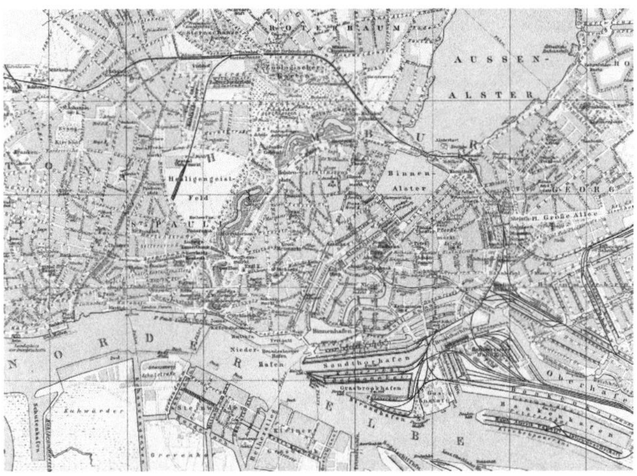

FIGURE 3.2.2 Hamburg – early configuration, 1897.

Source: iStock.

Background and Origins

The Free and Hanseatic City of **Hamburg** is both a city and a federal State lying on the River Elbe in northern Germany with its own natural harbour. On the confluence of the rivers Alster and Bille, the city centre is around Binnen-alster and Ausen-alster lakes that were formed by the damming of the River Alster. From its original castle in the days of Charlemagne, under a confederation of market towns and merchant guilds in northern Germany, in alliance with Lubeck, its port developed to become a major player in the Hanseatic League. In its role also as a free imperial city under the Holy Roman Empire, it became a sovereign state of the German Federation.

In the 1870s, Hamburg refused to join the Zollverein, but when it did finally agree in 1888, it negotiated its right to attain a freeport zone within its harbour. With the construction of the Spiecherstadt district, this enabled it to continue its significant role as entrepot port for Northern Europe. With improvements along the River Elbe, it became a riverine waterway inland up to Dresden and Prague. With its shipping lines and shipyards, it became a major maritime centre.

As an industrial city and port, it had its setbacks with a fire in 1842 and with poor sanitation and was hit by a major cholera outbreak in 1892 and then suffered considerably during World War I. The city was over-crowded and suffered from inadequate shelter with 'rent-barracks' and poor sanitation.

It was devastated and totally destroyed by Allied bombing in World War II. However, post-war it has been rebuilt and developed into a significant and successful modern commercial and industrial city. The city comprises seven boroughs, with Hamburg-Mitte and including Altstadt and Neustadt in the centre. Next to Mitte, Harberg contains the port area and Bergedorf lies to the south-east. The surrounding State is Schleswig-Holstein to the north and Lower Saxony to the south.

The cityscape is generally low-rise, with several churches on the skyline, and with its rivers and canals, there are many bridge crossings and several parks, such as Stadt Park. The city is surrounded by a peripheral 'Wall-ring' in several segments, which is a ring road now built up with buildings on one side that follows the line of the original old city wall. The outer perimeter is surrounded by a string of city parks with the western Wall-ring running alongside the Wallanlagen parks, which crosses the Alster River at the Lombardsbrucke and then the eastern Wall-ring runs along the railway tracks to the Elbe River.

A dominant building in the Atlstadt is the city hall, in Rathausmarkt, which is the seat of government of both the Free and Imperial city and the Federal State, and the tall St-Nikolai Church spire. In Neustadt near Grossneumarkt there is the rebuilt St Michaelis Church and tower. To the north-east along Lange Reihe is St Georg District beyond the Hauptbahnof station. To the west is St Paul's District with the harbour and its markets by Elbpark, and Fischmarkt beyond along the banks of the Elbe. As an important banking centre with its original merchant banks, such as Berenberg and Warburg, it has its own stock exchange.

Modern City – Buildings and Arteries

There are several significant buildings, such as the unusual angular Chilehaus, an expressionist building by Fritz Hoger from the 1920s near Burchardplatz. In the Speicherstadt district of warehouses, Hafen City is a waterfront area of urban regeneration recently realised with residential and commercial areas on Grasbrook Island in the former Freeport area, now with enhanced flood protection. In the Neustadt district the large commercial Hanseviertel scheme has been built.

The very modern Elbe Philharmonie concert hall has been elevated above warehouses (designed by Herzog de Meuron), standing tall at the harbour entrance, and the striking commercial Marco Polo centre (designed by Behnisch) and Unilever-Haus are nearby.

With the elevated Hochbahn, U- and S-Bahns, all inter-connecting at the Hauptbahnof, urban rail connections are very good across the city, with both rapid transit and regional rail services run by HVV Transit authority. Of the highways, there is a dense central network of streets and bridges with wider connections of the four main regional Autobahns outside the city. The cable stayed Kohlbrand Bridge crosses the Elbe (north and south) branches at Wilhelmsburg, as well as the newly enlarged Elbe Tunnel on the A7 Autobahn with its four bores and reversible lanes.

As an active industrial city with shipbuilding also, it acts as a deep river port and a major container port, effectively a seaport in the Elbe estuary. It has also now become a centre of scientific research and development with its universities and research institutions and aerospace industry.

With a city mayor it has seven boroughs around Central Harburg–Mitte with Altona and Harberg to the west and Bergedorf to the east. The population of the city is approximately 2 million, with approximately 5 million in the metropolitan area.

Transactional Forces and Typology

With its location in the north of Germany on the estuary of the River Elbe, between the ports of Bremerhaven and Lubeck, Hamburg developed as a pivotal city-state and played a key role in the Hanseatic League. With its role as a freeport, it developed as a significant entrepot port in the corner of north-west Europe by Schleswig-Holstein and Jutland. Now as a federal 'Stadtstaat', it has benefitted from its geographical location upstream in the deep River Elbe and connected with the waterways that run deep inland across northern Europe.

The city of Hamburg developed around the conjunction of the rivers Alster and Bille with the Elbe estuary, where a circular walled city was built. With the channelling of the waterways, the city grew outside beyond Altstadt and Neustadt in a concentric manner, with roads aligned radially from the city centre. Whilst dominated by its marine trade with its waterways south of the Norderelbe, with improved flood control, its railways and highways have complemented this and continued to serve the Port of Hamburg as a key logistics centre. With water courses to the south-east, development has focused northwards in corridors as growth has continued stretching towards Kiel and Lubeck. As a riverine city, its radial infrastructure is characterised by multiple bridge crossings over this extended catchment area of the River Elbe.

3.2.3 HELSINKI, FINLAND

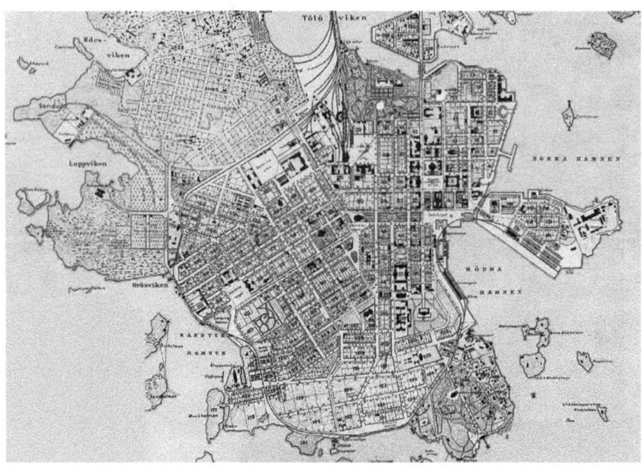

FIGURE 3.2.3 Helsinki – early configuration, 1899.
Source: Maphouse.

Background and Origins

Helsinki, as the capital of Finland, is located in the south of the country on the coast of the Gulf of Finland. Although part of Scandinavia, Helsinki in Finland feels more Baltic, located as it is between Sweden around the Gulf of Bothnia and Russia in the Gulf of Finland, 'the Pearl of the Baltics'.

Rugged with its lakes and forests, Finland encapsulates Lapland, which is beyond the Arctic Circle. With the retreat of the ice sheets, the first evidence of habitation here was during the Iron Age after the Ice Age. Strongly influenced by the 'Kalevala' epic poetry, the country has been guided by the aspirations of national romanticism.

The region of Tavastia around Lake Vanajavesi was populated with individual settlements near the mouth of the River Vantaa. It was taken over by Swedish colonisers around 1550 CE who started to develop the trading town of Helsingfors on a rocky peninsula that was to compete with Tallinn in Estonia on the opposing coast. However, in 1640 the centre of development had been moved to the Vronniemi peninsula to benefit from better siting for a harbour. The city accordingly developed trading in Hanseatic times from both its inner North and South Harbours separated by the area of Katajanokka.

With the 'Northern Wars' at the beginning of the 18th century between Russia and the Swedish Empire, the Russians prevailed and the Swedish retreated. Subsequently, despite development of a naval fortress on the island of Suomenlinna and with fortifications around Ullanlinna, Russia again defeated Sweden in the 'Finnish War' in 1809 during Napoleonic times and annexed Finland as a Grand Duchy. It was then developed as national capital having been moved from Turku to Helsinki.

After the Great Fire in of Turku in 1827, a new plan designed by Engel was prepared, with neoclassical buildings around Senate Square and the government palace that houses the state council, the Lutheran cathedral and the university around it. With the Presidential Palace near Aleksanterinkatu Street, the Orthodox Uspenski Cathedral (designed by Gornostayev) with its onion domes was consecrated in 1868 overlooking the harbour and Market Square.

The construction of a new rail line into the centre of the city was also a significant project, as the adopted route of the railway into a new station was for the railway to run from Hameenlinna into Helsinki with a route across Kluuvinlahti Bay. This station was opened in 1862 as a through station. As its population and traffic grew, a new larger terminal station was required, and after a competition won by Eliel Saarinen it was developed and subsequently completed in 1919 with its semi-circular style as the new Central Railway Station that dominates central Kluuvi.

Following World War I and the 'War of Independence', Finland gained its independence in 1918 and secured a new constitution. Eliel Saarinen updated the master plan for the city out from Senate Square with separation of land use and transport routes, with intervening green and forested areas.

During World War II and a very difficult period of winter war with Russia in 1940, at the end of the war, as peace was achieved, boundaries were reset and there was a movement of populations from rural areas to the larger cities. Around Helsinki the suburbs of Vantaa and Espoo grew, and as the economy diversified from forestry and paper products, the city modernised around the centre Kamppi.

Modern City – Buildings and Arteries

Helsinki Central Park stretches from the centre to the northern edge of the city. The city has main central areas with the districts of Kamppi, Kluuvi and Punavuori, North Helsinki with Toolo and Kallio, and Eastern Helsinki with Kruununhaka and Katajanokka. Using the local hard rock, Saarinen's station and clock tower near the botanical gardens were finished in dark granite in national romantic style along with the Finnish National Theatre and the National Museum of Finland.

There were also a significant number of Jugendstil-style buildings erected around the turn of the century. In 1952 the Summer Olympics was held in Helsinki with the stadium and Olympic tower built in the Toolo district.

Another building of particular interest is the Temppeliaukion Church of the Rock, which was constructed underground in the cut natural granite bedrock in a local square in 1969. As light streams through the upper windows, inside the circular rock-hewn building has a mystical presence.

Other modern buildings are Finlandia Hall (designed by Aalto), opened in 1971, which in contrast to the Tempppeliaukion Church of the Rock is a bright white undulating building. The Kiasma Museum (designed by Holl) was opened in 1998 along Mannerheimintie near Parliament House (designed by Siren).

The economy grew post-war from its fishing, timber and paper industries to design, electronics and media with major Finnish companies headquartered there, such as Nokia, Fortum and Neste. Finland joined the European Community in 1995 and groups of tall commercial towers were developed in the centre of the city at the beginning of the 21st century.

Having a comprehensive tramway system since 1900, the Helsinki Metro was opened in 1982 and operated by HKL. Commuter and inter-city rail is operated by VR with the main line to the north to Tampere and Oulu, and the coastal rail line to Turku.

There are now also three semi-circular ring roads I, II & III, that run around the city with a corresponding network of essentially radial highways leaving the city. Helsinki-Vantaa

Airport (HEL) is located in the north adjacent to Vantaa. In the port area, besides the fishing industry, it has a very active harbour for ferries travelling across the Baltic and with cruise ships and winter ice-breakers.

The long Paijanne water tunnel was bored through rock over 120 km. to provide the city with a supply of drinking water. The city is also unique in having a network of underground pipes that provide District Heating for 90% of its inhabitants. Helsinki also is the only capital city to have its own Underground Master Plan (UMP) for all realised urban development beneath 6 m. in depth.

The population of the city itself is approximately 0.6 million. The Finnish Capital Region that comprises the cities of Espoo, Vantaa and Kaunianen has a population of approximately 1.3 million. Greater Helsinki now includes the cities of Hyvinkaa, Jarvenpaa and Tuusala and others in the wider metropolitan area.

Transactional Forces and Typology

With its location in the Baltic region of Northern Europe, the city of Helsinki was to evolve in this Nordic area essentially around maritime trade. With its rugged coastline and islands squeezed between former Swedish and Russian empires, it is situated between the corresponding centres of Stockholm, Tallinn and St Petersburg. In the Gulf of Finland, from fishing and timber, fur and forestry, its long icy winters often result in frozen waters. With its north, south and western harbours, it is surrounded externally by multiple small islands whilst internally the land is populated by intervening lakes and forested areas. As a primate city, after tensions between competing neighbouring empires, it has only recently urbanised and expanded into its own larger independent metropolitanised area.

As a coastal city with rugged coastline and adjoining small islands, the centre has grown around its earlier core. However, extension has been linearly to the north and radially in the available land areas to the west and north-east around the protected lake and forest areas. This has been strengthened by new rail and metro lines to connect suburban areas and new growth poles from central Kamppi to Vantaa in the north and with new garden cities, such as Tapiola in the west. This is resulting in poly-nodal centres developing all within the broader urbanising area of outer Helsinki.

3.2.4 STOCKHOLM, SWEDEN

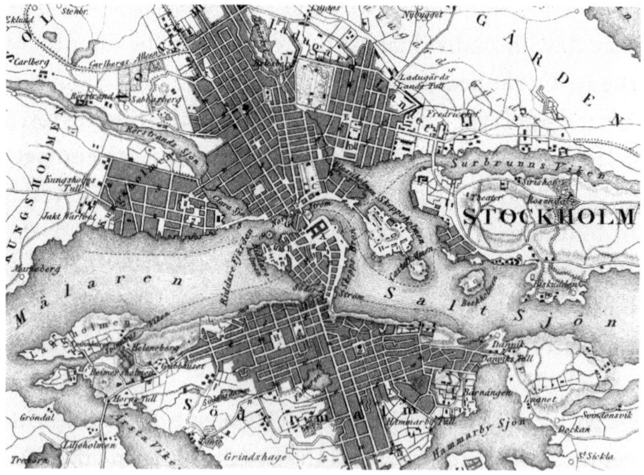

FIGURE 3.2.4 Stockholm – early configuration, 1851.
Source: iStock.

Background and Origins

Stockholm is located within the complex archipelago of the nearby Baltic Sea and the freshwater Lake Malaren and stretches over a group of islands in Riddarfjarden Bay, now interconnected by a series of roads and bridges. It has a long history within the waters around the central island known as *Gamla Stan*, the Old Town, which has been an active marine trading centre since mediaeval times.

First settlements were developed by the Vikings on the small island of *Helgeandsholmen*, and there is evidence of the first fortified castle around 1250 CE. The development of adjoining *Gamla Stan* came from commercial trading in the Baltic region as the city competed with the Hanseatic League. With its varying earlier forms of association from the Kalmar Union with its neighbours Denmark and Norway, Sweden consolidated under King Gustav Vasa, who was crowned in 1523 and inhabited Tre Kronor Castle. From that time onwards Sweden with its empire grew into a significant Scandinavian regional power.

Since its former defensive walls, the Stockholm Royal Palace, Kungliga Slottet, on *Stadsholmen* was designed by Nicodemus Tessin, around a courtyard after the castle was burnt down. The royalty in addition had their own residential country palace separately on *Drottningholm* on Lovon with its historic court Slottsteater outside Stockholm. With its churches, museums and national institutions around the main Stortorget Square, the long street Vasterlanggatan runs through the Old Town. The whole of *Gamla Stan* retains its former mediaeval street pattern and lanes.

The Great Church is situated on the smaller *Riddarholmen* and the House of Nobility. The newer Parliament building, Riksdagshuset, was built later on *Helgeandsholmen* nearby. New districts were developed around the centre at *Norrmalm, Sodermalm* and at *Kungsholmen*.

At the end of the 19th century the economy developed with industrialisation and with its university, museums and housing developments, the city's infrastructure was expanded and an early narrow-gauge electrified railway was built out to Roslagen in 1885. The Strandvagen waterfront was developed for the World Fair in 1897, and other more residential areas were developed farther out. The Summer Olympics were held in Stockholm in 1912 hosted at the Olympiastadion. Landmark buildings such as the imposing tall brick city hall (designed by Ostburg) were realised in 1923.

Modern City – Buildings and Arteries

With its varying island geography, each area has differing characteristics, with *Skeppsholmen* once a naval base and *Kungsholmen, Norrmalm, Södermalm* and *Östermalm* as distinct commercial districts.

Norrmalm is the commercial heart of Stockholm dominated by the Central Station and its bus and metro interchanges around Segels Torg. Farther north is *Vasatan*, with the circular Stockholm Public Library Stadsbiblioteket (designed by Asplund).

To its east is *Östermalm* with the Birger Jarlsgatan axis running down to Nybrohamnen and the waterside Strandvagen. Parallel streets run diagonally down towards the harbour as Karlavagen via Karlaplan and Valhallavagen.

Just to the west is *Kungsholmen* with its brick City Hall Stadshuiset (designed by Ostberg), opened in 1923 with its Golden Hall and Tower and adjoining park overlooking Riddarfjarden.

To the south of *Gamla Stan*, connected at Slussen, is *Södermalm*, with the open Medborgarplatsen central square bounded to the south by the Ringvagen.

A separate island farther to the east is the smaller *Skeppsholmen*, with Moderna Museet and also *Djurgarden*, with its botanical gardens and park at Skansen.

The city expanded through annexation and after the last world war, a general plan for Stockholm was developed by Markelius. A deliberate policy of land-banking and provision of new public housing was also pursued. In terms of land use this led to enabling the redevelopment of peripheral urban areas and a plan for new satellite towns.

The commercial area of *Norrmalm* was modernised under an urban renewal plan with five high-rise but ill-fitting Hortoget buildings (designed by Markelius) along Kungsgatan in the 1950s that completely transformed the original character of the former historic Klara area.

These were followed by the residential Gardet neighbourhood, the Grondel housing and suburban developments of new towns beyond at Vallingby and Farsta. In more recent years, several tall modern commercial developments have been realised also at Kista and Norra Tornen. With its development Kista has become the centre of a technology hub and houses Ericsson headquarters and the Victoria tower.

In terms of communications, a new metropolitan subway network T-banan to interconnect a hierarchy of area, district and neighbourhood centres was developed. This resulted in the concentration on construction of inner apartment blocks close to stations and having traffic-free neighbourhoods. However, this rather contradicted the generally higher demand for single-family houses to be built farther out. During the 1970s, further suburban residential developments were undertaken also outside the city, for example at places such as Tensta and Skarholmen, which were unsuccessful and became blighted areas.

Further commuter rail lines were developed, managed by Swedish State Railways (SJ). Due to the city's composition of individual groupings of islands, the realisation of the metro that evolved from 1950 onwards assisted in extending development to outer areas. New lines were built to connect in particular with the new satellite towns of Farsta and Vallingby, south and west respectively. There are also a number of tram lines. Stockholm Arlanda Airport (ARN) is located some 40 km. to the north of the city.

In 1966 a new Greater Stockholm County Council was created, which took over responsibility for the development and expansion of Stockholm's Tunnelbana. Due to local geology, however, several of the stations were created by excavation and by directly exposing underground bedrock. The hub station is at T-Centralen where all of the lines interconnect, and a new City Line was built beneath the centre in 2017 in order to increase urban capacity to Stockholm Central Station.

In terms of highway infrastructure, the city has a partial ring road around the city which has four sections that cross several sections of water and which is incomplete on the eastern side. In order to fund this, the Stockholm congestion charge was introduced in 2007 so as to control traffic at different hourly periods and is being used to contribute to overall funding in the transport sector.

Overall, with its varied geographical topography, mixed colourful and historic buildings and dominant city hall by the waterside, these forms create an interesting townscape and patterns around the waterfront edges of individual islands.

The population of Stockholm is approximately 1 million, with 2.4 million in the metropolitan area.

Transactional Forces and Typology

With its location within the natural archipelago in the Riddarfjarden, communications were by water as trade developed across the wider Baltic region. From early fishing, forestry and agriculture, Scandinavian peoples developed alliances in order to counter the influence of the Hanseatic League. As Sweden grew and expanded, influence of the Swedish Empire was at its peak during the mid-17th century. Moving from an agrarian economy to industrialisation during the 19th century and with its scientific achievements, modernisation has led to the evolution of several major manufacturers in Sweden, with Stockholm developing as a capital and financial centre. With its educational and cultural institutions, and media and design skills, it has become a vibrant modern Scandinavian city in an environmentally attractive area.

Stockholm is a most unusual composition as a coastal city comprising a series of interconnected islands and promontories. From its beginnings as a settlement on three small islands in Malaren Lake with its maritime trade, its infrastructure has gradually been developed. Connected by early road and bridge crossings and supported by subsequent underground tunnels, the insular characteristics of the city have been overcome as it has been knitted together in modern times. Whilst retaining outward connections by ferries, the road and rail routes define connections with new outer suburbs that have extended radially, but especially northwards, as reflected in the T-Bana network of Red, Blue and Green metro lines.

This has led to more peripheral development and housing districts, mainly linearly northwards along transit lines, with poly-nodal centres developing around intervening waterways and forested areas. As such, it retains its distinctive Scandinavian characteristics and benefits from its mixed natural features of both land and water.

3.2.5 OSLO, NORWAY

FIGURE 3.2.5 Oslo – early configuration, 1897.
Source: iStock.

Background and Origins

The city of **Oslo** was founded originally by the Vikings and established by King Hardrada of Norway in 1080 CE under the name 'Anslo' with the Akershus fortress at its heart. After different periods of conflict and long Kalmar Union with Denmark, in 1624 the mainly wooden city was destroyed by fire and a new city named 'Christiana' was rebuilt nearer the fortress. An area called Christiana Torv was developed as a quadrilateral, Kvadraturen, with a market square, buildings and old town hall, Gamle Radhus, beside the walled Akerhus Festning at the promontory.

Geographically most of Norway is a varied rocky country, having been shaped by geological forces. The capital of Oslo is located at the waters' edge at the north of the Skagerrak with its archipelago of fjords and inlets. This has led to the evolution of maritime outward-looking traders rather than more inward-looking mountain-based rural population from Lillehammer and beyond.

After Napoleon's defeat and with independence from Denmark in 1814, during the 19th century Norway's independence became increasingly under the influence of neighbouring Sweden. During this time new buildings of state were built in the city, which included the Royal Palace, Slottet, and the Parliament, Stortinget (design Langlet) and the neoclassical university. Karl Johans gate (boulevard) in the heart of the city and with adjacent Studenterlunden was built as the main thoroughfare, alongside which were new state buildings, the National Gallery and the National Theatre.

Eventually, in 1905, with final independence and a new king, the city's growth began to consolidate, and the name of the city was restored as Oslo in 1925. With its occupation by Germany during World War II, the country mobilised considerable resistance, but post-war required rebuilding. As parts of the neighbouring Aker area were incorporated into the city, this was formalised in 1948 to create a much-enlarged city. New garden suburbs were also built, such as Ulleval garden city and Torshov.

As a kingdom, the trappings of power are located in the centre of the capital around the royal palace and Slottsparken. With its parks and waterfront by the Oslofjorden, it has a unique urban setting with its wharves and marine activities. To the west of the centre also around the bay is the former island of Bygdøy, now the reclaimed peninsula of Bygdoynes with its small harbour and landmark Dronningen building, which complements the city.

Modern City – Buildings and Arteries

A significant addition to the urban landscape was the realisation of the new city hall, Radhuset (designed by Arneberg and Poulsson), an imposing brick building with its twin dark brown towers planned pre-war but not opened until 1950, where the Nobel Peace Prize is awarded annually.

Beyond Aker Brygge the site of the old shipyard has been converted into an active commercial and retail area. At Bjorvika, the new concrete and glass inclined Opera House (designed by Snohetta) is sited dramatically, sloping into the sea, and at Tjuvholmen the Astrup Fearnley Modern Art Museum (designed by Piano) is sited with the new Munch Museum (designed by Abalos and Herreros). With its maritime outlook and relatively small population, the economy has been strong since Norway's exploitation of the offshore oil and gas discovered in the North Sea. The city retains its active ferry business at the Port of Oslo and has an active marine industry.

As a compact Scandinavian city it has good public transport, with acomprehensive metro and tramway lines, with a connecting ring line and commuter rail services from the main Central Railway Station. The Drammen Rail line crosses the city centre underground in the Oslo Tunnel, enlarged in 2010, running between Skøyen and Oslo Station and the high-speed Gordermoen Rail line runs to Lillestrom via the Romerike Tunnel (14.5 km.) opened in 1999 and on out to Oslo Airport (OSL).

As a result of political agreement, a series of 'Oslo Packages' was agreed starting in 1990 when an urban toll ring was created around Oslo in order initially to raise funds for the construction of new road tunnels. This led to the realisation of a number of underground tunnels, notably in the centre where the Opera Tunnel runs from the western side of the city under Aker Brygge past the old fortress Festning Tunnel under Bjorvika Bay in immersed tunnel to connect with the ring road running up the eastern side of the city with tolls collected by Fjellingen A/S. With the northern sections of Granfoss, Smestadt, Tasen and Loren tunnels realised, these effectively connect the urban ring. Subsequent packages were to focus on improvements in public transport, rail and metro, and operations.

A significant suburban area has been developed around the Grorud Valley that runs north-eastwards in an ancient geomorphically formed valley. Under a protected continuous park plan, a number of satellite towns have also been developed that include Ammerud and Tveita, bounded by the Marka Forest and which are attractively integrated into the regional landscape. The population of the city is approximately 0.7 million and double that in the metropolitan area.

Transactional Forces and Typology

With its rugged coastline and deep fjords, multiple islands and Norse ancestry, geographically Oslo is protected at the head of Oslofjorden. Besides herring fishing and the timber industry, the seas and forests have shaped the terrain around the city. With its long coastline and mineral-rich North Sea, Norway is endowed with abundant resources for its small population. Despite its various past conflicts and coalitions with its neighbouring countries, it has benefitted since industrialisation from having its own natural resources and with skilled resolve, such as harnessing hydro-power also to make it a fine city with its own cultural traditions of writing, art and music.

As a coastal city at the head of the fjord, urbanised Oslo from its inverted 'U'-shaped waterside is changing with radial axes and sectors extending outwards as the terrain and forest lands dictate. With a central axis dictated by the line of Karl Johans gate, the fort and headland look out to the south-west with intervening islands and the southern promontory of the Nesodden peninsula. The central green Akerselva River corridor runs northwards, flowing into the Bjorvika Fjord, and acts as an effective boundary between the east and west sides of the city.

The modern T-bana metro provides a ring service in and around the city, and the ferries still serve the coastal waterways. Beyond its central urban core, the periphery is being populated by locally connected centres such as Hellerud, Grorud and Ammerud and beyond to Lillestrom to the east and Sandvika to the west. With its complex topography, future development of the city and settlements is being shaped by new infrastructure, that is, underground road and rail tunnels across the bay and similarly radiating out along corridors to the north-east and west.

3.2.6 COPENHAGEN, DENMARK

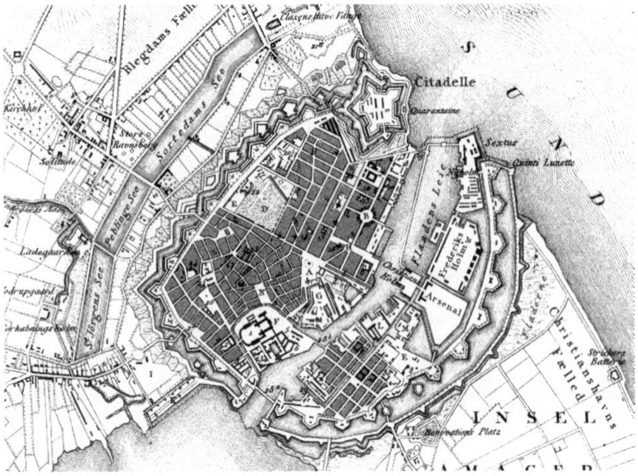

FIGURE 3.2.6 Copenhagen – early configuration, 1851.
Source: iStock.

Background and Origins

Copenhagen (København) is located on the Danish island of Zealand alongside the island of Amager by the Kattegat that connects with the Baltic Sea. Known in old Danish as Køpmannæhafn (merchant's harbour), evidence of the old city dates back to around 1050 CE with its early castle at Strandholmen.

With its emphasis on the sea and ships, the first habitation was around Gammel Strand, the old shore, and fishing village on the little island of *Slotsholmen*. With its defensive ramparts and walls, a fortress was developed as Copenhagen Castle (Kastellet). With its fishing industry, trade developed with other Baltic Hanseatic cities for several centuries. In 1426, the Danish king introduced 'the Sound Tolls' on ships passing through the Kattegat between Denmark and Sweden.

In 1479, as one of the earliest European universities, the University of Copenhagen was founded and during the 16th century after the Reformation, this led to the adoption of Lutheranism. In the 17th century, under Christian IV, the city was consolidated with new defences, such as the citadel fortress and arsenal, and *Christianshavn Vold* was built with its bastions and moat Stadsgraven. A shipyard was developed in *Nyholm*, and following land reclamation, groups of small islands around Holmen were consolidated to form Frederiksholm.

With two significant fires during the 18th century, one of the finest developments in the centre of the city is *Frederiksstaden*, with Frederik's Church and Amalienborg Square, which was developed as a group of elegant classical buildings with four angled royal palaces around an octagonal place and central equestrian statue.

After the Napoleonic wars, the city developed further in the 19th century and was expanded with new districts of *Østerbro*, *Nørrebro* and *Vesterbro*. The eastern area of *Amagerbro* was eventually incorporated into the city and the Freeport of Copenhagen was created in the northern part of the port in 1894.

Modern City – Buildings and Arteries

Post-war, a new 'finger plan' for the development of the city was prepared in 1947 with proposed development along the 'fingers' as five commuter rail transport axes with green areas in between. These green protected wedges were to guide development. More recently, in 1995, a plan for the development of the area of *Amager* with the new *Ørestad City* was proposed and has been substantially developed, due to its proximity to the Øresund Crossing to Malmö in Sweden supported by the realisation of a new extended line of the Copenhagen metro.

With its royal buildings, museums and newer town hall and the famous Tivoli Gardens, it is an attractive city to stroll around in, including the pedestrianised Stroget series of streets, one of the longest shopping streets in Europe, that encompass Gammeltorv and Nytorv squares. With the Inner Harbour and Frederikholms Canal, the separate quarter of Christianshavn is to the east, with its old homes and house boats, squares and gardens.

The main districts of the city of Copenhagen comprise:

Slotsholmen and Christiansburg Slot – from its fortified castle on the central island, destroyed by the Hanseatic League, this became the home of Christiansborg Slot, which houses the Danish Parliament and rebuilt palace, church and theatre and the Royal Danish Library (designed by SHL).

Frederiksstaden and Amalienberg – Frederiksstaden just north of Nyhavn Harbour was developed by Frederik V as a new district in the 1750s. Here, the unique Amalienborg Square was built with four individual Rococo palaces configured around a diagonal square, notably Christian VII, VIII and IX's Moltke, Levetzau and Schack palaces and Frederik VIII's Brockdorff Palace on the axis of Frederiksgade. On the waterfront directly opposite here, the dramatic modern Opera House (designed by Henning Larsen) was opened in 2004.

Christianhavn and Holmen – to the east of the Inner City, during the 17th century Christian IV built a fortification ring and canal around which the district of Christianshavn was created, and barracks at Christiania. Just to the north of this, further artificial islands were created at Holmen as a naval base in 1690 with its old dock (Gammel Dok) and subsequent warehouses.

Kongens Nytorv and Nyhavn – to the north of Slotsholmen, Kongens Nytorv was built by Frederik III with a new Nyhavn Canal to connect with Inderhavnen. Here in New Square, Charlottenburg Slot is sited on the eastern corner next to the Royal Theatre.

Inner City and Strøget – the Inner City (Indre By) contains a number of different districts, from Kongens Nytorv and the Latin Quarter linked by the main pedestrianised Strøget that connects the squares of Amagertorv, Nytorv and Gammel Torv through to the Radhus (town hall) and Tivoli Gardens.

Kongens Have and Rosenborg Slot – Rosenborg Slot stands in the extensive gardens of Kongens Have Park, built as a summer palace by Christian IV, north of Gothersgade Street.

Fredriksberg Have – to the west of the Inner City is Fredriksberg Slot, slightly raised in the extensive gardens of Frederiksberg Have, as an outer district separate municipality.

Østebro, Nørrebro and Vestebro – these areas lie just to the east, north and west on the opposite bank of Oster, Norre and Vester Sogade, around Osterbrogade, Sankt Hans Torv and Vesterbro Torv respectively, and comprise denser residential areas that are under regeneration.

Amagerbro – opposite the island of Amager is being developed as a separate centre north of the areas of Tarnby and Kastrup with its striking sloping Amager Bakke waste to energy power plant (designed by BIG) which incorporates a ski-run.

The modern city is now very active economically, with its financial institutions, major marine and shipping lines, design companies and a focus on the life sciences via a cluster of buildings that spans the Øresund region housing pharmaceutical and biotech companies. Additional land areas are being reclaimed, as for example Refshaleøen north of Margreteholm.

The urban public transport system is excellent and well-integrated with other transport modes where provision for pedestrians and cyclists is very good in the centre with extensive dedicated paths. The city metro is compact and comprehensive and was recently extended with the construction of the City Ringen circle line M3 and extension to Copenhagen Airport (CPH) at Kastrup and the new city of Ørestad.

The outer perimeter ring includes the O2 and O4, which pass the east and west of the city, with the E20 motorway connecting across Amager to the coast where from Zealand to the east the recently built Øresund Crossing of a major bridge and tunnel provides a fixed connection between Malmö in Sweden and Copenhagen, opened in 2000. This has had a significant impact on patterns of development.

Travelling west out of Copenhagen via the island of Funen, this route also has been connected by the realisation of another major project, the Great and Little Belt Crossings that ultimately connect Jutland with Zealand via Funen and Sprogø. Opened in 1998, this has completed the interlinkage of the whole country by road and rail now from Sweden to the western part of Denmark.

The population of Copenhagen is approximately 0.7 million and double in the metropolitan area. The Øresund region, which if combined to include the urbanised part of Malmö in Sweden, has a population of approximately 4 million in this greater urbanised metropolitan area.

Transactional Forces and Typology

Situated on the eastern coast of Jutland, Copenhagen is strategically located on the Øresund in a maritime context. As an active port in the Baltics during the period of the prevailing Hanseatic League, it played a defensive role variously in alliance with its Scandinavian neighbours Sweden and Norway. In modern times, with the realisation of both the Great and Little Belt Crossings and the Øresund Crossing, these have completely transformed land movements east-west by rail and road across the country. This axis, to be connected eventually with the southern Fehmarn Belt immersed tunnel crossing via Lolland, has physically integrated the larger centres across Zealand, and the adjoining Funen Island to the Jutland Peninsula.

With its location on the eastern coast facing the sea, the city of Copenhagen within almost feels like a peninsula between the two water courses of Sorterdams Sø and Inderhavnen with the castle (Kastellet) at its tip. The eastern fortified area of Christianshavn opposite the core of Slotsholmen and regal Frederiksstaden are at the centre, with Frederiksberg developed to the west. A flat waterside area, it is a distinctive, distinguished, older district surrounded by the denser adjoining Østebro, Nørrebro and Vesterbro districts.

Further substantial extension has developed by incorporation of the former island Amager area, with the realisation of the new city of Ørestad. This new south-eastern area is able to act as a relief centre and new pole for further city expansion. In addition, with the Øresund Crossing to Malmö, this has reinforced the southern and eastern development axis. As a coastal city with its historic core, besides

Amager, the city is also able to develop along its prescribed 'finger corridors' to the west. One of these is towards Roskilde and another benefits from connection with the Great Belt Crossing on the western axis across Denmark. Hence, these modern fixed-link infrastructure crossings have significantly strengthened the pattern of development of the outer perimeters of the city and accelerated communications to and with the city core. The following figures illustrate the modern configuration of the port cities as described and the evolution of their modern population..

FIGURE 3.2.7 Antwerp – modern configuration.

FIGURE 3.2.8 Hamburg – modern configuration.

FIGURE 3.2.9 Helsinki – modern configuration.

FIGURE 3.2.10 Stockholm – modern configuration.

FIGURE 3.2.11 Oslo – modern configuration.

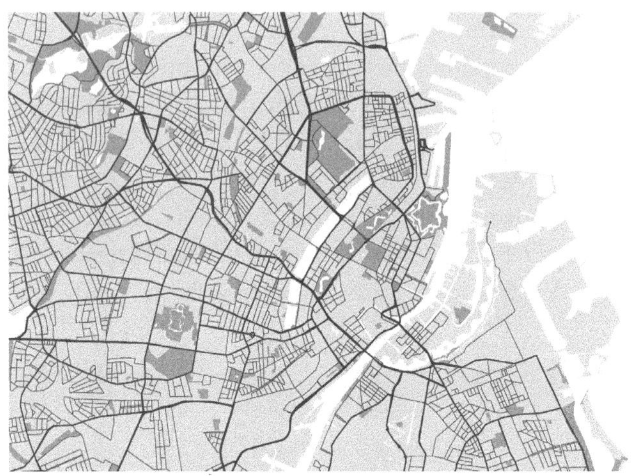

FIGURE 3.2.12 Copenhagen – modern configuration.

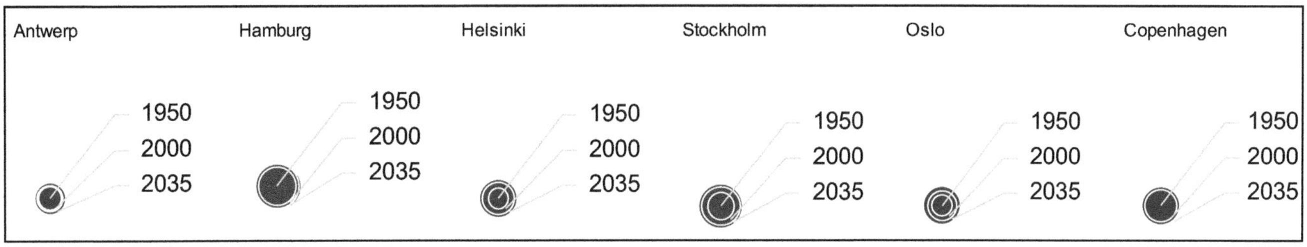

FIGURE 3.2.13 Lowland, Hanseatic, Baltic and Scandinavian port city populations.

Data source: Global Database of Metropolises 2020 – UN-Habitat (2020), UN DESA Population Division.

Further Reading

ANTWERP

de Goey, Ferry, ed. – *Comparative Port History of Rotterdam and Antwerp, 1880–2000*, Aksant Academic Publication, 2004.

Gaunt, William – *Flemish Cities*, Harper Collins, 1969.

Pye, Michael – *Antwerp, the Glory Years*, Penguin, 2022.

HAMBURG

Andreini, Laura – *Hamburg, on the Road, Architectural Guide*, Forma Edizioni, 2018.

Jefferies, Michael – *A Cultural and Literary History*, Signal Books, 2011.

HELSINKI

Amieel, Lieven – *The Narrative Turn in Urban Planning, Plotting the Helsinki Waterfront*, Routledge, 2020.

Drain, Kevin – *Gentrification in Helsinki*, Routledge, 2024.

STOCKHOLM

Griffiths, Tony – *Stockholm, a Cultural History*, Oxford University Press, 2009.

Magnusson, Lars – *An Economic History of Sweden*, Routledge, 2007.

OSLO

Hellerud, Synnove & Messel, Jan – *Oslo, a Thousand Year History*, Aschehoug, 2000.

Roe, Per G & Luccarelli, Mark, eds. – *Green Oslo, Visions, Planning and Discourse*, Routledge, 2016.

COPENHAGEN

Datz, Christian, ed. – *Copenhagen, Architecture and Design*, teNeus Publishing, 2005.

Stenogaard, Pernille – *Copenhagen, People and Plans*, Gyldendal, 2015.

3.3 Maritime Cities

Africa, Gulf of Guinea and the Cape – Americas – India

Freetown, Accra, Cape Town, Boston, New Orleans and Madras

3.3.0 INTRODUCTION – MARITIME CITIES

Straddling the equator and both tropics, the continent of Africa has the whole range of climatic and geophysical zones. As the home of ancient civilisations, Africa has been the source of a diversity of cultures and ethnicities over the ages. From the earliest settlements around Lake Tanganyika and Nilotic tribes up to the Mediterranean coast, with the West African empires of Ghana, Hausa and Benin, it has been the source of diverse groups of varying origins. With their own traditions, practices and languages, this has resulted in wide diversity across the continent.

With the colonial exploitation and the 'scramble for Africa' for both its human and natural resources during the 18th and 19th centuries, this led to divisions of land and created artificial boundaries that separated traditional tribal groups and individual groupings, especially along the coasts and around the Gulf of Guinea. With the so-named 'Gold Coast', 'Ivory Coast', 'Spice Coast' and 'Slave Coast', it became the source of exploitation of several foreign empires.

Since the main continental landmasses are separated by miles of oceans with deep and inhospitable waters, with advances in navigation and oceangoing shipbuilding capacity, European explorers were able to open up routes and locations that gradually became established routes for maritime trade and travellers.

Once the 'discovery' of lands both east and west of Europe were established, one of the greatest exploitations and transformations was the triangular trade across the Atlantic, with the transfer of slaves from Africa to the Caribbean, sugar and cotton and tobacco to Europe, and textiles and manufactured goods to Africa. This maritime trade was a strong force in the settlement and development of coastal communities and cities in the Americas, both north and south.

On the west coast of Africa, the coastal settlements of Dakar and Abidjan, Dahomey and Libreville were developed under French colonial occupation; Accra and Lagos under British occupation; Guinea Bissau, Porto-Novo, Sao Tomé and Luanda under Portuguese; Kinshasa under Dutch; Brazzaville under Belgian; Douala under German and Monrovia under American occupation.

It was not until after both world wars that the claims of nations to full independence were eventually heard, mainly during the 1960s after individual struggles and conflicts, that these countries gained their own independence as sovereign states. With the formation of the United Nations in 1945, this has since led to the formalisation of over 50 independent states in Africa, each having their own established or new capital city.

Besides selected African coastal cities, this chapter addresses certain principal continental ports between which maritime trade developed in America and India.

3.3.1 Freetown, Sierra Leone

Background and Origins

Freetown is located on the western coast of Africa on the Atlantic Ocean on the Sierra Leone estuary first named Romarong by the Temne, with the Lion Mountains beyond. Originally inhabited by the Krio-speaking Sierra Leone Creole people, in the tropical zone with rainforests, there are several other tribes, including Mende, Fula and Shebro. It was first settled by the Portuguese, who were seeking materials and labour and who built a basic fort there.

Around 1670 Bunce Island, named after Capt. John Bence, on the estuary was occupied by the Royal Africa Company, who built a defensive fort and yard as a base for the extensive triangular slave trade with North America, especially the southern plantations around Charleston and Savannah. From that time, it is estimated that approximately 50,000 persons had been shipped as slaves to the 'New World' of the Americas, men, women and children in bondage sold in return for guns, swords, rum and wine sponsored by state-backed companies that developed as a lucrative operation.

In 1787, a Committee for the Relief of the Black Poor, supported by abolitionist Granville Sharp, had been

DOI: 10.4324/9781003353348-14

established in London with the goal of seeking a home for freed slaves back in this country. He arranged for the transport of certain 'poor black' persons or indentured servants from London to be resettled in Sierra Leone by the St George's Bay Company, in an area known as Cline Town. These people in reality were persons of African descent who had sided with Britain during the American War of Independence, some of whom had been evacuated via New York and Canada and who may have escaped from southern plantations.

Hence the circle of their origin from Africa to the American South and then via Nova Scotia and on to Britain, was completed back to Africa as re-settlers into this area of Sierra Leone that became renamed Granville Town, in the 'Province of Freedom'. Whilst these new arrivals were resented by the resident Africans, their resettlement was fought over and homes and sites destroyed. New settlements were launched and established a few years later, when another wave of settlers from Nova Scotia founded a new colony at 'Freetown' in 1792 under the Sierra Leone Company on the same site of the former settlement. Under Clarkson as governor, a new town was set out on a grid pattern around the main street, Water Street.

Although these resettled Nova Scotians resented not being treated as 'freemen', King's Yard was used as a site for landing liberated Africans and the large Cotton Tree, a kapok tree was used as a ceremonial site of thanksgiving upon arrival.

At around the same time, a number of 'Maroons', or runaways, from plantations in Jamaica had similarly found themselves deported and resettled by the British in Canada. They in turn were transported from there to Sierra Leone and found themselves in Freetown but skirmished with the former arrivals from Nova Scotia. In consequence, they secured their own area of land, 'Maroon Town', just to the west of the original 'Settler Town'.

In 1808, the British had full control of the peninsula of Freetown and formed the Crown Colony and the Protectorate of Sierra Leone. Accordingly, the freed slaves established themselves in the area, and consequently their descendants formed a large group of Creoles, or Krios, resident in the country.

As the number of 'recaptives' increased, due to space limitations they were resettled by missionaries developing satellite villages around the peninsula. This led to the establishment of new towns such as Wilberforce and Kissytown as larger numbers of 'liberated Africans' were resettled. They in turn were gradually able to improve their own prosperity.

Freetown served as the capital of British West Africa for the next 80 years. However, malaria was always present and disputes arose, such as the levying of a five-shilling 'Hut Tax' in 1898 that caused a revolt with inhabitants, and there was no real interest in developing the interior of the country. A residential area also was built at Hill Station, essentially for Europeans but which increased separation from the Creoles.

Modern City – Buildings and Arteries

During the world wars, Freetown served as a significant port due to its natural harbour and acted as a base for naval operations across the Atlantic during World War I. During World War II, an air base was established for the Allies at Lungi on the other side of the Bay at Tagrin Point. This is now the location of Freetown International Airport (FNA).

Sierra Leone attained independence, becoming a dominion in 1961, and a republic in 1971. The main economy based on farming, forestry and fishing with food processing, tobacco and textiles, has been extended with the extraction of mineral resources such as bauxite and iron ore and with rarer diamonds.

Freetown is divided essentially into three regions, the East End, West End and Central Area. The East End is the most populated, with a large number of informal settlements east of Cline Town. The Central Area is the main area, with governmental buildings on Tower Hill and Victoria Park and the Queen Elizabeth II quay on Fourah Point. Kroo Town and PZ (after Paterson Zochonis) are in the Central Area.

The West End is more affluent beyond Kingtom and Murray Town to the peninsula of Aberdeen, connected by Aberdeen Bridge. With Lumley Beach, long Cockerill Bay leads south to Goderich. The higher Sugar Loaf Hill and Hill Station overlooks the city.

The Sierra Leone Government Railway, which had been built in 1897 with just one line of narrow gauge of two and a half feet (2'6"), eventually reached Makeni inland with Hunslet steam locos. With relatively steep gradients, it was not of great strategic value and was eventually closed.

The main harbour is the older Government Wharf to the north of the city. The main airport is on the opposite side of the bay at Lungi and accessed by the Kissy Ferry across the very wide estuary, with the main road access to the city via Highway 1 and Waterloo to the south of the city. With its busy traffic and tuk-tuks (kezes), 'ocadas and poda-poda' mini-buses, it is a very congested city. The population of the city is approximately 1 million, and 1.5 million in the urbanised area.

The country suffered greatly as a result of the coup in 1968 and the long civil war during the 1990s, with considerable destruction. With rival factions, considerable atrocities were undertaken during the prolonged civil war across the country that required UN intervention before final cessation of hostilities in 2001.

With another upheaval due to the Ebola outbreak in 2014, and in 2017 with the collapse of Sugar Loaf Mountain slopes and fatal mudslides, the city has had a challenging recent history.

Transactional Forces and Typology

Benefitting from its protected natural harbour, the first source of trade from Freetown was human resources, as the slave trade for labour exploited their sourcing and transportation from Africa to the Americas. Products such as palm oil, rice and groundnuts were found to be attractive for export also, as well as timber, as a railway was built farther inland (now closed) to support this. Other commodities such as iron ore and bauxite were discovered subsequently and to be of more value such that extractive minerals also became a source of exports.

Whilst the country was ruled traditionally by local chieftains, the city of Freetown has been significantly influenced by the resettled Creole population. With their origins as freed slaves and returning to settle in Africa, having particular motivation they acted as 'recaptive' administrators in promoting the spread of education across the country.

From its setting on the southern side of a deep natural harbour on the edge of the Atlantic Ocean, the city of Freetown benefitted from a series of bays around which a port could be developed outside the natural terrain of mangrove swamps. At the entrance to the estuary of the River Rokel, it is surrounded by a rocky mountain range that includes Sugarloaf Mountain and with the western spit jutting out into the bay at Cape Sierra Leone, Aberdeen is a hilly promontory with forests beyond.

All of this was developed as a downtown area on the southern land area of the bay with the main airport on the northern opposing side beyond Tagrin Point at Lungi. This natural inlet defines the harbour and the eastern river estuary, with its surrounding mangrove swamps. As a coastal city, it sits on a peninsula with a natural park and forest reserve to the south that results in development mainly sprawling in the limited available area on the eastern side towards Wellington. Accordingly, the city's characteristics are defined more as a peninsula coastal city having linear development eastwards rather than as a central port city developing internally around a natural bay.

3.3.2 ACCRA, GHANA

Background and Origins

Accra is located on the coast of the Gulf of Guinea in the dry Dahomey Gap where the forest separates into Upper and Lower Guinea forests. It is the capital of Ghana, just west of the River Volta Delta, almost at the centre of the world where the equator meets the meridian. The very earliest known settlement was of the Akan people around the 8th century CE.

Later, it became known as Bono State in the 11th century CE. Other adjoining states also developed around the peoples of the Ashanti, Akwamu and Dagombas tribes. The great ancient Empire of Ghana was actually situated to the north-west of the modern country of Ghana and at that time controlled the trade routes to and from the Sahel.

The Ashanti Kingdom developed inland around the centre of Kumasi. Early trade was north-south from across the Sahara with salt and cloth in exchange for gold, ivory and nuts. In the 15th century the Portuguese arrived and, attracted by potential gold deposits, established settlements at Elmina Castle and Cape Coast Castle. They were followed by the Dutch, who built Ussher Fort, which was used for the slave trade by the Swedes who built Christianborg, and other settlers each staking their claim, first Portuguese, then Danish and German with Brandenburger Gold Coast.

After these individual settlements, the British, who had built Fort James, started to acquire individual interests and sought to split coastal areas geographically to share with the Dutch. After war with the Ashantis, they took over and consolidated the British 'Gold Coast' in making Accra the capital in 1874. The scramble for Africa that was ongoing culminated in artificial boundaries of new states being created at the Berlin Conference of 1885.

As governor of the Gold Coast, Frederick Guggisberg in 1919 supervised the amalgamation of the districts of Osu, La and Jamestown and the creation of a deep-water harbour at Takoradi. After World War I, as Germany lost Togoland, which was incorporated into the east of the Gold Coast, it was renamed the Volta Region. After the revision of other regions, the reach of the city then stretched beyond Korle Lagoon in the west and Fort Ussher to the east and incorporated Christiansborg, Victoriaborg and Jamestown and the new Adrabaka to the north.

Areas were racially segregated, and when Korle Lagoon was bridged, this enabled further developments to be made to the west. Infrastructure was improved and the main railway was built from Accra to Kumasi by 1923 to connect with the main internal cocoa-growing region. The formal ridges, ministries and cantonment areas were developed in the city with many left-over spaces used for informal settlements by rural migrants.

Modern City – Buildings and Arteries

After the end of World War II, a new master plan for the city was prepared by designer Maxwell Fry around a central business district (CBD) and public squares with a grid layout and with a new parliament complex, but this was modified by Trevallion and never realised.

With the independence of Ghana in 1957, the new leader, Nkrumah, changed the emphasis of development of the city to focus on Independence (Black Star) Square, with the State House and new Organisation of African Unity (OAU) building and high-density development around the CBD. On becoming president of the new republic in 1960, Nkrumah had actively promoted the concept of Pan-Africanism and sought to position Ghana within the non-aligned movement.

The city that now forms part of the Accra Metropolitan Area with 12 districts in the Greater Accra region includes the historic centre and the CBD and subdivided Ablekuma, Ayawaso and Okaikoi municipal districts. Farther out along the coast more modern developments are focused around the forts of Jamestown, Ussherstown and Osu Castle. The city is now bounded by a ring road and encompasses the old districts and central area with the closest areas to the east and west being mainly residential.

With its earlier emphasis on the primary sector of agriculture, fishing and textiles, the city is now more focused on the tertiary sector and services especially the financial sector. During the 1960s, there was rapid industrialisation and increased migration to the capital. In addition to the University of Ghana, the educational sector created a number of new public and private universities and research institutes.

Despite political setbacks in the 1980s, after a period of structural adjustment, whilst being essentially an agricultural country, the economy has been modernised in the 21st century with focus on the exploitation of minerals and the development of oil and gas fields both on- and off-shore, although those benefits have not yet been fully realised. The large Akosambo hydro-electric dam was constructed where the Volta rivers meet and a deep-water harbour was built at Tema, which has developed as a major seaport with Tema New Town to the east.

The urbanised area of Accra has grown rapidly since the 1980s, in particular as some tribal lands that were administered by village elders on the periphery were sold, enabling individuals to benefit from developing these sites. In other prosperous areas, green enclaves are being created and gated communities built, for example at La in the north-east of the Accra Metropolitan Area and at Airport Hills.

There are many areas of low-income population and a large informal sector with low-income areas, which include Osu, La, Adedenkpo and Chorkor. These large areas of informal settlements are where rural migrants continue to seek work. There are huge areas of waste also, for example at Agbogbloshie, which has acted as a major garbage dump for international waste.

In the transport sector, urban public transport is mainly bus and minibuses, and ever-present 'tro-tros' and taxis. The M1 forms part of the Trans West African coastal highway which runs along the lowlands connecting the Ivory Coast in the west via Accra and with Lomé and Togo in the east.

The population of the city is approximately 2.5 million, with a high percentage of young people and double in the wider metropolitanised area. More recently under the Chinese Belt and Road Initiative, developments have focused on Cape Coast to the west and Kotokuraba Market.

A new sustainable HOPE City (Home, Office, People, Environment) is planned for outside Accra, with a cluster of six circular towers (designed by Paola Brescia OBR).

Transactional Forces and Typology

With its coastal location on the equator at the virtual centre of the world, Ghana is rich in natural sources beyond agriculture and cocoa, to include minerals and recently discovered oil. But it was the early discovery of gold that was to secure its attraction as the 'Gold Coast' as overseas prospectors who built coastal forts settled in Accra before venturing inland. Despite the dominant Ashanti, it became a British colony that sought to control the trade of key commodities of cocoa, timber and gold for export. Railways were built and infrastructure developed as a deep-water port in order to exploit this. As the country's administrative capital, Accra then became the prime regional centre of development.

In generally low-lying land on the coastal strip between the minor Densu River and the major Volta River to the east, Accra has grown inland on the axes of Kwame Nkrumah and Independence Avenues from Jamestown Lighthouse, as different districts have become populated. It has grown organically with both formal and informal sectors and reflects the form of radial development with localised growth poles as, for example, the newly planned sustainable urban Hope District. With its increasing growth of population and rural migration, this increased density is putting severe pressure on accommodation and the provision of public services.

3.3.3 CAPE TOWN, WESTERN CAPE, SOUTH AFRICA

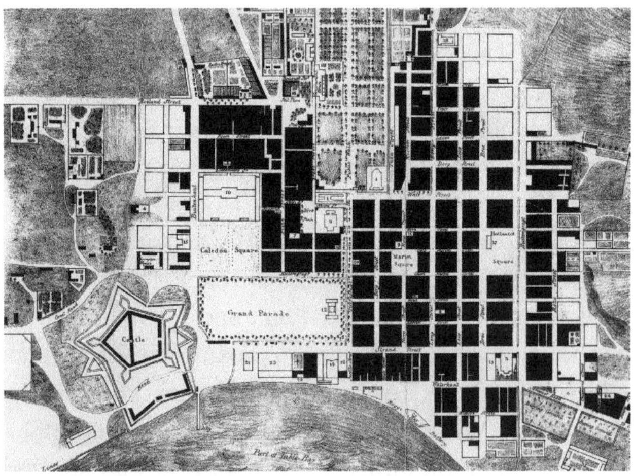

FIGURE 3.3.1 Cape Town – early configuration, 1827 (Thompson).

Source: Maphouse.

Background and Origins

Cape Town is located at the southern point of the continent of Africa, looking out westwards across the Atlantic Ocean. Overseen by Table Mountain and with vine-covered slopes, it is a most distinctive and attractive setting in southern Africa. Continuing down south along the rugged coast is the Cape of Good Hope, Cabo da Boa Esperanca, as named by Portuguese King Joao II after mariner Bartolomeu Dias' voyage in 1488, not quite the southernmost point of the continent but the place where the Atlantic and Pacific Oceans merge. It was considered as probably 'the fairest Cape in the whole circumference of the earth' by explorer Francis Drake.

Despite the indigenous Khoekhoe population, known for their 'click' language, in the early 1600s the Dutch settled and established the Verenigde Oostindische Compagnie (VOC) trading company. In 1674 the Dutch manned a bastion and fortress with a garrison there and the core of Cape Town was laid out. The importance of the Cape was recognised in strategic terms, as a revictualling point for ships to and from Europe and beyond, between India and Indonesia, via Madagascar and Ceylon.

With ongoing battles with the Dutch and French for territory, a British settlement was established in 1806 at Cape Castle. Other battles also occurred inland, for example between local Xhosa and Dutch trekkers (voortrekkers) that had headed east to Natal and the Transvaal in search of gold and diamonds.

The Municipality of Cape Town was established in 1840 as it started to be anglicised, and infrastructure was developed and new buildings established. An extensive breakwater was built to protect the harbour, and port facilities were improved. The land around Cape Flats was settled and developed. With the opening of Albert Dock, railways were introduced during the 1870s, with the first line from the Cape to Wellington. Subsequently, a railway line was built between Cape Town and Kimberley to connect with the diamond fields where prospecting was being undertaken. This was extended to Vryburg and beyond as the beginning of an African trunk line to Mafeking and on to Bulawayo, which was to constitute the southern part of Rhodes' unfulfilled aspiration of a Trans-African railway to connect Cape Town with Cairo.

Before the turn of the century, the population of Cape Town doubled with the arrival of prospectors, and the economy began to be transformed. The Cape Colony gradually grew into an exceptional mix of populations of different origins, besides African, to include Dutch, Malay, Malgache and Indian. In addition, with Dutch traders who had brought in enslaved people from Madagascar and Indonesia, this led to the development of a new ethnic group known as 'Cape Coloureds' outside the indigenous Zulu and Xhosa population.

The area was next affected by a series of Anglo-Boer wars, the first in 1880–1 and the second in 1900–02. This resulted in the creation of the Union of South Africa that was established in 1910, which unified Cape Colony with Natal, the Transvaal and Orange River, which was to be the forerunner of the eventual republic.

A significant programme of harbour improvements was undertaken with reclamation of a large area of foreshore land with Duncan Dock in Table Bay being realised that was of particular significance during World War II.

Under the original 'Native Location Act', a system of 'Pass Laws' had been introduced, which hindered the movement of local people and hence their place of habitation. With its continuation, it was after the world wars that racial divisions became further exacerbated, with introduction of the Group Areas Act of 1948, which assigned racial groups to different residential and business areas, a system of urban 'apartheid' (or separation in Afrikaans). The result of this was the forced removal of non-whites to denser, unfavoured areas and which led to the need for workers to commute much longer distances between their places of residence and work.

Whilst the country became an independent republic in 1961, it was not until after a long period of struggle that in 1992 the system of apartheid was finally ended, as eventually Nelson Mandela became president and a new constitution was formalised.

Modern City – Buildings and Arteries

Table Mountain is an extraordinary flat-topped sandstone plateau that stands above the city whose summit and Lion's Head can be reached up Signal Hill Road. It is a fine vantage point over the surroundings, with birds and wildlife. The coast and beaches extend outwards along with Robben Island, the offshore island where Mandela was held in confinement in a place of exile before his eventual release in 1990.

The Dutch had transformed Cape Settlement into Kaapstaad, which now known as Cape Town revolves around the Castle of Good Hope and the harbour, which is dominated over by Table Mountain. With its five main streets, Greenmarket cobbled square, Heerengracht canal, the reclaimed foreshore and waterfront, the older institutional buildings and city hall are located along Government Avenue. The national Parliament is situated here, and Cape Town acts as the legislative capital of the Republic as one of South Africa's three capital cities.

The commercial centre is downtown, and there are busy markets in each neighbourhood. With the white-fronted houses, verandas and tree-lined streets, colourful homes of the Cape Malay community are to be found in the Bo-Kaap district.

Passing west along Beach Road past Green Point and Sea Point from the city, there are rocky outcrops along the Atlantic seaboard down to Bantry Bay and Clifton beaches beyond. This route goes down the Cape Peninsula to the Cape itself. On the eastern side is Simons Town on the shore of False Bay with its naval base and historic mile, and Boulders Beach with unusual penguin colony.

Farther inland from Cape Town is the Boland (Uplands) with its lush winelands and which had been established at Stellenbosch and Cape Dutch buildings. With estates and wineries, it is a very productive area with some of the most iconic South African wines. With a unique geographic setting and warm climate, both Cape Town and its surroundings are able to benefit from being one of the most productive parts of the country. Cape Town has become the key economic hub of Western Cape Province, with modern business and financial services having overtaken the former agricultural, cattle, fruit and fishing sectors.

In terms of transport infrastructure, the main national roads N1, N2 and N7 originate from Cape Town in north-easterly and south-easterly directions and towards Namibia respectively. Cape Town railway station is in the centre of the city from where the luxury Blue Train connects with Pretoria over a 1,000 km. scenic route. Cape Town International Airport (CPT) is located to the east of the city at Bellville.

The city has no metro, but a series of dedicated busways run along the west coastline of the city, including the IRT, MyCITI service and numerous mini-bus services. The population of the city is approximately 0.5 million, but approximately 4.7 million in the wider metropolitan area.

Despite its chequered history and due to the natural resources from its geographic and strategic location, international mercantile forces have shaped its development and destiny. Its mixed resident population can now play a role within the reconstituted 'rainbow' country since elections under universal suffrage began in 1994.

However, whilst having a picturesque natural setting, the city has a fragmented urban form as a result of its earliest colonial origins, subsequent separate townships, apartheid planning and segregation, and modernist design that has created a fractured society and socio-spatial places.

One example of this is 'District 6' when, during the 1960s, an area of housing was completely demolished, the residents were forcibly relocated to Cape Flats. There are many such apartheid-type planned townships that remain, such as Kyetisha, a dense, highly populated, self-built neighbourhood that faces challenging conditions.

Transactional Forces and Typology

Cape Town's existence has been determined by its unique geographical location at the southern tip of Africa, with a natural harbour that has afforded protection to ocean-going mariners over the years. On the shore of Table Bay, since its discovery by the Portuguese in the late 15th century, it developed as a supply point in the maritime trading between Europe, the Indian sub-continent and the East Indies. Occupied and settled by the British as the 'Cape Colony', the primary agricultural activities were overtaken by mining and the exploitation of gold and diamond resources in the Witwatersrand region supported by the construction of railways. After the Anglo-Boer Wars, which resulted in the creation of the Union of South Africa, the city lost its dominance to the more industrial Johannesburg and with the installation of the executive in Pretoria in the Transvaal.

With its striking natural features of Table Mountain with the Devil's Peak and Lion's Head overlooking the bay, the city lies in a bowl surrounded by rugged mountains. It extends to be contained with the Cape Flats where the commercial and administrative centre of the city and Parliament are located in a laid-out gridded area alongside Government Avenue. The foreshore to the harbour has been considerably extended with the reclamation of a large area of additional land for use by the port.

With its geographical constraints between the sea and mountain, the coastal city is westward facing. From its central area it is surrounded by former townships now attaining different levels of development. With varying levels of provision of infrastructure and services, the city has a most fragmented nature, with greatly varying quality of residential provision in different districts.

With deficiencies in housing and transport facilities and with historical racial divisions, it remains a polarised South African city with all of the modern social challenges of health, education, poverty and security along with a fast-growing population, many of whom do not have access to basic or adequate facilities or services.

3.3.4 BOSTON, MASSACHUSETTS, USA

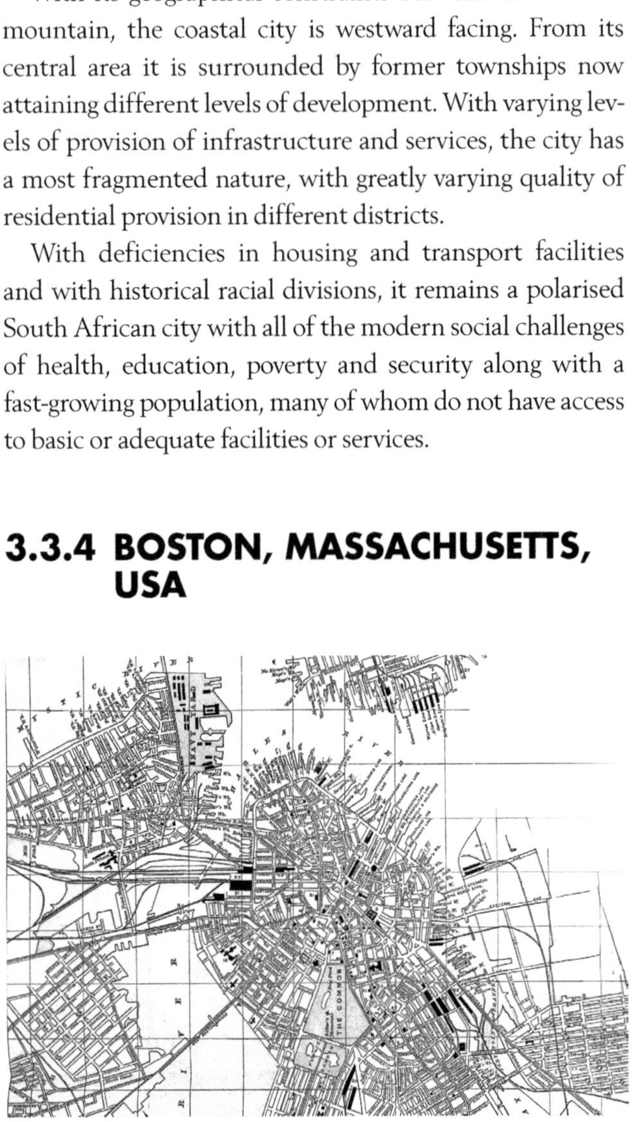

FIGURE 3.3.2 Boston – early configuration, 1900.

Source: iStock.

Background and Origins

The city of **Boston** was developed subsequent to the arrival of English Puritan settlers at Plymouth, with the foundation of Charlestown on the Charles River in 1629, which came to be a pivotal centre in New England. The original settlement around the Common on the Shawmut Peninsula of Massachusetts Bay and the Charles River was connected by a narrow isthmus on the eastern Atlantic coast of North America. At the centre on three hills, the Trimountain area with its flat side and slopes, and with its houses, churches and state institutions around the complex topography formed a compact centre. The city was gradually expanded westwards through extensive land reclamation of the tidal mud flats around Back Bay.

Across the Charles River, the settlers developed hilly Beacon Hill around Dock Square with Great Street, now State Street, leading up from the harbour. With its market square and meeting houses, this became the local centre of development with the Common beyond. Long Wharf extended Great Street well out into the harbour where stores and warehouses were built on either side, with the core acting as the centre of maritime activity.

In 1786, a toll bridge was constructed across the Charles River to Cambridge where Harvard University had already been established around its Yard. Mill Pond had been dammed and the State House was built adjoining the Common. Commonwealth Avenue was realised to strengthen the shoreline along the city perimeter, and streets were built with residential buildings of distinctive red-brick row houses on either side which came to be recognised as the New England vernacular style.

The port was active around Long Wharf and became renowned for the Boston revolt in 1773 against the colonists, as a result of the Tea Act in having 'no taxation without representation'. The following siege led to the Battle of Bunker Hill as the invaders resorted to the neighbouring Charleston peninsula.

With its seafaring background and activities of the 'Massachusetts Bay Company', the port was part of the triangular slave trade with Africa across the Atlantic and the Caribbean until its abolition. It was also a key destination of waves of immigrants from Europe, with the Irish during the potato famine to South Boston, and the Italians that settled in North End. There were also Polish and Russian Jews, such that by the end of the 19th century a large percentage of the residents of Boston was of immigrant extraction.

As the city developed progressively, new creative development plans were put forward. In 1844, Robert Gourlay proposed a significant scheme to enhance the city surrounded by a network of concentric highways with riverfront boulevards, park lands and river islands. In addition, an underground subway was proposed with a radial network spreading out from a circus around State House. Whilst radical, futuristic and unbuilt, it emphasised the importance of transport infrastructure.

Most significant was the actual development of outlying areas reached by horse-car and the growth of the streetcar suburbs, as described by Sam B Warner in the period of 1870–1900. Traditionally, most work was focused around the port, which was the hub and centre of activity. But in the second half of the 19th century, this was to change from the central walking city to the active larger suburban metropolis due to the realisation of new outer residential districts, as for example Roxbury, West Roxbury and Dorchester.

As a result of the construction of new streetcar lines, builders profited, and people were able to relocate to new residences in the more spacious suburbs, leaving the inner city for workers and lower-income housing. This changed the pattern of development of the city where as yet unregulated land was developed extending out radially to new areas.

At the centre, the gold domed State House had been built on Beacon Hill with its townhouses and the market at Faneuil Hall, Custom House, several meeting houses and taverns, and stores around Haymarket Square, many of which are still found on the 'Freedom Trail' that connects landmarks and cemeteries across the City and the Common. Besides local governmental entities, a number of educational institutions were also established that have now grown to become the great universities of Harvard and the Massachusetts Institute of Technology (MIT) in the neighbouring area of Cambridge, in addition to Tufts and Boston University. It is also home to the First Church of Christ, Scientist, known as The Mother Church in Back Bay.

Modern City – Buildings and Arteries

The city has evolved accordingly with the distinctive neighbourhoods of North End, West End and South End and has become a thriving metropolis. Around the Central Business District, considerable contemporary development has occurred, and a programme of urban renewal was undertaken in the 1960s by the Boston Redevelopment Authority. With the large austere Government Center overseen by Ed Logue and City Hall in Scollay Square, and with other commercial modern buildings and the tall John Hancock Tower, the central city has been transformed whilst retaining the open common.

With its history of streetcars and trolleys, good urban public transport was provided before the modern MBTA metro system was built. The overpowering elevated highway structure, the Central Artery, has been torn down and replaced by a tunnel, the 'Big Dig', which has led to considerable environmental improvement of several inner neighbourhoods. In East Boston, land has been reclaimed at Jeffries Point to extend Boston Logan International Airport (BOS). Here also Massport Port Authority is a major employer responsible for both the Port of Boston and the southern seaport with its container traffic and the airport.

The central area has been redesigned to encourage walking and cycling, and the redeveloped Quincy Market downtown, promoted by James Rouse at Faneuil Hall, is a favoured destination with its stores and restaurants. Boston remains a classic New England city, with fine red-brick buildings and row houses in the old town and larger buildings out towards Back Bay. From its colonial beginnings, it has been strengthened by its extensive educational institutions and attracts a young, creative population driving the legal, medical and science professions.

Outside the city is the Route 128 Corridor, which has been developing as a high-tech growth corridor promoting research in high-tech industries on the peripheral edge of the city. The population of the city is approximately 0.7 million and 4.8 million in the metropolitan area of Greater Boston in the Commonwealth of Massachusetts.

Transactional Forces and Typology

With its protected location on the Atlantic coastline, the city of Boston grew around its natural harbour, with trade emanating from the maritime activities of the Massachusetts Bay Company. From fishing and agriculture, seafaring communities have grown up and with a large number of European migrants that were attracted to and settled in the city, this has resulted in ethnic enclaves, such as Italians in the North End, Irish in South Boston, and Russians in the West End.

Many wharves were developed along the waterfront, the Charlestown Navy Yard was established and new areas of land were reclaimed. Beyond the centre, development of the city was extended outwards, following new streetcar lines as the residential streetcar suburbs were realised. Educational institutions were also developed, and local administration was centred around the State House on Beacon Hill as the state Capitol. With cross fertilisation between these institutions and business expansion that led to growth in commerce, new modern industries became established from spin-offs from the universities, forming high-tech industries.

As a marine port, the city developed organically around its coastal perimeter and riverine edges inland in a radial manner. Assisted by land reclamation beyond the historic centre, new areas have been consolidated from Back Bay and beyond into the residential suburbs. Around the riverside are the universities on the opposite bank of the Charles River in Cambridge, whereas on the eastern coastal side are the port and the expanding international airport on reclaimed land. The city of Boston developed in stages from its initial natural peninsula and shoreline to be subsequently extended with man-made wharves along its perimeter. Next, areas of land reclamation were undertaken with the filling of Mill Pond and the creation of the area known as Back Bay by the Charles River.

Beyond this, as new streetcar lines were developed outwards, new residential areas were built farther out and more land was annexed, which extended the size and reach of the city. Hence from its natural organic central city with its orthogonal reclaimed Back Bay grid, the city has stretched outwards, radially, north, west and south, linked by circumferential connections. Here now it is demonstrating 'edge city' characteristics with new businesses that are gravitating around the outer peripheral Route 128 Corridor. Whilst an historic state capital and marine harbour, from its early waves of immigrants, with trade and manufacturing, and now with established educational and research institutes, it has become a leading modern American city, with good communications pivotal to promotion of the developing digital economy.

3.3.5 NEW ORLEANS, LOUISIANA, USA

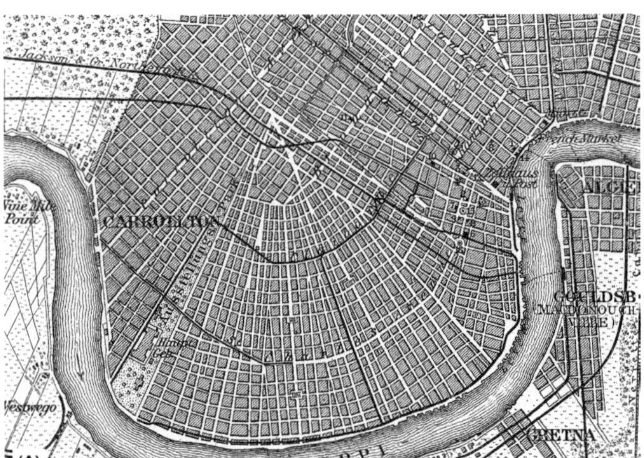

FIGURE 3.3.3A New Orleans – early configurations, 1897.

Source: iStock.

Background and Origins

New Orleans is located on the Mississippi River and by Lake Pontchartrain in the southern US state of Louisiana. Whilst earlier known waterways with natural levées were the Bayou Chapitoulas or Metairie, and Bayou Sauvage or Gentilly, the current course of the serpentine river was established around 1600 CE. An old site of Indian portage was known to be around Bayou St John.

First discovered by the Spaniard De Soto in 1540, land was cleared and a city was founded there named 'La Nouvelle Orléans' by French colonists, led by Sieur de Bienville in 1718 on land of the native Chitimacha Indians, named after the Duke of Orléans. With ongoing conflicts with the local Natchez native Indians, the French decided to turn also to the importation of African slaves in order to provide additional labour using their 'Code Noir'.

The city was planned and developed by Pauger in a classic grid style around a central square (Vieux Carré) as a feature near the long crescent-shaped bend on the left bank of the river. It was laid out and grew around the central Place d'Armes, now Jackson Square, in a pattern of blocks and streets. With its timber buildings and after two major fires at the end of the 18th century, St Louis Cathedral was founded and built.

Early fortifications were built by the French along with churches and the Catholic Ursuline Convent. It was ceded, however, to the Spanish under the 1763 Treaty of Paris ending the French and Indian War and remained so in a period under Spanish rule with its Cabildo (Council) until 1800, when it reverted briefly to French rule. The Spanish set about replacing buildings in the Spanish style with decorated windows, balconies and wrought iron grill work that have given colourful flavour to a number of streets in the centre.

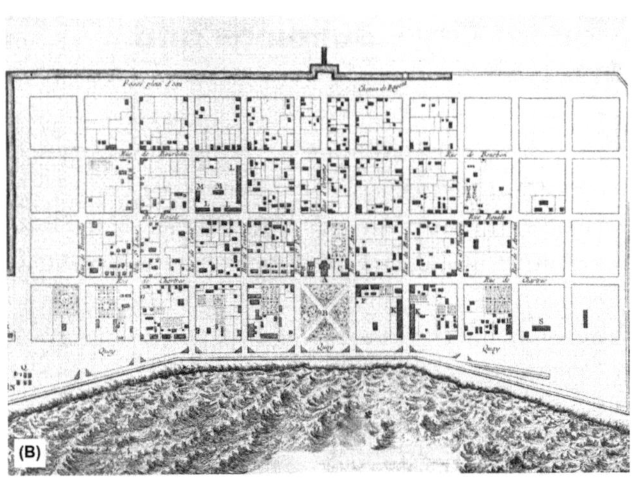

FIGURE 3.3.3B New Orleans centre – early configuration 1764

Source: Historic Urban Plans.

With its original fortifications, it was further developed later in 1815 to have new 'faubourgs' and boulevards around which new buildings were laid out. This enabled the grid measured in 'arpents' to be reorientated with a number of open malls or parade grounds as features. This gave the inherent character of the Vieux Carré, which still remains.

New Orleans was sold by Napoléon as part of the Louisiana Purchase, a significant area of Mid-America between the Mississippi in the east and the Red and Arkansas rivers in the west. This was to be followed by a period of intensive migration, especially with refugees and Créoles from the Caribbean after the Haitian Revolution. The colonial era was ended by defeat of the British at the Battle of New Orleans at Chalmette in 1815 by Jackson that served to cement the newly created American city and state.

The port on the river developed as significant trade and activities expanded in the region using the great Mississippi River. Many slaves had been brought into the Deep South to work on the local cotton plantations enhanced by the newly invented cotton gin (engine). As steamships were developed, they began to ply the river and were able to extend the range and volume of movement of commodities beyond flat-boats into neighbouring hinterlands where new wharves were built.

Although developed, the city remained at risk from its topographical siting and low elevation in the swampy bow of the river. With its high water table, marsh-land and general lack of sanitation, cholera and malaria were prevalent. Despite its importance as the 'Cotton Kingdom', there were other distractions.

In the 1850s, more working-class immigrants arrived from both Ireland and Germany, and tensions frequently developed between the different communities. The arrival of the railways was beginning to divert traffic, and the city was stripped of its status as the capital of Louisiana in 1849, being transferred to Baton Rouge. Its dominance as a port city began to wane.

With the US Civil War, Louisiana joined the Confederacy movement and Union troops captured New Orleans. With its eventual surrender in 1865, and the enforced end of slavery, planters became more impoverished and the economy started to decline. During the next 30 years, this decline continued with Jim Crow laws prevailing in the Southern states, and the Reconstruction Era evolved, ending in 1877, such that trade through the port of New Orleans was significantly lower than at its peak in 1860. In 1884, a World Cotton Exposition was held in the city.

Modern City – Buildings and Arteries

The distinctive Crescent City has a number of different neighbourhoods – the central French Quarter, the Warehouse District and the Garden District. In addition, it has the University District, City Park, Audubon Park and Carrollton. Mid-City is divided by north-south Canal Street, where a canal was never built, and the Pontchartrain Highway, with the city being bounded on the northern side by Lake Pontchartrain. The Bayou St John and other waterways cross the low-lying city. With both its French and Spanish cultural history and origins, it also reflects the distinctive Cajun and Creole flavours of Africa and the Caribbean. Canal Street, the main N-S street of New Orleans, divided the city between the Anglo-American and the Créole communities.

The Mississippi Riverfront contains the French Quarter running between tree-lined Canal Street with its streetcars and Esplanade Avenue with its historic Pontalba buildings on either side of Jackson Square with Saint Louis Cathedral and the Cabildo building, where the Louisiana Purchase was signed. Royal Street has its picturesque filigree cast iron facade buildings and also on Bourbon Street.

The Warehouse District to the south contains the Spanish Plaza and Riverwalk Marketplace and museums and art galleries with the ferry across the river to Algiers Point, which was the landing point for immigrants. The Garden District was built on what was once plantation land on the Livaudais Plantation having mansions and large gardens. As development intensified beyond bungalows and 'shotgun houses' or double tenements, townhouses were built of two to three storeys with elevated galleries. Since the river wraps around the central city in a bend, the buildings on each street side are either concave or convex facing.

Nearby is Audubon Park with private Tulane University and Catholic Loyola University. Other districts off the French Quarter are Treme, Marigny and Bywater, with Congo Square, the musical Frenchmen Street and the Piety Street area respectively. With its museums, art galleries, traditional, jazz and 'dixie' music and annual Mardi Gras festival, New Orleans remains a busy, dynamic and cultured centre, and with its unique characteristics the Vieux Carrée Quarter has been protected.

The current population of approximately 0.4 million has declined from its peak of 0.6 million since 1970, with 1.25 million in the metropolitan area.

In terms of transport, alongside the great river with its multiple barges, the railroad was used to support the movement of freight with trans-shipment at New Orleans as it became a significant railroad centre for the movement of cotton, coffee, bananas, grain and coal.

New highways were developed later, with the first major road and rail bridge across the river in the city, the Huey Pierce Long Bridge, built in 1935. The subsequent major crossing of the Mississippi, the Crescent City Connection, was built in 1958 and was doubled in 1988. The next main river crossing upstream is at Baton Rouge. The continuation of construction of the I-10 highway also led to improved accessibility of the city with new land developments.

During World War II, the lakefront at Pontchartrain served as a Naval Reserve Aviation base and was used by sea planes. Post-war, a new airport was developed at the former Moisant Stock Yards field to the west, now known as Louis Armstrong New Orleans Airport (MSY). The long Pontchartrain basic causeway across the wide lake was built in 1957 with a second parallel span built in 1969.

Mississippi River

The city of New Orleans is actually sited about 150 km. from the open sea and adjoining Lake Pontchartrain, on the long sinuous route of the Mississippi River from three separate marine entrance passes in the Gulf of Mexico, now protected by raised artificial levees. The city was formed in a marshy area in the 'back-swamps', where silt built up on the natural riverbanks and a Levée Board was established in 1890 in order to oversee the management and maintenance of these levées.

The Mississippi River acts as a huge drainage basin for much of the central North American continent, and has many tributaries, both large and small. In order to maintain parts of it suitable for navigation, this has to be managed. This has been undertaken partly upstream by the construction of tributary dams and spillways to provide protection against flooding, and partly downstream through the construction of connecting canals and higher levees.

In order to facilitate trade and shipping, an industrial canal was planned and opened in 1928 to connect the Mississippi with Lake Pontchartrain, which effectively separated New Orleans East from the rest of the city. In addition, along the Gulf Coast between the states of Florida, Mississippi, Louisiana and Texas, an Intracoastal Waterway was designed and built as a navigable inland waterway essentially for shallower barges connecting via Lake Borgne with the Mississippi coast. Completed in 1949, this connects with the industrial canal in New Orleans.

Increased offshore drilling for oil and gas in the Gulf has accelerated coastal erosion and saltwater intrusion preventing plant growth, exacerbating works required for extended and reinforced levees. These works in turn prevent the build-up of silt that previously replenished the adjacent land. In addition, the impact of dredging canals and piling up of silt on levees has the effect of such additional weight on swampy land, causing the marshes to sink.

With the emphasis now on containerisation, most of their original urban river wharves have been closed and activities transferred to the new larger outer facility at New Orleans East as a key trans-shipment centre.

In order to enhance traffic flows to the Gulf with deeper draft ships, a new connection to the Gulf Outlet of the Mississippi River was planned in 1956 and opened in 1968. This 122 km.-long channel, named the Mississippi River – Gulf Outlet Canal (MRGO), was realised by the US Army Corps of Engineers. Whilst beneficial for shipping, it had negative effects in serving to facilitate storm surges from the coast up this channel. As a result, a new storm surge barrier was eventually built across this in 2009 after the devasting effects of recent hurricanes.

Both the city centre and its surroundings suffered greatly because of Hurricane Katrina in 2005 when the artificial levees failed with breaches of the industrial canal, which led to major flooding, in particular of the Lower 9th Ward. Since then, a major programme of strengthening of the levees and enhanced protection has been undertaken under the auspices of the US Army Corps of Engineers with rebuilding and elevation of housing units in order to avoid future flooding after detailed reassessment of 'at risk' inundation zones.

Transactional Forces and Typology

Sited on a wide bow on the left bank of the grand Mississippi River, and bounded on the north by Lake Pontchartrain, the city of New Orleans was part of the 'Cotton Kingdom' across Louisiana. With its waterfront, Warehouse and Garden districts and the French Quarter, the great river was the main thoroughfare for people and goods. As a vital artery, whilst the development of railroads assisted in the transport of produce to and from plantations, waterborne transport has remained of prime importance. The most critical issue given the low-lying nature of the city is its vital dependence on protection from high water surges by long man-made levees along each side of the Mississippi River.

With general sinking of ground levels and subsidence, drainage and flood control has become the main concern for urban infrastructure in and around the city due to its riverine location and in the southern hurricane zone of

America. Despite this, built development of the city has evolved, with some taller buildings downtown in particular since the more recent development of the oil and gas fields in the adjoining Gulf of Mexico. Its whole livelihood, however, is now exacerbated by evolving climate change impacts.

As a near coastal city, New Orleans was planned and has been developed between a major river and lake in a very low-lying, swampy district. Its planned historic district was built around a tight grid parallel to the waterfront, with other areas platted orthogonally, except for the western Broadmoor district in the river bend that radiates out southwards.

Despite the downtown styled 'American Sector', it is generally dense but low rise, built on land that is gradually subsiding. Characteristically French and Spanish by origin, its Vieux Carré quarter is unique, as a southern American maritime city embellished with its Creole and Cajun cultural origins. Shaped both physically and economically by the majestic river and its trading and shipping activities, New Orleans is in many ways, not just metaphorically, 'Queen of the Mississippi'.

3.3.6 MADRAS (CHENNAI), INDIA

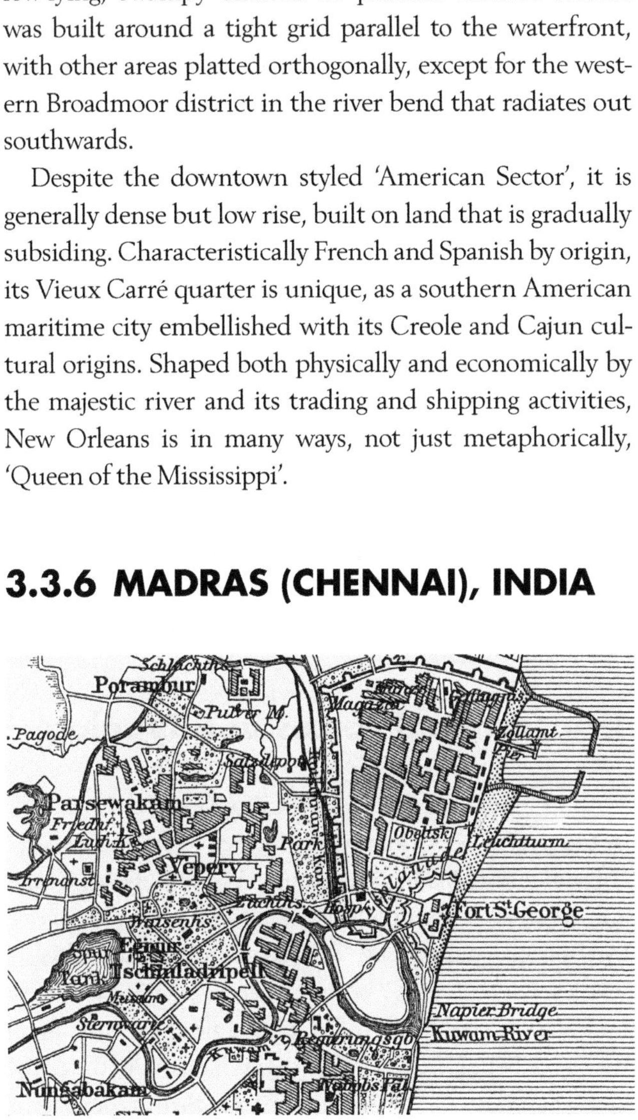

FIGURE 3.3.4 Madras – early configuration, 1897.

Source: iStock.

Background and Origins

Madras, now known as **Chennai**, is the capital of the modern State of Tamil Nadu, created in 1969, in the southeastern Coromandel coastal plain of India by the Bay of Bengal. Evidence of earliest megalithic settlement was known at Pallavarum, and parts of the region were ruled by the Chola dynasty. Then, under the early Pallava dynasty, the town of Mahabalipuram was built with its ancient Hindu temples around the 8th century CE until the subsequent Vijayanagara Empire dominated the region.

The first Europeans to arrive were the Portuguese in 1522, who built Fort São Tomé, followed by the Dutch settlers in Pulicat in 1612. Of more significance was the establishment of the British East India Company who, under Francis Day in 1639, negotiated the lease of a strip of land in order to build a warehouse and factory for its trading activities at the site of Madrasapatam. The following year, the company built Fort St George on adjoining land which, although lacking a natural harbour, became the nucleus around which the city grew by the meandering Cooum River.

The Corporation of Madras was established in 1688 under Royal Charter as municipal administrator that managed the development of the city and the neighbourhoods of Mylapore, Triplicane and Tiruvanmiyur.

With the eventual Seven Years War in Europe, the city came under siege from the French based in nearby Pondicherry in 1758. With the subsequent Anglo-Mysore War, this was settled by the Treaty of Madras, Fort St George was reinforced and under the Madras presidency Madras was established as the capital of the region. The city subsequently grew with a major naval base and became the headquarters for the British in southern India until 1772 when the headquarters of the East India Company was moved to Calcutta.

The city was adopted as baseline for the Great Trigonometrical Survey across India in 1802. It developed around Fort St George and Georgetown (Chennapatnam) and the Cooum River. As settlers gravitated to the area, there was a distinction between the fortified White Town for European residents and St Mary's Church and the more populous Black Town (Georgetown) for Indian residents.

With the Parade Ground and the Secretariat (designed by Chisholm) and Legislative Council Chambers, High Court, town hall and other administrative buildings, the British oversaw development, using a blend of Indo-Saracenic architecture. With the university Senate House, museums complex and other institutions, the Chepauk Palace, Banqueting and Memorial Hall, and the Triumph of Labour Statue on Marina Beach were key landmarks in the city.

A canal was built between the two rivers under the British, with the first section the Cochrane Canal and then extended to Pulicat Lake, that came to be named the Buckingham Canal, inset from the coastline that interconnects a series of marine backwaters and that was used for transportation. The Adyar River, with a wider estuary, is a physical feature farther to the south.

The earliest built infrastructure was the railways with the Southern Railway Headquarters and Central Railway

Station and suburban network initially developed using metre gauge track.

Georgetown, or Chennapatnam, is a busy commercial district with different trades-peoples where the Mint was located along with a mix of different trading communities, both Portuguese and Armenian.

Modern City – Buildings and Arteries

As the birthplace of India's National Congress Party, after India's independence in 1947, the city of Madras became the centre of political activities in the province. In 1996 Madras was renamed Chennai after a former Telugu ruler and was to become the capital of the newly created State of Tamil Nadu, mainly Tamil speaking. The Chennai Metropolitan area is now divided into three principal districts, North, South and Central Chennai, with the north being mainly industrial. The Chennai Metropolitan Development Authority (CMDA), which has responsibility for policy, developed a Master Plan to include the development of satellite towns around the city.

From its earliest textile manufacturing, the city now has a broad industrial base, especially in automotive manufacturing and heavy vehicles and railway rolling stock. It has diversified into financial services, electronics, IT and healthcare. One of the largest IT parks is Tidel Park, a special enterprise zone. Telecom and pharmaceutical companies also have research parks in Chennai, and a new commercial World Trade Centre Complex was opened in 2020.

In terms of infrastructure, the Grand Southern Trunk Road NH45 runs southwards to Theni, with the NH4 towards Mumbai, NH5 to Kolkata as part of the National Highways Golden Quadrilateral System that links the main State capitals, and there is a new highway around the city.

The railways radiate out from the centre with its broad-gauge Chennai Suburban Railway network with four main railway terminals, including Chennai Central. A programme of conversion of the original metre-gauge railway tracks to broad gauge has been ongoing to be compatible with the other electrified national Indian rail systems. The electrified Circular Line was opened in 2019.

The Chennai Metro opened in 2015 with the Blue Line and the Green Line and a second phase is planned. There are many bus companies and numerous maxi-cabs, taxis and auto-rickshaws that ply the streets of the city. Part of the elevated MRTS runs along the route of former Buckingham Canal. Chennai International Airport (MAA) is located to the South West of the city on the site of the original Madras aerodrome.

Chennai Port is a major container port that is pivotal in the region and to the north, Ennore Port, which handles bulk cargo, coal and minerals. The city is also the landing stage for a number of significant undersea fibre optic cables that connect India with Europe and South-East Asia between continents. There are a number of power plants around Chennai, which include gas turbine, thermal coal and atomic power generation.

Clean water is a major challenge for the city – it receives water from the Red Hills Lake and the Krishna River via the Telugu Ganga Project; however, additional water is also required from modern desalinisation plants. The city being in a cyclone region, it receives high annual rain during the monsoon season (between October and December). The main rivers in the city are generally heavily polluted and remain a considerable environmental challenge. The city has a national park in the midst of the city, Guindy National Park and the Adyar Eco Park around the River Adyar.

The population of the city of Chennai is approximately 8.6 million, and approximately 12.4 million in the metropolitan area, the city limits having been increased in 2011 to 426 sq. km.

It is a highly active and densely urbanised area with a highly diverse ethno-religious mix, although predominantly Hindu, and acts as the 'Gateway of the South'.

Transactional Forces and Typology

With its location on the eastern Coromandel coast of India by the rivers Cooum and Adyar, formerly a fishing and agricultural region, Madras was occupied by the Cholas, Cheras and Pandyas. Within the Vijayanagara Empire it had a rich local history and culture, with its temples and trading of textiles, salt and spices. With its marine environment, boat building was established to assist travel along the coastline of the Bay of Bengal. Much of this was to change with the arrival of the East India Company and period of British occupation.

With construction of forts and new breakwaters and a naval base, new industries and warehouses, railways were introduced and the meandering rivers were connected by a canal. The Chennai Port Trust was established in 1881, and areas were developed to the north and south of each river's islands and linearly along the northern coast towards Ennore Port. This has been renamed Kamarajar Port and is now functioning as a major industrial coal-handling bulk facility, which complements the Port of Chennai as a major container port.

Land was developed around the coast and river estuaries, which became the pole for subsequent development. With dependence on trading, the development of railway infrastructure facilitated transport and communication with central and northern India before the highway network was developed. With its increasingly educated workforce, this served as an alternative location for the growth of new industries, from textile to automotive, and more recently modern IT and service industries, with heavier industries located in northern areas.

As a coastal city with long outlying beaches, the core was settled around the two central rivers and artificial harbour that was created with new breakwaters. As a maritime centre, this was strengthened with the construction of the railways by the British, with the Southern Railway Headquarters of India established here. With its other administrative and institutional buildings, formerly the capital of the region managed by the Corporation of Madras, it is now the capital of the State of Tamil Nadu.

With its modern industry, automobile manufacturing, labour outsourcing, education, health and media, its perimeter has developed and grown, supported by suburban railways and new road infrastructure. Whilst possessing radial characteristics, it is developing more as a poly-nodal structure with intermixed industrial and residential areas, with informal settlements in and around the outer by-pass and service road.

The city of Chennai developed with characteristics of a coastal/riverine city having radio-concentric patterns of development on the western landward side of the city. Its burgeoning population has led to the creation of satellite districts with specialist business parks and IT districts that have housed modern international 'call centres'. With its improving infrastructure and new metro system, this is channelling development into planned new areas. However, significant areas of informal settlements still remain that need to be serviced, given the sparse water supply and inadequate sanitation . The following figures illustrate the modern configuration of the maritime cities as described and the evolution of their modern population.

FIGURE 3.3.5 Freetown – modern configuration.

FIGURE 3.3.6 Accra – modern configuration.

FIGURE 3.3.7 Cape Town – modern configuration.

FIGURE 3.3.8 Boston – modern configuration.

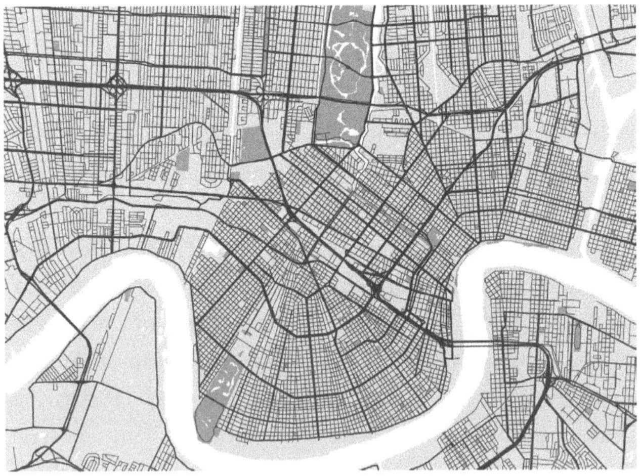

FIGURE 3.3.9 New Orleans – modern configuration.

FIGURE 3.3.10 Madras – modern configuration.

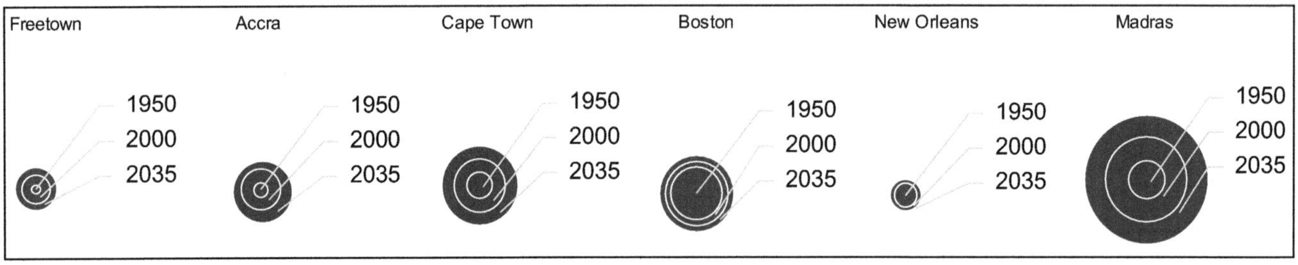

FIGURE 3.3.11 Maritime city populations.

Data source: Global Database of Metropolises 2020 – UN-Habitat (2020), UN DESA Population Division.

Further Reading

FREETOWN

Banton, Michael – *West African City, Study of Tribal Life in Freetown*, Routledge, 2018.

Fyfe, Christopher – *A Short History of Sierra Leone*, Longman, 1975.

ACCRA

Hart, Jennifer – *Making an African City*, Indiana University Press, 2024.

Ward, WEF – *A History of Ghana*, Routledge, 2023.

CAPE TOWN

Callinicos, Lili – *The World That Made Mandela*, STE Publishers, 2004.

Fairbridge, D – *Historic Houses of South Africa*, Oxford University Press, 1922.

Trotter, Henry – *Cape Town, a Place Between*, Catalyst Press, 2020.

BOSTON

Seasholes, Nancy – *The Atlas of Boston History*, University of Chicago Press, 2018.

Warner, Sam B – *Streetcar Suburbs*, Harvard University Press, 1962.

Whitehill, Walter & Kennedy, Lawrence – *Boston, a Topographical History*, Harvard University Press, 2000.

NEW ORLEANS

Cable, Mary – *Lost New Orleans*, Houghton & Mifflin, 1980.

Fraisier, Jim – *The French Quarter of New Orleans*, University Press of Mississippi, 2003.
Garvey, Joan & Widmer, Mary Lou – *Beautiful Crescent, History of New Orleans*, Pelican Publishing, 2013.
Guenon-Lelle, Dianne – *The Story of French New Orleans, History of a Creole City*, University Press of Mississippi, 2016.

MADRAS (CHENNAI)

Barlow, Glenn – *The Story of Madras*, Classic Books, 2021.
Krishan, Nandita & Doshi, Tishani – *Madras Then, Chennai Now*, Roli Books, 2014.
Srinivasachari, CS – *History of the City of Madras*, Varadachary, 1939.

3.4 Island Cities

Caribbean Sea – Indian Ocean – Makassar Strait – Pacific Ocean

Havana, Port-au-Prince, Kingston, Colombo, Makassar and Auckland

3.4.0 INTRODUCTION – ISLAND CITIES

Early trade developed terrestrially and across seas, long before modern trade expanded considerably and took also to the air. Typically, inter-continental trade by land and sea developed as modes of transport improved, and longer journeys could be made beyond using animal and wheeled transport on land having waypoints for caravanserai over longer land journeys, and progressively in sailing boats and steamships by sea. The latter was influenced predominantly by geography, and navigation of ocean currents and prevailing winds towards coastal estuaries and protected harbours, with the siting of islands used as stepping points for longer journeys.

In this respect a number of islands have performed a strategic role in the development of trade across waters in different geographical regions. Of varying scale, in different areas these comprise the islands and centres of the:

- Mediterranean Sea: Sicily (Palermo), Sardinia (Cagliari), Cyprus (Limassol) and Crete (Knossos).
- Caribbean: Cuba (Havana), Hispaniola (Santo Domingo), Trinidad (Port of Spain) and Jamaica (Kingston).
- Atlantic Ocean: Newfoundland (St John's), Bahamas (Nassau) and Cape Verde (Praia).
- Indian Ocean: Sri Lanka (Colombo), Sumatra (Medan) and Java (Jakarta).
- South China Sea: Borneo (Sibu), Celebes/Sulawesi (Makassar), Philippines (Manila) and New Guinea (Port Moresby).
- Pacific Ocean: Fiji (Suva), Samoa (Apia), Hawaii (Honolulu) and New Zealand-Aotearoa (Auckland and Christchurch).

With their coastal perimeters and as a result of forces of erosion, some islands have natural harbours which have provided safe haven to maritime travellers. Accordingly, they became points of arrival and departure for travellers carrying different types of trade and commodities, and centres around which settlers gravitated. From these, several centres developed into key cities or capitals of island states, such as Havana, Kingston and Port-au-Prince in the Caribbean and Suva and Auckland farther afield in Oceania.

3.4.1 HAVANA, CUBA

FIGURE 3.4.1 Havana – early configuration, 1762 (Duc de Choiseul).

Source: Historic Urban Plans.

Background and Origins

Havana (La Habana) is the capital of Cuba, a port on the northern coast of the island of Cuba in the Gulf of Mexico. Entered through a narrow inlet, it has three finger harbour areas of Marimelana, Guanabacoa and Atares. With its original Ciboney then Taíno indigenous population, the island was the springboard for the Spanish conquest of the Americas during the early 16th century CE after Velazquez landed at Baracoa in 1511. Settlement developed around the natural bay called Puerto de Carenas, renamed by the Spanish as San Cristobal de la Habana. It grew as a

DOI: 10.4324/9781003353348-15

trading port, as a key base and stopping point for Spanish shipping in the Caribbean.

It was fortified with the Castillo de los Tres Santos Reyes Magos del Morro in East Havana with its lighthouse at the harbour entrance and Fortaleza de San Carlos de la Cabana castle. On the opposite side these castles face Castillo de San Salvador de la Punta at the entrance and Castillo de la Fuerza in the Canal de Entrada, when the harbour mouth could be closed by iron and log chains.

It continued as a Spanish colony during the 17th and 18th centuries despite brief occupation of Havana by the British in 1762 before it was returned in exchange for the peninsula of Florida. The port grew in importance with its dry dock and with its Royal Shipyard, trade expanded between the Caribbean islands and particularly with the southern states of North America.

Plantations were developed and sugar production was significant form the 1820s. A railway line was built across the island in 1837 to link with Bejucal and the port of Batabano for the export of significant volumes of sugar cane, the majority of which went to America. This was facilitated on the basis of slave labour, which was planned on a 'golden circle' of states where slavery was legal, centred on Havana, although there was other diversified agriculture. After the Ten Years War (1868–78) between Cuba and Spain, this resulted in its eventual abolition, although repression of much of the rural population remained.

In 1898, as a result of the subsequent Spanish-American War, the USA gained control of Cuba as a protectorate, putting an end to a long period of Spanish occupation. A number of significant infrastructure improvements were made and with this and after the end of the earlier American Civil War, America's economic interest in the whole of the Caribbean increased, including to Central and South America beyond.

But in 1902, the Republic of Cuba formally obtained its independence. A period of significant building and infrastructure improvements was undertaken in and around the city. With Havana as capital, later in 1934 after the Great Depression, however under the Treaty of Relations, Cuba was obligated to give preferential treatment of its economy to the USA.

Tensions remained, and after World War II and a period under the dictatorship of Batista, in 1959 he was deposed during the Cuban Revolution by Fidel Castro. Here began a period of domestic communist rule in Cuba, with nationalisations and leadership of the international non-aligned movement, in direct conflict with the USA, which led to a mass exodus of professional Cubans from the island.

The situation was exacerbated with the installation of Soviet missiles on the island and the Cuban Missile Crisis confrontation in 1962. Whilst this was ended without conflict, later came the eventual collapse of the Soviet Union in 1991. The collapse of the Soviet empire had significant impacts, the Special Period when the US dollar was legalised and tourism increased, not least with a real shortage of food, as the country had relied upon cheap imports of food and fertilisers from the USSR, which was followed by a period of austerity. Subsequently, promotion of agriculture became a national priority. Despite some areas of liberalisation, the country has remained a one-party socialist republic.

Modern City – Buildings and Arteries

The city itself comprises three main areas, *Habana Vieja*, *Centro Habana* and *Vedado*.

Old Havana has the oldest square, Plaza de Armas, and the Plaza de la Catedral, with its large cathedral, and also contains Empredado (Paved Street) and Plazuela de la Cienaga (Little Swamp). With Palacio de los Capitanes Generales, the former president's palace and the old mansions of Casa de la Obra Pia, Casa de Africa (with Santería relics) and Casa de los Arabes, it has a number of attractive colonnaded streets and uniform building facades with upper half-moon stained glass windows.

The Prado promenade, Paseo de Marti, runs northwards from Parque Central to the sea and separates Old from Central Havana just outside the old city wall. The Gran Teatro de Havana here is the home of the Cuban National Ballet. The Parque Infantil runs alongside the Avenida del Puerto on the water's edge. The pedestrianised Calle Obispo is crossed by Mercaderes Street in the centre of Old Havana with restored Plaza Vieja and Plaza de San Francisco and its convent on the eastern edge. The Alameda de Paula promenade is at the southern corner of the harbour.

The grand Capitolio Nacional building, which is modelled on the US Capitol in Washington DC, dominates the centre with adjoining Parque Central and Parque de la Fraternidad (formerly Campo de Marte with four grand gates) and Palacio de Aldama.

Central Havana, with its dense network of streets, is an active residential neighbourhood (barrio) of the city with Calle San Rafael as its main commercial street. New Havana was developed after the city walls were razed and developed along the Malecon coastal promenade with the seawall towards Miramar, which has been developed in the west.

The area between Central and Vedado is separated by the diagonal area that houses the Palacio de la Revolucion by the Plaza de la Revolucion and memorial obelisk. Beside the military Castillo del Principe, Avenida Carlos III (formerly Paseo de Tacon) was built, which acts as a distinctive avenue across the district with the adjoining Quinta de los Molinos. Also in the centre is Barrio Chino.

Vedado is a planned area laid out on a diagonal grid with numbered streets (calle) and lettered avenues loosely modelled on Miami. To the south, the major Colon Cemetery is described on a Greek cross; built in the 1880s, it dominates this neighbourhood. The principal street is Calle 23, La Rampa, that runs north-east as the main business district which, with Linea, runs towards the water's edge and which is crossed perpendicularly by Avenida de los Presidentes.

Beyond the Rio Almendares are the new suburban districts of Miramar and Playa running out to the west. On the other side of Havana to the east are the beaches of Playa del Este towards Cojimar. During the 1920s and 1930s, the city had a growing tourist industry, which led to the development of hotels and casinos such as the Hotel Nacional de Cuba and several Art Deco buildings such as Edificio Bacardi and the Radiocentro CMQ Building. However, during the Castro regime, most of the privately owned businesses were nationalised. The Parque Deportivo José Marti with its pioneering concrete structures was also built here then.

During the 1950s, a new master plan for the City of Havana (Plan Piloto) was developed by Sert and Weiner, but these restructuring plans were not implemented. There is still much Spanish-influenced and colourful architecture, such as the old San Carlos and San Ambrosio Seminary. The Lopez Serrano building (designed by Mira) was the tallest residential building before the tall Edificio Focsa 'Y'-shaped, reinforced concrete, Corbusian-style apartment building (designed by Sampora/Esteban) was completed in 1956.

The modern economy remains based on the traditional sugar industry, light manufacturing and food processing, fishing, textiles, alcohol and tobacco, with the famous Habanos cigars. The port remains a significant area of activity around the wharves. The main railway station at La Coubre is located in the southern corner of the city and runs out towards the south and east.

The population of the city is approximately 2.1 million with an expanding metropolitan area. There is a large Chinese community in Barrio Chino who had originally immigrated to Cuba in the mid-19th century via the Philippines. Whilst the city grew extensively during the early 20th century, it has not grown significantly since. It now has 15 municipalities and is administered by the CCPCC, Comité Central del Partido Comunista de Cuba. Housing conditions are often very poor, with many families sharing occupancy in 'cuarteria', as subdivided mansions, and there are many areas of inadequate housing.

The extensive early railway network was developed by the Spanish. Now it is mainly commuter rail run by FFCC to the suburbs from the Central and La Coubre stations. The old Hershey Electric Railway runs from Casablanca station on the eastern side of the harbour to Matanzas parallel to the Via Blanca highway. The statue of Cristo de la Habana in white Carrara marble overlooks the city from Casablanca on La Cabana Hill.

The highway network is extensive, with the A2 ring road which connects with the Via Monumental. This now passes under the harbour entrance in the Tunnel de la Bahia de la Habana to connect with East Havana, an immersed tunnel built by the French and opened in 1958. The A1 connects with the Carretera Central, which is the main east-west highway axis that runs the length of the island across Cuba. Roads are generally in poor condition, and urban public transport is mainly buses, minibuses and bicycle taxis.

Transactional Forces and Typology

With its pivotal position in the Straits of Florida in the Greater Antilles, the natural harbour of Havana on the northern coast of Cuba was a natural point of call and 'carenage' for early maritime shipping. With the land well suited for the growth of sugar cane and tobacco, all that was needed was an abundant source of labour in order to exploit this resource. These factors combined to make Havana a place of settlement, exploitation and trading as the triangular Atlantic trade took hold.

From American support and protection at the beginning of the 21st century, the island eventually secured its own independence. With the challenges and national fragility during the 1930s, much was to change with the Cuban Revolution and the creation of the socialist republic in 1959. With its focus on Marxism-Leninism, this resulted in a state-controlled planned economy that despite its abundant resources failed to remain competitive internationally. After the loss of Soviet subsidies, its economy declined and with austerity and rationing in place, many areas of poverty exist.

With its coastal location, early settlement was to develop around the natural harbour that was afforded protection on either side. As the Spanish established the first development with its formal squares within the compact walled and fortified perimeter, Habana Vieja, the area expanded organically westwards with Habana Centro and a planned neighbourhood in Vedado, with Cerro south of the centre. Connecting infrastructure, both of rail and road, was introduced essentially to support the exploitation and exportation of sugar cane that expanded rapidly.

Havana generally has the characteristics of a tropical island port with radial extensions and linear development on either side along the coast. The urbanised city had newer planned districts to the west of the original Spanish colonial centre. They were enlarged to support growth in the rural economy and extended radially in sectors with individual districts surrounded by an outer ring road to connect with the Cross Island Highway. As growth stalled, limited and localised satellite areas were developed in order to rehouse inner-area inhabitants from areas of inadequate housing.

3.4.2 PORT-AU-PRINCE, HAITI

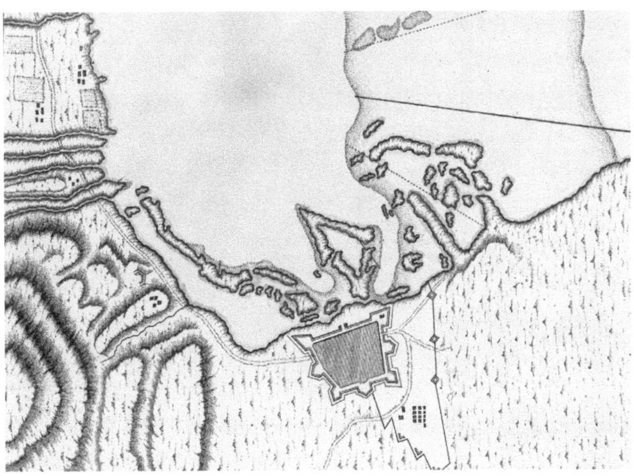

FIGURE 3.4.2 Port-au-Prince – early configuration, 1764 (Bellin).

Source: Maphouse.

Background and Origins

Port-au-Prince is the capital of Haiti, located on the country's west coast in the Gulf of Gonave in the Greater Antilles archipelago of the Caribbean. A natural harbour, the island was originally inhabited by the Taíno tribe from South America with Caribes on neighbouring islands. It is located in the western region of the country in the Plaine du Cul-de-Sac, which is a depression between the saline Lake Trou Caiman and the large, brackish Lake Azuei.

With Columbus's discovery of the islands under the Spanish expeditions, first settlements were established at La Navidad and Isabella in 1500 and created at Santo Domingo on the southern side of the island. Named Hispaniola or Isla Espanola, the Spanish, however, lost control to French 'filibustiers' (buccaneers) who gained control and established a presence at Turgeau Heights at a region that became known as l'Hôpital and developed a port town at Léogane.

The French authorities aimed to retain control with the arrival of a senior administrator on the ship *le Prince* in 1706, hence the name Port-au-Prince. With settlement also at Cap-Francais, Port-au-Prince hence was named the capital of Saint Dominique in 1770.

With ongoing disputes between France and Spain, the island was divided between the two, with the western portion (3/8) being ceded to France, called Saint Dominique, and the eastern part (5/8) to Spain, which came to be known as the Dominican Republic under the Treaty of Ryswick in 1697. With the importation of slave labour from Africa, sugar cane plantations were established across the island and tensions arose between the black-dominated north and Cap-Haitien- and mulatto-dominated south.

In the middle of the French Revolution, the slaves on the island rebelled against the occupying French army in the Haitian Revolution, eventually attaining independence in 1804 with the native name Haiti under leader JJ Dessalines, becoming the first independent nation in the Caribbean to have abolished slavery. For a brief period the country was split in two between the north and the south but was reunified in 1821. The Haitians also aspired to unify the country east and west, but there was strong resistance from the Dominicans such that the Dominican Republic itself became independent in 1844 after the Dominican War of Independence, with its capital at Santo Domingo.

During the 19th century, Haiti suffered several periods of political instability and a period of German influence following the former French influence. The USA, opposing foreign interference under the Monroe Doctrine, intervened and in 1915 occupied Port-au-Prince, taking over and introducing a new constitution.

This was the time to upgrade the basic infrastructure in Haiti with new roads, irrigation and social facilities. Agriculture was improved and cash crops of sisal, cotton, sugar and coffee became significant exports. However, the Haitians resented American occupation, and the USA eventually left in 1934. Haiti retained its independence as a republic and under a new constitution.

Modern City – Buildings and Arteries

After World War II, there was further turmoil within Haiti, and in 1957 Dr Francois Duvalier was elected as president. Known as 'Papa Doc', with his private militia of Tontons Macoutes, people were terrorised and corruption was widespread. Succeeded by his son Jean-Claude ('Baby Doc'), the economy continued to decline until his departure in 1986.

In 1994 the USA intervened again after a coup d'état, with the goal of upholding democracy. Later that year the country was devastated by Hurricane Gordon and in 2004 and 2008 Haiti was hit again by more severe tropical storms. Then in 2010 a massive earthquake struck, killing many and leaving 1.5 million people homeless. All of these severely impacted the whole of the city of Port-au-Prince.

The cathedral, parliament, Palais de Justice and presidential palace in the Champs de Mars was severely damaged, along with the basic infrastructure, main port and airport, schools and hospitals. With its westward-facing port in the bay facing Gonave Island, the city stretches form Cité Soleil in the north to Bicentenaire in the south. The old city is laid out on a grid around the Rue Champ de Mars, with the outer Route de Delmas running to Pétion-Ville. The developed area extends to the Rivière Grise in the east to Carrefour in the south-west.

The city has very high levels of unemployment and a large informal sector. The most notorious areas are Cité Soleil and Bel Air with extremely high levels of poverty. In addition, there are several 'tent cities' and temporary areas for internally displaced persons camps due to natural catastrophes.

The national highways radiate out from the city of Port-au-Prince, but all modern infrastructure is in very poor condition. The UN subsequently retains a strong presence in assisting at all levels of government. The population of the city is approximately 1.2 million with double in the metropolitan area and an increasing number of displaced persons.

Transactional Forces and Typology

With its geographic coastal characteristics in La Baie de Port-au-Prince, and with the mountains of La Chaine de la Selle to the south, sugar cane and cotton have been cash crops of fundamental value to the area. From its historic struggles to its modern geophysical challenges, the capital has been at the centre of disputes exacerbated by occurrences of tropical storms and earthquakes around the fault zone of the Gonâve microplate. Whilst deforestation and erosion have also contributed to difficulties, as the largest city in Haiti, Port-au-Prince's main challenges now are basic infrastructure, accommodation and underemployment of the growing volume of local and displaced persons.

With the west-facing coastal perimeter between the Fort Dimanche and Bicentenaire, the planned centre of the city straddles Blvd Jean-Jacques Dessalines with the harbour to the north at Cité Soleil and Airport Toussaint Louverture. South-east towards Pétion-Ville, there are many unplanned informal areas, and these are growing significantly with the increasing volume of displaced people. As a Caribbean coastal city, Port-au-Prince stretches eastwards into a depressed saline plain to an inland lake bordering the Dominican Republic. Both its natural topography and areas of intense human habitation are under constant climatic challenge as well.

3.4.3 KINGSTON, JAMAICA

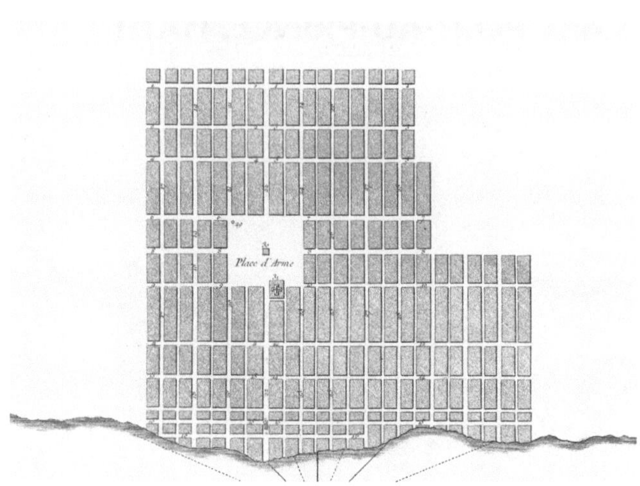

FIGURE 3.4.3 Kingston – early configuration, 1764 (Bellin).
Source: Maphouse.

Background and Origins

Kingston is the capital of Jamaica, one of the largest islands in the Greater Antilles, located on the south-east coast of the island. A large natural harbour protected by a long outlying sand spit, Palisadoes, extends out to Port Royal at the mouth of the Kingston Harbour. The promontory was developed with Fort Charles at the tip with its battery, but Port Royal was badly impacted by an earthquake in 1692 when much of the sand spit liquefied and was destroyed.

The city sits on an alluvial plain alongside Hope River. Since the centre of the island of Jamaica is quite mountainous, Kingston lies in the rain shadow of the central ridge of the Blue Mountains. It is also in the hurricane zone of the Caribbean.

Jamaica was originally inhabited by Taíno peoples and came under Spanish rule after the arrival of Columbus and his followers. In 1655 it fell under the possession of the British, who developed a plantation economy for sugar, based upon the labour of indentured African slaves, which continued until their eventual emancipation in 1838. Other crops, such as cotton and coffee, were also developed. The island had also sustained a large population of 'Windward Maroons', escaped slaves that sought to develop their own free communities.

There are essentially two districts in Kingston, historic Downtown and new Uptown Kingston, each with their own individual neighbourhoods. Historically, after the earthquake, a central gridiron plan between North, East Queen and Harbour streets and Orange and High Holborn streets was drawn up by John Goffe with wide main streets and brick buildings.

New areas of housing were developed after 1848, and in 1872, the government was transferred from nearby inner Spanish Town on the Rio Cobre to Kingston as the new capital. In 1923 the adjoining parishes of Kingston on the water's edge and St Andrew were amalgamated.

The city has a number of large parks that include Emancipation Park and Hope Gardens, which is a large area of botanical gardens. Bordered by the Hope River to the east, the ground rises to the north-east and the Blue Mountains towards hilly Gordon Town and Irish Town.

In 1907 there was a substantial earthquake, subsequent to which building height restrictions were introduced. During the 1930s, the country suffered severely economically under the Great Depression, and it was not until after World War II, with Jamaica having joined the Federation of the West Indies, that the country attained full independence in 1962.

Modern City – Buildings and Arteries

The central Parade Square is now named St William Grant Park, with its individual central neighbourhoods of Rae Town, Denham Town, Jones Town and Trench Town and surrounded by other individualistic quarters and Emancipation and Independence parks. Bounded in the north by Old Hope Road and with Spanish Town Road running north-west towards Washington Boulevard, the city is a densely developed area from the waterfront to the surrounding hillsides.

The city of Kingston has grown extensively but because of its increasing population density, it suffers from poor areas of housing and has a large informal sector and many areas of inadequate accommodation.

Having previously been heavily dependent on the sugar trade economically, bauxite began to be exploited and the tourism industry now prevails. With increasing volumes of visitors, the commercial and financial headquarters of a number of international companies have become established in Kingston and the port and airport have expanded.

With roads radiating out from the centre, some of the infrastructure has been improved, with the main A1 Highway that leads in and out of Kingston across the island to join with Highway 2000 to connect with Montego Bay and northern Ocho Rios. This was developed by Trans-Jamaica Highway (Bouygues and CHEC), built in phases and with the north-south and east-west sections extended to Montego Bay. There was formerly a railway between Kingston and Montego Bay across the island, but this is now closed.

The Norman Manley International Airport (KIN) formerly Palisadoes was developed on reclaimed land in the bay of Kingston harbour where the port has also been enlarged. The island has a mixed economy with agricultural, mining (bauxite), financial, commercial and service industries. Kingston itself has a current population of approximately 0.7 million out of the overall island population of 2.8 million.

As a centre of reggae music, the annual Kingston Carnival is a colourful event. The small offshore islands just to the east of Port Royal that include Gun Cay, Lime Cay and Maiden Cay are small attractive and scenic islands.

Transactional Forces and Typology

After early Spanish rule in Jamaica, one of the largest Caribbean islands, the British focused on the development of sugar cane plantations with local and indentured labour. With its natural harbour, activity developed at Port Royal, around which merchants gravitated and which grew as a centre of trade in the Caribbean. Whilst sugar predominated, cotton and coffee were also produced – bauxite production was a later development. As inhabitants gravitated to the city, adequate accommodation has failed to keep pace with denser settlements, resulting in inadequate housing provision.

With its coastal location and its unusual rocky spit that afforded protection to the harbour, the city of Kingston developed from its coastal edge northwards inland. From a small historic core, its growth has extended outwards in each direction, in a less planned manner, to the outskirts and the foothills. Despite some land extensions for the port and airport, the newer western area around Portmore has been extended around the bay in the direction of former Spanish Town. This has resulted in patterns of expansion radially on the periphery where there are many outer areas that are becoming more populous.

3.4.4 COLOMBO, SRI LANKA

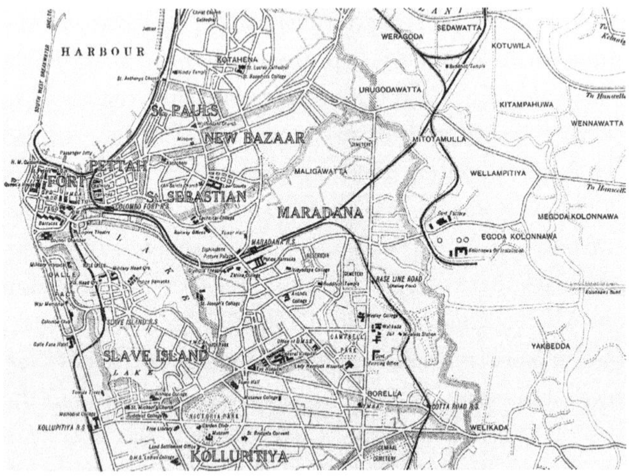

FIGURE 3.4.4 Colombo – early configuration, 1925.

Source: Survey Office.

Background and Origins

Colombo is located on the west coast of the island of Sri Lanka and is the commercial capital of the country. With a large natural harbour, it is strategically placed on the east-west shipping routes across the Indian Ocean. The first settlers arrived from Northern India around 377 BCE and established a settlement on the island at Anuradhapura. It next came under the influence of the earlier Chola dynasty, which was a thalassocratic empire in Southern India that established a city at Polonnaruwa until it was overcome by the Sinhalese king Vijayabahu.

Tamil invaders followed and took over the northern area of the country around Jaffna. Known to earlier traders, the island was then occupied along the coast by Portuguese explorers in the early 16th century, who built a fort at Colombo and annexed Jaffna, taking command of much of the coast and negotiating the sale of cinnamon and other commodities with the Sinhalese king.

In the next century, the Dutch United East India Company (VOC) gained control of coastal Jaffna and also Colombo in 1656 and traded successfully until the island was ceded to Britain in 1815 after the Napoleonic wars. Colombo was made capital of British Ceylon, with new settlements developed around the port peninsula and Oute Stad (outer city), now Pettah market, and the infrastructure of the city was developed in and around the Clock Tower, a former lighthouse. The north-east of the city is bordered by the Kelani River, and with Beira Lake in the centre of the city, a number of canals were built.

The trading economy expanded after the Suez Canal was built, the port was expanded and new buildings were developed, with Victoria Park, now Viharamahadevi Park, and the town hall and other buildings. After the two world wars, when Trincomalee was a naval base, the country gained its independence in 1948 as a dominion. Renamed Sri Lanka with Colombo as its capital, a complete change and mix of cultures resulted, with Sinhalese, Tamils and Moors. Independence Memorial Hall was built in Independence Square surrounded by the denser development of other buildings nearby.

During the 1960s, growth faltered, and a new Democratic Socialist Republic of Sri Lanka was created in 1972. Tensions grew between the Tamils and the Sinhalese, and in 1983 civil war developed that lasted over 24 years, until 2009, and with conflicts over sacred Hindu sites.

The Asian tsunami that occurred in 2004 had significant impact and caused great devastation around the southern coast of Sri Lanka, especially at Galle.

Modern City – Buildings and Arteries

In 1980, the administrative capital was moved from Colombo to neighbouring Sri Jayawardenepura Kotte with the parliament, president and secretariat, Supreme Court and main ministries relocated there. This inland site of Kotte, an old fortress on a triangular site surrounded by two streams of water, was the former capital of the Kingdom of Kotte during the 15th century. Now built on reclaimed land around the Diyawanna Lake, this is the site of a new central parliamentary complex, with only limited government institutions still remaining in Colombo.

The heart of Old Colombo is the Fort District, with its lighthouse, opposite which is the Pettah commercial bazaar area. Along the coast is Galle Face Green, a long promenade along the sea front, with Slave Island and a number of temples jutting into Beira Lake. There is a mix of different colonial-style buildings, such as the general post office, the national museum and the old parliament building.

Farther south is Cinnamon Gardens and its surrounding residential area. The main road axes running to the south are Galle Road and the north-south Baseline Road that eventually connects with the Mount Lavinia suburb in the south and Bandaranaike International Airport (BIA).

Rail is the main form of transport, with the Fort Railway Station connecting the Maradana and Dematagoda stations, as well as urban buses and autorickshaws. The former tram lines have now been discontinued. A new light rail transit (LRT) system has been built and is being expanded across the city.

New businesses have developed in the commercial areas of central Colombo, with the World Trade Centre towers and maritime activity around the busy port of Colombo. With its temples and mosques, significant new commercial buildings include the slender, tall Lotus Tower, Altair straight and sloping towers and One Galle Face residential towers. The new Colombo International Financial City (CIFC) is also under development nearby as a port city under Chinese finance, but it is not yet fully established.

With its importance on east-west trading routes, the Port of Colombo had been extended as a major new area was created on reclaimed land for a new container port protected by breakwaters. As the busiest port before the northern ports of Trincomalee and Jaffna, the port is now impacted by the newly realised Hambantota Port competing under China's international Belt and Road Initiative.

The population of the city is approximately 0.75 million and 5.6 million in the wider metropolitan area.

Transactional Forces and Typology

As a substantial island in the Indian Ocean, from its early settlers to its later colonisers, Colombo has found itself on the active east-west maritime trading routes between Europe and from South-East Asia via Suez and the Malacca Strait. As early Europeans sought a toehold, the Port of Colombo became a regular stopping point for shipping travelling around the Indian Malabar and Coromandel coasts and beyond. From its original harbour to its now extended port city, cinnamon, tea, textiles and other commodities have been exchanged for modern imported goods. As a capital city with a mixed population of Sinhalese and Tamils, Indian and Moors, it has an impressive coastal perimeter and inland forestry resources.

As a westward-facing coastal port in the Indian Ocean, Colombo is defined by its central core and harbour, from which all communications radiate out eastwards inland. With its roads and railways, and urbanised area now complemented by an LRT system, it has evolved from a colonial capital to a large modern city. With the new core around Kotte City and parliamentary centre by Diyawanna Lake, Colombo is evolving into having a bi-polar centre with other active development nodes in annular locations around the city.

3.4.5 MAKASSAR (UJUNG PANDANG), SULAWESI, INDONESIA

Background and Origins

Makassar (Ujung Pandang) is the capital of South Sulawesi, formerly Celebes, one of the four great Sunda islands of Java, Sumatra, Borneo and Sulawesi. It is located on the southern peninsula of the island, a port on the Makassar Strait of eastern Indonesia. Low lying but with mountainous geography, having very early prehistoric settlements, its indigenous population was organised under the Gowa, Luwu and Bone kingdoms. The Gowa was based at Sunggaminasa, just south of Ujung Pandang.

The Portuguese settled and established a naval base around 1525 CE, which was followed by Dutch occupation under the United Dutch East India Company (VOC), becoming Sulawesi's principal port and naming it Makassar.

It became a leading centre in the 17th century for trading spices, cloves and nutmeg with the 'Spice Islands', or Moluccas (Maluku). In 1669 the VOC developed Fort Rotterdam and a walled town called Vlaardingen as a trading centre. It developed also as a fishing centre and as a source of fine veneer woods like ebony, sandalwood and rattan, as well as makassar oil for hair treatment.

With its monsoon climate, Makassar acted as a strategic port in South Sulawesi in the Java Sea between Asia and Australia, despite continuing conflict between the VOC and the Makassarese, both Gowa and Bone. In 1905 it was to become eventually a colony of the Netherlands East Indies until it was occupied by the Japanese during World War II as a key military base.

Modern City – Buildings and Arteries

After the war, initially part of East Indonesia, Sulawesi became part of the unitary Republic of Indonesia in 1950. Then, in 1964, it was divided into four separate physical provinces – North, Central, South-east and South – with Makassar as the capital of the province of South Sulawesi. Briefly known as Ujung Pandang again, the name reverted to Makassar in 1999.

The Port of Makassar, Soekarno-Hatta, has been developed as a major facility here with its new port and container facility which dominates the coastal area and with new breakwaters that have been built offshore. Sultan Hasanuddin International Airport (UPG) is located to the north-east of the city.

Economically the service sector has grown and areas such as the Makassar Industrial Estate have been expanded. In addition, a major new area is being developed, known as the Centre Point of Indonesia (CPI) or Citraland, opposite Losari Beach with the 99 Domes Mosque and the twin towers development on reclaimed land, which acts as a major new coastal landmark.

With its 'becak' (pedicabs) and 'pete-pete' (minibuses) for transport there is also a small BRT network across the city as new highways have been developed in the growing city and expanding metropolitanised area.

The population of the city, mainly Muslim, is approximately 1.4 million and double in the metropolitan area of Mamminasata, which includes the Regency areas of Maros, Sunggaminasa (Gowa) and Takalar.

Transactional Forces and Typology

The southern peninsula of Sulawesi was occupied by Makassar tribes who, due to their marine situation, became skilled sailors and set up trading networks with other islands in the Indonesia archipelago, with the Malay peninsula and the northern coast of Australia. These included fishermen who collected 'trepang', or sea cucumber, from the seabed. In addition, with rice farming, communities were both focused on agriculture and fisheries, and subsequently forestry. The region has very fine fauna and flora; however, it is suffering the loss of mangrove swamps, lowland forests and extensive deforestation as population pressures have increased.

Sited on the coast between the meandering Tallo River in the north and its wetlands and the Jeneberang River in the south, by the Western Cape point at Barombong, this is a generally forested region. The core of the city developed around the original Fort Rotterdam, around which has extended the significant and modern Port of Makassar. With its main axes running north and south, Jalan Pangeran Pettarani and Jalan Veteran Selatan parallel to Canal Panampu, development has extended eastward towards the river and north-east towards the airport and Jalan Insinyur Sutami toll road towards Maros. The new modern central area of Citraland is a feature that is being developed with its reclaimed land and landmark twin towers on the coastal point known as the Centre Point of Indonesia. There are many built areas having Indonesian characteristics and district neighbourhoods that house dense areas of population and more extensive informal areas of individual 'kampungs' in low-lying areas.

3.4.6 AUCKLAND, NEW ZEALAND

Background and Origins

The city of **Auckland** (Tamaki Makaurau) is located on the North Island of New Zealand (Aotearoa), with the Hauraki Gulf to the east and the Pacific and Manukau Harbour to the south-west opening out to the Tasman Sea. Unusually situated on a narrow isthmus, the city looks out with its harbours on two separate ocean sides. To the south is a range of volcanic mountains, the Auckland Volcanic Field, that is generally dormant, but with geothermal springs. Located out in the South Pacific on the International Date Line, Auckland is the most populous Polynesian centre.

The area has been occupied since around 1000 CE with the first Polynesian Kupe tribal settlements. It was not until 1769 that Europeans, under Captain Cook, discovered and mapped the islands with their indigenous Maori population. In 1840, the British government claimed sovereignty over the island, which was agreed with the Moris at the Treaty of Waitangi in Northland, and the city of Auckland was declared the capital initially, but it was subsequently transferred south to Wellington in 1865. Despite early local disputes, an agriculture economy was developed based essentially on sheep and cattle, wool and meat, and an exporting trade developed.

The infrastructure and roads were developed towards the south of the island and railways were introduced. In 1907, the country's status changed to a dominion rather than being a British colony. The Australian and New Zealand (ANZAC) regiments performed influentially in the world wars, especially World War I, to whose memory the Auckland War Memorial Museum is dedicated.

Modern City – Buildings and Arteries

The sea and the waterfront are always evident on either side of the city. Waitemata Harbour is the principal access from the sea, and with the construction of the Auckland Harbour Bridge in 1959 enabling across to the North Shore, this has become a dominant feature of the Harbour. Farther east is Mission Bay and Rangitoto Island beyond. With its harbours, marinas and yachts, sailing is a dominant activity in the region.

The city is focused around the north-facing waterfront, with the Old Customs House and Ferry Building between

Queen's and Princes wharves. At the centre is the town hall, with Victoria Park and Albert Park on either side of central Queen Street and the Old Government House. Auckland Domain and the Winter Gardens are located beyond. Now the tall Sky Tower, opened in 1997, dominates the skyline with its observation deck and illumination at night.

The terminus of the North Island main trunk railway is at the Britomart Transport Centre, which feeds the Western, Southern and Onehunga railway lines via Newmarket, now each electrified. The City Rail Tunnel Link is under construction to connect Mount Eden Station to Britomart so as to turn this into a through station.

The main highways pass through and around the city that connect with State Highway #1 to the north across the Harbour Bridge. The Central Motorway Junction (CMJ) is a major complex 'X' junction of State Highway #1 with Highway #16 near the centre of the city. The two main arterial roads, the Great North Road and the Great South Road, lead outwards from the city.

In terms of public transport, there was formerly a tram network across the city, but it has been withdrawn. The Ferry Terminal handles a large volume of coastal traffic. The city has a high rate of car ownership, with the suburbs typically spread out at low density, generally single-family homes on quarter-acre sites. Auckland Airport (AKL) is located on the south-western harbour side by Mangere.

Auckland is now the major economic and financial hub in New Zealand, and its manufacturing sector has also increased. In recent years in-migration has increased from the Pacific islands and South Asia and beyond. The population of the city is approximately 1.5 million.

With its complex geography, twin harbours and rugged coastline, it has separate inlets and districts, fine beaches and forests, natural areas, parks and scenic outlying areas, a true 'City of Sails'.

Transactional Forces and Typology

Given its geographical setting with water on either side of Auckland in the Pacific Ocean and the Tasman Sea, early marine activities revolved around sealing and whaling. Inland, the rich and undulating land was suited to sheep farming, forestry and fishing. As settlers took advantage of the location from 1840 onwards, development occurred on the eastern side of Hauraki Gulf and as the local economy developed with trade around the harbour protected by the land of Narrow Neck. With its outer Barrier Islands and the Coromandel Peninsula, the region attracted new migrants. With the export of wool and frozen meat, this was then extended to butter. Subsequently, at the beginning of the 20th century, vineyards were also established, which have further contributed to the regional economy.

With its most unusual natural setting, Auckland's main port is on the eastern side via Waitemata Harbour. With its rugged coastal features, the terrestrial infrastructure connections run diagonally across the city, now shaped by the Auckland Harbour Bridge, which connects the city with Takapuna and the north beyond Northcote Point. As a generally low-density city, with its marine perimeters, it is outward looking with its scenic surroundings and natural environment around its centre by the northern wharves and harbour. Its physical constraints as an isthmus restrict over-development. This is being contained effectively within the isthmus and on the northern harbour side and at North Shore opposite outside the wider and more rural Auckland region beyond. The following figures illustrate the modern configuration of the island cities as described and the evolution of their modern population.

FIGURE 3.4.5 Havana – modern configuration.

FIGURE 3.4.6 Port-au-Prince – modern configuration.

FIGURE 3.4.7 Kingston – modern configuration.

FIGURE 3.4.8 Colombo – modern configuration.

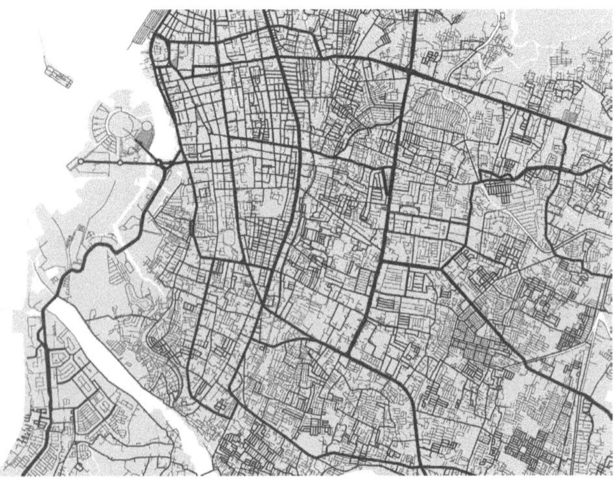

FIGURE 3.4.9 Makassar – modern configuration.

FIGURE 3.4.10 Auckland – modern configuration.

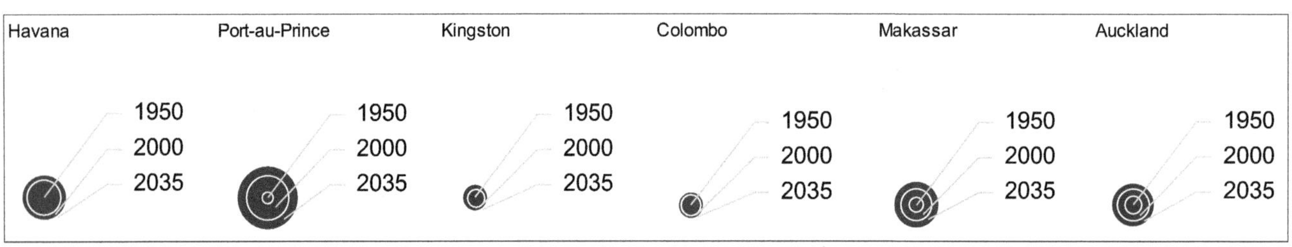

FIGURE 3.4.11 Island city populations.

Data source: Global Database of Metropolises 2020 – UN-Habitat (2020), UN DESA Population Division.

Further Reading

HAVANA

Estrada, Alfredo J – *Havana, Autobiography of a City*, Palgrave Macmillan, 2007.
Garcia, Guadalupe – *Beyond the Walled City, Colonial Exclusion in Havana*, University of California Press, 2016.
Hugh, Thomas – *Cuba, or the Pursuit of Freedom*, Eyre & Spottiswoode, 1971.
Montalvo, Maria & Hugh, Thomas – *Havana, History & Architecture of a Romantic City*, Monacelli Press, 2000.

PORT-AU-PRINCE

Duval, Carline – *Mobilité et Transport dans l'Aire Metropolitaine de Port-au-Prince*, Bibliothèque Nationale de Haiti, 2023.
Orner, Peter & Lyon, Evan – *Lavil: Life, Love and Death in Port-au-Prince*, Verso, 2017.

KINGSTON

Ehrengardt, Thibault – *The History of Jamaica 1494–1838*, Dread Editions 2015.
Howard, David – *Kingston, a Cultural and Literary History*, Ian Randle Publication, 2004.

COLOMBO

Corca, Ishvari – *Glimpses of Colombo*, Municipal Council of Colombo, 1988.
De Silva, KM – *A History of Sri Lanka*, Penguin Books India, 2005.

MAKASSAR (UJUNG PANDANG)

Ricklefs, MC – *A History of Modern Indonesia*, Stanford University Press, 1993.
Sutherland, Heather – *An Indonesian History, Personalised Politics in Makassar and South Sulawesi*, NUS Press, 2024.

AUCKLAND (TAMAKI MAKAURAU)

Reed, AW – *Illustrated Account of the Founding and Development of the City of Auckland*, 1955.
Stone, Russell – *From Tamaki-Makaurau-Rau to Auckland*, Auckland University Press, 2013.

3.5 Mountain Cities

South American Cordillera – Himalayas – Arabian Plateaux – African Rift Valley

Bogotá, Quito, La Paz, Kathmandu, Tehran and Addis Ababa

3.5.0 INTRODUCTION – MOUNTAIN CITIES

Many countries, and particularly landlocked countries, have been determined by their surrounding natural topographical features which may act as barriers to accessibility. From plains and savannah to forests and hillsides, mountain ranges often respected historically for their grandeur also separated human habitation, land uses and animal husbandry.

This led to the search for passages through inhospitable terrain and the development of way stations for travellers as points of refuge. In different continents there are imposing mountain ranges around which cities developed:

- South America: Andes Cordillera (Bogotá, Quito, Cusco and La Paz).
- Americas: Rocky Mountains (Denver, Salt Lake City, Albuquerque and Santa Fe).
- Asia: Himalaya Mountains (Kathmandu and Lhasa).
- Africa: Great Rift Valley (Addis Ababa, Nairobi and Arusha),
- Europe: Alps (Innsbruck, Grenoble and Trento) and Pyrenees (Andorra).

Each of these encompass not only their own ecological regions, but also contain different areas for human habitation. In each continent, settlements have located either on mountain passes or plateaux, which have their own microclimate, around which particular cultures with their own crafts have gravitated, depending on the available local resources, materials and soils.

This has been determined by the surrounding natural environment with different forms of settlement and types of habitat, as exemplified particularly by Bogotá, Quito, Cusco and La Paz in South America and Kathmandu and Lhasa in the Himalayas, as well as individual places such as Addis Ababa in Africa.

As lines of communication were established and inland roads developed, exploitation of mineral resources was facilitated, with previously inaccessible resources extracted. Of high value, these became of increasing interest for trade and foreign exploitation, from precious silver and gold to mineral ores and more recently rare earth materials.

3.5.1 BOGOTÁ, COLOMBIA

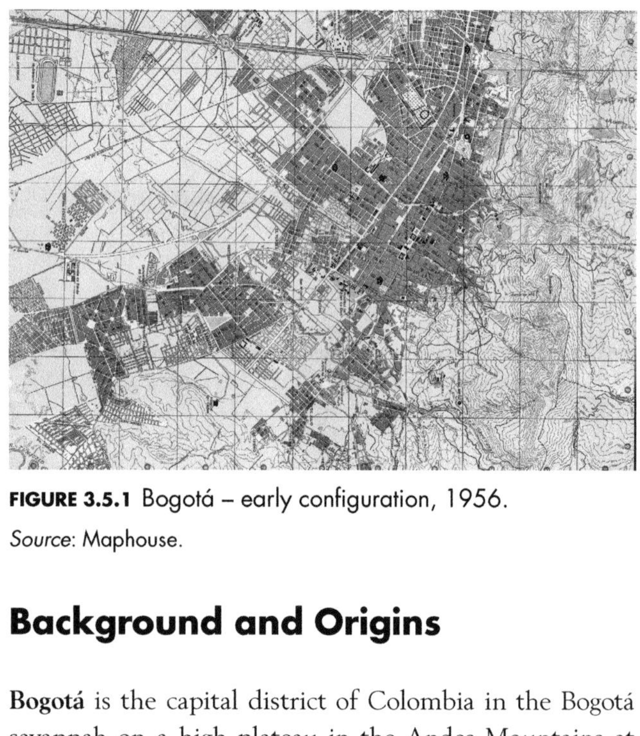

FIGURE 3.5.1 Bogotá – early configuration, 1956.
Source: Maphouse.

Background and Origins

Bogotá is the capital district of Colombia in the Bogotá savannah on a high plateau in the Andes Mountains at 2,640 m., bordered to the west by the Bogotá River, a tributary of the Magdalena River, and to the east by the Cordillera Oriental. Originally inhabited by indigenous peoples speaking the Chibcha language, the Muisca settled in the Altiplano in an agriculture-based confederation of settlements. They were active in exploiting local mineral sources, in particular gold and silver, carbon and copper, in addition to the salt mines in the area.

In 1533 CE the Spanish conquistador Jimenez de Quesada launched an expedition by land and river into the interior

DOI: 10.4324/9781003353348-16

of the highlands and up the Valle del Rio Magdalena. They reached a place named Bacatá where he overcame the local tribes and 'zipa' and set up a military encampment. The settlement of Nuestra Senora de la Esperanza was extended and renamed Santafé within an expanding part of Spanish colonisation in the region known as the Audiencia de Santafé in the New Kingdom of Granada. This grew around Plaza Mayor and Calle Royal and was gradually extended.

With abdications of the House of Bourbon in Spain and of the Spanish oversees administration, local juntas were created in the Latin American colonies. Accordingly, the Junta de Santa Fe was established in 1810 with the Declaration of Independence by Colombia. This led to tensions, and in 1819 Simón Bolívar, under a campaign of resistance and following the Battle of Boyacá Bridge, succeeded in liberating Santa Fe, which he renamed Bogotá, which then became capital of the new State of Gran Colombia that comprised Colombia and Venezuela, Ecuador and Peru.

After periods of ongoing civil war, in 1863 the United States of Colombia was formed with the federal states of Colombia and Panama. However, after the 1,000 Days War in 1902, Panama eventually seceded to leave Colombia solely as an independent republic. At the end of the 19th century, rail infrastructure and trams were introduced, Bogotá grew as the political centre of the country and in-migration developed from rural areas, especially the Eastern Highlands. Whilst the economy gradually expanded, there were ongoing territorial disputes with Peru in the highlands of the Amazon basin.

Modern City – Buildings and Arteries

After World War II, there were further disputes and periods of violence, La Violencia (1948–58), which included the El Bogotazo riots in 1948. From the 1960s onwards, conflicts continued with guerrilla groups, paramilitaries and drug cartels, such as in Medellín in 1993, and asymmetric battles were ongoing across the city. It was not until 2016 that this was eventually settled to bring an end to the ongoing conflict under the Colombian peace process.

Subsequently, in the 21st century, the economy has diversified from its emphasis on agriculture to light industry and the service and financial sectors. Being rich in natural resources and minerals, chemicals and manufactured and scientific goods have increased and financial services have grown. Much of the country's energy is now being produced from hydro-electric schemes constructed in the high mountains.

The historical centre of the city is La Candelaria, with its old colonial buildings and Capitolio Nacional (Congress), Palacio de Justicia and Palacio Lievano (Alcadia – city hall) located around Plaza de Bolívar, formerly Plaza Mayor and dominating Catedral Primada. The oldest churches in the city are Iglesia de San Francisco, de la Vera Cruz and La Tercera.

The areas north-east towards the hills are the more affluent areas, with Chapinero containing Zona 'T' – Rosa and Zona 'G' – Gourmet and Parque de la 93. Residential Usaquen is beyond and contains a market and leafy park. To the south are poorer areas where extensive informal settlements and barrios are found, as well as many migrant and displaced workers in and around Cuidad Bolívar, Kennedy and Bosa.

Based on the original grid layout at the centre of the city, the main avenues (carrera) run north-south and the streets (calle) east-west. A comprehensive network of cycle paths, 'ciclorutas', has been built across the city, with traffic-free zone 'Ciclovia' held each Sunday on the main streets. The city mayor has targeted vehicle congestion at the centre and improving public services in the suburbs.

Within the city of Bogotá, public transport is provided by buses, now mainly electric. The Trans-Milenio dedicated rapid bus transit network of articulated buses was introduced to increase capacity on the main north-south avenues (Caracas, Americas and Jimenez) and radials. A new metro is also planned. The main highway network is connected with the Autopista Norte and Sur. The main International Airport, El Dorado (BOG), is sited in the north-west.

The City Business Centre is around Calle 19 and Carrera Septima (7th Avenue), often known as Santa Fe. Centro Internacional is an adjoining business district, north of Calle 26 with its major new commercial developments. Nearby is the Maracena residential district and the Bullring. Many modern shopping malls have also been built. The tallest stepped BD Bacatá tower is under construction here, surpassing the earlier Colpatria building in height.

The industrial zones are on the outskirts and the main university is located at University City. The centre of the city contains many museums, libraries and the national archive, national park and botanical gardens. To the east the high Cerro de Monserrate hill overlooks the city.

The city is divided into 20 localities and is administered as the Capital District of Colombia. It is surrounded by the growing Department of Cundinamarca. The population of the city is approximately 7.5 million and 10 million in the metropolitan area with a large volume of migrant workers.

Transactional Forces and Typology

Trade was the main activity of the Muisca in this region with salt and coca, maize and potato, cotton and textiles. The subsequent discovery of emeralds and gold encouraged new prospectors, but coffee, palm oil, sugar and rice became additional staples. In this mountainous enclave, it was the Spanish who were to change the environment with their settlement and development of the rectilinear city around Plaza Mayor in accordance with Spanish settlement principles. With the attraction of valuable minerals to new prospectors, the city's destiny was to change with liberation under Simón Bolívar and subsequent consolidation of Bogotá as the capital of independent Colombia.

From the organised Spanish central plan of Santa Fe, Bogotá has grown into a substantial capital city nestling in the Cordillera mountains. With its centre of old Santa Fe and Centro Administrativo, it has evolved linearly running north to Chapinero and El Retiro. New districts are emerging to the south and west where densely populated barrios have been developing along new diagonal axes. With its growing districts having individual characteristics, it has a civic centre and dynamic commercial core along with distinctive outlying neighbourhoods driven by the 'Vital Neighbourhoods Programme'. Constrained by its eastern range of mountains, its extensions are of an axial nature, with linear development along the main thoroughfares. Cosmopolitan and dynamic, it retains its colourful Latin American culture.

3.5.2 QUITO, ECUADOR

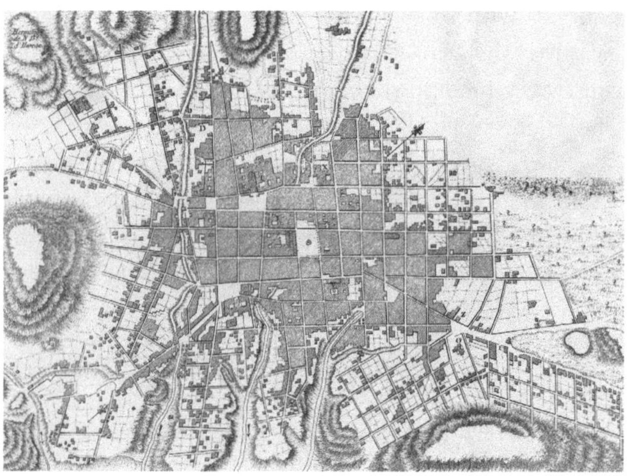

FIGURE 3.5.2 Quito – early configuration, 1764.

Source: Maphouse.

Background and Origins

Quito, as the capital of Ecuador, is located in the highlands close to the equator. At an elevation of 2,850 m. by the slopes of the Pichincha volcano on a plateau by the Guayllabamba river valley, it lies in the 'Avenue of Volcanoes', which includes nearby Cotopaxi that rises to 5,900 m. Quito was the site of ancient settlements with peoples of the Cara and Quitu culture, who extracted obsidian from the igneous volcanic rock.

In the 15th century CE the Inca civilisation extended over the area from the Peruvian Highlands and Cusco with their Quechua language and sun worship. Local roads were developed and stonework, textiles and animal pastoralism pursued. However, this was interrupted by the Spanish conquistadors under Francisco Pizarro, who invaded the coast around Guayaquil. In 1534, the Spanish, under Sebastián de Benalcázar, founded the Villa de San Francisco de Quito and developed it within the Real Audiencia de Quito.

The central area was laid out along with churches and convents of the different monastic orders, which was the beginning of several years of colonial development. Subsequently Quito became an artistic centre of the New World with its School of Quito–style in both art and architecture.

In 1822, seeking independence, José de Sucre, under Simón Bolívar, led local troops into a successful rebellion at the Battle of Pichincha on the slopes of the volcano. Becoming part of Bolívar's Gran Colombia, with neighbouring parts of Colombia and Venezuela with Ecuador, it was to attain its own independence in 1831. The 19th century was characterised by considerable political upheaval and with a series of different, often ex-military, rulers and border disputes with neighbouring Peru in the Oriente around the source of the Amazon. A number of significant earthquakes also occurred during that century.

Modern City – Buildings and Arteries

The Ecuador-Peruvian War was ended with the Rio Protocol in 1942, which aimed at reaching agreement on the borders. However, this was prolonged and not finalised until 1998 with the Brasilia Presidential Act. With a number of intermittent civilian and military dictatorships, the emergence of the Amer-Indian populations' influence and the search for land reform, later governments sought austerity in the economy. The discovery of oil in the Amazon basin and the development of a pipeline in 1972 enabling its exploitation led to the beneficial development of the economy, which had been primarily agriculture based with bananas, potatoes, coffee, cocoa and sugar, as well as textiles and weaving.

The city is divided into the old historic city, Centro Historico, with 48 blocks, and the new city and southern populous area. In the historic city centred around Plaza de la Independencia is Palacio de Gobierno and Palacio de Carondolet together with Palacio Arzobispal and the new city hall (Administración Urbana).

Just adjacent is the Cathedral of Quito and the exquisite church of La Compania de Jesu, La Concepcion and La Merced, and also the church and convent of San Francisco and the fine churches of El Sagrario and Santo Domingo. There are also the monasteries of San Agustín and del Carmen Alto. Off Plaza Santo Domingo is the active old La Ronda with its white colonial houses.

The New Town radiates along five avenues out to the north-east past the Parque La Alameda to the large Parque El Ejido. Here is the Palacio Legislativo and more commercial areas and the Mariscal Sucre area and museums.

Overlooking the city to the west is Cerro Panecillo Hill at 3,000 m. with the statue of the Virgin Mary and with its TeleféricQo cable car to the top of Cruz Loma at 4,270 m. to the north-west. The Parque Bicentenario and Parque Carolina and larger Parque Metropolitano are sited to the east.

In terms of urban transport, there are numerous buses, taxis and MetrobusQ rapid transit lines, and the new Metro MDQ is under construction from Quitumbe in the south to El Labrador in the north with intermediate stations at El Recrio, La Magdalena and Universidad Central. Railways were developed early in Ecuador now run by EFE and with the main station at Estación de Chimbacalle.

There are two main north-south highways running on either side of the city, Avenida Occidental and Avenida Oriental (Periferico), and the Panamericana Highway by-passes the city to the east.

There are high levels of poverty, however, in the city and it has a large informal sector. The population of the city is approximately 2 million and 3 million in the metropolitan area.

Transactional Forces and Typology

From its indigenous Quitu origins, the city was first shaped by the Incas in their expansion beyond Cusco. Subsequently with colonisation by the Spanish, the settlement was developed with a central core of square blocks and the introduction of monasteries and convents. As a valley in the Cordillera mountainous region, with cinnamon forests, it was Pizarro who discovered a route through the mountains to the east to reach the Napo River that runs eastwards into the Amazon. With its animal grazing and horsemen, the area was taken over by the Spanish seeking mineral riches in the region.

The city of Quito is elongated, running north-south between natural valleys within which the historic core was settled by the Spanish. Narrowly constrained, it has individual districts shaped by the terrain with one area of urbanisation stretching out beyond to the east at Tumbaco and the Los Chillos Valley. There are interspersed green spaces within the city, and the southern area is more industrialised and contains many areas of informal settlements. As a narrow mountain city, it is constrained by its development just to the north and south along trading routes. With developable land at a premium, settlements are evolving erratically where service provision of utilities is challenging.

3.5.3 LA PAZ, BOLIVIA

Background and Origins

La Paz is located in the centre of Bolivia south-east of Lake Titicaca in a canyon in the Rio Choqueyapu surrounded by the Altiplano Mountains. Occupied by indigenous Aymara people, the first settlement was at Laja on the trading route from Potosi and Ouro to Lima, before being relocated to Chuquiago Marka. At an elevation of 3,650 m. in the mountains and overlooked by the triple peaked and snow-capped Mt. Illimani, it was developed by the Incas in the 14th century CE.

In the 16th century, with the arrival of the Spanish conquistadors, the region known as Charcas was developed and a city named Nuestra Señora de la Paz was established by conquistador Alonso de Mendoza in 1548. The main square was laid out with its cathedral and public buildings around La Plaza de los Espanoles, under the firm control of the Spanish.

In the region, mining became most significant with the finding of gold nearby and silver at Potosí, where people were forced to work under the mit'a tax. In the 19th century, the freedom forces grew during the Spanish-American Wars of Independence in Alto Peru led by Pedro D Murillo and eventually becoming an independent republic in 1825. Then, in 1898, La Paz became the seat of government of the pluri-national State of Bolivia with the constitutional capital remaining at Sucre (Chuquisaca).

Modern City – Buildings and Arteries

The city of La Paz has developed physically around a narrow canyon through which the Rio Choqueyapu has been channelled and built over, now very polluted and which divides the city.

There are seven main districts of La Paz, the principal ones being Zona Centro and Cotahuma District. The former contains the old quarter, Casco Viejo, with its rectangular blocks around Plaza Murillo, around the long avenue of El Prado.

To the east of El Prado beyond Plaza Murillo is San Juan and Miraflores, a wider area with Parque Urbano Central and the university. Here are the two cathedrals, San Francisco Basilica and Metropolitan Cathedral, and also the government Palacio Quemado (burned palace). Farther east is the long Avenida Busch, running north-south between Plaza Triangular and Plaza Gualberto Villaroel and out along the axis of Avenida de las Americas to the north-east.

To the west of El Prado is Rosario and the Plaza San Francisco and the large Mercado Negro. After San Pedro and Plaza Sucre is the main neighbourhood of Sapocachi with the Plaza Eduardo Avaroa, Kantutani and San Jorge areas.

El Alto is a district to the west, a fast-growing commercial centre comprising La Ceja District and Mercado 16 de Julio and with international airport nearby. Three road axes run south-westwards which define new sectors of development and Ciudad Satelite.

There are also two separate northern and southern districts – the north, which is more elevated and mainly industrial, has a large national brewery. The south is the lower and more residential Zona Sur, with Calacoto by the Rio Irpavi bounded by major highways.

Downtown is densely developed and with increasing commercialisation and many international businesses located there. The tallest commercial buildings are Torre Girasole, del Posta and Azul.

The modern industrial economy is based on mineral processing with its extensive tin deposits, and on light manufacturing and food and agro-processing.

In terms of the residential sector, there is a real deficiency in adequate housing and with many self-help developments, an extensive informal sector.

The city has a developed highway network with the main tolled Autopista La Paz – El Alto that connects with the fast-growing neighbourhood at altitude of El Alto where El Alto International Airport (LPB) is located.

Public transport is mainly buses which run from La Paz bus station (designed by Eiffel), which was originally designed as a combined bus and train station. With its complex topography, the city benefits from a modern 30 km. network of eight aerial cable car lines, 'Mi-Teleférico', supplied by Doppelmayr, with the first red, yellow and green lines erected that cross less accessible terrain.

Fresh water supply to the city is fed from the neighbouring glaciers of Zongo, Tuni and Condorri in the Cordillera mountain range. However, due to climate change and generally receding glaciers, this is becoming a yet scarcer and limited resource.

The population of the city is approximately 0.75 million with approximately 2 million in the metropolitan area.

Transactional Forces and Typology

Traditionally La Paz developed as a mountain trading centre with agriculture and textile products. On the route between Lima and Potosí, it was minerals that were to shape its development with the discovery of tin at Ouro and silver deposits at Potosí and also gold. With its rich mines, the road network was developed to connect with them as the Inca civilisation extended over the Aymara. As Alto Peru was taken over by Spanish conquistadors who were granted encomiendas, the landscape was changed. However, tensions continued to the present day between farming and mining interests until the nationalisation of the mines in 1952 and gas in 2006. From tin and silver, attention is now turning to exploitation of another scarce natural resource – rare lithium – recently discovered in the region.

As a city nestling in a mountainous region in a river canyon, its topography constrains its development. From its central planned Spanish colonial core, Casco Viejo, Sapocachi and Miraflores are planned gridded areas. Elsewhere, however, the city has developed organically in upper residential areas of challenging hilly terrain before lower industrial areas in flatter terrain. With a large informal sector, new satellite areas have been planned, for example Zona Mariscal Santa Cruz in its hexagonal blocks on the south-western periphery of the city. A high and densely populated city, it is shaped by its enveloping natural rocky terrain as an unusual mountain capital high in the Andean Altiplano.

3.5.4 KATHMANDU, NEPAL

Background and Origins

Kathmandu is the capital of Nepal, located in the Kathmandu Valley on the Indo-Gangetic plain, at the confluence of the Bagmati and Vishnumati rivers, the gateway to the Himalayan mountains. The city was known initially as Kantipur, from the Sanskrit words of 'kanti' – beauty – and 'pur' – place – the 'city of light'. Its elevation is approximately 1,400 m. There is evidence of very early settlement here at two places named Yambu and Yengal, and of the first Buddhist monasteries in the region. Its location served as a transit point for trade, especially of salt between India and Tibet. The city developed and was laid out in the form of a sword with a series of eight surrounding barracks. However, setbacks have occurred due to periodic earthquakes.

From 1200 CE onwards, during the Malla dynasty, considerable developments were undertaken in the city with the three Durbar Squares, fine temples and palaces, built like pagodas. Education came to be based essentially around the monasteries. There was a sophisticated local culture amidst the three fortified cities in the region of Kathmandu, Patan and Bhaktapur. This era was superseded by the Gurkha era after the conquest and unification of Nepal after the battle of Kirtipur and taking over of the Kathmandu Valley. The Shah dynasty established the Kingdom of Nepal. However, conflicts continued and with a series of massacres, plots and assassinations during the 19th century, it was an uncertain period.

This was exacerbated by the Nepalese invasion of Tibet and the Sino-Nepalese War. The country suffered at that time in being effectively a buffer state between China and the provinces of British India, with the Anglo-Nepali War, during which the strength of the Gurkhas became evident. Eventually, in 1923, a Nepal–Britain Treaty of Friendship was agreed upon. Following India's independence in 1947, Nepal established a parliamentary democracy in 1951 with the ending of the 100-year-old Rana dynasty.

This was modified to a tiered national 'Panchayat' system of fixed assemblies without political parties in 1960. This enabled some progress in development and modernisation to be made under a series of five-year economic plans and resettlement plans and with the building of the new East-West Highway to improve access. At this time new building developed on the outskirts that started to change the pattern of development.

Modern City – Buildings and Arteries

In the late 1980s the government started to favour China over India in its Treaties of Trade and Transit, causing considerable friction. Accordingly, in 1990, the national Panchayat was dissolved, the People's Movement ended the monarchy, challenged by the Maoist party, which led to a period of conflict with the Nepali civil war. Finally, the country became a secular state and declared a Federal Democratic Republic in 2008 with a new constitution in 2015 and divided into seven individual provinces.

The main axis of the city is north-south with the Kanti Path leading down from the former Narayanhiti Palace with the Durbar Marg in parallel, which run on either side of the Tundikhel central parade ground with the Rani Pohkari Lake (Queens Pond). To the east along Prithvi Pass is Singha Durbar, which houses Nepal's Parliament with its extensive courtyards and ministries. Juddha Sadak (New Road) connects west to the old city via Ganga Path into Durbar Square. The city retains a fine series of monuments despite terrible destruction in the catastrophic 2015 earthquake. These comprise Durbar Square and Basantapur Square with its palaces and courtyards, Hanuman Dhoka Durbar Royal Palace, Kumari Bahal and Kasthamandup (House of Wood) and Basantapur Tower and Royal Palace.

The Hindu temple complex at Pashupatinath is beside the Bagmati River, and outside the city are the splendid Buddhist stupas of Boudhanath to the north and Swayambhunath to the west. The city is a most decorative place with temples and religious art developed over the centuries, with sculptures, carvings, Paubha scroll paintings and colourful prayer flags and wheels in evidence. This is elaborated upon in its many annual religious festivals and fairs. Nepalese culture is very strong with its multi-cultural origins and religions and active artisanal markets.

Nepalese architecture also is distinctive with its stupas and wood and brick pagodas and temples, with brick cores and multi-tiered roofs with timber struts often elaborated with carvings, clay tiles, copper and brass. With its individualistic streets, arcades, courtyards and terraces, the cityscape is a vibrant mix of Hindu 'shikhara' and Buddhist 'vihara' and occasional 'dharmasala'.

As an important trading centre off the Silk Route, it is now a significant commercial centre in the region. With its handicrafts and carpet-making supplemented by manufactured garments, it is an economically active city. Improved highways, such as the BP Highway and others across the region, are assisting in the transport of a large number of seasonal visitors out to the mountains also facilitated by cycle tracks and ropeways. Tribhuvan International Airport (KTM) has a tabletop runway with frequent flights to Lukla. The population of the city is approximately 1 million people and 3 million in the metropolitan area, with a large number of people gravitating to the centre. It has a high density of approximately 29,000 pers./sq. km.

Transactional Forces and Typology

It was during the Malla dynasty that the three city-states of Kathmandu, Patan and Bhaktapur with their Durbar Squares dominated this Himalayan region during the 13th century. With corresponding trade routes from Tibet to India through the valleys, these acted as centres around which local produce and crafts could be exchanged. Despite its often harsh climatic environment, these places with their bazaars and temples have provided bases around which communities could gravitate and grow spiritually and artistically.

Located in the Kathmandu Valley by the Bagmati River, the city grew around the early mediaeval core and compact Durbar Square with its palaces and courtyards. Along the commercial east-west New Road to the open Tundikhel parade ground and then the Prithvi Path leading to the Singha Durbar Palace, and with the north-south Kanti Path and Durbar Marge axes, the city is now extending eastwards. As a mountain city, close to the Himalayas, it has unique characteristics.

Over time as palaces have been rebuilt and restored after fires and earthquakes, the Nepalese style of architecture is being retained. Accordingly, the organic centre, with its lanes, alleys and bazaars, is supplemented by its later straight paths that have opened long axes across the city, such as Kanti Path and Prithvi Path. With its peripheral ring road, the city is sprawling outwards in each direction in the hilly terrain as settlements have filled remaining available buildable sites into the east past the airport and south towards Patan (Lalitpur) with its old palaces and temples just south of the river.

3.5.5 TEHRAN, IRAN

Background and Origins

Tehran is the modern capital of Iran located in the north of the country by the Alborz mountains and its northern borders with Mount Damavand that runs along the south-east coast of the Caspian Sea and with the Great Salt Desert to the south. The city itself has a great variation in elevation, varying from the south to the hilly north at around 1,700 m. From an ancient civilisation, the earliest Median people had their capital at **Ecbatana** with the first Persian Achaeminid Empire under Cyrus the Great that was founded in the 6th century BCE.

Succeeded by Darius, he established **Persepolis** (Takt-a-Jamshia) in Shiraz Province some 1,000 km. south of Tehran as the capital of his empire with the fine Palace of Parsa and its tall, decorated columns. He was succeeded by the Sassanians, who were the last dynasty having a capital at **Ctesiphon** before the Muslim conquest in 650 CE, which was an Islamic Golden Age. There was a period of rule by the Umayyad and then the Abbasid caliphs. Next the Seljuk Turks established their capital at **Isfahan** in 1051 and later at **Rey**, which is now a suburb of modern Tehran.

This changed with the Mongol invasion in the 13th century and the country was fragmented under Tamarlane, until in the 16th century the Safavids reinvented the country, which was a turning point, with its capital at **Tabriz** through conversion to Shia Islam. Shah Abas subsequently moved the capital to Qazvin and later to **Isfahan**.

During the 16th century, under the Safavid King Tahmasp, the city of Tehran was considerably developed, fortified with walls and series of towers, and laid out gardens and construction of a royal palace. The masterpiece Shahnama, 'the Book of Kings', was created at this time with its illustrated narratives. During the mid-18th century, under Nader as Shah of Iran, the country exerted considerable power in the region over the Ottomans and Mughals and occupied the Caucasus and Trans–Oxiana, the former Sogdiana who moved the capital to **Shiraz**. Once deposed, under the following Qajar dynasty, a new capital was created in **Tehran** in 1795.

The first oil fields were discovered at Masjid- i -Suleiman in 1908, which were to be of great significance. Despite the country's independence in World War I, during World War II the Persian campaign was fought against the Ottomans. In 1923, after the war, Reza Shah came to power in establishing a new Pahlavi line who set about modernising the country.

During World War II, the Persian corridor supply route was maintained by the Allies through use of the Trans-Iranian Railway to Russia via the Caspian Sea and Volga River. The newly functioning oil fields, where Abadan was refining oil, were protected. In 1943, after the Tehran Conference of the three Allies, Iran secured its independence that evolved under the new Pahlavi dynasty.

Modern City – Buildings and Arteries

South Tehran contains the original city of Tehran developed around Rey, which is enveloped now to the south of the centre. Here the former royal Golestan Palace and gardens and reflecting pool, which is off Panzdah-e Khordal Avenue, was modified during the 19th century with its main halls and galleries. Adjacent is the Grand Bazaar with its myriad of lanes and selling a multitude of commodities.

The previous Toopkhaneh Square, now Khomeini Square, is just to the north along with the museums and the site of the former Tekyeh Dowlat Dome Theatre before its demolition. Besides Park-e Shahr, the Iranian Majilis (Parliament) are nearby.

Central Tehran contains the Abbas Abad and Mosalla districts and is bounded by Enghelab Avenue in the south. As the main commercial area, it is an active commercial area north of the Grand Bazaar and Soleimaniyeh.

North Tehran was more recently constructed off the main north-south Valiasr Avenue with the White Palace and Green (Shahvand) Palace and the newer Niyavaran Palace as the last modern palace of the Pahlavis. During the 19th century, new master plans were prepared for the city but were never realised.

It was during the 1930s, under Reza Shah, that significant changes to the urban fabric were first made with the demolition of buildings and new streets constructed and the Grand Bazaar was divided in half. Under Mohammed Reza, the last Shah, changes continued during the 1960–70s with significant new building, which was undertaken with a focus on industrialisation and the 'White Revolution' as modernisation continued. Besides agriculture, this modern economy focused on industrial manufacturing of electrical and chemical industries and automobiles. The city developed many commercial areas and new shopping malls, notably the huge Iran Mall, and the city's infrastructure was heavily modified such that new expressways and streets resulted in having high traffic levels.

Whilst the Shah held lavish celebrations for the 2,500th celebration of the Persian Empire in Persepolis, he was subsequently deposed, with a new Islamic Republic formed in 1979 in its place. To the west the modern Azadi Tower (designed by H Amanat) had been built in the form of a Sassanian arch to commemorate the anniversary of the Persian Empire, echoing a blend of Sassanian and modern architecture.

Greater Tehran

Modern Tehran has been shaped by new bordering expressways, namely in the north the Chamran, Sadr, Kordestan and Modarres Expressway running north-south, and the Haqani, Hemmat and Resalat Expressway running east-west.

These complement the original avenues, namely Valiasr (Pahlavi) towards Darband, Afriqa (Jordan) and Shariati Avenue running north-south, and Abbas Abad Street, Motahari Avenue, Keshavarz Boulevard and Enghelab Avenue running east-west.

To the north by the Hemmat Expressway is the tall Milad Tehran Telecommunications Tower and observation deck. Crossing the Modarres highway between Park-e Taleghani and Park-e Abu-Atosh is the unusual modern enclosed truss twin-level pedestrian Tabiat Bridge (designed by L Araghian).

An extensive public transport system has been developed with metro and BRT and articulated trolley buses connecting with the central railway station. A new modern underground metro system was opened in phases since the year 2000 and extended since with seven lines and over 100 stations. Tehran Imam Khomeini International Airport (IKA) is located to the south-west of the city.

The population of the city, which is divided into 22 districts, is approximately 9 million people with 16 million in the metropolitan area. The city lies in a seismic zone with very cold winters and semi-arid climate.

Transactional Forces and Typology

As an early city of the Medes peoples established at Rey around the 7th century BCE, the capital of the Persian Empire has been variously located throughout its history. From the sites of Susa, Ecbatana, Persepolis, Isfahan and Shiraz during the former Achaemenid, Sassanian and Safavid empires and subsequent dynasties, it has become modern Tehran within the Islamic Republic of Iran.

From the role of the central Silk Roads passing through Rey and Marshad, and with its central location and proximity to the Caucasus territories and the Caspian Sea, Tehran has contributed to the strategy of extending trading routes across Central Asia and beyond. Whilst the exploitation of oil and gas has modified the intensity of development, this has led to it becoming a more industrialised region.

From its ancient Persian roots and despite Mongol invasions, the original city comprised a walled citadel with its bazaar and narrow streets, and neighbourhoods from mediaeval times. In the mid-19th century, outer city walls were built in essentially an octagonal shape, but it was during the 1930s that significant changes were made to the urban fabric, with street widening and division of the Grand Bazaar, and construction of a series of new buildings.

With its broadly quadrilateral centre around Golestan Palace, the Grand Bazaar and Khomeini Square, the main axes of the city have been redeveloped from Azadi Square and the east-west Keshavarz Boulevard (Bolvar) with the modern expressways extending deep into and surrounding the central area of the city.

Broadly oriented east-west with the mountains as a backdrop, there are a number of differing neighbourhoods of the city, which has expanded mainly southwards and south-west past the airport towards Eslamshahr. With new residential and commercial areas acting as poly-nodal centres of development supported by its seven newly extended metro lines since 2000, its expansion is continuing. As an Islamic city that still retains some original features, its modernisation and densification has transformed it into one of the largest Central Asian metropolises.

3.5.6 ADDIS ABABA, ETHIOPIA

Background and Origins

Addis Ababa is the capital of Ethiopia located in the centre of the country at the north of the East African Rift at an elevation of 2,350 m. at the foot of the nearby mountains. The capital was previously located at Entoto, a high-table mountain, but this was considered too high for the Oromo tribe. In 1886, Menelik II chose to relocate the capital of the Shoa Kingdom to a lower location in the valley and moved it to Addis Ababa, building an imperial palace known as the Great Ghebbi near hot water springs.

Very early human occupation was known to have existed here with the discovery of Australopithicus female skeleton remains in the Awash River Valley dating back approximately 3.2 million years. Much later an Ethiopian empire was established, with the Kingdom of Aksum around 400 BCE which eventually adopted Christianity and became the early Orthodox Church with its featured rock-hewn churches built in the region. The Habesha people, which became known as Abyssinians, settled in Ethiopia and Eritrea and developed agricultural practices trading with Arabia via the Red Sea.

Islam was introduced into the Horn of Africa, and in the 16th century CE there were wars between Ethiopia and the Adal Sultanate, a Somali kingdom, especially in the Ethiopian Highlands. During the 17th century, the capital was moved to Gonder near Lake Tana, which was the new seat of government of the emperor, and in the next century, after various disputes during the Era of Princes, new wars with the khedive of Egypt developed.

As colonial expansion across Africa developed, war followed with Italy, which was concluded with the Treaty of Addis Ababa in 1896 that recognised the independence of Ethiopia under Emperor Menelik II. Thereafter came a prosperous period of growth and development as new neighbourhoods were built with new settlements.

Subsequently Emperor Haile Selassie came to power, and as the city consolidated, coffee production increased. But in 1936 under Mussolini, Italy reconquered and created Italian East Africa and Selassie was forced into exile. With a grand plan for the aggrandisement of the city, there followed a period of infrastructure development of roads, railways and dams and reinforcement of the earlier Ethio-Djibouti metre-gauge railway.

With their eventual defeat after World War II, Hailie Selassie returned. He was to preside over the formation of the OAU in Addis Ababa, and the permanent UNECA for Africa was set up in 1958, located in Africa Hall in Menelik II Avenue. New master plans were prepared by Abercrombie for the city. Selassie ruled until he was deposed by the 'Derg' in a coup in 1974, the Provisional Military Administrative Council (PMAC) backed by the Soviets who established a socialist government that remained until 1991 but which was followed by an intense period of civil war. Eventually a new constitution for the country was adopted in 1995 and with an elected mayor and council for the city directly responsible to the federal government.

Modern City – Buildings and Arteries

In the centre of the city are the main axes of dual Menelik II Avenue that runs north-south from Unity Park down to Meskel Square, with another axis of Churchill Avenue to Gambia Street. These are crossed by Sudan Street and Ras Mekonnen Avenue by the central station.

The Grand Anwar Mosque and Ethiopian Orthodox Cathedral are principal religious buildings in the city and the imperial palace, with its clock tower, is the seat of government. Meskel Square is the populous circumscribed major square where public gatherings are now held. Just to the west, the Piazza District with St George's Cathedral shows Italian influence, and the Mercato District has a large open outdoor market.

With its essentially rural economy and production of coffee, light manufacturing started to be developed around the capital, along with new financial services. A large number of new commercial buildings were realised that has included the African Union Hall and Conference Centre and office complex and corporate headquarters buildings as the city has been modernised.

To the south-east, the Bole Road (Africa Avenue) area is more densely occupied and leads down to Addis Ababa Bole International Airport (ADD) and the Outer Ring Road, newly built around the city in phases financed by China Road & Bridge Corp. (CRBC). An elevated light rail system built across the city opened in 2015 with two lines financed by China Ex-Im Bank. A new Addis Ababa to Djibouti standard-gauge electrified railway has also been built with Chinese finance parallel to the old Ethio-Djibouti Railway.

Overall Addis Ababa has a large informal sector and poor 'kebele' areas and inadequate housing provision as a result of a high rate of urbanisation and migration into the capital. The population of the city is approximately 3 million and 5 million in the metropolitan area, the main ethnic groups being mainly Amhara, Oromo and Guraje.

Transactional Forces and Typology

With its location by the nearby mountains in the Rift Valley not far from the equator, with evidence of very early human occupation, Addis Ababa has had an interesting history of religious peoples and multi-ethnic groups before the subsequent Islamic presence. From its hot springs to its productive land, it has proximity to the Blue Nile, Omo River Valley and Awash River, each flowing in different cardinal directions.

In modern times the city has been impacted by its colonial occupation and tensions in the Horn of Africa, which, with a major famine in 1985, have impacted its development, along with population growth pressures. It is now home to the African Union Commission and its directorates.

Addis Ababa is an elevated, more recently established capital which is surrounded by the heavily populated districts of the Shewa region and the Oromo people. With the former Ghebbi Palace at the centre, and with the Great Mosque and Ethiopian Orthodox Cathedral, its differing origins are evident, especially with the Italian Piazza district. With the government buildings, Africa Hall and new commercial buildings, Addis Ababa University and the market area of Merkato, there are districts developing differentially that are surrounded by mountains on most sides.

The main axes run north-south and east-west and development has occurred in sectors, generally infilling available land and radiating out from central Unity Park, and with industrial zones to the west and by Bole Airport to the south-east. It is a characteristically African tropical modern capital, having increasing population pressures around the periphery and with localised centres of growth. The following figures illustrate the modern configuration of the mountain cities as described and the evolution of their modern population.

FIGURE 3.5.3 Bogotá – modern configuration.

FIGURE 3.5.4 Quito – modern configuration.

FIGURE 3.5.5 La Paz – modern configuration.

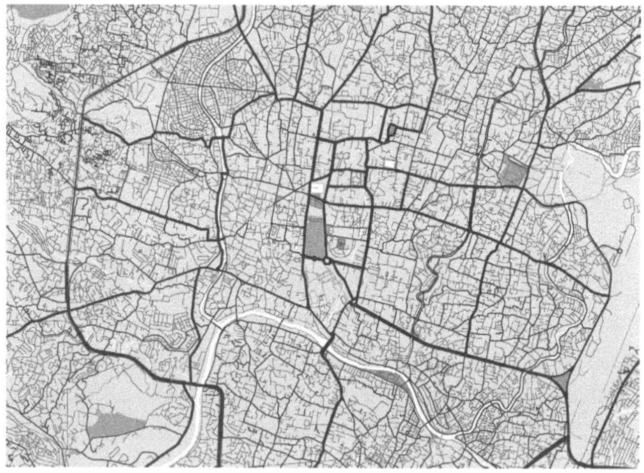

FIGURE 3.5.6 Kathmandu – modern configuration.

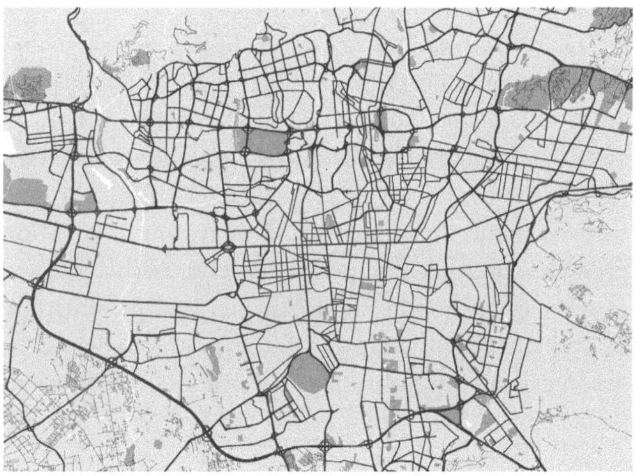

FIGURE 3.5.7 Tehran – modern configuration.

FIGURE 3.5.8 Addis Ababa – modern configuration.

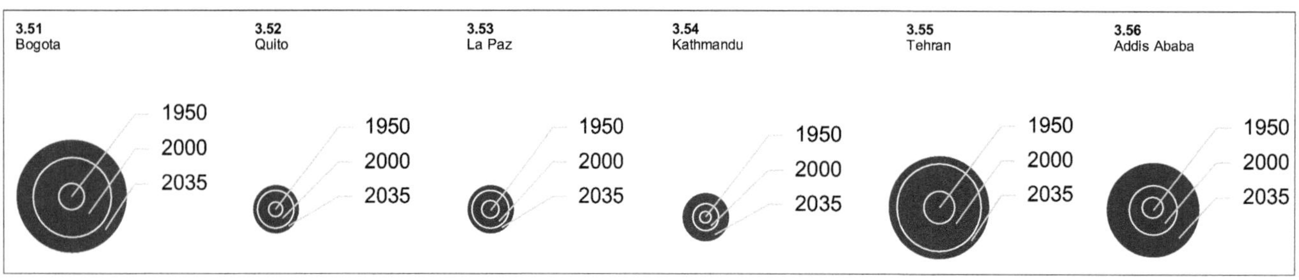

FIGURE 3.5.9 Mountain city populations.

Data source: Global Database of Metropolises 2020 – UN-Habitat (2020), UN DESA Population Division.

Further Reading

BOGOTÁ

Berney, Rachel – *Learning from Bogotá*, University of Texas Press, 2017.

La Rosa, Michael & Pavony, German – *Colombia*, Rowman & Littlefield Publication, 2023.

Molina, Diego – *Planting a City in the Tropical Andes, Bogotá 1880 – 1920*, Routledge, 2024.

Zeiderman, Austin – *Endangered City, the Politics of* Security *and Risk in Bogotá*, Duke University Press, 2016.

LA PAZ

Klein, Herbert – *A Concise History of Bolivia*, Cambridge University Press, 2021.

Kohl, James – *Indigenous Struggle and the Bolivian National Revolution, Land and Liberty*, Routledge, 2020.

QUITO

Mincham, Martin – *The People of Quito, 1690–1810, Change and Unrest in the Underclass*, Routledge, 2019.

Parreno, Christian – *Modern Architecture of Quito, Global, Local and the In-between*, Bloomsbury Visual Arts, 2024.

Phelan, John – *The Kingdom of Quito in the Seventeenth Century*, University of Wisconsin Press, 1967.

KATHMANDU

Bell, Thomas – *Kathmandu*, Haus Publishing, 2017.

Hosken, Fran – *The Kathmandu Valley Towns*, J Weatherhill, 1974.

Rademacher, Anne – *Reigning the River, Urban Ecologies and Political Transformation in Kathmandu*, Duke University Press, 2011.

TEHRAN

Madani-Pour, Ali – *Tehran, the Making of a Metropolis*, John Wiley, 1998.

Mehan, Asma – *Tehran, from Sacred to Radical*, Routledge, 2024.

Naraghi, Ashkan – *A Social History of Modern Tehran, Space, Power and the City*, Cambridge University Press, 2023.

ADDIS ABABA

Garretson, Peter – *A History of Addis Ababa from Its Foundation in 1886 to 1910*, Harrassowitz, 2000.

Marcus, Harold – *A History of Ethiopia*, University of California Press, 2023.

Part IV

Planned Cities

4.0 Introduction – Planned Cities

In contrast to those cities that have developed organically and grown around natural settings that have their own configuration, in recent centuries man-made planned cities have been developed with quite specific purposes. Some have been planned as new capitals as new nation-states have emerged, and in securing their own symbolic capital that is purpose designed, others for a particular industry or function.

Exceptionally, planned developments were laid out under particular regimes, notably under the Soviet empire, where in a repetitive manner, virgin sites were developed for collective and industrial purposes.

Other cities have been established as individual industries have taken hold in locations around which 'company towns' have been developed in order to house the workforce and their families.

In some countries, new towns have been planned as either 'garden cities' or residential areas as 'overspill' from congested urban centres, aimed at affording a more amenable environment and where space was less of a premium.

Besides such built developments, there are some 'divided cities' which became and remain divided, due to either political or religious tensions, which have become segregated, often forcibly, resulting in physical separation, with walled areas physically dividing formerly homogeneously inhabited areas.

More recently, newer contemporary areas are being planned and targeted as 'sustainable cities' that are more compact, integrating land use and transport considerations with renewable resources, having greener credentials, and smart systems.

As developable sites have grown in their scale and coverage, new 'edge cities' have also emerged which, due to development constraints, have sought expansion at the edge of existing urban realms, so forcing new development yet farther out beyond previously built-up urban areas. All of these different patterns of planned urban development are prevailing across different regions and continents.

This **Part IV** addresses the different types of planned cities, from former Soviet and divided cities to planned industrial and modern American cities and new capital cities that have been established across different continents (see Table 4.0.1).

DOI: 10.4324/9781003353348-18

Table 4.0.1
Planned City Types

		City	Topography	Plan	Trans. Forces	Type
4.01	**PLANNED CITIES**					
4.1	***Former Soviet and Eastern European***					
		Moscow	Riverine	Radio-centric	Fed. Capital	Soviet
		St Petersburg	Coastal/bay	Axial	Reg Capital	Neo-classical
		Kyiv	Riverine/isles	Linear/axial	Capital	Ukrainian
		Warsaw	Riverine	Radio-centric	Capital	Rebuilt-modern
4.2	***Divided***					
		Belfast	Coastal	Divided	Adm. Capital	De-industrial
		Berlin	Riverine	Axial/sectoral	City-state	Global
		Beirut	Coastal	Organic	Capital	Mediterranean
		Johannesburg	High veld	Poly-centric	Minerals	Witwatersrand
		Jerusalem	Hill	Organic/divided	Religious	Cultural
4.3	***Industrial and De-industrialised***					
		Chicago	Lake	Grid/radial	Transport hub	Modern
		Detroit	Lake/straits	Grid/axial	Industrial	De-industrial
		Cleveland	Lake/riverine	Grid/linear	Industrial	De-industrial
		Pittsburgh	Riverine/confluence	Grid/axial	Industrial-coal	De-industrial
		Manchester	River/canal	Core/sectoral	Industrial	Modern
4.4	***Modern American***					
		Philadelphia	Riverine	Grid	Services	Metropolitan
		Baltimore	Estuarine/bay	Grid	Maritime	Port
		Miami	Coastal/barrier	Ortho. grid	Maritime	Global
		Atlanta	Hilly	Grid	Sta. Capital	'New South'
		Houston	Estuarine	Grid/radial	Petro-chemical	Global
		Seattle	Bay/lake	Linear/axial	Maritime-services	Port
		Vancouver	Coastal inlet	Grid	Maritime	Global
4.5	***New Capital***					
		Washington DC	Riverine	Grid/axial	Fed. Capital	Global-admin
		New Delhi	River	Grid/axial	Nat. Capital	Global-admin
		Chandigarh	Plains	Regular sectors	Sta. Capital	Punjabi
		Islamabad	Foothills	Recti-linear	Isl. Capital	Admin
		Canberra	Lake	Radio-centric	Aus. Capital	Modern
		Brasília	Highland	Curved axial	Fed. Capital	Modern
		Abuja	Forest/savanna	Axial	Fed. Capital	Nigerian

Table 4.0.2
Planned Cities Population Growth (thousands)

City	Country	1950	2000	2020	*2035**	Area
Moscow	Russian Federation	5,356	10,005	12,538	*12,823*	City proper
St Petersburg	Russian Federation	2,903	4,719	5,468	*5,643*	City proper
Kyiv	Ukraine	815	2,606	2,988	*2,991*	City proper
Warsaw	Poland	768	1,666	1,783	*1,802*	City proper
Belfast	UK	446	571	631	*697*	Urban agglomeration
Berlin	Germany	3,338	3,384	3,562	*3,639*	City Proper
Beirut	Lebanon	322	1,487	2,424	*2,304*	Urban agglomeration
Johannesburg	South Africa	911	3,046	5,783	*7,461*	Urban agglomeration
Chicago	USA	4,999	8,315	8,865	*9,830*	Urban agglomeration
Detroit	USA	2,769	3,899	3,548	*3,837*	Urban agglomeration
Cleveland	USA	1,392	1,786	1,763	*1,932*	Urban agglomeration
Pittsburgh	USA	1,539	1,753	1,704	*1,862*	Urban agglomeration
Manchester	UK	2,422	2,345	2,730	*3,028*	Metropolitan area
Philadelphia	USA	3,128	5,156	5,717	*6,378*	Urban agglomeration
Baltimore	USA	1,168	2,079	2,325	*2,597*	Urban agglomeration
Miami	USA	622	4,933	6,122	*6,952*	Urban agglomeration
Atlanta	USA	513	3,522	5,803	*6,889*	Urban agglomeration
Houston	USA	709	3,847	6,371	*7,569*	Urban agglomeration
Seattle	USA	795	2,720	3,433	*3,909*	Urban agglomeration
Vancouver	Canada	556	1,959	2,581	*2,957*	Metropolitan area
Washington DC	USA	1,298	3,949	5,322	*6,122*	Urban agglomeration
New Delhi	India	-	-	-	-	
Chandigarh	India	40	790	1,148	*1,572*	Urban agglomeration
Islamabad	Pakistan	36	569	1,129	*1,672*	City proper
Canberra	Australia	20	344	457	*540*	Metropolitan area
Brasília	Brazil	36	2,932	4,646	*5,345*	Metropolitan area
Abuja	Nigeria	19	833	3,278	*6,071*	Urban agglomeration

*Note: *Forecast.*
Source: Global Database of Metropolises 2020 – UN-Habitat (2020), UN DESA Population Division.

4.1 Former Soviet and Eastern European Cities

Former Soviet – East European

Moscow, St Petersburg, Kyiv and Warsaw

4.1.0 INTRODUCTION – FORMER SOVIET CITIES

During different ages, groups of peoples gathered and assembled around towns and cities that had grown and become organised into nation-states. Such amalgamations of peoples have come under the power and direction of different leaders from varying backgrounds, many autocratic, driven by differing religious and political aspirations, each having varying degrees of impact.

Historically, typical form determinants of organic towns and cities were predominant natural features as cited earlier, around rivers and estuaries, natural harbours and islands. Planned developments were formed around other features – more likely as a result of the availability of suitable land and sited on routes of communication. Such lack of special distinctiveness has led to the repetitive nature of new settlements, with fewer constraints, enabling a looser, more sprawling type of development.

This became prevalent in areas of Asia, noticeably Russia, when after the Revolution, the extensive Russian Empire was transformed from its former imperial regime into the Soviet Union. As a socialist state, it had strategic needs both to house its growing population and to expand its industrial capabilities. The Union expanded to comprise 15 republics whose goals were focused equally both on the collectivisation of agriculture and industrialisation. These objectives led to the construction of planned developments on a significant scale, using new methods and materials, quite distinct from vernacular types of built form found locally.

Based upon an extreme socialist philosophy, the reach stretched beyond the physical borders of the Soviet Union for 70 years from 1919 as the Union of Soviet Socialist Republics (USSR) encompassed neighbouring states in Central Europe. This Eastern Bloc* of satellite states came under the hegemony of the Soviets.

These republics fell under a regime with common rules, which led to authoritarianism and standardisation of practices that restricted creativity and led to the construction of repetitive housing and factory blocks that destroyed the traditional urban fabric. Initially focused at main urban centres, these were extended out across other regions under Soviet control, which became heavily industrialised with rigid planned developments. Pre-fabricated panel and new reinforced concrete construction enabled the repetition of apartment blocks and factories assembled intensively on virgin sites.

The Soviet empire's ascendancy before World War II, after a period of consolidation, reached its apogee and before eventual decline and rapid collapse at the end of the 20th century. With the sole remaining Russian Federation, other countries transformed into a number of Newly Independent States (NIS) and the Caucasus countries.**

Characteristically each has its own primate city, such as Kyiv and Warsaw, Bucharest and Bratislava, as state capitals that dominate over other secondary cities. The modern state capitals are the main cities of the generally smaller Baltic states with capitals of: Estonia (Tallinn), Latvia (Riga) and Lithuania (Vilnius), and also the Balkan (former Yugoslav) states of Croatia (Zagreb), Slovenia (Ljubljana), Serbia (Belgrade), Bosnia-Herzegovina (Sarajevo) and Montenegro (Podgorica).

* Comprising the People's Socialist Republics and their capitals: Albania (Tirana), Bulgaria (Sofia), Byelorussia (Minsk), Czecho-Slovakia (Prague and Bratislava), German Democratic Republic (Berlin and Leipzig), Hungary (Budapest), Poland (Warsaw), Romania (Bucharest) and Ukraine (Kyiv).

** The Caucasus and the NIS and their capitals: Georgia (Tbilisi), Armenia (Yerevan), Azerbaijan (Baku), Kazakhstan (Astana and Almaty), Kyrgyzstan (Bishkek), Tajikistan (Dushanbe), Turkmenistan (Ashgabat) and Uzbekistan (Tashkent and Samarkand).

DOI: 10.4324/9781003353348-19

4.1.1 MOSCOW, RUSSIAN FEDERATION

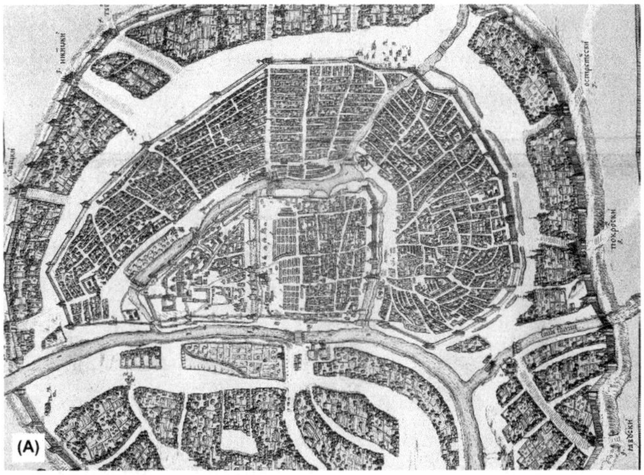

FIGURE 4.1.1A Moscow – early configuration 1660.

Source: British Library.

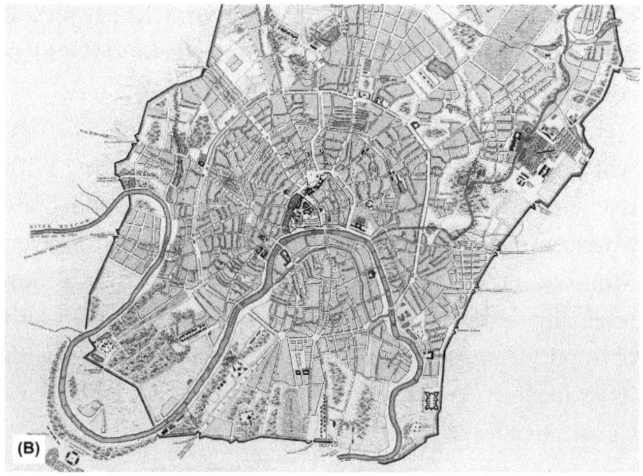

FIGURE 4.1.1B Greater Moscow – early configuration 1808.

Source: iStock.

Background and Origins

From its Slavic origins in 1150 CE, the city of **Moscow** (Moskva) in the centre of the Russian steppes developed on the Moskva River from being a wooden fortress ('kremlin') to a walled mediaeval city to protect itself from Mongol invasions. From an area of Russian principalities, the nucleus of the Grand Duchy of Moscow, or Muscovy, developed connected by the Russian river network. After the former khanates, the first real tsar of Russia was Ivan IV, or 'the Terrible', crowned in 1547 in the Dormition Cathedral. The Muscovite State expanded absorbing other principalities and a series of buildings of the Kremlin were built and 'russified'.

It was not, however, until after the death of Peter 'the Great' who had established the capital of Russia in St Petersburg, and subsequent Catherine 'the Great' that power returned to Moscow under Tsar Alexander I, effectively after the role in defeat of Napoleon and retreat of the French.

The focal point of the city is the Kremlin, the citadel of the tsars, and the centre of absolute power. By the Moskva River, with its tall walls and 19 towers, built around 1500, it encloses the Great Kremlin and the Faceted and Patriarch's palaces, the Cathedral of the Annunciation, of the Assumption and of the Archangel, the State Armoury and the Ivan the Great Bell Tower. The cathedrals and bell tower are distinguished by their external golden onion domes and with their internal embellished frescoes and icons. In the 17th century, further defences and walls of the city were built.

During the mid-19th century parts of the city were replanned and parks were introduced, along with sewerage and railways which enhanced the city and with the Garden Ring around the former city ramparts beyond the boulevard horseshoe.

The major upheaval in the early 20th century following World War I, with the Russian Revolution and the overthrow of the Romanovs, was the taking over by the Bolsheviks. With the transfer of the capital back to Moscow in 1918 and followed by the long period of Stalin's rule up until and during the Great Patriotic War of World War II. There was a wave of rebuilding and buildings such as the Basilica of Christ the Saviour were demolished to be replaced by the huge Palace of the Soviets, although this was eventually abandoned.

After victory Moscow continued to grow into an extensive city. There were new developments and expansion of the city and with the construction of multiple high-rise Soviet apartment blocks across the city and any sense of a former green belt was lost.

Adjacent to the Kremlin is enormous Red Square, originally a market square, now used for annual parades and celebrations. At one end is St Basil's Cathedral, the Cathedral of the Intercession with its colourful ribbed and faceted onion domes. Alongside the walls of the Kremlin is the mausoleum, built of cubes of red granite and black labradorite, where the embalmed body of Vladimir Lenin lies, plus those of former Soviet leaders Stalin and Brezhnev.

Behind the square is the District of Kitay Gorod, a commercial area with several churches. The infamous headquarters of the Cheka and the KGB is also to be found in Lubyanska Square.

As a large metropolitan city, the heart of the commercial area is Tverskaya, which also features Theatre Square with the Bolshoi Theatre, home of the Bolshoi Ballet, the Moscow Arts Theatre and also the Maly Theatre. Tverskaya

Ulitsa is the main shopping street around which several grand hotels are located, as well as museums and the Moscow Music Conservatory.

To the west of the Kremlin is Arbatskaya District, which is a pedestrianised area around Ulitsa Arbat. With many restaurants and cafés, it comprises the Pushkin State Museum of Fine Arts, the Stalinist gothic-styled Foreign Ministry and the rebuilt Cathedral of Christ the Saviour, originally destroyed under Stalin's orders. The area south of the city Zamoskvoreche (beyond the Moskva River) with its main street Bolshaya Ordynka crosses the Vodootvodnyy Canal, an older area, with a number of distinctive churches.

Modern City – Buildings and Arteries

A General Plan for New Moscow, which was developed by A Shchusev in the 1920s, did not progress and a new Moscow Plan was developed in a competition won by V Semenov in the 1930s. Based on a plan of radial and concentric ring boulevards, it aimed at overall urban containment with a peripheral green belt and forests. The new metro system was planned and started in 1933 with its neo-classical stations, and Soviet housing blocks were arranged in districts (rayony) with neo-classic facades. A new central city Gorky Park was built, and a canal system was developed to link up the adjoining rivers.

In terms of infrastructure, a series of ring roads has been planned and built around the city, with the outer MKAD Ring Road and intermediate 3rd Ring Road. The city has continued to expand with its 5th Ring Road, which has been built striving to accommodate the increase in highway traffic with a further external CKAD planned. Although the 4th Ring Road was cancelled, the surrounding green belt has come under significant pressure, as land has been lost to both infrastructure and building.

The city metro network is very comprehensive with deep stations, several in different styles that reflect different national characteristics, such as Mayakovskaya, Komsomolskaya, Teatralnaya, Belorusskaya, Novokuznetskaya and Kievskaya stations built during the 1930s and 1950s. Of the railways there is a circular route, the 'Little Ring' of Moscow, renamed Moscow Central Ring (MCC), and also the Metro Bolshaya-Koltsevaya large circle Line 11 which interlocks with the MCC.

There are four major international commercial airports in Moscow: Sheremetyevo (SVO), Domodedovo (DME), Vnukovo (VKO) (Andrei Tupolev) and Zhukovsky (ZIA) (Alexander Pushkin).

With the collapse of the Soviet empire in 1991, Moscow's role was transformed as the capital of the new Russian Federation. Since then, the city has been beset by several scandals in terms of corruption under different mayors. The city's skyline is continuing to be modernised, with a large number of new commercial developments in the Moscow International Business Centre as the Moscow and Naberezhnaya Towers, the Federation Twin Towers (Vostok and Zapad), Neva Towers, Eurasia and Capital Towers. The Ostankino TV Tower, at 540 m. in height, is also located here. Permits for building construction are issued by the central Moscow Construction Committee that oversees the release of land. With no real market signals as to the actual 'demand' for development, this has resulted in residential densities conversely increasing with distance from the centre.

The whole city has a certain serenity but with its freezing winters, heavy industrial production, air pollution and restrictions, it has a severely challenging environment. The population of the city is approximately 12.5 million with approximately 20 million in the whole metropolitan area. The population density of the whole city region is approximately 16,700 pers./sq. km.

Transactional Forces and Typology

Moscow is a centralised metropolitan city with an historic core around the river that has developed in surrounding rings since the mid-19th century. With mixed residential, commercial and industrial zones, there is a green Moscow Forest and Park Protection Belt that surrounds the city. With groups of socialist high-rise apartment blocks developed in clusters, it was not until after the ending of the Soviet system that the desires for more rural living were enabled, leading to the suburbanisation of different areas.

With individual aspirations for small family 'dachas' and with privatisation of the housing stock in 1991 and housing reform, there was an increase in private ownership. Since historic protection of western areas from industrial development upwind to Moscow's west, areas such as around the Rublevo Uspenskoe Highway, or Rublevka, with its green space have attracted residents and increased real estate values.

Despite this, it has been challenging to protect the city's 'green lungs' with ongoing expansion of sets of further outlying peripheral ring roads. This has resulted in ongoing urbanisation across Moscow but with more haphazard multi-nodal development around the periphery.

As a major riverine city developed around its initial central fortification, Moscow has grown in a typical radio-concentric manner. Over the decades progressively it has extended itself with the annexation of new suburbs and successive circumferential ring roads. It is now developing nodes at and around the periphery that has more resemblance of modern 'edge cities'. Of a lesser density, this has led to a wider sprawl in all directions still in a concentric pattern. Across this, it is traversed by major road and rail infrastructure axes radiating out from the centre into the extending region of Greater Moscow.

4.1.2 ST PETERSBURG (LENINGRAD), RUSSIA

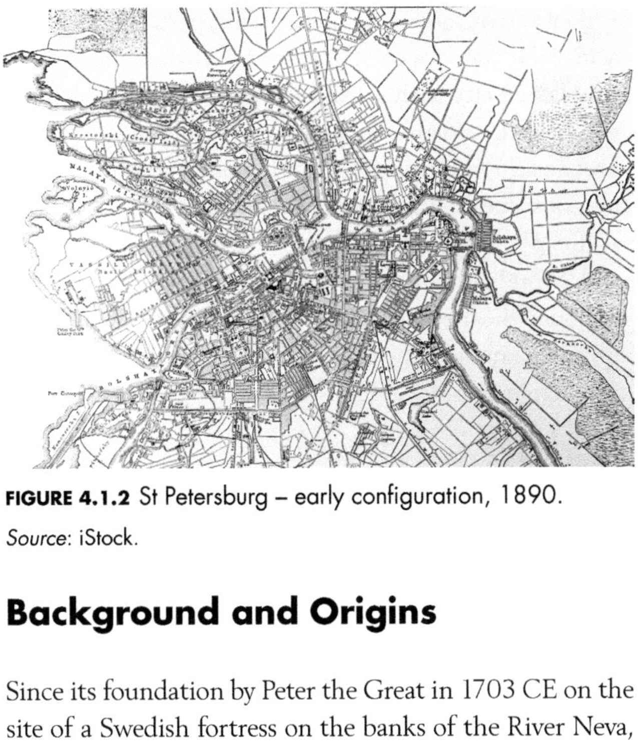

FIGURE 4.1.2 St Petersburg – early configuration, 1890.

Source: iStock.

Background and Origins

Since its foundation by Peter the Great in 1703 CE on the site of a Swedish fortress on the banks of the River Neva, by Vasilevsky Island, the city of **St Petersburg** has been steeped in significant historical developments. In a strategic maritime location in the Gulf of Finland and the Baltics, with a defensive fort on Kronstadt Island, it became a leading port for Russia.

With the Peter and Paul Fortress built on Zayachy Island across the River Neva, the Admiralty building, with its shipyard and slipway, was developed as a pivotal maritime centre. The spit of Vasilevsky Island, known as the Strelka, is where the Academy of Sciences and the naval building were located with the two Rostral Columns that have served as distinctive lighthouses to the port.

The Peter and Paul Fortress on Petrogradskaya was first built in wood, then in stone, with two entrances to the bastion surrounded by water, of St Peters Gate and Neva Gate, or 'death gate', at the riverside level where prisoners were brought in directly to the fortress. Within is the elegant Cathedral of Saints Peter and Paul, with its distinctive soaring golden spire, where the Russian tsars were buried.

As a fortified outpost it became the home of the tsars, who whilst normally resident at Tsarskoe Selo, had the Winter Palace designed by Rastrelli for Tsarina Elizabeth on the banks of the Neva and a railway built in 1837 to connect St Petersburg with Tsarskoe Selo and the Imperial Palace at Pavlosk.

However, the makings of the Revolution broke out here on Bloody Sunday in January 1905, as the Romanovs were losing power. In 1917 strikes led to the advancement of the Russian Revolution, the tsar abdicated, the Bolsheviks took over, and a provisional government was installed at the Smolny Institute and the city was renamed Petrograd. When in October the Red Guard stormed the Winter Palace after a signal from the cruiser *Aurora*, the Cossacks were defeated and Lenin returned from exile to lead a new government.

This eventful period was quite short-lived and with Lenin's death in 1924, the city was renamed Leningrad, as the capital was moved to Moscow and Stalin took over. As he exerted greater control during the 1930s, in the following period during World War II, the city was besieged and surrounded by German troops. For a period of over two years, the city was totally isolated and deprived of food except for that which could be smuggled across frozen Lake Ladoga. With its eventual relief in 1944, it came to be honoured by Stalin later as 'Hero City' after victory in the Great Patriotic War, with a commemorative obelisk erected in Vosstanya Square.

There followed the subsequent Communist period in Russia, which lasted from 1945 until 1990. With the eventual introduction by President Gorbachev of the policies of 'glasnost' and 'perestroika' (openness and restructuring), this was to prove to be the beginning of the end of the Soviet Union. Nationalist forces intervened and in the rump of Russia under Mayor Sobchak, the people voted to restore the original name of the city as St Petersburg.

Flat and low-lying in the mouth of the river Neva, the city developed in different areas – separately on Vasilevsky Island, and the areas of Petrogradskaya, Gostinyy Dvor and Sennaya Ploschad. With its rivers and canals, bridges and river crossings have become features of the urban landscape.

On the other side of the Neva River from Vasilevsky Island, along the granite embankment, is where the three grand avenues of Nevsky Prospekt, Gurokhovaya Street and Voznesensky Avenue meet by the distinguished Admiralty building with its slender spire.

By St Isaac's Cathedral in St Isaac's Square is the Winter Palace, now comprising one of the most famous museums, the Hermitage. The facade is bold along the embankment, and on the other side it opens out to Palace Square, which faces the curved General Staff Headquarters Building opposite with its triumphal arched entrance with the tall Alexander Column that celebrates victory over Napoleon, placed in the centre of the square. Within the Hermitage, the galleries and replacement staircase contain the comprehensive art collection.

Adjacent is Gostinyy Dvor, the Great Bazaar, that encompasses the Griboedova Canal, a thriving neighbourhood of shops and markets around Arts Square and the Church on Spilled Blood, where Tsar Alexander was assassinated, with its onion-shaped domes. Mikhaylovsky Palace is another major Russian museum, and straddling this district is Nevsky Prospekt, the long, wide and most famous street in St Petersburg with its shops and stores, invariably packed with pedestrians and several lanes of fast-moving traffic. Anichkov Bridge, with its wild horse statues, is one of the elegant bridges that carries Nevsky Prospekt across the Fontanka Canal.

The western part of St Petersburg contains Sennaya Ploshchad (Haymarket) with St Nicholas Cathedral. In this neighbourhood is Mariinsky Theatre, with the former Kirov, opera and ballet company, and the Rimsky-Korsakov Conservatoire in Theatre Square near the tree-lined canal embankments.

Modern City – Buildings and Arteries

Besides its historic centre and fine buildings, with the development of the city post–World War II, a large number of contrasting concrete slab apartment blocks were built in different neighbourhoods across the city, having the same general Soviet-style characteristics, that replaced former communal apartments. As a feature of many other Soviet cities also, this has changed the profile of the city. A generally low-rise city, the skyline has been disrupted more recently by the enormous Lakhta Centre tower in the north-west district of the city and construction of the St Petersburg TV Tower.

In terms of infrastructure, the city has a dense network of metro lines that have opened since 1955 with many deep stations that provide public transport across the city. The main central railway stations are Finland Station, at which Lenin arrived upon his return, and the Moscow, Vitebsk and Baltic stations.

The Neva River is an important marine access route that freezes over in the winter when access for shipping is restricted. Generally low-lying and with a propensity to flood, the St Petersburg Flood Protection Scheme and complex of dams was built and completed in 2011. It is 25 km. in length, acting as a levee and causeway that protects the city during cyclones from storm surges from Neva Bay. It extends from Lomonosov to Kotlin Island and the island city of Kronstadt and then to Cape Lisy Nos. It has two lock openings for the shipping channel and also acts as carrying the city's new ring road, the tolled Western High-Speed Diameter Highway with its complex bridges, that connects with the main urban highways across the city.

With its frenetic traffic in the modern city, recent development activity appears quite anarchic, partly being driven by invisible modern-day oligarchs. It remains a fascinating and historic city with its canals – the city is known as the 'Venice of the North' – and its notable buildings and parks, it retains important cultural and musical centres, such as the Conservatoire and Mariinsky Theatre.

The population of the city, which grew appreciably post–World War II, is approximately 5.5 million and 7 million in the metropolitan area. Farther out beyond in the suburban areas are the fine parks of Peterof, Tsarskoe Selo and Oranienbaum.

Transactional Forces and Typology

Strategically located in the Baltics at the head of the Gulf of Finland by the Neva Bay, and linked with the Volga-Baltic Waterway, the Imperial City of St Petersburg became a significant port. Despite its freezing winters, with its low-lying islands, as capital of Russia under the Romanov Dynasty, St Petersburg acted as the 'Window to Europe', as the northernmost metropolis in Eurasia. With a maritime outlook and symbolic fortress, it served as a base for the large Russian Navy.

With the Moyka River and Fontanka Canal, and bordered to the south by the Obvodny Canal, the city is characterised by a number of man-made canals that were constructed in and around the city along with their elegant crossing bridges. As a major trade gateway and also administrative centre, the regional economy has been strengthened by the industrial sector and ship-building. More recently this has become more specialised with the oil and gas trade, with its pipelines across the west of Russia and beyond.

The city was first established and shaped around the original fortress on the banks of the Neva River. With the rebuilding of the Peter and Paul Fortress, the city developed around Trinity Square on the right bank. Initial plans were for the development of Vasilevsky Island and with a grid of canals around the city, but it was the shaping by the three grand avenues, in particular Nevsky Prospekt and the bridge crossings that connected the main squares such as Palace Square and the Admiralty, that defined the historic centre of St Petersburg.

As an estuarine and riverine coastal city, the city has developed with radial characteristics, having circumferential roads around the eastern side of the city. Now connected by the new Western High Speed Diameter Highway, this also contributes to supporting the modern flood protection scheme of the city with dedicated lock openings for maritime shipping, which is critical for protecting the entire urbanised area.

4.1.3 KYIV, UKRAINE

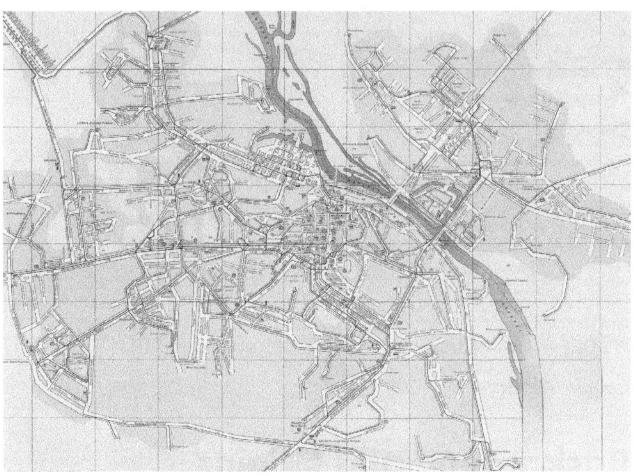

FIGURE 4.1.3 Kyiv – early configuration, 1976.
Source: Maphouse.

Background and Origins

Kyiv, the capital of Ukraine, is located in the centre of the country, the cradle of Slavic civilisation on the banks of the wide meandering River Dnieper. It has seven hills and varied topography. In mediaeval times it was capital of Kievan Rus' when original river crossings consisted of floating wooden pontoon sections.

One of the oldest areas is Lavra where the first monastery was built in underground caves of St Anthony in 1051 CE. In the lower Pecherska area a number of subterranean churches were built around the catacombs and contained within an extensive walled area. In the upper area, the monastery has an entrance via Trinity Gate Church by the Dormition Cathedral where the later great bell tower and cluster of gold domed churches were constructed.

With its Byzantine origins and sacked by Mongol raiders of the Golden Horde in 1240 CE, in the centre St Sophia Cathedral remains defiant, standing adjacent to the Royal Palace. With its later additions of onion domes and its tall bell tower, it is striking both externally and internally, with its frescoes and mosaics. Facing it is St Michael's gold domed monastery and St Andrew's Church nearby (designed by Restrelli) as a five-domed cross, along old Andriyivsky Street that curves to the upper town.

In the commercial Podil District the city revolves around the central area where assemblies of people coalesce in the 'Maidan', Nezalezhnosti, which is intersected by Khreshchatyk Street. Here is St Volodymr's Cathedral, a more recent addition with its seven blue domes.

After the Bolshevik Revolution in Russia came the destabilising Polish-Soviet War that was ended with the Treaty of Riga in 1921. At this time both the Ukrainian and Byelorussian Soviet socialist republics were established, the beginning of the Soviet era of Ukraine. During the following period, the city was divided into 'rayons', and many Soviet-style neighbourhoods of apartment blocks were built.

There are two significant landmarks in the city – firstly, the memorial at Babyn Yar to commemorate the Jewish people from Kyiv who were massacred by Nazis in 1941 during World War II and left in a ravine to die. Secondly, the giant Motherland Monument, now transformed with the Ukrainian symbol that has replaced the former Russian symbol, in the park looking out across the River Dnieper. The city suffered considerable loss and destruction during this war.

Modern City – Buildings and Arteries

During Soviet times, Kyiv had a significant volume of heavy industry and factories, in particular specialising in the manufacturing of armaments and Antonov aircraft. Modern Kyiv is very different from its ancient beginnings with its complex modern existence seeking to escape from the Russian yoke. However, Ukraine is still a major wheat producer and specialises in agro-processing.

One major feature of the city's infrastructure is the very deep metro lines and characteristic stations dating from 1960, similar in design to the Moscow subway, that traverse the city along with the long bridges that span across the wide River Dnieper. The city also has one of the original tramway systems in Europe.

Since becoming an independent republic in 1991 after joining the Commonwealth of Independent States, the economy has undergone a significant transformation with the closure of many former heavy industries, and modernisation has been undertaken.

The very wide River Dnieper has always been an obstacle to crossing the city. In 1850, a multi-span suspended chain bridge, 'Nicholas', was built for road traffic, and in 1870 a box trussed bridge 1 km. in length was built for the railway (designed by A Struve). These operated successfully until 1920 when they were both destroyed by retreating Polish troops after the Kyiv offensive against the Soviets. Whilst the Nicholas Bridge was replaced by the Soviets by a floating pontoon bridge, this was also destroyed in World War II. The principal bridges in Kyiv and their date of realisation are:

- Petrovski rail truss bridge (1929), Paton road and trolley welded bridge (1953), Metro road and rail bridge (1965), Pivnichnyi cable-stayed road bridge (1976), Pivdennyi road and metro suspension bridge (1990), Darnytskyi road and rail bridge (2012) and Podilsky road and metro bridge (2020).

The centre of the city revolves around Independence Square and the Maidan. With the government institutions, one strange modern building is the House of Chimeras with its roof of gargoyles which is used as residence of the president.

The population of the city is approximately 2.5 million people and 3.5 million in the wider metropolitan area. Modern events have been traumatic – subsequent to the Orange Revolution and protests in the Maidan the government fell but further disruption was to occur with the attempted Russian invasion in Spring 2022 and the subsequent occupation of the Luhansk and Donetsk regions and separation from main Ukrainian territory, which has resulted in population exodus from these areas.

Transactional Forces and Typology

Located in the midst of continental Europe, the city of Kyiv has developed by the Dnieper River, a broad, meandering river with a series of islands and rapids that flows south to the Black Sea, long used as an original trade route. Whilst a riverine means of transport, this has also acted as a barrier across the country east and west. Whilst predominantly a rich agricultural region and major grain producer, it was industrialised by the Soviets during the 1930s and enhanced with its railway connections. As a strategic location during World War II, it has become a centre of resistance in ongoing territorial disputes with Russia.

A city dominated by its major river and islands and lake reservoir to the north, it has developed on either side, predominantly the left bank with residential and industrial zones, linked by a series of long bridges forming interconnections. Much affected by developments during the Soviet regime, it was highly industrialised and developed with neighbourhoods of dense residential housing blocks.

As a riverine city it is bi-polar, with settlements on either side of the Dnieper River, with the historic centre and governmental area on the western side and surrounded by forested areas. Whilst dominated by the river, its development has followed the road and rail axes radiating out from the centre.

4.1.4 WARSAW, POLAND

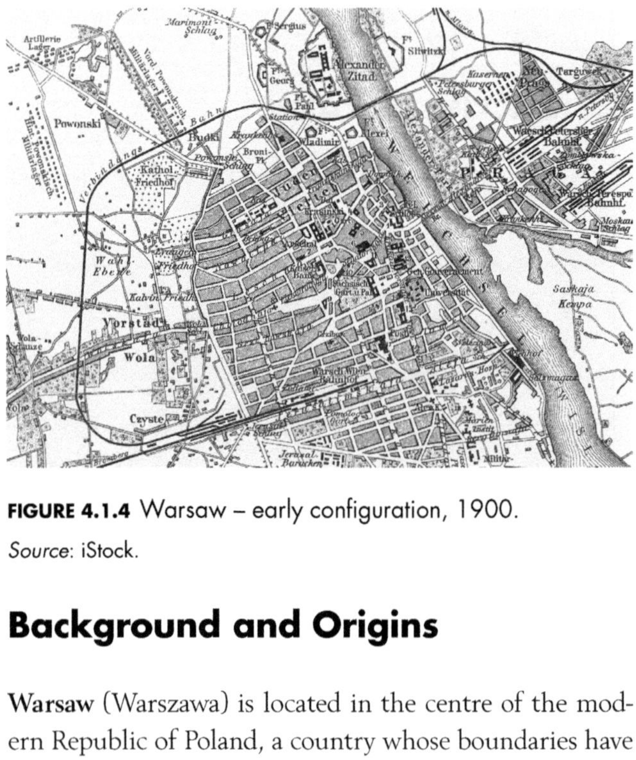

FIGURE 4.1.4 Warsaw – early configuration, 1900.
Source: iStock.

Background and Origins

Warsaw (Warszawa) is located in the centre of the modern Republic of Poland, a country whose boundaries have ebbed and flowed over the centuries. From its earliest settlement as a fishing village on the escarpment of the Vistula River, the Kingdom of Poland was founded in 1025 CE, and fortifications were built in the 13th century as the city became the capital of the Duchy of Masovia in 1413, which was subsequently incorporated into the Kingdom of Poland having the seat of Parliament (Sejm).

A New Town was built to the north of the Old Town ('Stare Mesto'), and the city was strengthened with a Barbican and Royal Castle. On the intersection of the north-south and east-west trading routes, a permanent bridge across the river was realised in 1549. A commonwealth was created after the Jagiellonian Dynasty with the Federation of Poland and the Duchy of Lithuania. With the capital and

Royal Court established here in 1596, transferred from Krakow, the city grew in influence during the 18th century.

However, with earlier partition, it was annexed by the Kingdom of Prussia in 1795 and after further partition was reduced to being just the capital of South Prussia. After Napoleonic invasion, it came under the influence and dominance of Imperial Russia during the 19th century when a new citadel was built.

During the latter half of the 20th century, the city urbanised and manufacturing developed in and around the district of Mlnow, and the population grew on the east bank of the district of Praga from which Warsaw grew to become one of the largest cities in the region.

At the end of World War I, whilst the Polish Republic was re-established, the Polish had to defeat the approaching Bolshevik Red Army in the Battle of Warsaw in 1920. Further districts were developed in the inter-war years as Mokotow, until all was to be overturned again with the overwhelming Nazi invasion of Poland in September 1939.

Warsaw suffered terribly during World War II, being completely overrun, razed and destroyed. With the Jewish people herded into the Warsaw Ghetto, and subject to deportation by rail from Umschlagplatz, the population fiercely resisted the Nazis, even with the failed Warsaw Uprising in 1944. The concentration camp and Pawiak Prison is evidence of this. Despite final liberation, there was total destruction of the city, including the Krasinski Palace, National Museum and Archives, Royal Castle, churches, monuments and apartments from which the momentous task of rebuilding remained after the war.

Modern City – Buildings and Arteries

Alongside the Vistula River, new bridges had to be rebuilt to reconnect with the old eastern district of Praga. The old town, Stare Miasto, the palace and royal apartments, churches, museums, Castle Square and walls were completely rebuilt as originally designed. The original Chamber of Deputies (Sejm) and other institutions around newly cobbled streets and alleyways were all rebuilt.

Subsequently, it had to endure a long period post-war under the Soviet yoke. The large Palace of Culture, a gift from the Soviets in the 1950s, is one of the central old-style Stalinist buildings that still remains.

With the fall of communism and since independence as a democratic republic again in 1990, Poland has transformed into a market economy and flourished, having joined the EU in 2004. Modern Warsaw in the 21st century has been completely transformed with a series of modern skyscrapers such as the commercial Rondo 1 (designed by SOM) and residential Zlota 44 (designed by Libeskind).

The commercial centre has been reformed around the Central Station and Al Jerozolimskie, Swietokrzyska and Marszalkowska boulevards where a number of high-rise commercial buildings have been constructed. It is a mixed city with generally indistinct buildings in Mokotow, although the rehabilitated Old Town retains its original designs and facades.

With many memorials and cemeteries and rebuilt landmarks, there are a number of places for remembrance of the atrocities of the past. The population of Warsaw is approximately 1.8 million with 3.2 million in the metropolitan area.

Transactional Forces and Typology

From its early history, under the Dukes of Masovia, the more recent history of the region around Warsaw has been heavily influenced by continuing conflicts. Initially these were between Prussia, Russia and Austria, followed by the Bolshevik Red Army after World War I and then Nazi Germany and Soviet Russia during World War II. With its moving national borders, the Vistula River served as a formidable barrier during World War II, despite the annihilation of many parts of the city.

With its rebuilding, the city remained under the socialist yoke for 45 years until the collapse of the Soviet empire in 1991, with many residential brutalist slab blocks constructed that reflected the Soviet model. Now independent, Warsaw has regained its prominence as the national capital, consolidating the Warsaw Voivodship and with enhanced industry and re-population has developed into a modernised city with a new metro system and upgraded infrastructure.

The centre of the city was traditionally compact, developed around the central area and the Old Castle that overlooked the river. With a number of large parks, development has extended along transport corridors, reaching out to both the east and the west. This riverine city is divided by the Vistula River that runs north-south through the city, resulting in a bi-polar city on either side.

With the castle, the Historic District and the Governmental Centre on the western bank, more residential areas are on the eastern side around Praga. With its denser urban core, suburban developments are emerging peripherally in concentric zones around the city in each direction, since the effective rebuilding of the city and restitution of historic landmarks has been completed especially around the Royal Route. The following figures illustrate the former Soviet cities as described and the evolution of their modern population.

FIGURE 4.1.5A Moscow – modern configuration.

FIGURE 4.1.5B Greater Moscow – modern configuration.

FIGURE 4.1.6 St Petersburg – modern configuration.

FIGURE 4.1.7 Kyiv – modern configuration.

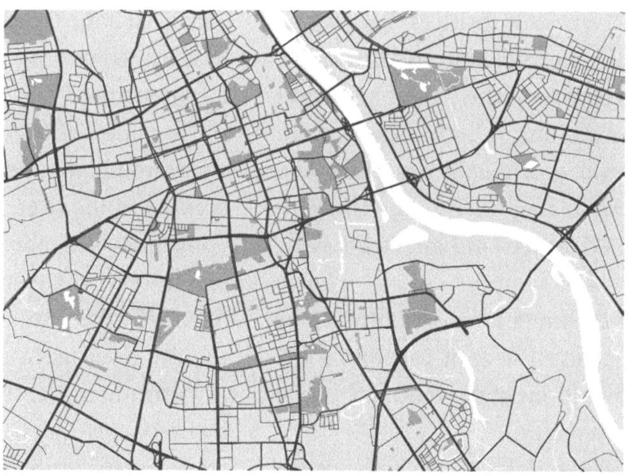

FIGURE 4.1.8 Warsaw – modern configuration.

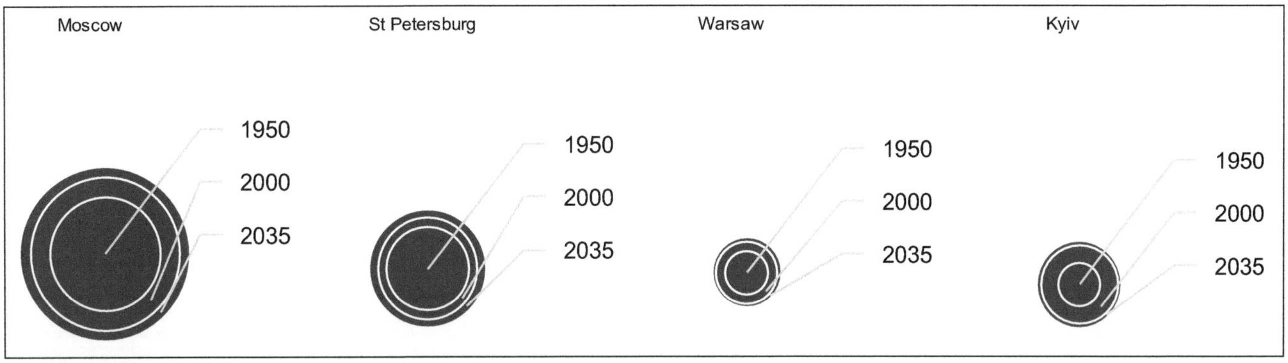

FIGURE 4.1.9 Former Soviet and Eastern European city populations.

Data source: Global Database of Metropolises 2020 – UN-Habitat (2020), UN DESA Population Division.

Further Reading

MOSCOW

Brooke, Caroline – *Moscow, a Cultural History*, Oxford University Press, 2006.

Merridale, Catherine – *Red Fortress, History and Illusion in the Kremlin*, Metropolitan Books, 2013.

Zubovich, Katherine – *Moscow Monumental, Soviet Skyscrapers and Urban Life in Stalin's Capital*, Princeton University Press, 2021.

ST PETERSBURG (LENINGRAD, PETROGRAD)

Butter, Prit – *Hero City, Leningrad 1943–4*, Osprey Publishing, 2024.

George, Arthur & Elena – *St Petersburg, Russia's Window to the Future, the first three Centuries*, Taylor Trade Publication, 2003.

Giroud, Vincent – *St Petersburg, a Portrait of a Great City*, University Press of New England, 2004.

Lincoln, W Bruce – *Sunlight at Midnight, St Petersburg and the Rise of Modern Russia*, The Perseus Press, 2001.

Ometov, Boris & Stuart, John – *St Petersburg, Portrait of an Imperial City*, Cassell, 1990.

KYIV

Bilanky, Serhiy – *Imperial Urbanism in the Borderlands, Kyiv, 1800–1905*, University of Toronto Press, 2018.

Plokhy, Serhii – *The Gates of Europe, a History of Ukraine*, Penguin, 2016.

WARSAW

Davies, Norman – *God's Playground*, Oxford University Press, 2005.

Davies, Norman – *Rising '44, the Battle for Warsaw*, Pan Publisher, 2018.

Ritchie, Alexandra – *Warsaw 1944*, William Collins, 2014.

4.2 Divided Cities

Europe – Middle East – Africa: Religious and Political Divides and Partition

Belfast, Berlin, Beirut, Johannesburg and Jerusalem

4.2.0 INTRODUCTION – DIVIDED CITIES

From the earliest settlements, towns and cities have developed organically from their natural setting and as a result of particular human influences on their built form. With increased organisation, regulation and direction, urban areas were developed based upon their administrative and regulatory arrangements. Dependent on the type of ruling power prevailing, such political orientation determined the planned built developments that ensued, which generally overruled earlier organic developments.

Hence divisions have been shaped with demarcations and boundaries determined in accordance with different types of rule or ownership. This has been evident throughout the centuries into modern times, often leading to the source of conflict between peoples. Depending on the strategic value of a region, areas have become subdivided and invariably fought over for either religious or political reasons and for individual or collective gain.

From original city-states to our modern recognised nation-states, predominantly current borders are now generally respected, although many of these may have been determined after long territorial disputes. Significant modern conflicts, however, are still ongoing over land, territory and resource rights on several continents.

Certain cities where such divisions were manifested by the physical erection of dividing walls, have significantly shaped the form of these cities as they exist today. Others have only comparatively recently been reconciled, such as the re-unified German capital of Berlin. In many other areas, tensions still remain, such as in divided cities like Jerusalem, which stand out in particular in urbanised environments today.

4.2.1 BELFAST, NORTHERN IRELAND, UK

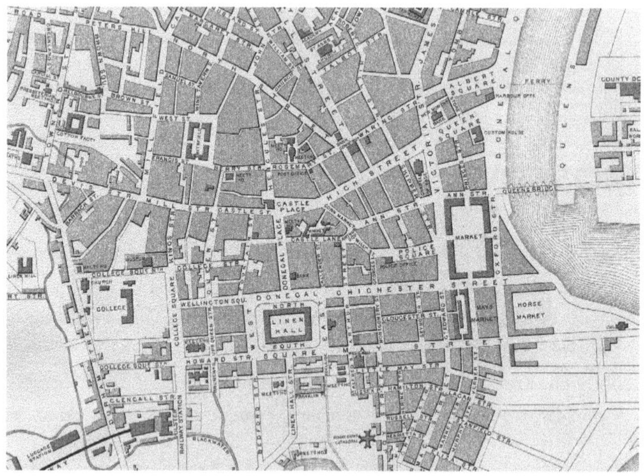

FIGURE 4.2.1 Belfast – early configuration, 1851 (Tallis).
Source: Maphouse.

Background and Origins

Belfast is the capital of Northern Ireland, which was created after partition in 1921 with its six counties forming Ulster, one of Ireland's former traditional kingdoms. From the 16th century CE onwards a process of 'plantation' of Ulster had occurred with the taking over of parcels of land by English Protestants. Following England's break with the Catholic Church, tensions between Catholic and Protestants developed. Local clans resisted foreign rule of the Tudors and Stuarts, who had adopted a policy of persuasion through implantation of local rectangular or diamond-shaped garrisons, often with London livery companies. Oliver Cromwell was more persuasive, but when William of Orange defeated James II at the Battle of the Boyne, these new settlers resulted in Irish Catholics being increasingly marginalised, which sowed the seeds of several hundred years of conflict.

DOI: 10.4324/9781003353348-20

From early Iron Age hillforts, the city of Belfast, or Béal Feirste, 'the river mouth of the sandbar', stands on the banks of the River Lagan. The sandbar was formed from the confluence of two rivers, the Lagan which flows into Belfast Lough and the Farset now underground, at Donegall Quay where the original settlement developed. It was, however, not until 1994 that a weir was built across the river by the Laganside Corporation which raised the water level so as to cover unsightly mudflats. This also enabled a larger area of land to be regenerated and developed.

In the 17th century, the port was developed and with the growing of flax in the province, linen and rope-making industries were established. From its market town beginnings, as further industrialisation developed, migration of the rural Irish population occurred, mainly settling in the west of the town, which attracted more capital. Heavier industries developed, such as shipbuilding, which became well established at the Harland & Wolff shipyard, building ships for the White Star line and including the infamous RMS *Titanic* in 1912. The city grew by agglomeration as outlying districts were incorporated, along several arterial roads, such as Newtownards and the Falls Road.

Modern City – Buildings and Arteries

With its inherent historical tensions between the Catholic and Protestant inhabitants, this sectarian conflict erupted in the 1970s into the vicious 'Troubles' between nationalist and unionist parties. With the formation of rival paramilitary groups, the city became divided between republican and loyalist areas. Segregation took place with the creation of fortified walls to separate different neighbourhoods, as along the Protestant Shankill Road and the Catholic Falls Road. Many political murals still remain on the buildings in these areas of either republican or loyalist persuasion, with the western parts being predominantly Catholic and the eastern parts Protestant.

After long deliberations under a challenging peace process, a multilateral agreement was finally reached, the Good Friday Agreement of 1998, which aimed to address the inherent tensions between the resident populations. A bilateral agreement was also made between the UK and Ireland which aimed at further devolution to Stormont and constitutional changes with improved processes of governance, which was agreed to in a referendum. Physically some of the former barriers came down but not removal of all the 'Peace Walls' that separated different communities.

The original townlands of Belfast comprised those in County Antrim on the north bank of the River Lagan and those of County Down on the south bank. The city contains different quarters – the Cathedral Quarter around St Anne's Protestant Cathedral and the southern Queen's Quarter around Queen's University and the botanic gardens. The centre of the city revolves around Donegall Square and city hall, which connects along the 'golden mile' to Queen's University. The Entries is a network of narrow alleys with pubs and taverns near the clock tower.

The Titanic Quarter is a waterfront area that has been regenerated around former Belfast Harbour and Queen's Island, now becoming a new ICT hub for innovation. The Gaeltacht Quarter is an area to the west of the city around the Falls Road. The tallest new commercial building is the Obel Tower, and there are a number of parks and gardens across the city.

Since the Peace Agreement, as part of the 'peace dividend' there has been considerable redevelopment and regeneration of real estate, but the Peace Walls that recognise this divide still exist. With its chequered history, there is still evidence of the Troubles, with areas remaining separated and having different majority populations. Slowly some convergence is beginning to develop assisted by a narrowing of economic differentials between the separate neighbourhoods and communities.

In terms of infrastructure, the West Link Highway connects with the main motorways and Translink is the main railway company providing commuter rail services. The population of the city is approximately 0.3 million, with broadly 40% each brought up under either Catholic or Protestant traditions, and approximately 0.6 million people live in the metropolitan area.

Transactional Forces and Typology

The city of Belfast developed around the River Lagan and the natural inlet of Belfast Lough from the Irish Sea. As the 'river mouth of the sandbar', it was a natural location for a harbour as trade developed with the production of linen and rope from locally grown flax. After the construction of the Lagan Canal, the creek was developed significantly with the creation of new wharves on reclaimed land, and around 1850 a deepened Victoria Channel was straightened in order to improve marine access. It became a key gateway for the movement of raw materials and has shaped the immediate coastline.

Soon after, new shipyards were constructed for shipbuilding, notably Harland & Wolff, that grew in importance during the subsequent world wars. Aircraft manufacturing was also developed, with Short Brothers building military aeroplanes. Post-war, as these industries declined, attention focused on ship repair, oil tankers and supporting the offshore oil business and the newer renewables sector which is becoming increasingly significant.

As a coastal and estuarine city around an enclosed bay, the city has grown initially along the river and outwards radially. However, it has been shaped and constrained into different quarters since the time of 'plantation' as separate neighbourhoods have developed, particularly the western Catholic areas around the Falls Road and eastern Protestant areas around Shanklin Road. These arterial roads have served to shape the segregation and division of different wards of the city of Belfast rather than to create nucleated or integrated settlements. Accordingly, this sectarian undercurrent has acted as an inherent regulator of the pattern of development of the city and its ongoing division.

4.2.2 BERLIN, GERMANY

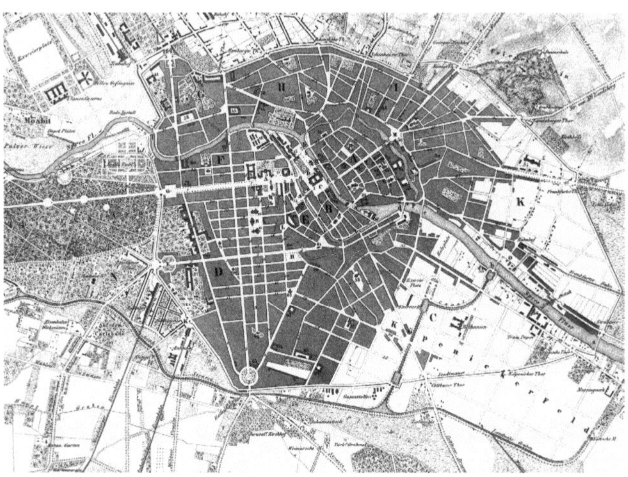

FIGURE 4.2.2 Berlin – early configuration, 1897.

Source: iStock.

Background and Origins

Berlin developed around an island on the River Spree that flows into the River Havel in Spandau in a region of forests and lakes in the north of Germany. Inhabited by Slavic tribes, the first records of settlement date from the 12th century CE of development at Cölln on the Fischerinsel and Berlin at Nikolaiviertel on the Spree, with walled perimeters. These settlements were strategically located on the Holy Roman Empire trade routes, old 'Via Imperii' running north-south towards the Baltic coast.

As the city developed, the combined Berlin-Cölln of both the right and left banks became the capital of the principality of Brandenburg, known as Prince's Capital, and the royal palace, which became the permanent residence of the Hohenzollerns, was built in 1450.

During the 17th century there was a setback and devastation with the 30 Years War and after the Edict of Potsdam in 1685, a large wave of mainly Huguenot immigration occurred. With evolution of the Duchy of Prussia, Berlin was to become the new capital of the Kingdom of Prussia in 1701, succeeding Konigsberg and with the integration of new areas of Friedrichstadt, Fredrichswerder and Dorotheenstadt aligned with the Altstadt.

These new districts were laid out on the edge of the existing mediaeval city. By the district to the south-west of Friedrichstadt a new city wall was planned that would act both as a line of defence and a customs wall. The wall was to contain three gates, Brandenburg, Potsdam and Hallisches gates, that would overlook three eventual great public spaces respectively: the square Pariser, the octagonal Leipziger and the circular Mehring Platz. These carefully selected forms of a square, octagon and circle were to echo the planned design of the royal squares created in Paris some years earlier.

This resulted in the creation of the main axes of avenues that connect these three geometrically constructed open spaces: the southern rondell of Mehring Platz (or Belle Alliance), the western octagon of Leipziger Platz and the central square carré of Pariser Platz, which together define a triangular area of Berlin, all built around 1735 under King Friedrich Wilhelm I.

The octagonal 'Achteck', with its eight-cornered marketplace, was subtle in its configuration, abutting the wall and the four main roads radiating outwards from Potsdamer Platz. Here the space was the subject of significant plans for development, firstly as a monument to Frederick the Great by Gilly and subsequently to the Wars of Liberation in 1815 by Schinkel, neither of which were realised. Rather, Potsdamerstrasse was to become a street defined by its rail tracks and streetcars bordered by large public and private buildings.

Under Frederick the Great, the city grew and consolidated during the 18th century. However, in 1806 the city was taken over temporarily by Napoleon, but after his defeat and the Congress of Vienna in 1815 it became part of the new Province of Brandenburg. With the eventual unification of Germany in 1871, under Kaiser Wilhelm I, Berlin became the capital of the Second Reich of the German Empire. New plans were developed with the Prussian Building Commission under KF Schinkel, and the 1875 Prussian Planning Act enabled new urban areas and infrastructure to be designed and built. During the 19th century, the city developed as a centre of learning with the founding of the university and technical university.

At that time there were many scientific and industrial advances in electrical and mechanical industries with the introduction of new steel. In 1862 James Hobrecht prepared a physical plan for developing new apartment blocks, five storeys high across the city, which were developed but without adequate planning for open spaces and ventilation, known as Mietskasernen ('rental-barracks').

Towards the end of the 19th century, industrialisation of the city really evolved, with growth in population and production leading to the development of the adjoining suburbs of Wedding, Moabit and Kreuzberg beyond Mitte connected by urban tramways. Manufacturing grew from light to heavy industry and steelworks and with the significant development of new railways and factory installations having their own dedicated housing settlement, as at Siemensstadt.

With the defeat and upheavals of World War I, in 1920 the new Weimar government created the Greater Berliner Act to considerably expand the size of the capital and to include incorporation of several surrounding towns of Charlottenburg, Kopenik, Lichtenberg, Neukölln, Rixdorf, Schoenberg, Spandau and Wilmersdorf with the city of 'Alt Berlin'.

This was to lead to a virtual doubling of the urban population between 1890 and 1910 and there were new modern built schemes from the legacy of Neus Bauen, either 'siedlungen' or 'zeilenbau'. As an example, Taut and Wagner's Hufeisen-siedlung (horseshoe housing estate) cooperative was built at Britz in Neukölln as a three-storey building around a lake with adjoining flat-roofed terraced housing. The public housing agency GEHAG was also formed to develop a significant volume of subsidised public housing.

After the decadent Roaring '20s with further industrial development, there was even the construction of the Flieguder Hamburger, the streamlined diesel-powered railway that ran between Berlin and Hamburg. The first motorway, AVUS (Automobil – Verkehrs und Ubungsstrasse), was built through the Grunewald forest.

But all was to change in the late 1930s with the coming to power of Hitler and the National Socialist Party. Albert Speer had developed a grandiose plan for Greater Berlin in 1939 with a monumental north-south axis and new Runder Platz crossing Potsdamerstrasse, leading via a triumphal arch to a gigantic domed Kupferhalle. None of this, however, was built, except for a northern section that ran east-west that was completed post-war by the Russians and named Stalin Allée.

During World War II, after a continuous period of intensive Allied air raids, most of the city was destroyed. At the end of the war and the Third Reich, after 1945 the Allies divided the city of Berlin into four sectors, to be managed individually by the victors, the USA, the UK, France and the USSR.

Scharoun had a plan for rebuilding the city rejected before a broader plan by K Bonatz was approved in 1948. However, that year the Soviets, under Stalin, opposed currency reform measures and started to blockade access to its sector. This led to immediate division between the three western and one eastern sector. With the western sectors entirely surrounded by that of East Germany under Soviet control, their inhabitants were marooned and required supplies to be air-lifted into their own parts of the city to maintain their livelihood. With the resulting creation in 1949 of the Federal Republic of Germany (FRG) of West Germany and the German Democratic Republic (GDR) of East Germany, an 'Iron Curtain' had fallen that divided both the country and the city in two.

Divided City

With an increase in Cold War tensions, in 1961 the East Germans decided to build a wall to physically separate their sector from the others, the so-called Antifascist Protection Rampart. This was physically divisive of neighbourhoods and communities and remained for nearly 30 years. Huge housing blocks had been built along Stalin Allée and elsewhere in the Soviet sector, which totally changed the urban fabric in their area, using prefabricated concrete 'platten-bau' systems.

West Berlin post-war was a very different place, having survived the blockade under its separately administered sectors and thanks to the Berlin airlift in 1948. Being attached to the western sphere of influence, it had managed to remain economically active. The main activity focused around Kurfurstendamm, which had an intensity of stores and shops as well as less salubrious places and activities. Very plain, with an absence of historic culture, the real character of the place had been lost, and the Federal

German administrative capital had been transferred to Bonn.

The area of Hansaviertel was rebuilt around Hansa Platz during the Interbau 57 international building exhibition that resulted in individual innovative modern housing designs being undertaken. With the city fragmented, in the early 1960s a plan was outlined that aimed at creating a new centre for West Berlin, with the development of a Kulture Forum. This area was effectively filled with the realisation of the Philharmonie (designed by Scharoun), the National Gallery (designed by van der Rohe) and the Berlin Science Centre (designed by Stirling), unrelated modern buildings realised during this period of partition.

The city was physically divided from 1961 by the walls and a controlled no-man's land, which created an enormous barrier between East and West Berlin extending between Potsdamer Platz in the west and Alexanderplatz in the east and beyond. The wall, which crossed Unter der Linden at the closed Brandenburg Gate, was a 3.6-m.-high, 'L'-shaped precast concrete wall capped with a rounded concrete pipe and barbed wire. Progressively more sophisticated, the fourth generation of barriers was actually two rows of walls with electric fences and separated by a no-man's land that was overseen by watchtowers with ditches and tank traps.

In East Berlin there was constant surveillance and a certain curiosity as to occurrences on the other side. In the West, there was outward-looking interest, but as an isolated area within surrounding East Germany, they were determined to retain their democratic principles and economic system, although being totally dependent on western support for security. Barriers and the extended wall were all pervasive, with occasional entry points, such as 'Checkpoint Charlie' between sectors at Friedrichstrasse. Both the 'U' Bahn and the 'S' Bahn continued to run, but certain 'ghost' stations were closed where no passengers were allowed to alight or disembark. There was not a curfew, but the general atmosphere was restrictive across the divided city.

In 1986, the East Berlin Planning Office presented its plan for the reconstruction of Berlin 'Mitte' on the 750th anniversary of the founding of the city. However, during the late 1980s there was a growing move across Europe for greater democracy, and in Russia, where the Soviet socialist doctrine was waning, under President Gorbachev movements towards 'perestroika' (restructuring) and 'glasnost' (openness) were facilitated in seeking political and economic reform in the east. The laws on state enterprises and cooperatives were changing and decentralisation was being encouraged.

Gradually, a whole wave of people seeking liberalisation swept across Europe as the Cold War thawed, with the resulting pressure for movement of people out of the east into the west. In early November 1989 this came to a head when after pressure in Hungary and Poland, the East German GDR General Secretary Krenz allowed checkpoints in the wall to be opened in order to allow the free movement of people across their border.

Not only was this the beginning of a new freedom of movement, but this led to the re-unification of Germany in October the following year. With the eventual transfer of the capital back from Bonn to the unified capital of Berlin in 1991 under the new Federal Republic of Germany, Berlin became one of the three city-states of the 16 federal states of Germany.

Modern City – Buildings and Arteries

With re-unification, a huge amount of rebuilding was required, especially infrastructure in reconnecting city roads and streets, bridges and communication networks. A city forum was established, Stadt Forum, in order to canvass opinion on rebuilding, although this tended to exacerbate differences rather than coalesce opinion.

Much was to change during the next 20 years. With the divisive wall being pulled down and previously abandoned areas being restored, new neighbourhoods began to become re-established. Here was the chance to rebuild everywhere, with older buildings being acquired, pulled down and rebuilt, and derelict sites being redeveloped. A huge amount of money flowed back into the city as districts and infrastructure were upgraded; apartments were modernised, and re-integration and renovation occurred in many areas.

Instrumental in the large-scale redevelopment were the state railway corporations from both east and west that controlled all the original railway land and which were combined into a new corporation named Vivico to manage their property portfolio.

As a very large city with many different neighbourhoods, extensive numbers of new central developments occurred, especially in Mitte and around Freidrichstrasse, formerly part of East Berlin. Quite extensive in-migration also took place. Here the eastern neighbourhoods quickly regained their stature as economic activity was restored; with large stores and commercial outlets returning, property values rose within a reinvigorated centre again.

With the developed public transport system of the 'U' and 'S' Bahns, a new central station, Hauptbahnof, was built with a huge redevelopment at Lehrter Bahnof where east-west and north-south rail routes crossed over each other at split level. This area has been completely transformed now with new developments along the adjacent River Spree.

The centre of gravity of the city has moved back from Kurfurstendamm and Zoogarten to Mitte. With the redevelopment of the new Reichstag government (designed by Foster) and new commercial developments by the Brandenburg Gate around Pariser Platz, and with a revitalised Unter der Linden, upgraded Museum Island Quarter and new Humboldt Forum (designed by Stella) replacing the old concrete Palast der Republik of the GDR, which had been built over the historic Berlin Palace of Brandenburg-Prussia, this area has become totally transformed.

After much controversy, consultation and competition, Potsdamer Platz now features major new commercial real estate around the former square, with the high-rise buildings of the Sony Centre (designed by Jahn) and Daimler-Benz HQ (designed by Piano), alongside the original octagonal Leipziger Platz, many people have moved back into the centre of the city around Mitte. With major companies headquartered in Berlin, such as Siemens, and with its role as federal administrative capital and chancellery, it has become a significant European capital and economic hub.

The renamed Karl-Marx Allee remains a major boulevard to the east overlooked by the tall Fernsehturm TV tower in Alexanderplatz. Schoneberg retains the high-rise Sozialpalast as a relic built by the West German government as social housing. The new main airport has been opened to replace the four former separate French, British, American and Russian airports of Tegel, Gatow, Staten and Tempelhof by one large international Berlin Brandenburg Airport (BER) in the south-east located beside the former Schönefeld airport, which has been converted into a public park.

Popular inner areas such as Kreuzberg are growing and have consolidated, as have outer areas, such as Prenzlauerberg, Spandau and Potsdam over the infamous Glienicke Bridge across the Havel River where formerly spies were exchanged between east and west.

Now integrated, Berlin is an active and united modern city again. Despite this, whilst it may have lost some of its former intrigue, with its earlier divisions, it is becoming a pivotal global city with a significant developing hinterland in the region. The population of the city is approximately 3.8 million with 6.2 million in the metropolitan area.

Transactional Forces and Typology

The city of Berlin developed at the crossing of historic trade routes running north-south following rivers to the Baltic coast and east-west between areas of Frankish and Slavic realms. With settlements around the islands of Cölln and Alt Berlin on the River Spree, these served as the basis of ties and trading alliances between neighbouring cities that were to develop across the region.

Subsequently, as Berlin became the capital of Prussia, and with immigration from other regions, the city developed militarily, administratively and commercially. Most significantly, this was to be transformed during the 19th century when, as a result of the Industrial Revolution, new manufacturing and technological processes were introduced that generated increased economic activity.

With the construction of the railways, new industries were developed and the population increased greatly. Greater Berlin was created in 1920 by incorporating adjoining towns and communities into a substantially larger metropolitan city. Of a much larger area, it was divided into separate boroughs incorporating previous towns and outer areas, such as Neukölln, Köpenick, Pankow and Reinickendorf, and is now a substantial region.

All was to change under the German Reich and the devastation of World War II, during which the majority of the city's buildings and infrastructure were destroyed. It also left the city divided firstly under four occupied zones of the Allies, which was to lead to the separation of the Soviet sector and the creation of East Berlin and physical barriers between east and west. It was not until the fall of the Soviet empire in 1991 that the city was to be reunified again.

During the subsequent 30 years, this has led to rebuilding and substantial strengthening of the economy, leading to the city's reverted dominance as the leading and largest German city-state. As a result of increased homogenisation, the city has coalesced again to become a formidable and productive integrated modern city within the Greater Berlin conurbation, yet a poly-centric city with distinctive neighbourhoods (kiez).

As a riverine city centred around the natural Fisherinsel, the city is characterised by a series of concentric zones with radial routes emanating outwards in all directions. From its early classic core with three central geometric squares with their interconnected axes, the city has grown generally in a planned manner with individual areas as extended neighbourhoods have been developed.

Comprising a number of different quarters – residential, commercial and industrial – its density is reasonably balanced and extends beyond into the suburban areas and the lakes on both the eastern and western side. Its comprehensive urban and suburban U-Bahn and S-Bahn transport system provides good access to and from most of the outlying areas.

Previously subdivided sectors have been reintegrated now, although the eastern area still contains many dense concrete residential apartment blocks as a remnant of the Soviet era. Since reunification, as new areas have been replanned and reconstructed, the central federal administration area has new significance and commercially Mitte has re-established itself as the commercial centre of the city.

4.2.3 BEIRUT, LEBANON

FIGURE 4.2.3 Beirut – early configuration, 1897.
Source: iStock.

Background and Origins

Beirut is one of the ancient cities in the Levant, located on a peninsula on the eastern Mediterranean coast. As a Phoenician port it was known as Laodicea during the Seleucid Empire whose civilisation developed around city-states such as Tyre, Sidon and Byblos. During the Roman Empire, it was one of the Roman colonies to the east of the empire. Since it was conquered by the Muslims in 635 CE, it became the Principality of Mount Lebanon, the basis of the current country.

After a period of Turkish Ottoman rule, in the mid-19th century, under French influence, it became an active port and commercial and cosmopolitan centre producing silk and agricultural products. Its harbour and railway station were a place of entry for pilgrims en route to Makkah.

After World War I and with independence after the end of the French mandate in 1946, in recent times the city became divided between the Christians predominantly in the east and Muslims in the west, with a no man's land in between, around a nominal 'Green Line'.

Modern City – Buildings and Arteries

With civil war in 1975, after constant struggles and years of war, the centre of the city started to be reconstructed in the 1990s. Under development company Solidere, set up by the then prime minister Hariri, aggressive rebuilding started, until the conflict with adjacent Israel was re-ignited in 2006.

Since then, periodic attacks have disturbed the trend towards stability, which has caused significant setbacks on each occasion. All of this has contributed to huge devastation and waste areas around the Jdeideh suburbs and along the northern coast. Memories of these remain, like the scarred concrete building known as the Egg. New areas have been targeted for development, such as the Bourj Hammoud landfill and the Linord coastal reclamation.

However, the disastrous August 2020 explosion in the port area in Hangar 12 where volatile material had been stored caused huge devastation across the city followed by an enormous fire.

The main city is divided into a dozen quarters, where different population groups reside. With a number of diverse religious groups, mainly Muslim and Christian communities, these comprise different Muslim sects (Shia, Sunni and Druze), Christian sects (Catholic, Maronite and Orthodox), Assyrians and Copts.

The Beirut central business district has been extensively reconstructed, somewhat controversially, under Solidere's expropriation and intervention, with much cultural heritage displaced and former Ottoman and mediaeval artefacts lost. The unique local sense of place of previous souks, small private houses and stores has disappeared, replaced with large public concrete structures. The role of the former banking and financial centre has also lost most of its trade and significance.

Whilst the centre is now largely pedestrianised, Hamra Street leads along the Corniche and on down to the Raouche limestone rocks, which remains an attractive area. Once known as the 'Paris of the Middle East', sadly few areas of real architectural merit remain in the city except Gemmayzeh and the Badaro green village district, with its cafés and restaurants.

Besides these, there are the central government buildings, the American University of Beirut, and many educational establishments and museums. Despite its recent troubled history, its Middle Eastern spirit remains, and thanks to its Mediterranean location and good climate, it is a fascinating, yet complex city.

The population of the city is estimated at approximately 0.4 million and 2.2 million in the metropolitan area (with approximately 50% Muslims and 40% Christians).

Transactional Forces and Typology

As an historic city under earlier Roman, mediaeval and Ottoman periods, Beirut was a significant trading centre in the eastern Mediterranean. Coming under different spheres of influence, its agricultural produce, silk and other export materials led to the expansion of its port and the development of the interior railway line to Damascus. It also developed as a financial and banking centre under the French mandate, evolving into a thriving regional business centre.

However, with instability in the region, the commencement of the Lebanese Civil War in 1975 led to the division of the city between the Muslim West and Christian East around a north-south demarcation line. After a long period of division, despite intermittent setbacks, this has been followed by periods of reconstruction, although much of the local heritage has been lost. Tensions still remain.

As a coastal city, the natural configuration of Beirut is with its northern coast and the port and a western coast with cliffs and the international airport. To the east beyond is the range of rising mountains running in a north-south direction. It has mixed linear and radial features despite having lost much of its former determining characteristics in individual quarters. Defined by its coastlines, it continues to retain its mixed Middle Eastern, Mediterranean and Arab cultures.

4.2.4 JOHANNESBURG, SOUTH AFRICA

Background and Origins

The city of **Johannesburg** in the modern province of Gauteng in South Africa is located on the High Veld, on the southern side of the prominent ridge named Witwatersrand (white waters ridge), which features springs arising from rocks in the watershed between the Limpopo and Vaal rivers. The first settlements were of Sotho-Tswana-speaking inhabitants in the Transvaal region before the Mfecane wars between different indigenous communities and the Ndebele.

Everything changed in 1884 with the discovery of gold in this mineral-rich area, and with the establishment of miners' tented camps, such as Ferreira's Camp, the gravitation of people to this area began. It was a tough time as Dutch settlers from the Cape, Afrikaaners, foreign migrants and newcomers, 'uitlanders', were all seduced by potential riches and there was tension between the Boers and the British occupiers in the Transvaal and Orange Free State that led to the ongoing Boer Wars.

Whilst resolved at the Treaty of Vereeniging in 1902, the city started to be industrialised. However, there were several negative outcomes of these wars at the time with scorched-earth tactics in clearing the countryside and the policy of internment and concentration camps, which had a lasting impact and profound consequences on future demographics.

The city of Johannesburg developed in multiple nuclei, with differing characteristics in different areas. Downtown was established having a regular grid pattern with developments in the eastern and western ridges spreading out from the centre.

The Anglo-American Corporation was established by Oppenheimer in 1917 to manage both gold and diamond mining in the region. Expansion occurred during the 1930s after South Africa left the gold standard and as the economy started to grow.

Modern City – Buildings and Arteries

Post–World War II, the system of apartheid (apartness) racial segregation was introduced in 1948, which led to the separation of inhabitants into distinct black and white areas. But the new Land Acts reserved the best land for the white population. As a result, in the 1950s the government started the Western Areas Removal Scheme and the policy of building townships for black people outside Johannesburg was created in order to provide housing for workers for the city and the mines.

This included Meadowlands and the township of Soweto (South-West Townships), which was originally a collection of settlements on the outskirts of the city. This area grew enormously as rural migrants moved towards the city, but who were not allowed to reside in the city proper.

This policy of apartheid also led to the creation of 'Lenasia', a dedicated area for those South Africans of Indian descent. All of this had a significant impact on the overall subsequent development of the city, in different clusters around Sandton and Rosebank, Randburg and Braamfontein.

Subsequently, high-rise blocks were developed in the denser commercial central district downtown, such as the Carlton Centre, with new retail areas and residential areas in Hillbrow, along with more expensive residential areas in Hyde Park and Sandhurst.

During the following period of international sanctions in the 1980s, the economy contracted, security had deteriorated and many businesses downtown were abandoned and relocated to more secure areas such as Sandton and Randburg. Inner-city areas deteriorated as buildings were taken over and occupied by squatters. The suburbs sprawled in every direction as the urbanised area of Greater Johannesburg expanded. As crime increased in the inner cities, flight to the suburbs occurred, with fewer people wanting to live downtown, and whites in particular retreated to gated communities. There were some drastic renewal examples, such as the decaying then refurbished cylindrical tower block Ponte City in Hillbrow.

Huge political changes were to come, however, once a ban on the African National Congress (ANC) was repealed and in 1991 eventually the Group Areas Act, which had enshrined in law the system of urban apartheid, was scrapped by premier De Klerk. The ANC was legitimised, Nelson Mandela was released from confinement on Robben Island and elected president in 1994 and took over from the old National Party. The entire previous regime, including its rules and regulations, had finally been dissolved. All of this was to have an impact on land policy, housing and settlements across the country.

Originally, the city of Johannesburg had a significant downtown core district that was supplemented by a number of peripheral small towns along the gold-bearing reef. However, gradually a number of satellite areas developed, and the city has fragmented into differentiated clusters of corporate businesses with separate individual gated residential estates surrounded by a patchwork of former townships. These spatial developments have been significant, completely changing the nature of the city into a polycentric district in the midst of the sprawl of Greater Johannesburg.

Gauteng Region

Gauteng, meaning 'place of gold' in the Sotho-Tswana languages is a new province of South Africa. Originally it was part of the old Transvaal province, named Pretoria–Witwatersrand–Vereeniging (PWV). Now it has become one of the most urbanised areas surrounding the city of Johannesburg. The Witwatersrand basin is a long geological gold-bearing area featuring north-flowing rivers form waterfalls and comprising a conurbation from Randfontein in the west to the town of Nigel in the east, encompassing the East, Central and West Rands.

With population pressures and a high rate of urbanisation around Johannesburg, this entire Gauteng region has become a magnet for new settlement. This has resulted in major strains due to high population densities and a large informal sector.

The area contains national roads that run through the region and which are the responsibility of SANRAL, as well as bus rapid transit and railways. The new Gautrain express train on standard gauge between Johannesburg Park Station and O.R. Tambo International Airport (JNB) connects via Mid Rand and Centurion with Pretoria, operated under the Bombela Concession.

As the principal city in a modern reforming South Africa, Johannesburg acts as its main business and financial centre with a population of approximately 1 million in the central city, 8 million in the metropolitan area and 16 million in the Gauteng region.

Transactional Forces and Typology

The origin of the city of Johannesburg was as a mining town as gold was sought and extracted along the geological reef. From its central core, residential clusters developed farther out as suburbanising pressures led to lower density sprawl during the 1950s. This was strengthened with the formation of office and retail centres at Randburg, Rosebank and Sandton that developed into independent municipalities north of the central core, around which further development coalesced.

Even farther out, rings of development emerged with concentrations of commercial businesses and new centres such as Mid Rand and Centurion, and those on the northern axis between Johannesburg and the capital Pretoria. With the later realisation of the fast Gautrain railway, this has reinforced the attractiveness of the Gauteng region. However, this has strengthened the segregation between affluent suburbs and enclaves, and new shopping malls, office complexes and casinos, with the less well-serviced township areas where the majority of the population are poorly housed in peripheral areas.

The city of Johannesburg was originally set out on an orthogonal grid on a dominant east-west axis around which development took place. Its localities of Braemfontein and

Doomfontein were laid out by the Dutch founders with roads and railways to support early mining exploration activities in the Witwatersrand area.

Outside the central area, the land has been shaped by the creation of the peripheral townships such as Soweto, where the black community of mainly miners was contained and segregated under the Urban Areas Act of 1923. With the restrictive Pass Laws, this resulted in significantly shaping the patterns of occupation and land use across the city and its outlying areas. Now as intervening areas have been developed, a poly-centric and hybrid nature of development has resulted, with clusters of groups housing different extremes of wealth aggregated together distinct from many more segregated and poorer areas. With Johannesburg as part of the Gauteng region, this now has the characteristics of a larger megalopolitan region.

4.2.5 JERUSALEM, ISRAEL

The background and history of this ancient city has been described in Chapter 2.2.1 and in particular its religious significance for Jews, Muslims, Christians and Armenians.

Modern City – Division

Despite its long, complex history and continuing symbolism for different religious faiths, Jerusalem is an evolving yet divided modern city. With its unicameral Knesset Parliament of the State of Israel, ministries and governmental institutions, it acts as the administrative centre of the country.

After the original formation of the State of Israel in 1948, this has led to the partition of the city between West and East Jerusalem along the Green Line separating the Israeli and Arab sectors. This has generally prevailed since its creation, leading to the ongoing separation of the city.

After the Six-Day War in 1967, Israel took over the West Bank and annexed the eastern part of the city. This was a challenging period with the sensitivity of the Western Wailing Wall and the Dome of the Rock, Al Aqsa Mosque on Temple Mount. The Old Arab Mughrabi Quarter was destroyed and the shrine of Haram e-Sharif was returned to control of the waqf.

Since the declaration of the independence of Palestine and subsequent Palestinian Intifadas – the First Intifada between 1987 and 1993 and the Second Intifada in 2000 – parts of the city have been separated by the Israeli administration physically by newly constructed sections of very high concrete walls woven through different neighbourhoods with a no-man's land in between.

Now over 400 km. in length, this coarsely separates land under Israeli jurisdiction from the Palestinian West Bank. This has segregated the local environment and forced division between communities, in some areas straying beyond the agreed-upon Green Line. Such non-contiguous areas have severely divided the West Bank, with new settlements also often in elevated clusters overlooking neighbouring surroundings in a defensive posture.

The essential forces that are endemic across the city are both strongly religious and political. With established built features having a long history of representation and symbolism, individual monuments reflect deeply embodied meanings for different sects. Absent tolerance and understanding, conflict and tension have ensued, which has resulted in a deeply divided city of segregated resident populations (see also Support Material: Annexe 4.2A – Destruction of Cities). The following figures illustrate the modern configuration of the divided cities as described and the evolution of their modern population.

FIGURE 4.2.4 Belfast – modern configuration.

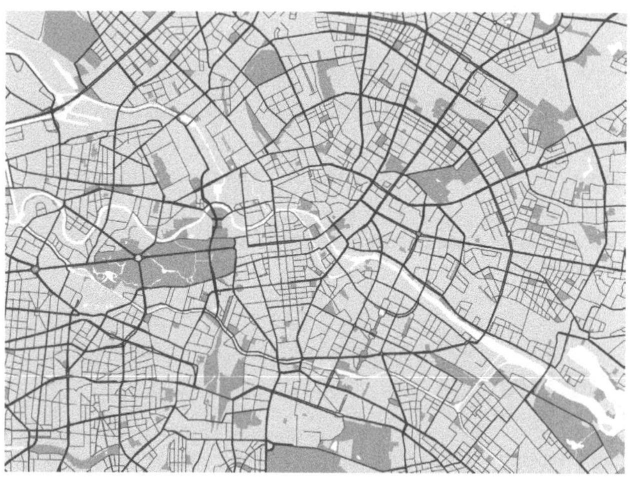

FIGURE 4.2.5 Berlin – modern configuration.

FIGURE 4.2.6 Beirut – modern configuration.

FIGURE 4.2.7 Johannesburg – modern configuration.

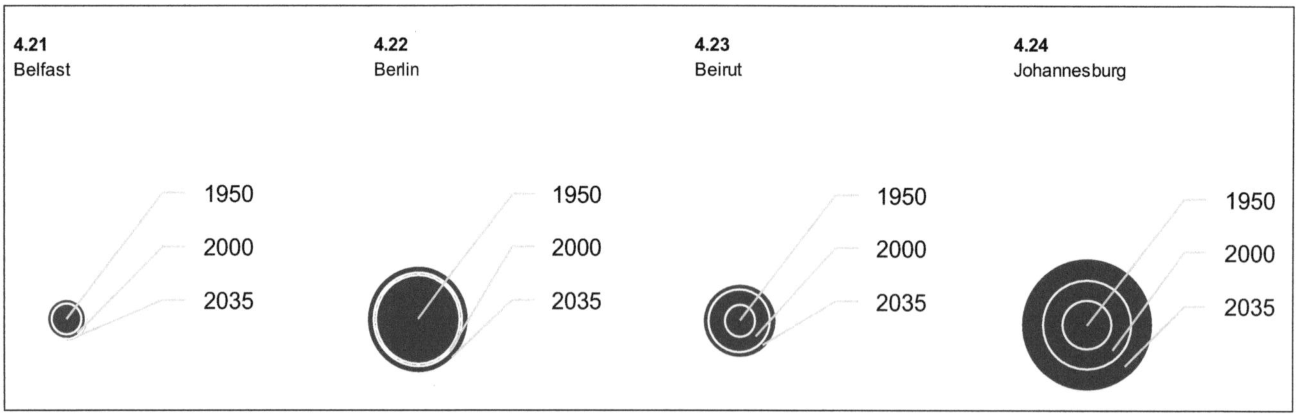

FIGURE 4.2.8 Divided city populations.

Data source: Global Database of Metropolises 2020 – UN-Habitat (2020), UN DESA Population Division.

Further Reading

BELFAST

Cochrane, Feargal – *Belfast*, Yale University Press, 2024.
Connolly, SJ, ed. – *Belfast 400: People, Place and History*, Liverpool University Press, 2012.
Johnston, Wesley – *The Belfast Urban Motorway*, Colourpoint, 2014.
Maguire, William – *Belfast, a History*, Carnegie Publishing, 2009.

BERLIN

Balfour, Alan – *Berlin, the Politics of Order*, Rizzoli International Publications, 1990.
Beevor, Anthony – *Berlin the Downfall*, Viking, 2002.
Kempe, Frederick – *Berlin 1961*, Penguin Books, 2012.
Maclean, Rory – *Berlin*, Weidenfeld & Nicolson, 2014.
McKay, Sinclair – *Berlin*, Viking, 2022.
Read, Anthony & Fisher, David – *Berlin*, Pimlico, 1994.
Richie, Alexandra – *Faust's Metropolis, Berlin*, William Collins, 1998.

BEIRUT

Excoffier, Guillaume & Salvaing, Matthieu – *Modernist Beirut*, Editions Norma, 2023.
Kassir, Samir – *Beirut*, University of California Press, 2010.
Khalaf, Samir – *Heart of Beirut, Reclaiming the Bourj*, Saqi Books, 2013.

JOHANNESBURG

Callinicos, Luli – *A People's History of South Africa*, Ravan, 1985.
Kruger, Loren – *Imagining the Edgy City, Writing, Performing and Building Johannesburg*, Oxford University Press, 2013.
Murray, Martin – *City of Extremes, the Spatial Politics of Johannesburg*, Duke University Press, 2011.

JERUSALEM

Chiodelli, Francesco – *Shaping Jerusalem, Spatial Planning, Politics and the Conflicts*, Routledge 2018.
Lemire, Vincent, ed. – *Jerusalem, History of a Global City*, University of California Press, 2022.
Sebag Montefiore, Simon – *Jerusalem, the Biography*, Phoenix, 2011.

4.3 Industrial and De-industrialised Cities

America – England

Chicago, Detroit, Cleveland, Pittsburgh and Manchester

4.3.0 INTRODUCTION – INDUSTRIALISATION AND DE-INDUSTRIALISATION

Beyond their earliest activities, humans gravitated to agricultural practices in seeking to secure their own subsistence, either from the propagation of animals and fisheries or the cultivation of soils and forests, through exploitation and better utilisation of the soil and natural resources. From grazing and stock rearing to subtropical agriculture, it was possible to develop different produce from the land depending on the soils, resources and means of cultivation. From this primary source of activity, transformation using methods developed from the creation of tools and invention of machines to assemble different man-made products through the shaping of timber, stone, tin and other found materials came the secondary processes.

As migrants moved around different regions, centres of attraction developed around which new 'making' activities were able to be developed. Shaped largely by the availability of mineral or lumber resources, these became localised centres of production. With the combination of natural resources and human labour and ingenuity, the mining of coal led to new methods of production and fabrication and ensuing manufacturing processes and 'industrialisation' of the labour force.

With gravitation to these centres, industrialisation and urbanisation went hand in hand, especially during the 19th century as the Industrial Revolution took hold in Western Europe and North America, facilitated by the extraction of mineral resources, which led to the extraordinary growth of new industrial cities. Here new inventions, tools and machinery were to totally transform their economies with steam power, the production of iron and steel and then electricity, enabling basic resources to be processed and manufactured that fundamentally altered the ways of life of subsequent generations.

It was essentially this presence of coal and other minerals in the ground that, once mined and processed, acted as new sources of power, which led to the growth of large manufacturing plants and the coalescing of communities around centres of production. These developed, for example, in Manchester and Sheffield in the Pennines in Britain, around the Ruhr Valley in Europe and around the Allegheny Mountains near Pittsburgh in America.

Such developments were quite transformative right up until the end of the 20th century when the reverse process of de-industrialisation came into play, which in its own way began to transform these cities and built-up areas into quite different communities with altered patterns of work and habitation. This came as a result of the decline in the comparative advantage of particular regions such that with changing conditions, manufacturing moved to other more favourable areas of resources of both labour and materials.

In consequence, whole areas of industry were closed down with the loss of employment that has necessitated in cities having to re-invent themselves with the creation of entirely new sectors in the service industry, typically in the fields of health and education, defence and information technology.

4.3.1 CHICAGO, ILLINOIS, USA

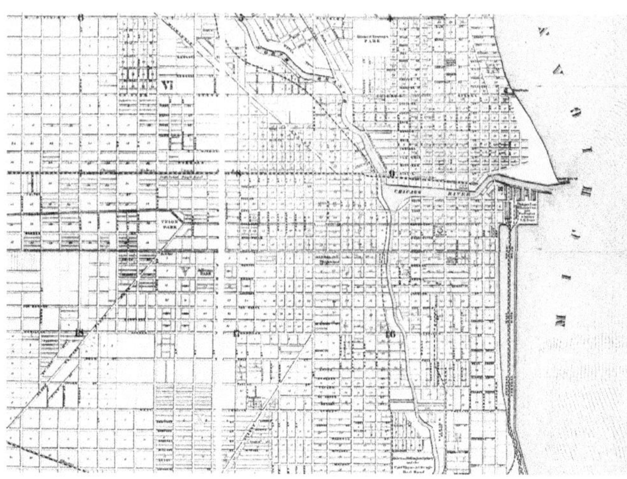

FIGURE 4.3.1 Chicago – early configuration, 1857 (Mendel).
Source: British Library.

DOI: 10.4324/9781003353348-21

Background and Origins

The city of **Chicago** is located on Lake Michigan in the American Midwest where with its continental climate, the wind and snow can swirl around in the coldest of winters. The city was established by French settlers where the north and south branches of the Chicago River converge into Lake Michigan at Wolf Point and where a section of historic portage was developed, as the river, along with a mud lake, was too shallow to be navigated across. This was in order to connect with the Des Plaines River, which flows south to the Illinois River and eventually into the Mississippi River, that was originally discovered by the French explorers Marquette and Joliet in 1673.

Native American tribes named this swampy land 'Checaugou' or 'Chicagoua', from which the name has stuck. After early settlement, eventually a large area of Native American–held land was ceded by the indigenous Illinois Potawatomi under the Treaty of Chicago in 1833, subsequent to the first foreigners establishing settlements in 1795 and trading as the American Fur Company and with Fort Dearborn built in 1803. The first grids of streets were platted (approximately 330 x 660 feet) and initial plank roads were constructed.

At the lakeside a channel was cut through the sand bar and two long piers were built out into the lake at the mouth of the river. With its key geographical location where the Chicago River meets Lake Michigan, river transport of freight became pivotal, especially as other lake ports developed when the Great Lakes started to be interconnected by the Erie Canal and the Illinois & Michigan Canal built from Bridgeport in 1848 to link with the Mississippi River.

With water transport becoming a significant means of transport, plans were made for a new canal into the city. With the appointment of commissioners in 1829, works were approved and commenced in 1836. Development occurred around the neighbourhood of this canal and land speculation increased as the canal was eventually opened in 1848 and building developed on both the north and south sides of the city. During the 1850s, civic buildings began to be constructed around the main square under local associations, and balloon frame–type wooden residential buildings with clapboard siding were developed across the platted grid.

New economic activities developed from the abundance of grain and cattle in the prairies and of lumber in the region. These commodities were each to require their own form of infrastructure with large grain silos for storage and stock-yards for handling cattle. These materials and products were to lead to the colloquial naming of Chicago as the 'Great Reaper City' and 'Porkopolis'. Active markets developed accordingly for trading these commodities in the 'Pit', with 'bulls' and 'bears' selling long and short.

In parallel, the first rail line was opened, followed by a network of rail lines radiating out from the city centre. Chicago became the 'Big Junction' with the Union Pacific having met with the Central Pacific at Promontory in Utah in 1869, thus providing a connecting trans-continental rail line between east and west. The regional Chicago & North Western, Galena & Chicago Union and Chicago & Rock Island and Pacific regional rail lines, meeting with the major Michigan Central (MC) and Illinois Central (IC) line that runs south to the junction of the Ohio and Mississippi Rivers, created so as to avoid St Louis, formed the St Louis cut-off. One area to be protected in approaching the city centre was the lakefront where the Illinois Central was built on trestles just offshore, thus enabling a lakefront park to be realised and retained.

This was to shape the extension of the city and its suburbs with the overall development of the railroads in the 1860s and their individual terminals. The huge Union Stock Yards for the handling of cattle were also extensive. Besides meat processing, resulting in Chicago becoming the meat packing capital of the world, the city became a major centre for grain handling and processing in the region. Both of these activities led to a significant increase in demand for employment.

With an increase in immigration, socio-economic conditions in the city deteriorated, resulting in over-crowding and poor housing and sanitary conditions, especially in the West Side during the second half of the 19th century. In 1854, a major cholera outbreak occurred, and in order to improve general sanitation in the central city, around 1860 in the central low-lying area, a programme of lifting entire blocks of buildings by jacking above the ground was undertaken in order to enable new sewerage pipes to be installed and run underground, requiring building foundations to be totally rebuilt. In 1866, a new dedicated intake was also constructed in a caisson out in the lake for the water supply, which was connected by tunnel to a new pumping works for distribution.

In 1871 a major fire occurred in the core of the city that decimated the North Side, after which plans were laid for the Great Rebuilding. New streets were planned and developed across the city, with several swing bridges across the waterways and a network of streetcars. The centre of the city was orientated around the axes of State Street (north-south), Lake Street (east-west) and Washington Boulevard.

The Great Conflagration of 1871 was a precursor to a consolidated period of growth and development in

Chicago, which benefitted from a series of pivotal technical inventions that was to result in the city expanding horizontally along new railroad tracks and vertically with tall steel skyscrapers and internal elevators. From the prairies to the lakefront, the city was shaped by a number of significant local entrepreneurs, such as Charles Yerkes, Marshall Field and George Pullman, and designers and builders such as Louis Sullivan, Daniel Burnham and William Le Baron Jenney.

Previously all of the water from the Chicago River ran into Lake Michigan. However, at the turn of the 20th century, with the construction of a new drainage canal, the direction of flow was reversed. Instead of stock-yard waste flowing into the lake, under the auspices of the Sanitary District of Chicago, this was improved by cutting through the watershed to the Des Plaines River now running outwards and eventually into the Gulf of Mexico instead. This new Chicago Sanitary & Ship Canal (CSSC) was created under a huge earthworks operation and connected the South Branch of the Chicago River to the Des Plaines River at Rockport, opening in 1900. This was further extended in 1907 to Joliet, and the additional canals of the North Shore Channel and the Calumet-Saganashkee Channel were opened in 1910 and 1922 respectively with management and navigation subsequently taken over by the US Army Corps of Engineers.

In 1892, an elevated tramway known as the 'El', centred around downtown, was built in a loop around which much development occurred. The World's Columbian Exposition, overseen by Burnham, was realised the following year, which was to celebrate the 400th anniversary of Columbus's journey to the Americas. A site was chosen at Jackson Park and with Olmsted acting as landscape architect, the marshy land by the lakeshore was channelled and zoned, with dedicated buildings sited by man-made lakes having a pier by the railroad terminal.

This comprised the great buildings of the large Manufactures and Liberal Arts Building; individual machinery, agriculture and transportation exhibition buildings; and an art gallery and administration building. It was an eclectic site of impressive but temporary buildings, mainly in white stucco, around the Grand Basin, hence the name 'White City' as Chicago came to be called.

Despite individual setbacks, such as the major fire and cholera outbreaks, the city had attracted several waves of immigrants both from South and Central European countries, which having different ethnicities, has led to conflicts between different groups of migrants.

In 1909 a Plan of Chicago was prepared by Burnham that proposed a new master plan for development, aimed at improving the lake front, introducing a regional highway system with diagonal boulevards, improving the rail terminals and streets and introducing new outer parks and the Chicago Civic Center. With a strong vision and taking into consideration the regional implications of metropolitan growth, it led to the creation of the Chicago Plan Commission. Whilst they did not formally adopt the Burnham Plan, a number of development plans and regulations going forward were prepared for the benefit of the city.

A significant plan in the transport sector, the Chicago Transportation Study, was developed later in 1956 that applied new models to calculate the trip generation, distribution and assignment of vehicles and which integrated consideration of land use with transport planning. This was a progressive approach for infrastructure planning, as the city had grown to become America's second city, and one of the largest regional transportation hubs.

Modern City – Buildings and Arteries

It was within the Loop that the skyscraper was born, with raft and caisson foundations for several significant buildings developed in the downtown area, by new bascule bridges crossing the Chicago River. The earliest tall steel-framed buildings were the Rookery and Reliance buildings, the Sullivan Center, the Monadnock masonry building, the Fine Arts Building, the Auditorium and Santa Fe buildings, all of which influenced subsequent built development (see Support Material: Annexe 4.3A – Chicago Significant Buildings).

By the Near North Side, the Astor Street District was developed comprising a number of elegant residential buildings, with the Gold Coast Historic District nearby where old money gravitated, and Oak Street Beach off North Lake Shore Drive. In the South Loop area and Near South Side, areas of former warehouses have been converted more recently to residential use by upgrading a rundown industrial area.

Farther south, in the South Side, is the University of Chicago, with FL Wright–designed Prairie School–type buildings in Oak Park in the West Side and adjoining Washington Park. The modern Illinois Institute of Technology (IIT) engineering campus (designed by van der Rohe) with its set of horizontal glass and steel-framed buildings, was realised here in the 1950s. The Museum Campus is in the south by Grant Park.

Downtown has the stepped Willis Tower (previously Sears), which was the world's tallest building in 1997, having over 100 storeys. The other major tower having as many storeys is the cross-braced John Hancock Center along the Magnificent Mile (North Michigan Avenue). The Trump Tower and the twin circular towers of Marina City are located in the Near North Side, where there are other tall modern buildings, such as the Aon Center and 333 W. Wacker Drive. In the 1980s, a revitalisation strategy was undertaken of the riverfront zone of the Loop and around Navy Pier, and with the subsequent Lakefront Millennium Park Project. Here now is the Pritzker Pavilion and reflective elliptical sculpture in AT&T Plaza.

Whilst having localised areas of affluence, there are, however, today many vacant land plots and abandoned homes that are found across the city and in suburban areas. The population of the city is approximately 2.7 million, from a previous peak of 3.6 million in 1950, and approximately 9.6 million in the whole metropolitan area.

As a major hub in the Midwest, whilst overtaken in size and scale by Los Angeles horizontally, in terms of verticality, Chicago remains the most striking and denser city architecturally of impressive scale, with a different zoning regime from New York.

In the highway sector, centrally Wacker Drive is double-deck, with the outer Chicago Skyway and network of freeways and inter-state highways around the city. Each district is linked by the CTA Rail Transit System with its seven rail lines that circle around the Loop and radiate out in north-south and westerly directions to connect with the populated suburbs.

Transactional Forces and Typology

With its strategic location on the lake, waterborne transport triggered the development of Chicago. With grain from the prairies, cattle and lumber in the region, these commodities each necessitated their own silos and stockyards for storage and trade. As major canal, rail and road routes were constructed and expanded to facilitate their exchange, the city became the 'big junction' of the Central Pacific and Union Pacific railways.

The core developed around the Loop, the central business area around which all trading, business and commercial activities took place. With land at a premium, new technologies were developed at the beginning of the 20th century that enabled taller buildings to be constructed in steel with elevators and fire protection and equipped with electricity and telephones. From the 'Windy City', 'White City', 'Reaper City' to 'Skyscraper City', Chicago has continued to reinvent itself as different transactional and market forces have emerged to shape the developing city and its lakefront, assisted by areas of land reclamation.

Since the city is located on the north-south lake waterfront, its development has radiated westward on the land side in a semi-circular manner with its axial roadways and railways radiating out from the inner loop. In a pattern of concentric rings, different zones have evolved, such as those under transition or deterioration as highlighted by Park and Burgess's assessment of the city 100 years ago. This model, along with Hoyt's assessment of land values and distance from the centre, has reflected the evolving forces in the pattern of development and expansion of the city (see Support Material: Annexe 4.3.B – Chicago – Urban Economic Theory).

As a radio-centric city, both its commercial and residential central areas have risen vertically in the third dimension, as this has enabled more attractive financial returns to be generated. Farther out, however, there are many pockets of land that have been vacated in deteriorating neighbourhoods, where older industries have declined and moved out, leaving vacant plots and decaying buildings requiring regeneration.

4.3.2 DETROIT, MICHIGAN, USA

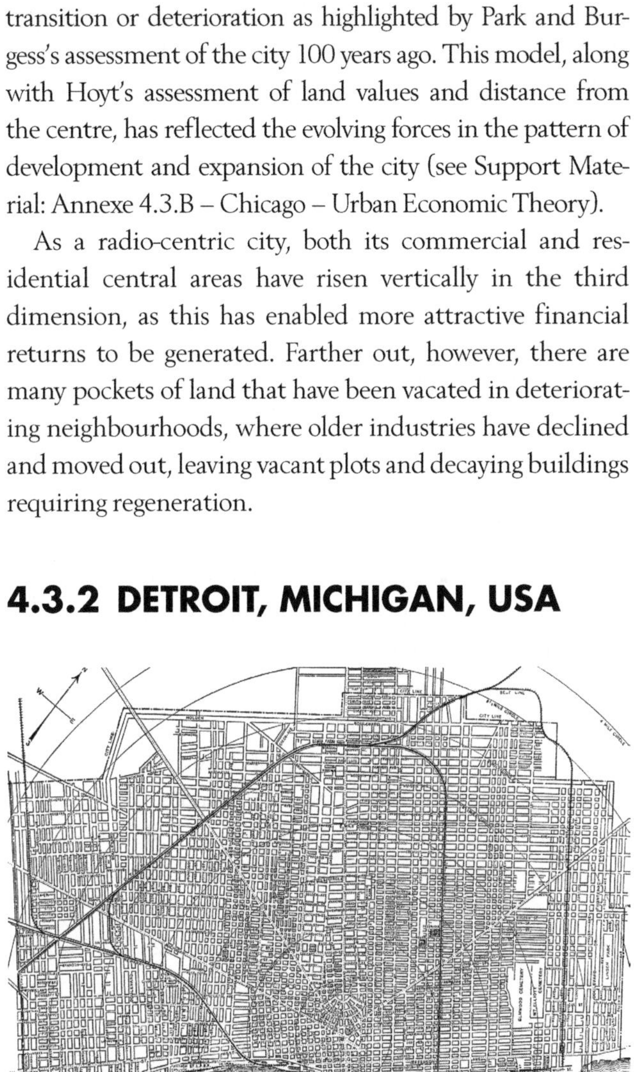

FIGURE 4.3.2 Detroit – early configuration, 1899.

Source: iStock.

Background and Origins

Detroit, in the state of Michigan, was founded originally by French traders on the western bank of the river straits by Belle Isle island. The first settlement was developed with a plan by Antoine Cadillac in 1701 on the straits that connect with Lake St Clair and Lake Erie. With a fortress built on the north bank Fort Pontchartrain du Détroit (straits), the waterfront was a key location in acting as a trading post and also for controlling water transportation.

The region was known as the League of the Iroquois by the French who defended the territory. In 1760, the British gained control of the region with the end of the French and Indian Wars. However, in 1805, the settlement was completely destroyed by fire and required complete rebuilding.

At the Battle of Lake Erie in 1813, the Americans defeated the British and retook Frenchtown, and under the auspices of Judge Augustus Woodward surveys to prepare a new plan for the town of Detroit were commissioned. Unusually, this was developed based on a series of equilateral triangles bisected by perpendicular lines to create right-angled triangles, known as 'sections'. With wide avenues running north-south and east-west, the intervening spaces were sub-divided and at the intersection of the avenues, Grand Circus was laid out, featuring a rectangular park named Campus Martius (Field of Mars) along with Cadillac Square.

Along the riverfront, streets were to be laid out in a grided and hexagonal system. It was an unusual plan with residential squares and tree-lined avenues and boulevards. However, it did not receive full acceptance with the proposed divisions and sale of lots. Accordingly, during the 1820s, an adjusted plan was developed partly on a gridiron system, with only the features of Grand Circus and Campus Martius remaining and Washington Boulevard extending north-south towards the riverfront.

There Detroit faces the city of Windsor in Ontario, Canada, on the opposite bank of the straits. That city has also developed, but less than the metropolis of Detroit. After the 1880s, the city grew steadily with the rise of lake shipping, shipbuilding and manufacturing. With the construction of the Welland Canal between the lakes and the Erie Canal to the east, its role developed as a major port on the straits that eventually connect via the Great Lakes to the St Lawrence Seaway and the Atlantic.

In 1889, the Detroit International Exposition Fair was held on the lakefront when a causeway was built to connect the city across the rail tracks. With industrialisation, it grew significantly at the beginning of the next century to become the centre of the American automobile industry with Ford, General Motors and Chrysler, each headquartered there. Henry Ford had founded the Ford Motor Company, with the Model-T produced at the Piquette Avenue plant in 1908, eventually transferred to Highland Park and later to River Rouge. This was followed by the actions of Messrs Dodge, Packard and Studebaker that led to the city of Detroit and Dearborn becoming America's great automotive capital.

With the increasing demand for industrial labour, migration came both from the South and in waves escaping from events in Central and Eastern Europe, such that the tonnage shipped on the Detroit River expanded considerably, reaching 60 million tonnes in 1907. This led to it becoming the 'greatest commercial artery on earth' and making Detroit the fourth-largest city in the USA in 1920.

During World War II, production had turned to armaments and defence manufacturing to support the war effort, making Detroit the 'arsenal of democracy'. Post-war the major auto manufacturers began to amalgamate into larger companies, as for example General Motors. Despite the increasing volume of jobs, employment was skewed with mainly black Americans facing lower-paid factory jobs and discrimination in housing opportunities with red-lining of districts relating to mortgages.

As a result, the city extended and sprawled outwards beyond 8 Mile Road in suburban areas with poorer neighbourhoods, having displaced residents and with less-effective transport routes until the introduction of electric streetcars towards the more urbanised centre.

Modern City – Buildings and Arteries

The older residential districts are located along the spines of the city, on the Woodward and East Jefferson corridors and in areas such as West Canfield and Brush Park. Central neighbourhoods comprise Downtown, Midtown, New Center and the Southwest District. One experimental area, Lafayette Park, was designed by M van der Rohe and L Hilberseimer on the east side in a landscaped park with a mix of high- and low-rise apartment buildings. Older downtown Art Deco buildings include the fine Guardian Building, the Penobscot and the Fisher Building with its internal arcade. Midtown contains a group of landmark commercial/office and residential buildings (designed by A Kahn) to act as the headquarters of General Motors.

Due to the intensive automotive activity, the city retained the name of 'Motor City' but post-war, with the music industry of jazz and rhythm & blues, it was also known as 'Motown'. The city is home to a number of universities, including Wayne State University and the University of Detroit Mercy, with the University of Michigan initially established in Detroit in 1870 but which moved to Ann Arbor in 1837. Several major medical centres were also developed there.

At the lake entrance of the city straits is Belle Isle Park, located on an island in the Detroit River between Detroit and Canada which is connected by the multi-span MacArthur Bridge. Farther south also is the larger Grosse Isle island in the straits.

Longstanding tensions in the city, however, led to urban conflicts during the 1960s. In 1963 Dr Martin Luther King Jr carried out his 'walk to freedom' down Woodward Avenue and spoke at Cobo Hall. With the subsequent July 1967 12th Street Riot, houses were destroyed and businesses closed and/or relocated to other areas after the 'Devil's Night' of fires and riots. This led to the beginning of an extended period of dereliction of neighbourhoods which was to last for several decades. Such urban decay was exacerbated by poverty and the issue of racial segregation in schools. Many buildings were burnt down, which led to many blocks remaining empty or torn down in outlying areas.

During the 1970s, with a newly elected mayor, Coleman Young, an attempt was made at revitalising the Downtown district. An elevated People Mover loop was constructed and further reinvestment in commercial areas was sought downtown, such as the Renaissance Center.

But after businesses had been lost to the auto-dependent suburbs, the People Mover was inadequate in itself to stabilise the city, the tax base declined as unemployment rose and crime remained high. With the continuing white flight to the suburbs, the downward trend continued.

Ultimately, in 2013, the city filed for bankruptcy, which was its final low point. New efforts were made after this to reinvigorate the city, and the Detroit Institute of Arts monetised its art collection. Further initiatives were undertaken to regenerate areas of the city and its infrastructure, including street lighting, landscaping and upgrading the Central Station and its reconstruction. At Wayne State University there was emphasis on the development of new medical and health facilities and medical colleges.

Downtown, more modern 21st-century commercial buildings were developed, including One Detroit Center, the Renaissance Center, Campus Martius Park, Grand Circus Park, Detroit International Riverfront and Hart Plaza.

With an urban population that once reached 1.85 million in 1950, it has declined to approximately 0.6 million, with Downtown mainly 80% African American, and approximately 5 million in the surrounding metropolitan area. Detroit also contains several ethnic neighbourhoods, such as Greektown, Little Italy and areas of Polish, Slavic, Ukrainian and other Central European descent. Beyond, in the surrounding Wayne County, there is a growing Asian population. In addition, Dearborn suburb is an area having one of the largest proportions of Arab population in North America.

The city has four border crossings with Canada. The Ambassador Toll Bridge was built in 1929, privately owned by the Detroit International Bridge Company, and the sunken tube Detroit-Windsor Road Tunnel opened in 1930. The Michigan Central Railroad Tunnel opened before in 1910, and there is a separate Detroit–Windsor truck ferry by Zug Island. The new cable-stayed Gordie Howe International Bridge is under construction between Detroit and Windsor, financed by Canada.

The city is circled by highways, with the lateral east-west Grand Central Boulevard and the I-94 highway running east-west to Ann Arbor via Dearborn and Willow Run. I-75 runs north to Troy and south to Toledo, Ohio, and I-96 runs from the west and connects with the 401 King's Highway in Canada. I-375 to the east is being rebuilt as an urban boulevard. The Q-line is a modern streetcar that runs up and down the north-south axis of the city on Woodward Avenue.

Major efforts have been made to regenerate downtown with new buildings, especially around the central zone with its new elevated People Mover. However, farther out, around 8 Mile Road, many lots still remain vacant, with sites of previous residences burnt out and decayed and not replaced. While some communities are being rebuilt, many others have been abandoned or have been taken over, for example as urban farms. The huge Packard Automotive Plant remains abandoned. One area of regeneration is the former Michigan Central Station, which has been converted and reopened as a technology and community hub.

Transactional Forces and Typology

With its location on the western side of the Straits of the Detroit River, passing between Lake Erie and Lake St Clair, the American city of Detroit directly faces the city of Windsor in Ontario, Canada, on the other side of the straits. From its origin as a key trading post, it is strategically placed to control waterborne transport between the lakes and beyond. As a result, as trade and commerce developed in this region, it became one of the busiest commercial centres at the beginning of the 20th century.

Due to both its location and the availability of materials and labour in Michigan, the metropolitan area expanded considerably. With the active neighbouring city of Windsor, this also became a significant international crossing point between the two cities. Supported by the gravitation of migrant labour, especially from the South, manufacturing developed with complementing industries such that Detroit became a major manufacturing hub for the automobile industry. With this and its armament manufacturing during World War II, this expanded both in Detroit and neighbouring Dearborn.

However, post–World War II, as manufacturing declined, the city entered a period of rapid deterioration with racial conflict, population flight and residential dereliction that led to eventual bankruptcy of the city in 2013. Since then, the city has had to reinvent itself and regenerate many areas and its infrastructure, which it is still in the process of undertaking. One symbol of renaissance in the downtown area is the circular loop of the elevated People Mover track that was erected in order to regenerate parts of the commercial business district for business purposes and visitors. As a small beginning, much still remains to be done to restore the economic health of the city.

A flat riverside city, Detroit developed across the straits from Windsor, initially along the waterfront. The city had taken shape in response to the 1820s Plan with its different equilateral triangle sections that were bisected with boulevards. The unique formal layout of Grand Circus and Woodward Avenue led to Campus Martius Park next to Cadillac Square. Whilst not fully developed or completed, the core of the city remains characterised by the three original major avenues – Michigan, Woodward and Gratiot – that run diagonally and radially from the centre.

From this rather classical layout, the city has spread outwards to the suburbs and beyond 8 Mile Road in generally an orthogonal manner and through many mixed, older suburban districts which have decayed and are derelict pending renewal. This is the current stage of aspiring recovery of the 21st-century de-industrialised city.

4.3.3 CLEVELAND, OHIO, USA

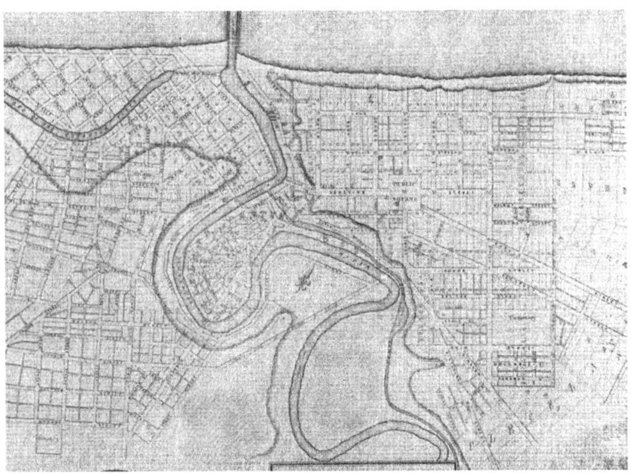

FIGURE 4.3.3 Cleveland – early configuration, 1835 (Merchant).
Source: Historic Urban Plans.

Background and Origins

Cleveland, in the state of Ohio, lies on the waterfront of the southern shore of Lake Erie that developed around the oxbow loops of the outflowing Cuyahoga River. It was first laid out by General Moses Cleaveland in 1796 as a former reserve named Cleaveland with a grid of streets and the grand Public Square. Divided into a number of lots, settlers arrived from across the Appalachians and established themselves, some as individual proprietors and some under company sponsorship.

As the infrastructure was developed, the land to the east of the river grew around Lakeside and Superior, Chester and Euclid avenues and the grid system. Between the loops of the river, the topography determined the shape of the lots and sub-divisions and development in the south-west, radiating out from Gravity Place, although little evidence of this remains. It developed further with the addition of the new canal infrastructure and the arrival of the railroads, each having its own separate terminal.

The significance of the city lies in its strategic location on the Great Lakes. In 1832, when the Ohio and Erie Canal was constructed, a link between the Ohio River and the Great Lakes was created, subsequently connecting with the St Lawrence Seaway out to the Atlantic Ocean, and also via the lakes and the Erie Canal to the Hudson River.

It was constructed as a city in 1836 and renamed Cleveland; it later annexed the neighbouring Ohio City to the west in 1854. John D. Rockefeller established Standard Oil in Cleveland in 1870, which marked the beginning of the oil-refining industry. With its pivotal location, Cleveland developed as a transportation hub for both water and rail transport as the railways were developed in the region, and around which manufacturing developed.

This required labour and materials, processing and warehousing, which attracted the migration of labour both from the South and from several Eastern European countries, particularly Italy and Russia. As new iron and steel facilities were developed around the abundant local coal resources and with the Bessemer steel-processing technique, automotive manufacturing also developed with petrol-powered engines.

The 1903 Group Plan outlined a plan for the commercial centre of the city with its civic buildings of city hall, the county courthouse and public library each clustered around the mall. During the 1920s, the population grew and the Terminal Tower skyscraper was built in the centre of the downtown area in 1926 that adjoined the multi-level Cleveland Union Terminal railroad and station. During the 1920s, many areas became very over-crowded, and areas of poverty were evident in many different ethnic areas, especially of Central European migrants. Shipbuilding was prevalent around the port area, which developed into an important role for inter-lake shipping services across and around the Great Lakes.

The city prospered during the 'Roaring Thirties', but this was to collapse with the Great Depression after the Wall Street crash in the 1930s. In 1936, the Great Lakes Exposition was held to celebrate the centennial of the city on the North Coast Harbour, which was an opportunity to connect the city with the lakeside shore that had been cut off by rail tracks.

The Cleveland Union Terminal and Tower Complex acted as a central landmark with the Old Victorian Arcade nearby, modified in 1939, and the prestigious Euclid Avenue with grand mansions running east from Public Square. The lakefront has several parks and reservations, with a group of metroparks that encircle the city.

In terms of further built development, the residential streetcar suburbs of Shaker Heights, connected by a rapid transit line, and Cleveland Heights and others were developed along the irregular bluffs along the lakeside.

During the world wars, Cleveland acted as a significant manufacturer of machinery and necessary armaments. In particular, United Steel Corporation and others built aircraft, and Sherman tanks were built during World War II. Goodyear produced rubber tyres in nearby Akron. Cleveland was a significant contributor in the industrial sector.

Modern City – Buildings and Arteries

After the interruption of World War II, the city's growth resumed, and in 1950 the population was approximately 1 million people. But in the 1960s the city deteriorated with decline in the iron and steel industries due to competition from overseas manufacturers, and with residents' white flight from the city, leaving a declining central area.

With high unemployment, discrimination in housing and mortgage red-lining, there was social unrest and a series of riots across the city. During the 1970s, the city defaulted on federal loans and the recession deepened. It was not until the 1980s, after a programme of urban renewal, that recovery eventually began and reconstruction and regeneration of urban areas downtown recommenced. With construction of the Key Tower (designed by Pelli), 200 Public Square skyscraper and the Gateway Sports and Entertainment Complex, the character of the centre of the city changed.

On the lakeside with the Great Lakes Science Centre, the city began to emerge from default in 1987 and the economy diversified, with high-tech companies being formed and emphasis on the health sector. Environmental clean-up was undertaken along the Cuyahoga River, after the polluted river had actually caught fire in 1968 and the emphasis on clean-up and regeneration was occurring.

The city now has several distinctive neighbourhoods downtown – Civic Centre, Central 9/12 District, the former Financial District, the Warehouse District and Playhouse Square, and the Campus District. Several of the residential districts, both on the east and west side of the Cuyahoga River, are beginning to be revitalised and residents are returning from the suburbs. To the west is Edgewater and Lakeside, and farther to the east is Euclid.

In terms of transport, historically the city has been a railroad hub with several intermodal freight terminals for Norfolk Southern and CSX Railroads, each railroad having its own terminal. The Lakefront Station for Amtrak passenger services and commuter rail has been redeveloped.

Mass transit is run by Greater Cleveland Rapid Transit Authority (RTA) and has three lines, Red, Blue and Green, as well as the bus rapid transit Health line on Euclid Avenue and a BRT line out to the airport. The east-west rail corridor, however, has divided the city and separated it from its lakefront.

The city has a good network of major highways, with named avenues running east-west and numbered streets running north-south. The city has the interstates I-71 and I-77 running north-south and I-80 and I-480 running east-west, as well as I-90, which turns into the Inner Belt downtown.

The Port of Cleveland at the river mouth acts as a major bulk freight and container terminal. The meandering Cuyahoga River has necessitated numerous bridges to be built across the river, including the high-level bridge and Main Avenue Bridge, and characterised by a series of major arch bridges and lower swing or lifting bridges, many of which are now archaic and obsolete.

The city has changed significantly from the former industrial period but is re-inventing itself into a modern high-tech hub with a major health centre around University Circle. This has contributed to a significant transformation from an industrial to a service economy, which is needed in order to reinvent and revitalise the city. Case Western Reserve University and Cleveland State University are also major educational centres, and Wade Park has several museums and business centres.

The population is approximately 0.4 million, down from a peak of 0.9 million in 1950, with 2 million in the metropolitan area. The Greater Cleveland MSA includes Cleveland, Akron and Canton, a combined statistical area (CSA) with approximately 3.6 million people. Burke Lakefront Airport is sited along the lakeshore on the edge of the lake with the main Cleveland Hopkins International Airport (CLE) located southwest of downtown adjacent to Glenn Research Centre.

Transactional Forces and Typology

The city of Cleveland was to evolve as a result of its strategic location on one of the Great Lakes, Lake Erie, where the meandering Cuyahoga River Valley meets at a junction with the lake. Here, inter-lake shipping services were the first means of communication before railways and highways were developed. With the discovery of local minerals and coal, iron and steel manufacturing facilities were developed, which necessitated transporting heavy bulk materials and freight. This was followed by oil discovery and with the development of oil refining, the industrial activities of the city were consolidated.

With developments along the lakeshore around the harbour, manufacturing and warehousing districts were developed on both banks of the Cuyahoga River with many multiple bridge crossings, both raising and swing bridges. The centre of the city developed around Public Square, which was to act as the core of the urban transport system with Ohio City on the western side of the river. As the city de-industrialised post-war, the stockyards and warehousing districts have been transformed and replaced with newer technologies and service industries.

As a lakeside city, initial development was linear along the shore and also inland on either side of the Cuyahoga River. With a central orthogonal grid that now accommodates downtown, both commercial and residential developments extend eastwards into the suburbs along the main boulevards of Lakeside, Superior and Euclid avenues. Development around University Circle now acts as a further pole of new industries, especially in the health sector with its large medical centres that have replaced areas of the de-industrialising city. Although hybrid, the city retains characteristics both of a radial and orthogonal layout as its sprawls beyond the city eastwards and westwards and south of the lakeshore.

4.3.4 PITTSBURGH, PENNSYLVANIA, USA

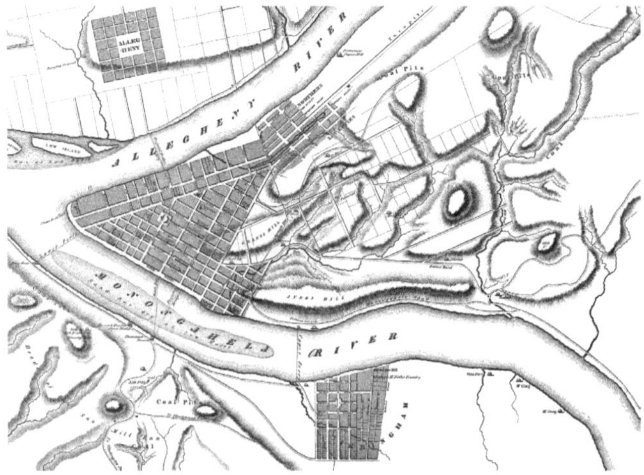

FIGURE 4.3.4 Pittsburgh – early configuration, 1815 (Darby).
Source: Historic Urban Plans.

Background and Origins

Pittsburgh is located in the west of Pennsylvania, west of the range of Appalachian Mountains at the confluence of the Allegheny and Monongahela rivers that become the Ohio River. It was inhabited originally by the Shawnee Native Americans before the arrival of European settlers, firstly the French who built Fort Duquesne at the confluence of these two rivers.

However, the British continued to contest their presence under several wars until at the end of the Seven Years War, in 1758, General Forbes then declared the place 'Pittsburgh', a borough to be named after William Pitt, British prime minister, and constructed a new fortress named Fort Pitt to replace the old one burnt down by the French. Next, in 1768, under the Treaty of Fort Stanwix, the boundary lines between the Indian and English territories were redrawn to include the 'New Purchase' of lands and the definition of the purchase areas that was to constitute Pennsylvania.

After the American Declaration of Independence in 1776, Pittsburgh continued to develop under a new ground plan with boatbuilding by the rivers due to the abundance of lumber. A baseline was established parallel with the Allegheny River with two principal streets, Penn and Liberty streets, in essentially an east-west direction. With the abundance of mineral resources in the neighbouring Allegheny Mountains of coal, iron and tin, new industry began to develop that attracted immigrants arriving from the 'Old World', especially Central Europe and the expansion of mining in the 1830s.

Aided by escaping slave labour from the South via the Underground Railroad, manufacturing developed and new factories were established, for example the Carnegie Steel Company, founded in 1875. Further development gravitated around this and with blast furnaces and ironworks and construction of new railroads, further in-migration arrived to provide labour for the steel mills. The railways were pivotal to development of the city, and with different geographic lines, six different clocks were needed to be displayed at Central Station to reflect the different American time zones through which the various lines were routed.

The Pittsburgh Survey – considered a social science masterpiece – was undertaken in 1907–8 by the Russell Sage Foundation to record the living and industrial conditions at this time. The area became a major steel producer during both world wars and formed part of the heavy industrial sector in the regional triangle of the cities of Pittsburgh, Cleveland and Chicago.

Modern City – Buildings and Arteries

Post-war industrial production continued; however, by the 1970s and 80s, with the 'grand re-adjustment' as more competitive manufacturers developed in South-East Asia, the city suffered from a period of de-industrialisation when large numbers of workers were laid off, residents moved out and the city had to strategically reinvent itself. With a programme of urban renewal and development in the inner Golden Triangle, it started to be transformed into a centre for biotechnology and healthcare, with the University of Pittsburgh Medical Centre and Allegheny Health, as well as new R&D (research and development) centres attracting new technology companies into the centre of the city.

In the past, Pittsburgh benefitted from a few rich main benefactors such as Carnegie, Mellon and Frinck. From its former strength in individual materials of steel (US Steel), aluminium (Alcoa) and glass (Pittsburgh Plate Glass), the headquarters of these manufacturing industries have dominated the city. Other major modern companies headquartered in Pittsburgh include Kraft Heinz, PNC Financial Services and Arconic.

As a modern city it has a large number of cultural assets, museums and institutions and is home to three major universities – the University of Pittsburgh, Carnegie Mellon University and Duquesne University – constituting strong educational and health sectors.

Adjoining Mount Washington, Downtown is within the central area, the Golden Triangle between the two main rivers, the Allegheny and the Monongahela. With a large number of bridge crossings, the city has distinctive areas, North Side & North Hills, South Side & South Hills, and East and West Ends.

In the centre were the old historic wharves and the North Side, formerly Allegheny City, which was mainly residential. Beyond East Side at Oakland and Shady Side and Squirrel Hill, are the universities, with parks and green neighbourhoods. West End encompasses Mount Washington with its two funicular railways of the Duquesne and Fort Pitt inclines, and South Side is the site of former steel works with extensive rail yards.

There were several railway stations that served different railway lines, the Baltimore & Ohio, Pittsburgh & Lake Erie, Pennsylvania Railroad and elegant Wabash Pittsburgh terminal. Formerly, Union Station provided mainline services to connect with the major cities of Chicago, New York and Washington DC. Freight rail lines were very active with the Pittsburgh line of the Norfolk Southern Railway, running in Keystone Corridor as the main line across the state of Pennsylvania.

In terms of transport, the city is well provided for and has a large number of distinctive bridge crossings painted yellow over the rivers that include Fort Duquesne Bridge and Fort Pitt Tunnel & Bridge into the city. At the Conference of the Three Rivers, the Ohio Fork, historic Point State Park has been developed as a green area at the apex of the Golden Triangle.

Modern highways of Parkway East connect with I-76 and Parkway West with I-79 and I-80. Cross town is I-579 that connects with the Liberty Tunnels and the three triangular interstates that form a type of beltway around the city and out to Pittsburgh International Airport (PIT).

Today, the city is divided into different neighbourhoods, with the Cultural Centre, the Financial Centre and Downtown. Overall, from its peak of 0.68 million in 1950, the population of the city has fallen to approximately 0.3 million, most of this since the 1960s–70s, and has 2.4 million in the surrounding metropolitan area. It has a large proportion of Central European Slavic, German and Polish populations.

Particularly distinctive historic and modern buildings downtown include the Oliver Building (designed by Burnham), Gulf Tower, Alcoa Tower, US Steel Tower, One PPG Place (designed by Johnson & Burgee), BNY Mellon Center and PNC Plaza.

Transactional Forces and Typology

Located at the confluence of two major rivers, these provided access routes around which Pittsburgh developed from its initial emphasis on boatbuilding and lumber and in a region of productive agricultural land. As mineral resources were discovered in the neighbouring hillsides, labour was transferred into mining activities in the extraction of tin, iron and coal, each of which was to radically transform the region, with its added economic attraction for migrant workers.

New industries and technologies were developed to transform these raw materials into steel, with coal used to power blast furnaces and steam-driven railways to haul heavy materials to and from the Allegheny Mountains to link with other growing centres. Hence it was to become the industrial powerhouse of the region in the late 19th and early 20th centuries. However, post–World War II, with changing world markets it had to de-industrialise and has now been transformed with modern health, defence and IT companies that have been developed to replace former industrial activities.

With its mineral resources and industrial origins, the centre of Pittsburgh is typified by its Golden Triangle at the conjunction of the two rivers where wharves were developed along its riverbanks. With a number of bridge crossings built to connect with Allegheny City and its Commons to the north and Mount Washington to the south, and with Point State Park, the central area was platted and areas laid out with orthogonal grids parallel to each of the two rivers.

As the railways were developed, new termini were built and road and rail axes routed through and around the city. The core of the city is the triangular central business district with its modern commercial tower buildings, with university and residential quarters farther east in the suburbs of Oakland and Squirrel Hill respectively. Unusually, it is a riverine city on a fork, typified by its triangular core between the Allegheny and Monongahela rivers and having distinctive yellow bridge crossings. Its former riverside heavy plants are now deindustrialised and under redevelopment with new service businesses as older manufacturing areas are being regenerated, which is changing the former industrial characteristics of the city. Anchored around the universities, this is developing as new educational and health hubs for 'Eds and Meds'.

4.3.5 MANCHESTER, ENGLAND

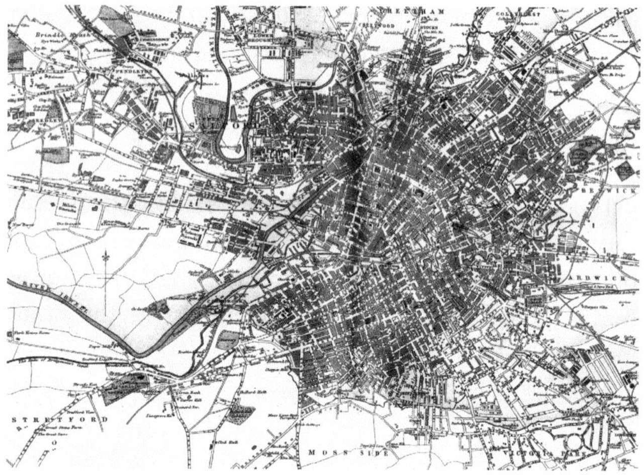

FIGURE 4.3.5 Manchester – early configuration, 1880.
Source: iStock.

Background and Origins

Manchester is located in the north-west of England, bounded by the Cheshire Plain and the Pennines to the east, on the River Irwell, inland from the Mersey estuary. It was first inhabited by a Celtic tribe, the Brigantes. In Roman times, known as Mancuniam, a fort was established around 79 CE in the confluence of the Rivers Irwell and Medlock, now Castlefield, to support the Roman interests at Chester (Deva Victrix) and York (Eboracum). With the Saxon conquest and the Roman withdrawal, the focus of development moved and with the coming of the Normans, development occurred in the Salford Hundred and the County of Lancashire.

The town was chartered in 1301. With the original manor house, marketplace and parish church by the reinforced Hanging Ditch, these became the premises of the subsequent Manchester Cathedral and Chetham's Hospital and Library, founded in 1421. Several churches were also built under the patronage of the Earls of Derby.

From the market and with an influx of Flemish weavers in the 14th century, came the beginnings of the textile industry of both woollens and fustians (linen). With the increase in trading of quantities of cotton, the emphasis was on creating cotton fabrics and new methods of fabrication were invented, starting with cotton spinning in the 18th century, which led to an increase in dwellings and workshops around Market Street and Deansgate.

The population began to grow around these settlements, and as trade expanded in the region new forms of communication were required. Roads to Stockport, Oldham and Ashton-under-Lyne were developed as turnpikes, and links to Liverpool were excavated as canals between the rivers Mersey and Irwell. Parcels of land around the centre were sold for the development of housing, and the city was transforming rapidly.

The earliest development of infrastructure was in 1765 when the Bridgewater Canal was built with the Giant's Basin at Castlefield for the transport of coal, designed by Bradley, and was extended subsequently to Runcorn and the River Mersey. With the construction of the Manchester to Bolton, Oldham, Ashton and Rochdale canals, these greatly facilitated the local movement of goods by water. The Rochdale Canal joined the Bridgewater Basin in 1806, which completed the whole canal route between the east and west coasts of Britain.

Cotton and Textiles

The development of these transport routes led to an increase in cotton mills with their tall chimneys and steam engines used in spinning and manufacturing, but also in the need for specialist storage warehouses for commodities for the home trade, for storing and packing and were built both in the Ancoats and Chorlton areas. The design of the warehouses evolved as the brick buildings were enhanced with cast iron beams and columns and sprung brick arches that were better fireproofed. Wet docks were also constructed that enabled barges to enter buildings directly from the canals via shipping holes and to be directly unloaded inside, for example at the Grocer's Warehouse built in 1770.

In terms of cotton production, the Flying Shuttle was invented by Kay in 1733 and in 1768, at Shude Hill, Arkwright patented a spinning frame which was further developed in 1771 at Cromford Mill in Derbyshire using a water-powered system, for both rotary carding and spinning. The machinery within the cotton mills became more advanced as steam-powered machines were invented for mule spinning and manufacture using a multi-spindle Spinning Jenny designed by Hargreaves in 1765.

The cotton was prepared on lower floors, with spinning mules on the upper floors, and the mills were lit by gas. The principal mills then were McConnel & Kennedy Mill, Murrays' Mills, Sedgwick Mill and the Royal Mill. In 1801, the population of Manchester was only 75,000, but by 1851 it had reached 0.3 million as workers began to be attracted to the new business in the city, which was becoming known as 'Cottonopolis'. Later, in 1876, the Clayton Aniline factory was built for the manufacture of synthetic dyes used in calico printing. Technical innovation continued and companies such as Rylands developed textile manufacturing with the 'ready clothing' trade and was a major employer with many mills and warehouses.

Railways, Ship Canal and Industrialisation

Development of the first railway between Manchester and Liverpool was a challenge, with tunnels and cuttings, a viaduct over Sankey Brook and the crossing of Chat Moss. Stephenson, supported by Rennie, developed this project and the construction of a double line track, with the operation of the first trains powered by Stephenson's 'Rocket' steam engine.

The Liverpool and Manchester Railway (L&MR) was opened in 1830, with separate stations built at Manchester Liverpool Road, Manchester Central, Manchester Exchange and Manchester Mayfield. The latter three stations, Central, Exchange and Mayfield, are now all closed, with just the Midland Hotel remaining. Manchester London Road was opened in 1840 as part of the Manchester and Birmingham Railway (M&BR).

The Sheffield, Ashton-under-Lyme and Manchester Railway (SALMR), which connected Sheffield under the Pennines with a blasted tunnel at Woodhead with Manchester, was opened in 1845, later to become the Manchester, Sheffield and Lincolnshire Railway (MS&LR). The MS&LR and the M&BR then amalgamated to become the London and North Western Railway (L&NWR). They had each grown and due to overcrowding were divided into two parts with separate platforms on the north-west and south-east sites in Manchester.

With further subsequent amalgamations, the London Euston to Birmingham and the Grand Junction on to Warrington became L&NWR along with the LM&S (London, Midland and Scottish) respectively, and the MS&LR became L&NER (London and North East Railway), having the station terminus at Manchester Piccadilly. The other main station that developed was Manchester Victoria on the Manchester and Leeds Railway (M&LR) and which served other cities farther north, and the Great Northern Railway goods warehouse at Deansgate.

Goods had been transported by water on the combined Mersey and Irwell Navigation and on the canals until they fell into disrepair. With the Liverpool and Manchester

Railway, in order to avoid Liverpool dock and rail charges after the Long Depression in the 1870s, a plan was made to construct the Ship Canal so as to give ocean-going vessels direct access to Manchester.

After passing the Manchester Ship Canal Act in 1885 and various financing challenges, its route was designed that incorporated a series of large locks. It was built in eight sections and was eventually completed in 1894, 58 km. in length, only slightly shorter than the Panama Canal. It had several engineering challenges besides the lifting locks and the swing bridge crossing and swing aqueduct at Barton. But it did enable fairly large ships to sail directly up the Mersey estuary to Runcorn via Eastham Locks and then to run directly into Salford and the edge of Manchester at Pomona–Manchester Port. Whilst beneficial, it never quite attained the commercial success anticipated, since after disposing cargo, often ships had to return back down the canal empty. Between the wars it carried 5–6 million tonnes and post-war peaked at 20 million tonnes in 1958.

With the extension of steam, another energy source – hydraulic power – was developed, produced through placing water under high pressure and extending it in a network of underground hydraulic pipes across the city for powering and lifting systems.

Engineering and industrialisation developed with iron foundries that were able to cast gear wheels and machine tools. These were used to develop and power machinery and locomotives with bodies and either fixed or moving reciprocating steam engines. Machine engineering was refined by engineers such as Whitworth, with his standardised screw thread, micrometres and planing machines. This led to the establishment of several manufacturers of carriages for the new railways.

Given this industrialisation in the area around Manchester and Salford, a number of clay pits were excavated and many kilns built for the firing of bricks and tiles. With the Bradford and Clayton collieries producing coal, a large volume of bricks was required for the extensive warehousing and buildings. Glass works were also developed for glass blowing.

Housing, Civic Design and Infrastructure

The cotton industry continued to flourish, and as more workers moved into the city pressures increased on housing them, which led to congested back-to-back building. As the Industrial Revolution took place, it also brought poverty and squalor here as living conditions deteriorated, as infamously cited by Engels in *On the condition of the Working Class in England*, published in 1844. The tall chimneys and columns of smoke signified this new industry, but with poor environmental air and polluted water conditions. Cotton mills peaked around 1850 and expanded in the adjacent areas of Bolton and Oldham, making it a heavily industrialised region along with its warehousing and storage facilities.

Overcrowding had developed in stages, with the initial construction of row houses having attic floor workshops and basements, with ground and first floors subdivided for tenement housing. With a rear yard and outside privy, these yards then became built over with one-up, one-down cottages facing the rear alley. With restricted access, poor lighting and sanitation, these rooms housed multiple occupants with inadequate space and facilities. With cholera and typhoid epidemics, the building of new back-to-back buildings was banned in 1844, and cellars were later made illegal in 1853.

As the city developed, its municipal infrastructure was also developed, as well as civic buildings such as the Exchange, Waterhouse's town hall and the Portico Library. With some earlier Greek Revival–style buildings, the predominant style was red-brick Victorian architecture, the Assize Courts, Free Trade Hall, Owens College, the Rylands Library and the Mechanics Institute. New buildings were constructed along Princess Street and Deansgate, with the Barton Arcade, and many new churches were built.

Owens College grew into the University of Manchester and a branch of the Bank of England was built in King Street. As the city grew as a financial centre, banks, insurance companies and large hotels were also developed. At the turn of the 20th century, the Royal Infirmary was demolished to make way for the newly laid-out sunken Piccadilly Gardens on the site of former clay pits.

World War I and the Great Depression impacted export markets, and cotton processing started to develop in other areas of the world. The first aeroplane designed by Roe flew in 1909, followed by bi-planes that were manufactured by AV Roe, which was to become a significant aero manufacturer. During World War II, manufacturing was focused on armaments, with AVRO manufacturing Lancaster aeroplanes and Dunlop producing barrage balloons, and other machine industries were developed. A large area of the city was destroyed in the Christmas Blitz of 1940, both industrial and residential buildings.

In business terms, Trafford Park Industrial Estate had been developed in the building of cars alongside the Ship Canal. Cycle manufacture began in the 1880s with motorcar bodies at the Belsize Works and Crossley Motors. After the war, whilst trading continued, the movement patterns changed with the introduction of containerisation, which led to a reduction in the use of the Ship Canal, with larger container ships docking at Liverpool on the Mersey that led to the eventual closure of Manchester Port.

With the increase in population, housing conditions had deteriorated in Hulme and Moss Side and during the inter-war years people began to move to the suburbs. The local authority set about the construction of a large number of areas of council housing to try to clear the old back-to-back housing, especially in the individual suburbs of Chorlton-on-Medlock, Ancoats and Hulme.

The satellite town of Wythenshawe was planned by Parker in 1927 with an adjoining green belt. Ardwick had developed around Ardwick Green and the Pin and Paper Factory was built in a loop on the River Medlock at Pin Mill Brow with its classic saw tooth–shaped roof. Cheetham contained the Jewish quarter and synagogues, and at Strangeways, Waterhouse had developed the six-wing radial Victorian prison in 1868 on the Pentonville model. Charlton-on-Medlock, just to the south of the city, with its mills and Macintosh Works, was intensively developed, having the major landmark of the Roman Catholic Holy Name of Jesus Church modelled on earlier French cathedrals.

Ancoats was an area of very intensive activity around the mills, as an industrial hub in the 19th century with its cotton textile mills, developed around the McConnel and Kennedy Mill and the Murrays' Mills and with new technologies that were emerging at that time. Connected by man-made canals, the brick and iron-built mills were able to produce manufactured materials for sale and export. Around these mills dense rows of back-to-back brick housing were developed along with terraced houses to accommodate the workforce, but housing conditions failed to keep up with sanitary requirements.

The manufacturing tradition continued right up until World War II when production was to be converted to uniforms and parachutes. Post-war, with competition from cheaper producers in the Far East, the area declined and ran down and became subject to a programme of slum clearance. With this decline, the manufacturing sector fell into disrepair and the area became a 'sink estate' typical of inner-city physical and social decay. Most of the cotton mills were closed during the 1960s and 1970s.

It was not until the 1990s that attention turned to regeneration, conservation and redevelopment of the Northern Quarter of Manchester with change of use of older buildings. Ultimately, with trade developers reinvesting in the area with new housing and occasional preservation, the area began to regenerate itself with gentrification along refurbished streets and canals, the ultimate de-industrialisation cycle.

Hulme was developed with dense streets and terraced housing back in the mid-19th century, without backyards or privies. In recent years this became very run down and was subject to major urban clearance and renewal also. However, the first stage of this modern rebuild was unsuccessful and this had to be entirely redeveloped.

The major rebuilding scheme here had consisted of four linear blocks, 'The Crescents', designed by Wilson and Wormersley in 1971, with 'Radiant City' aspirations, with elevated decks in concrete tower blocks which became dysfunctional. With technical design flaws, this led to social delinquency that evolved into a modern-day elevated slum. It was so severe that the whole scheme had to be demolished after only 20 years in 1991. More recently this has been subject to yet further redevelopment under the 'Homes for Change' municipal strategy at the beginning of the 21st century, with the aim of producing better-suited community architecture.

Modern City – Buildings and Arteries

Since the 1960s and 1970s, across Manchester there was a period of economic decline which led to the need for fundamental transformation of the city. Businesses closed with the downturn of heavy industry and regeneration of the city was required, especially after a major terrorist bomb devasted a large part of the central area in 1996.

During the 1980s, under the Trafford Development Corporation and newly created Enterprise Zone, with the Ship Canal closure Salford Docks was closed and redeveloped as Salford Quays as a major urban regeneration project. With a new waterfront and redeveloped piers, this enabled the realisation of landmark buildings such as the Lowry theatre and the Imperial War Museum-North as well as new buildings, like MediaCityUK. Several major commercial and residential building projects were also developed, and a new vertical lift Millennium Bridge was built over the quays.

Post-war the canals had lost their usefulness. The Bridgewater Canal became a cruising canal, and the Ashton Canal was rehabilitated. The Bridgewater Canal and Locks were repaired and restored by British Waterways.

After the Commonwealth Games were held in 2002, the stadium was converted for football and other new developments were realised, such as the Arena, G-Mex, Beetham Tower and four tower blocks around Deansgate Square. The University of Manchester, UMIST and Manchester Metropolitan University are large university campuses in the centre of the city. As the home of Granada Television, BBC Northern Studios and Salford Media City, there is also a significant media presence.

The population of the city is approximately 0.6 million with approximately 2.75 million in the metropolitan area of Greater Manchester. This includes the metropolitan boroughs of Salford, Stockport, Tameside and Trafford. The city is governed by the Manchester City Council with 32 electoral wards. The Greater Manchester Combined Authority, GMCA, with a directly elected mayor was formed in 2011.

From the original Liverpool and Manchester Railway, the two mainline railway stations now are Manchester Piccadilly and Manchester Victoria that connect with the electrified West Coast Main Line to the north. The two other main remaining stations are Oxford Road and Deansgate.

In 1992, a new urban tramway, Manchester Metrolink, was opened which runs on urban street tram-lines and former commuter rail track, and now has a network of over 100 km. These eight lines were developed in phases radiating out to Altrincham, Ashton, Bury, Rochdale, Manchester Airport and Trafford commercial centre, overseen by GMPTE.

In terms of highways, the city is surrounded by motorways with the outer peripheral M60 and connection to the M61 and M62, and M56 and M66 and M6 north-south, and the central obtrusive M602 elevated Mancunian Way that traverses the city, opened in 1967. The aerodrome at Ringway, where Fairey Aviation built aircraft in World War II, has been developed with twin runways and terminals as an active international airport, Manchester Airport (MAN).

Transactional Forces and Typology

With its historic origins, the city of Manchester has benefitted from its natural resources and location and continuously harnessed its human potential to create its own realm of comparative advantage. By developing firstly its waterborne infrastructure of canals, it was able to support the growth of cotton mills, and subsequently its railway infrastructure facilitated the movement of bulk materials and locally found coal that led to the industrialisation of the region. With technological developments and human endeavour, this led to continuing advancement in manufacturing and industrial production.

In consequence, the city and the adjoining city of Salford continuously reinvented itself from an early linen market to a weaving and cotton-spinning town. With its subsequent industrialisation and radical growth of manufacturing, this eventually, in turn with changing markets, has been superseded as Manchester has de-industrialised. Now through regeneration it has converted its centre into a service-orientated and media-focused city with research and technology centres, as evolution of its inherent adaptability, industriousness and aspirations as a 'northern powerhouse'.

Although located just inland around the small River Irwell, after development of canals in the 18th century, it was the realisation of the railways and construction of the Manchester Ship Canal to connect with River Mersey and the Atlantic that this infrastructure contributed to enhancing the commerce of the city as it industrialised during the late 19th century. Powered by locally sourced coal seams, it was development of the iron and steel foundries that facilitated fabrication of machine tools and machinery that were to lead to the growth of new factories. Around these, settlements grew up of increasing density to house the resident workforce. From its centre, the elongated city has grown to incorporate Salford and now forms part of the larger metropolitan region of Greater Manchester, with peripheral growth poles around the perimeter bounded by modern highways. The following figures illustrate the modern configuration of the industrialised and de-industrialised cities as described and the evolution of their modern population.

FIGURE 4.3.6 Chicago – modern configuration.

FIGURE 4.3.7 Detroit – modern configuration.

FIGURE 4.3.8 Cleveland – modern configuration.

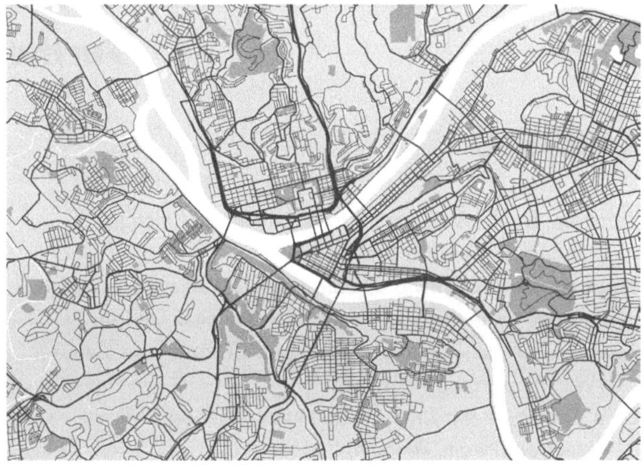

FIGURE 4.3.9 Pittsburgh – modern configuration.

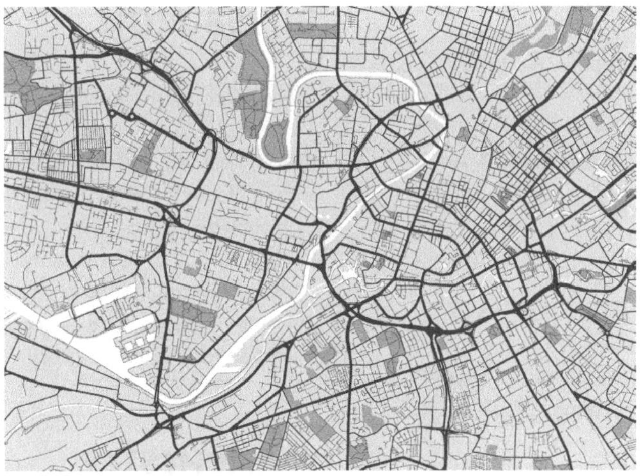

FIGURE 4.3.10 Manchester – modern configuration.

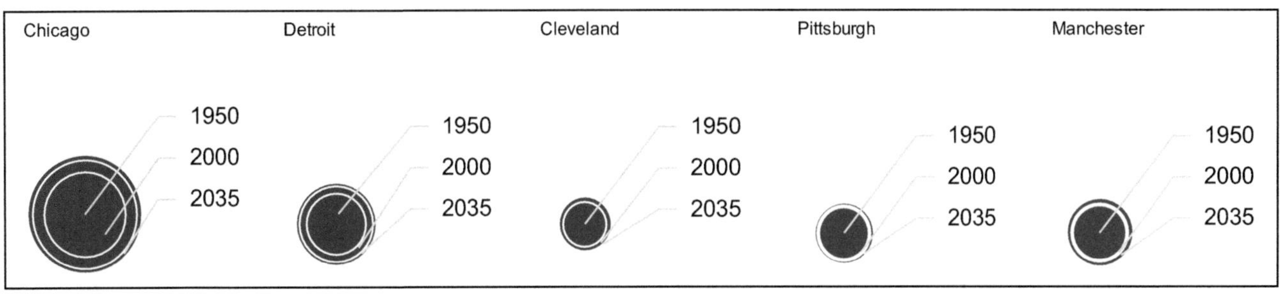

FIGURE 4.3.11 Industrial and de-industrialised city populations.

Data source: Global Database of Metropolises 2020 – UN-Habitat (2020), UN DESA Population Division.

Further Reading

CHICAGO

Burnham, Daniel & Bennett, Edward – *Plan of Chicago*, Commercial Club Chicago, 1909.
Busquets, Joan & Crawford, Christina – *Chicago, Two Grids Between Lake and River*, Applied Research & Design, 2017.
Cronon, William – *Nature's Metropolis, Chicago and the Great West*, New York, 1991.
Dyja, Thomas – *The Third Coast, When Chicago Built the American Dream*, Penguin, 2015.
Miller, Donald – *City of the Century, the Epic of Chicago*, Simon & Schuster, 1996.
Sampson, Robert – *Great American City, Chicago and the Enduring Neighbourhood Effect*, University of Chicago Press, 2012.
Zuchowsky, John, ed. – *Chicago, Architecture & Design, 1923–93, Reconfiguration of a Metropolis*, Prestel-Verlag & Art Institute of Chicago, 1993.
Zuchowsky, John, ed. – *Chicago, Architecture & Design, 1872–1922, Birth of a Metropolis*, Prestel-Verlag & Art Institute of Chicago.

DETROIT

Cohen, Jean-Louis & Crawford, Cristina – *Detroit – Moscow – Detroit, an Architecture for Industrialization, 1917–45*, MIT Press, 2023.
Gay, Cheri – *Detroit, Then and Now*, Pavilion Books, 2016.
Jay, Mart & Conklin, Philip – *A People's History of Detroit*, Duke University Press, 2020.
Maraniss, David – *Once in a Great City, a Detroit Story*, Simon and Schuster, 2015.
Martelle, Scott – *Detroit, a Biography*, Chicago Review Press, 2014.

CLEVELAND

Love, Jeannine – *Cleveland Architecture 1890 – 1930, Building the City Beautiful*, Michigan State University Press, 2020.
Miller, Carrol & Wheeler, Robert – *Cleveland, a Concise History 1796 – 1996*, Indiana University Press, 2009.
Van Tassel, David & Grabowski, John – *The Encyclopaedia of Cleveland History*, Indiana University Press, 1996.

PITTSBURGH

Fifield, Barringer – *Seeing Pittsburgh*, University of Pittsburgh Press, 1996.
Kellog, Paul, ed. – *The Pittsburgh Survey*, New York, 1914.
Lorant, Stefan – *Pittsburgh, the Story of an American City*, Doubleday & Co., 1964.
Pittsburgh, ASCE – *Engineering Pittsburgh, a History of Roads, Rails, Canals & Bridges*, History Press, 2018.
Zurowski, L, Jacob, M & Succop, C – *City Steps of Pittsburgh, a History and Guide*, History Press, 2024.

MANCHESTER

Hartwell, Clare – *Manchester*, Yale University Press, 2002.
Hylton, Stuart – *A History of Manchester*, Phillimore & Co., 2010.

4.4 Modern American Cities

Early Settlements – Atlantic Coast – Gulf of Mexico – West Coast

Philadelphia, Baltimore, Miami, Atlanta, Houston, Seattle and Vancouver

4.4.0 INTRODUCTION – BACKGROUND AND ORIGINS

The North American continent is a land of very different geographies from Alaska, down the Rocky Mountains, eastwards across the central plains and prairies to the Appalachians, and southwards on to the Gulf of Mexico. As early human indigenous civilisations ventured across the Bering Strait and down the west coast, on the one hand, later colonial explorers on the other hand reached the eastern coast from another direction of Europe.

Coastal and estuarial developments were the first places of settlement as 'planted sites' at places on the East Coast having natural harbours; around estuaries such as the St Lawrence, Hudson, Charles and Delaware rivers that acted as places of refuge; and likewise, around Vancouver Island and Seattle on the West Coast. As key places on local trading routes, they grew as a result of their strategic location.

The Great Lakes and their eventual inter-connection into the St Lawrence River facilitated the development of the northern American cities of Chicago, Detroit and Cleveland and Toronto, Montréal and Québec in Canada.

Along the Eastern Seaboard that comprises the harbours of Portland, Boston, New York, Philadelphia and Baltimore, Norfolk and Wilmington and Charleston and Savannah down to Miami, early settlements developed both in New England and the Tidewater colonies around Chesapeake Bay and the James River (see Support Material: Annexe 4.4A – Colonial Settlements in North America).

Around the long Mississippi and Missouri rivers, at places that had been populated by the pre-Columbian tribes such as at Cahokia, these became prime transport routes that led to the development and expansion of the southern and central cities of Houston, New Orleans, St Louis, Kansas City and Minneapolis.

To the west in the Rocky Mountains, water was scarce except around the Rio Grande, Colorado, Columbia, Snake and Yellowstone rivers, which encouraged interspersed settlements at Salt Lake, Denver, Boise and Cheyenne. With its highly varied geology, topography and climate, these have been key determinants in the differing settlements and demographic pattern of developments across the whole American continent.

On the western side the Pacific Coast had its own contrasting forms of development from the Yukon down the Canadian coast to Vancouver and along the Californian shoreline to Mexico.

A number of planned cities are highlighted in order to illustrate the evolution and development of distinctive cities across this continent in becoming significant cities, which in their own way encapsulate typical American modern urban typologies. These constitute city centres generally having gridiron layouts and with downtown areas, mostly being automobile focused, with high-density cores and sprawling outer suburbs, and more recent poly-nodal 'edge cities' developing as new neighbourhoods are formed on the urban fringes.

Modern American Urbanised Development

Post–World War II patterns of development changed locally as suburban development took off across America with the development of mass-produced 'tract housing', as for example promoted by William Levitt in Levittown, Long Island, New York. Suburban life was resulting from the separation of the activities of 'living' and 'working', which, led by mono-cultural zoning, encouraged the sprawl of development and extending streets and highways.

DOI: 10.4324/9781003353348-22

Early American primary cities up until the 1930s generally had characteristics of a 'mono-centric' city. Through specialisation, the central business district (CBD) was the focus of business, commercial and retail activity, whilst decentralised outer areas became residential areas with more space for housing. Commuter trains, streetcars and buses then typically provided transport services between home residence and work.

Metropolitan economies evolved, essentially into 'trading places' as the dynamics changed, with municipal areas becoming more open economies, as they tended to specialise in producing goods and services in fields where they had a comparative advantage. This changed the patterns of production and consumption and, consequently, the population location and residential densities in different urban areas.

Subsequently, this has led to trading places evolving beyond central areas to 'edge cities', bedroom suburbs and specialist hubs around which people gravitate in order to cross-fertilise skills and resources under newly developing technologies. Despite zoning constraints and some voluntary segregation into similar community and cultural groups, this has resulted in a considerable change in the morphology of modern American cities that has loosened the structure into 'poly-centric' areas and which is leading to greater variation in patterns of land use.

This trading place economy is also leading to the expansion of wider metropolitan areas that cross state and even national boundaries, leading to the formation of combined economic regions that are evident, as around Washington and Baltimore, Detroit and Windsor, San Diego and Tijuana, and Texas and Monterrey.

4.4.1 PHILADELPHIA, PENNSYLVANIA, USA

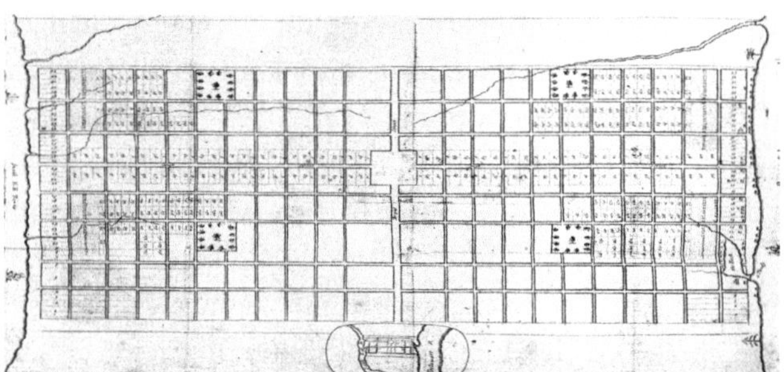
(A)

FIGURE 4.4.1A Philadelphia – early configuration 1682 (Holme).
Source: Historic Urban Plans.

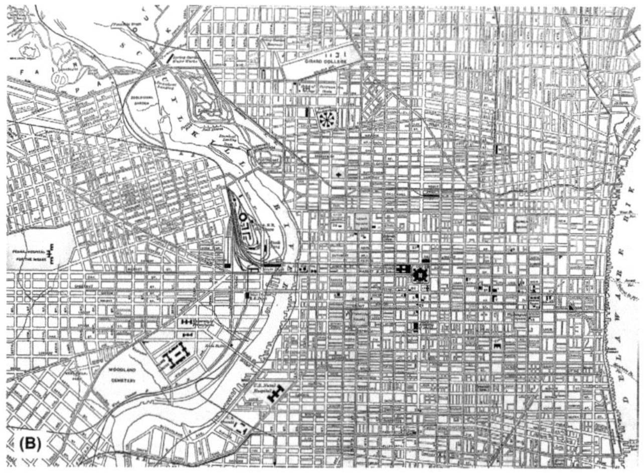
(B)

FIGURE 4.4.1B Philadelphia early configuration 1899.
Source: iStock.

Background and Origins

Philadelphia is located on the Eastern Seaboard of North America, built around its basic layout as established at its founding in 1682. With its original wards and gridiron layout designed by Thomas Holme, this stretched east-west between the Schuylkill River and the Delaware River waterfronts.

A site had been identified by the commissioners selected by William Penn between the Schuylkill and Delaware rivers of approximately 100 acres that was considered suitable for development. This flat area was developed according to a model with streets and wharves fronting each river, and connected by main cross streets Market, Chestnut, Walnut and Spruce that were intersected by 24 north-south numbered perpendicular streets.

At the intersection of Market and Broad streets was placed a new square, to house city hall, along with four other main public squares symmetrically placed across the city. The statue of the Quaker founder William Penn on top of city hall has been the landmark in Penn Square above whose height new buildings were not approved to be built. With these five fine characteristic squares of ten acres each, they became key features of the city, namely Logan and Rittenhouse squares, Washington, Franklin and Penn squares.

Between the main east-west commercial streets of Chestnut and Market, and 2nd and 6th streets, is the old Historic District around Independence Square where the American Declaration of Independence was drafted in 1776 and where the Liberty Bell is housed. Beyond the state house is Penn's Landing where William Penn first stepped ashore, with its original wharves and former naval facilities on the

Delaware River. It was this area around which most of the early development took place.

In 1802 PC Veerlé proposed an extension to the city on the west side of the Schuylkill River with a system of squares and diagonal avenues and a central oval place. Whilst this would have been an interesting counterpoint to the existing city on the opposite bank acting as an approach to it from the west, it was never built or developed.

The whole city is characterised by its vernacular red-brick buildings with a number of fine churches and buildings in early colonial style, especially around the old city. In the urban centre, this comprises mainly terraced buildings of three to four storeys in height, with individual backyards. The city received several waves of immigrants, notably of German, Irish and Italian origin.

Modern City – Buildings and Arteries

From its early origins and beyond the old city core, farther out the housing stock and row houses had deteriorated to generally poor conditions before a concerted effort was led by the former mayor Frank Rizzo and Edmund Bacon in the late 1960s before the American Bicentennial celebrations.

Centre City and the area around Market and Broad streets was also extensively modernised with the twin towers of Liberty Place constructed which were the first in 1987 to break the built height convention. Society Hill became a fashionable residential area near the waterfront with the Italian District and its markets and restaurants nearby and overlooking the Benjamin Franklin suspension bridge and industrial area of Camden opposite in New Jersey across the river. At the western end of the city Ben Franklin Parkway leads diagonally out past Logan Square alongside the Schuylkill River towards Fairmount Park with the Philadelphia Museum of Art and Barnes Foundation fine art museum.

Across the river past 30th Street Station, in West Philadelphia, University City developed, the home of the University of Pennsylvania that was founded by Benjamin Franklin, with Temple, Drexel and other educational institutions. Regenerated Locust Walk is now the hub of Penn University with its scientific, medical, engineering and business centres, and the Wharton Business School, interspersed with libraries and fraternity houses, sports field and football stadium.

The expansion of the city has followed the evolution of the transportation system, which with its former streetcars, was the first city to develop a comprehensive urban transport network. The early streetcars followed the privately developed omnibuses and the horse-car lines in the 1850s, which were expanded into transit lines that were eventually to be electrically powered.

These urban trams and trolleys became a distinctive urban feature and were extended across streets of the whole city. They connected with the steam-powered railroads and the Suburban Railroad Station, with the Frankford Elevated and the Market Street line developed in the central area. The linkage to outlying districts was significant and acted as important connectors for growing commuter traffic, now managed by SEPTA. The main bridges across the Delaware are managed by the Delaware River Port Authority. Besides the main marine wharves along the Delaware River, Philadelphia International Airport (PHL) is located to the South West alongside the interstate highway I-95.

A large amount of urban renewal has occurred also across the city, supported by the Philadelphia Housing Authority, and the downtown neighbourhoods have been extensively gentrified. So too the near suburbs have been renewed where previously North and West Philadelphia were not affluent places, with large areas of deprivation, crime and poverty. Changes have trickled down to most sections of the population such that recent suburban flight has been arrested and residents are coming back into the city.

The 'City of Brotherly Love' has grown significantly from its early origins but retains its character, as often found in older cities and seaports. With its trading background and strategic significance since the Founding Fathers and the symbolic Declaration of Independence, it has re-invented itself several times, industrially and commercially, around the port and railroad, and is now evolving into a modern bio-technology hub and with its universities, teaching and research institutions.

It has grown to approximately 1.6 million in the city, with the surrounding six counties in Pennsylvania, and around 6 million people in the outlying Delaware Valley metropolitan area, stretching between Trenton and Wilmington.

Transactional Forces and Typology

The Eastern Seaboard of America having been essentially colonised by the British, Philadelphia was the centre of one of the original Thirteen Colonies that had focused resistance during the American War of Independence. With its maritime role, and sited between two major rivers, the Delaware and the Schuylkill, it was a pivotal location which afforded good communications with neighbouring states. Accordingly, it was the meeting place of the nation's Founding Fathers and served as the nation's first capital. As an active industrial city, it developed commercially and administratively to extend its influence across the region. Now with its individual and specialist educational districts, its transformation is continuing, driven by urban renewal into a modern service-based economy, particularly in the health sector.

With its original layout as established by William Penn, the central city retains its orthogonal gridiron characteristics and its five main squares. Situated and bounded between the two major rivers, it has riverine characteristics with its waterfront and wharves on the Delaware River which border with industrial New Jersey on the opposite shore across the major Ben Franklin and Walt Whitman highway bridges. Whilst generally aligned virtually north-south, it also reaches out westwards along the suburban Main Line.

To the north, the city has been developed with dense residential districts and mixed populations, many from former migrant families. Population pressures have extended out to the suburbs also, which has led to growth especially in the western areas, as along the streetcar lines north-west towards Bryn Mawr. With its historic centre, it has emerging revitalised districts and features poly-nodal centres on the edges that are extending into the greater agglomeration in the Delaware Valley of the North-Eastern Seaboard.

4.4.2 BALTIMORE, MARYLAND, USA

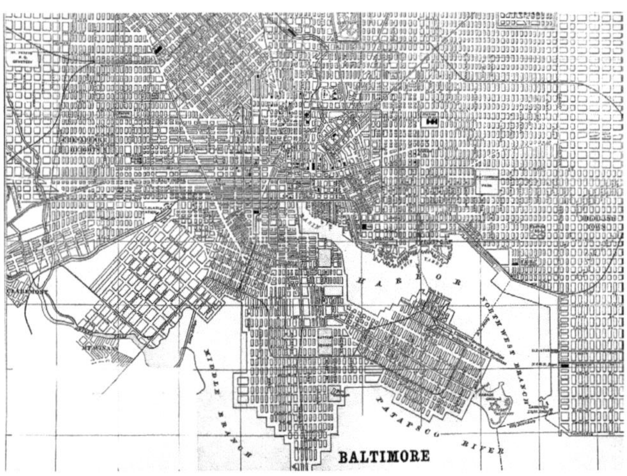

FIGURE 4.4.2 Baltimore – early configuration, 1899.

Source: iStock.

Background and Origins

Baltimore is located on the east coast of the USA on the Patapsco River, which flows into the Chesapeake Bay. Originally populated by Native tribes, Iroquoian-speaking from the Susquehanna River Valley, Algonquian-speaking tribes were also known farther south, along the Potomac River.

The region was sparsely populated when the first European settlers arrived around 1635, and as a British colony under Lord Baltimore, Jonestown was established in 1661 and Fells Point. The town and port of Baltimore was established in 1706, which acted as a port facilitating the triangular Atlantic slave trade and the export of tobacco. Further plantations of grain and tobacco were expanded and traded with the labour-intensive sugar plantations in the Caribbean. In the second half of the 18th century, public markets were developed in specific commodities, such as at Broadway Market, Centre Market and Lexington Market.

As one of the original Thirteen Colonies in the American War of Independence, the Province of Maryland was incorporated under the US Declaration of Independence that convened in Philadelphia in 1776. However, with the ongoing War of 1812, a sea battle resulted with the successful defence of Fort McHenry at the Battle of Baltimore and the Battle of North Point during the British Napoleonic Wars, to which a monument was erected subsequently at Battle Monument Square by the City Circuit Courthouses.

During the 19th century, infrastructure began to be developed with the construction of the federally funded National Road (now Route 40), and the Baltimore and Ohio Railroad, each of which enhanced communications and facilitated the shipping of products and materials, mainly timber, tobacco and textiles across the country to and from the Midwest, and the first telegraph line between Baltimore and Washington DC was also realised.

With the American Civil War in the 1860s between the Union and the Confederacy, Maryland, although part of the Union, had still permitted slavery until its abolition in 1861. With the subsequent period of the Long Depression, it was not until the turn of the century that the city started to grow in discrete areas under a formal grid pattern along with radial avenues.

Despite its segregation ordinance and the setback of the Great Fire in 1904, the city grew by annexation of portions of the surrounding counties, and streetcars were installed in several districts. With ongoing migration from the Deep South, the percentage of the black population in the city grew correspondingly.

Modern City – Buildings and Arteries

With its natural geographical configuration in having an Outer Harbor and an Inner Harbor, the port area developed, around which manufacturing activities gravitated. In the mid-19th century, the centre was consolidated and populated by a number of individualistic buildings, such as the Baltimore Basilica, the Brown Memorial Church and the Sun Iron Building that led to a series of other iron-fronted buildings downtown. In the residential areas the streets were laid out in a grid pattern and lined with 'formstone' faced row houses that typified the original housing stock in central Baltimore.

With a radiating form out from the centre, the city of Baltimore has different geographical districts, that comprise the following areas:

- Central Baltimore and north of the Inner Harbor and Downtown has commercial and residential areas that include institutions such as the University of Maryland Baltimore, the former Lexington Market and Baltimore City Hall.
- North Baltimore, which has older residential neighbourhoods, includes historic Charles Street and Mount Vernon Place with the George Washington monument, Old Goucher College and Charles Village beyond and Wyman Park, Jones Fall Valley and Johns Hopkins University farther north.
- South Baltimore contains the Old South Baltimore, Baltimore and Ohio Railroad sidings, and the Federal Hill and Park District beyond with its old Fort McHenry at Locust Point promontory.
- Eastwards, North-East Baltimore contains Morgan State University, an historically black college and university (HBCU), founded in 1867 as a biblical institution, and East Baltimore, which is home to smaller residential neighbourhoods and borders the harbour with Canton and Patterson Park and Fells Point and Brewers Hill beyond.
- North-West Baltimore comprises Pimlico and Druid Hill Park; West Baltimore contains the Old West Baltimore Historic District that includes Sandtown, Madison Park and Upton, and South-West Baltimore contains Ridgely's Delight, Gwynns Falls/Leakin Park and the Patapsco Valley.

With periods of civil unrest and riots in 1968, inner areas became run down and industrial warehouses were abandoned. During the 1970s, a period of urban renewal began with a large programme for redevelopment of the Inner Harbor promoted by the developer Rouse. The construction in 1977 of the Baltimore World Trade Center (WTC) pentagonal tower (designed by Pei), and Convention Center Harbor Place, plus the Transamerica Tower transformed the whole area around the Inner Harbor, which significantly changed the level of economic activity in the central area.

The main highway axes of Charles Street run north-south and Orleans Street run east-west (R 40), and the diagonal avenues, Falls Road and Liberty Heights run north-west, with Harford Road and Belair Road (R 1) north-east, and Wilkins Avenue and Washington Boulevard south-west.

The major through expressways are centrally the I-95 that crosses east-west under the harbour at the Fort McHenry Tunnel, opened in 1985. The peripheral I-895 crosses farther out in the Cross Harbour Tunnel, and then beyond the I-695 crosses over the bay at the former Francis Scott Key tolled bridge that forms part of an outer beltway, now under reconstruction.

The east coast Amtrak Rail lines run through the main Penn Station to Washington DC on one branch, which connects with Philadelphia and New York along the East Coast via the Baltimore Belt Line, and the major Baltimore and Ohio Railroad system radiates out westwards to interconnect with the cities of Chicago, Detroit, Cleveland and Pittsburgh, on another branch and with St. Louis and Cincinnati.

Baltimore Thurgood Marshall Airport is located in Maryland to the South West of the centre and with its proximity to the Washington metropolitan area is now known as Baltimore Washington International Airport (BWI).

During World War II, Baltimore was a vital port for shipping, and as the manufacturing centre of steel, Bethlehem Steel's Sparrows Point Steel Mill was one of the world's largest. The Port of Baltimore remains a very active port on the East Coast, carrying approximately 30 million tonnes of cargo annually.

Post-war the population of the city had peaked at nearly 1 million in 1950 but with de-industrialisation, it has declined to approximately 0.6 million in 2020 with 2.8 million in the metropolitan area. Despite a long period of white flight to the suburbs, the inner population has gradually begun to increase again since 2010 with gentrification taking place around the redeveloped Inner Harbor area and nearby Federal Hill.

Transactional Forces and Typology

With its location at the northern end of the Chesapeake Bay, the City of Baltimore has developed in and around its natural harbour that has evolved in activity since its early Atlantic and Caribbean maritime trade. With deep channels and its former wharves and warehouses, in recent times these areas have been expanded with land reclamation and modernised. They now comprise the established Ro-Ro and Container facilities with the Dundalk, Seagirt and Fairfield marine terminals that handle significant volumes of freight traffic each year.

Whilst maritime trade continues at these active port terminals, the Inner Harbor has been redeveloped commercially and successfully transformed around the World Trade Center with new business developments and associated real estate improvements around the inner-city area.

The city is characterised as a major port city with an enhanced inner harbour around which the central business district has been redeveloped. As a marine estuarine city, it has notable natural promontories with Fort McHenry at Locust Point and Fells Point opposite, as well as the industrialised areas of Fairfield and Dundalk and Sparrows Point grouped around the harbour bay on land which has been reclaimed.

Farther inland the city has been developed on an orthogonal grid, with Falls Road Creek acting as a divider in a north-westerly direction. Surrounded by circumferential highways with its different districts, the city has the outward characteristics of a typical modern American marine trading city along the North-Eastern Seaboard, but also inward-looking to its regional connections of the urbanised agglomeration in the post-industrial era.

4.4.3 MIAMI, FLORIDA, USA

Background and Origins

Miami is located in the south-east of the state of Florida on a plain between the Everglades and Biscayne Bay and the barrier islands. All of the land is very low lying in the Biscayne aquifer located on the Tropic of Cancer in the Gulf Stream ocean current with a tropical monsoon climate.

Inhabited originally by the Native American tribe the Tequesta, it was settled with a Spanish Mission in the mid-16th century CE, round the Miami River, named after the Mayaimi tribe. Florida ceded to the United States in 1821 and after the Seminole Wars, Miami was incorporated as a city in 1896. It was the construction of the Florida East Coast Railway (FEC) that assisted in the agricultural development of citrus crops, around which migration increased and a real estate boom followed.

During World War II, Miami acted as a significant port and naval base for the US fleet. Since the end of the war, the city has grown around the port, divided into different quarters, North-East, South-West and South-East and Downtown.

With its location close to the Bahamas and the Caribbean, the city has been shaped by separate waves of immigration coming from Cuba, Haiti and Latin America beyond. Modern Miami has been affected in particular by the US relationship with the nearby island of Cuba, resulting in major waves of Spanish-speaking migrants, in particular under the Mariel exodus from Cuba in 1980. These arrivals processed at the Freedom Tower settled around the expanding quarter of Little Havana in the middle of Miami.

The African American population also has become concentrated in central Miami, at Overtown, Liberty City and Opa-locka. Just to the north is Miami Gardens where this population has grown with West Indian and Haitian immigrants. From Lemon City, initially to Little Haiti around Notre Dame d'Haiti, the Haitian community has also established itself farther out now at North Miami.

Another smaller wave of arrivals came from Nicaragua as a result of the Sandinista Revolution in 1979. They were first housed by the Maduro Stadium with support of the Cubans, and then Venezuelan migrants who concentrated themselves at El Doral. More recently, Russians and Ukrainians have settled also at Sunny Isles, north of Miami Beach, in more upmarket luxury apartments.

Modern City – Buildings and Arteries

Since the turn of the 21st century Miami has developed as a major centre of commerce and now hosts the headquarters of many major global businesses, thanks to its proximity to Latin America and the Caribbean.

Downtown, on the eastern side, the Central Business District (CBD) has developed with the growing financial district and international banks situated long Brickell Avenue, the extension of the north-south US Route 1, south of the Miami River. Within the Brickell financial district, the highest buildings are Panorama Tower and One Bayfront Plaza.

Towards the south, the tree-lined Coral Way leads to Coral Gables and to the west of downtown are a number of different neighbourhoods, including Little Havana. Midtown is mainly high-rise commercial developments, Edgewater with its residential condominiums, and Wynwood. Farther north, Upper East Side has affluent residents and the separate district of Little Haiti.

The city itself has a population of approximately 0.5 million and the Miami metropolitan area, which includes the counties of Miami-Dade, Broward and Palm Beach, has a population of approximately 5.5 million people, about 70% Hispanic. Two-tiered Miami-Dade County was renamed in

1997 and has grown into a significant metropolitan county. Pan American Airport and terminal was opened in 1929 and has since grown into the present busy Miami International Airport (MIA) within Miami-Dade County.

Manufacturing, finance and media have each contributed to the real estate boom, assisted by the favourable climate and beaches, attracting new housing and condominiums. It did suffer significantly, however, during the 2007/8 global financial crisis, and there are several peripheral neighbourhoods of low income and poverty.

Tourism is a major factor with the magnet of South Beach and its Art Deco buildings; Miami Beach, Lincoln Road and Virginia Key and Key Biscayne, each connected by causeways; and Port Miami with its deepened harbour and major base for cruise ships. Much of the more recent architecture is styled as Miami Modernist, known as 'MiMo'.

Land has been reclaimed within Biscayne Bay and real estate developments have intensified on new land plots, with golf courses even on Normandy Shores and Indian Creek. Elsewhere new condominiums are being extended linearly along the coastal shoreline.

Due to its geography, with Miami River and its inner bays, the city's connections are defined by the crossings of the Tuttle and MacArthur Causeways and elevated highways. Due to the low-lying nature of the terrain, the city is generally built upon porous limestone. With rising sea levels, this is generating the challenge of protecting these built-up areas.

With the north-south roads and avenues, and the east-west streets and terraces, Miami is laid out on a grid system from downtown, with the North West and South West as the largest quadrants. Communications have been shaped by the waters' edges and the intersections of roadway crossings, such as at the mechanised raising bridges across the Bay and Miami River. The main Biscayne Boulevard and Brickell Avenue downtown are interrupted by frequent bridge openings to enable ship traffic to pass, causing traffic bottlenecks at peak periods.

The old Venetian and MacArthur Causeways built in the 1920s connect Downtown with South Beach and Miami Beach respectively. I-195 crosses into Miami Beach and I-95 goes to Miami International Airport (MIA). Florida's Turnpike, the unsigned SR 91, runs diagonally down across the state of Florida. A tunnel and rapid freight link connects the port with the airport. The long, elevated Rickenbacker Causeway connects Miami with the barrier islands down to Virginia Key and Key Biscayne across Biscayne Bay.

In terms of public transport, Amtrak and Metro Rail provide rail and rapid transit, links with the Tri-Rail commuter rail. The MetroMover serves the central areas with its Downtown, Omni and Brickell loops. Miami Central Station is a mixed-use station for the newly named Bright Line rail system, with Miami Intermodal Centre (MIC) as the transportation hub for different modes.

Transactional Forces and Typology

With its coastal and tropical location around the Miami River and by Biscayne Bay with its mangroves, the area thrived initially through agricultural production of citrus crops. With the development of the Florida East Coast Railroad, new waves of immigration and settlement were attracted, leading to Miami becoming the 'Gateway to the Americas'. With its propinquity to the Caribbean also, it became a refuge for Spanish-speaking migrants, especially from Cuba and Haiti, with the growing districts of Little Havana and Little Haiti, and its warm climate acts as a magnet for modern tourism.

As the real estate market developed, coastal developments have been built linearly all along the coast from Fort Lauderdale and North Miami down to South Miami. With its barrier islands, the Keys and Miami Beach, the city is afforded some coastal protection. However, the city is very low-lying and in a hurricane region with the increasing threat of rising sea levels from climate change.

With the growth in activity of its modern expanding port and airport, it has become a major dynamic hub in the region. With its developing real estate market, it has also become a thriving financial centre. Its increasing yet ageing population has also led to an increase in the construction of apartments and condominiums and a growing health sector. Alongside the developing financial district and associated businesses and media services, the city has a very active modern economy.

A linear coastal and tropical city with its outer barrier islands, Miami has an unusual geographical configuration. With its orthogonal built grid and neighbourhoods and swampy Everglades to the west, available developable land is a constraint for developments which are otherwise forced up vertically. Given these constraints of buildable land, the city has been stretched out linearly along narrow corridors both with dense localised commercial and residential condominiums becoming the archetypal development pattern. With land at a premium, and marine and environmental constraints in the south in and around Biscayne Bay, development is centred around the north-south I-95 linear business corridor, resulting in its elongated shape as a modern American metropolitan city, expanding outwards also into the surrounding Miami-Dade County region.

However, its proximity to the ocean brings risks with policy initiatives such as the 'Forever Bond' introduced in 2018 so as to start to finance climate-resilient initiatives.

4.4.4 ATLANTA, GEORGIA, USA

FIGURE 4.4.3 Atlanta – early configuration, 1847.

Source: Historic Urban Plans.

Background and Origins

Atlanta was effectively founded at the Zero Mile Post in 1847 on the intersection of two railroads as the terminus of the Western and Atlantic railroads where they crossed the Chattahoochee River to connect the Port of Savannah with the Midwest. These grew and developed into a classic hub-and-spoke system for the transport of key commodities of cotton and eventually steel in the region.

On a ridge in the foothills of the Appalachian Mountains, the land was originally occupied by the indigenous Creek people near where Peachtree Creek flows into the Chattahoochee River, but they were forced out with Indian removal around 1830 and were relocated to Oklahoma.

During the 1860s, at the time of the American Civil War, the region around Atlanta was the scene of several conflicts until the Confederates in Atlanta surrendered to the Union Army in 1865. A period of reconstruction followed, and slavery was eventually abolished with the Reconstruction Amendments to the Constitution. Atlanta became the state capital in 1868.

Water supply was guaranteed for the city by the construction of a dam in the Chattahoochee River and the formation of Lake Lanier. Its basic rail network, with the Terminal and Union stations, was consolidated and new buildings were constructed at the turn of the 20th century, such as the Flatiron and Empire buildings.

With the decline of the cotton industry, new industries developed around the city and were to provide valuable manufacturing capacity during World War II. The airport, which was developed at the original site at Chandlers Field, was expanded during the war.

Post-war, educational institutions were established, Georgia Tech University was founded and new research institutes gravitated to the city. With its industry and resources, Atlanta developed into becoming the leading city in the American South with its significant black population.

Modern City – Buildings and Arteries

Atlanta annexed the adjoining areas of Fulton and DeKalb counties, and a Plan of Improvement was outlined in 1952 for the area. The Atlanta Housing Authority (AHA) introduced a programme of public housing improvements also as infrastructure was gradually improved.

During the 1960s, Atlanta became the main organising centre of the civil rights movement and especially with the Supreme Court's ruling to desegregate public transport and public schools. Suburbanisation had increased, with 'white flight' continuing from urban areas, and fundamental demographic shifts were occurring. Significant programmes of urban renewal were also undertaken in different areas.

Delta Airlines had adopted a plan to build a major airport that would operate as a hub-and-spoke system for the whole Southern region of the USA so as to connect beyond with the major cities of New York and Chicago. After eventual agreement, this became Atlanta Hartsfield-Jackson International Airport (ATL), named after the former mayors.

As new businesses developed, growth of the economy expanded in and around the metropolitan area of Atlanta, which attracted the headquarters of several major corporations such as Coca-Cola, Delta Airlines, and AT&T. With its railroads, freeways and international airport, it developed logistic centres to act as a major hub in the region. The merged Norfolk Southern Railway is also headquartered here.

With CNN World Headquarters it became a centre for film and TV production companies and in actively promoting a multi-cultural city. Another factor in development was central downtown hotels that were developed by Portman under a new model of major skyscraper hotels such as the Regency Hyatt and Westin Marriott. Major commercial buildings such as the Bank of America Tower were developed as well as the Civic Center with a large separate parking area.

With the Georgia Institute of Technology located in Midtown, the city contains the Atlanta University Center, which comprises a contiguous consortium of HBCUs. With Piedmont Park, Centennial Olympic Park and Chastain Memorial Park, there is also the extensive Chattahoochee River National Recreation Area, and with the Botanical Garden, the city retains large green areas with extensive tree cover such as magnolias, dog woods and southern pines.

Atlanta is the seat of the state capital, and the governor and Georgia General Assembly are located at the Capitol here. The city also has a mayor and city council at city hall, and it contains two main counties, Fulton and DeKalb. The population of the city is approximately 0.5 million people, with 6 million in the metropolitan area where Atlanta MSA has been growing very fast.

In terms of infrastructure, its modern interstate highway system includes the I-20 that runs east-west just south of downtown. I-75 and I-85 combine to become the main N-S spine Downtown Connector and run southwards to Atlanta International Airport. The encircling outer I-285 Beltway acts as the perimeter or boundary of 'inside' and 'outside' the perimeter (ITP and OTP) of the city. The Georgia 400 Corridor runs north through Buckhead and via the Lenox Center regional mall to Sandy Springs.

The city's subway system was commenced in 1975, managed by MARTA, and serves the five main counties in the metropolitan area. The first line was opened east-west in 1979 then followed by the north-south line and is integrated with the central Atlanta streetcar that opened in 2014 with associated rail and bus systems. The rapid transit system has four lines, Red and Gold north-south, and Blue and Green east-west, with a very deep station in the rock at Peachtree Center Station, and the main interconnection at the Five Points Station at split level.

In 1996 the city was awarded the Summer Olympic Games for which substantial works were undertaken in the creation of the Centennial Olympic Park along with housing for the athletes in the Georgia Tech area. From the beginning of the 21st century, the demographics changed, with more residents reverting to central and downtown areas. The Martin Luther King, Jr. Center for Nonviolent Social Change is to the east of the city along with Little Five Points located farther east.

Recently the regenerated Atlanta Beltline has been redesigned to convert a former peripheral railway line into a light rail line with off-road trails under the PATH foundation in surrounding neighbourhoods and to connect with the Silver Comet Trail.

The city is now widely spread out with low density urban sprawl and localised nodes in different wards. It is dependent on the inter-connecting highway system with areas such as the Peachtree Center, individual office parks and retail centre nodes farther out and expansion at developing centres such as Roswell and Sandy Springs.

The larger Atlanta region now comprises the counties of Cherokee, Clayton, Cobb, Dekalb, Douglas, Fayette, Forsythe, Fulton, Gwinnett, Henry and Rockdale as well as the city of Atlanta itself.

Transactional Forces and Typology

The siting and growth of the city of Atlanta developed from its original role as a transit point with the intersection of two railroads, around which the transport of cotton and manufactured materials developed. This extended into heavy industries that then became supported by the highways and the subsequent interstate network. This was further reinforced by the development of Atlanta International Airport, which now plays a pivotal role as a major logistics centre in the midst of this expanding metropolitan region.

As a highly automobile-dependent city, its focus has been on the continued development of the extended NW Corridor Project. These economic forces have led to the development of the city as a linear metropolis in a corridor north and south along Peachtree ridge with a downtown stretching through Midtown from Five Points north to Buckhead uptown.

Since the original Downtown of Atlanta centred around Five Points, more recent real estate development has been focused around Midtown and farther north. Whilst the downtown area has a reorientated orthogonal grid, the city is bounded by its hub-and-spoke highway system and structured through the Connector and circumferential Beltway.

With its dominant central commercial areas, and with intervening less prosperous residential areas, the city has characteristics of typical modern automobile-dependent American cities. With its incoming population predominantly from the catchment area of the American South, its development is increasingly poly-centric in the outer and more peripheral areas in the wider urbanised region.

4.4.5 HOUSTON, TEXAS, USA

Background and Origins

Houston is located in the south of the state of Texas inland from Galveston on the Gulf of Mexico. Originally inhabited by the indigenous Karankawa peoples, now extinct, the city was named after Sam Houston, who was a general in securing the independence of Texas from Mexico after the Battle of San Jacinto. Most of the land is either on the central plain and low lying with marshy land and wet bayous and lakes.

The first settlers that arrived at Allen's Landing explored sites at Buffalo Bayou and White Oak Bayou near Galveston Bay in 1836. The transport of enslaved black people evolved with the triangular trade with Africa, and a commercial hub developed on the coast for the export of cotton. As the plantations grew, the railway was developed and trading expanded around the ports of Galveston and Beaumont.

The major development was the discovery of oil in the Spindletop oilfield near Beaumont. With plentiful resources in the region, expansion of the port was necessary for commercial exploitation. Accordingly, the Houston Ship Channel was dredged, and a deep-water port was developed in 1914 at Houston with its turning basin. Gradually an entire petro-chemical industry developed in the period between the world wars along with manufacturing and shipbuilding industries.

It had become a strategic location, and with developing industries a high volume of immigration ensued to service the new businesses. After World War II, the city annexed several unincorporated areas and grew significantly larger. With its low-lying outer land areas, however, being built up rapidly, the area remains susceptible to flooding.

From its early ward system, new districts and neighbourhoods were developed across the city. Never formally planned, Houston remains one of the few US cities not having statutory zoning regulations, although it does have restrictive covenants. The impact of this has led to generally lower density and outward suburban sprawl of the city.

Modern City – Buildings and Arteries

Populations moved south from the Rust Belt in the northeast into the oil and gas industry. In addition, during the 1960s, a new aerospace industry took off as the NASA Manned Spacecraft Centre was established in Houston. The city also became a centre for medical care and specialist hospitals.

The oil and gas industries are concentrated around the Houston Ship Channel and towards the coast, along with other ancillary industries of oilfield equipment and services. At the entrance Baytown is highly industrialised, as is the area around Morgan's Point.

With its location on the Gulf Coast, Houston is often impacted during the hurricane season with high tides and flooding as a result of inadequate protection and with excessive development having been undertaken in the flood plain with sand having been dredged for building purposes.

The skyline of downtown Houston now comprises a number of mid-rise commercial buildings along with the modern tall skyscrapers of Wells Fargo Plaza (designed by SOM), Williams (Transco) Tower (designed by Johnson and Burgee), Pennzoil Place (designed by Johnson), Bank of America Center (designed by Johnson) and JP Morgan Chase Tower (designed by Pei). With the advent of air conditioning, more companies relocated to the area and lower wage economy. A network of underground tunnels and skywalks that connect these buildings in the central area downtown was built also to provide a controlled environment for pedestrians in the humid and tropical climate.

The city benefits from a number of large parks such as Hermann Park, Memorial Park and Buffalo Bayou Park. There are a number of universities in the city, including the University of Houston, University of Houston–Clear Lake, Texas Southern University (TSU) and Rice University close to the Texas Medical Center (TMC). The eclectic Museum District and the Menil Collection are west of the city's centre, and the residential River Oaks neighbourhood is even farther west.

The population of Houston is approximately 2.3 million, with 7 million in the metropolitan area. The profile of the population is approximately Hispanic–Latin 45%, black 30% and white 20%, an ethnically diverse metropolitan region. With very high levels of immigration into the area, however, approximately 0.5 million persons are unregistered.

The metropolitan area extends beyond Houston into the residential Woodlands and Sugar Land areas. The adjoining Houston–Woodlands–Sugar Land metropolitan area is considered one of the most economically active regions. The Woodlands, first conceived by Mitchell under McHarg design principles as a residential area, was taken over by Rouse and then developed commercially by Howard Hughes in Montgomery County. Sugar Land, as the site of a former sugar plantation in Fort Bend County, has similarly grown extensively as a commercial district.

The inner city is contained within the I-610 highway loop and the I-45/69 downtown, surrounded by Uptown and other specialised districts such as the TMC and the Energy Corridor. The highways are extensive with an outer Beltway SH8, the Sam Houston Parkway and Tollway, and Grand Parkway SH99 yet farther out.

The city region is highly automobile dependent with a hub-and-spoke highway system and orbital roads managed under a Major Thoroughfare and Freeway Plan. It is crossed by three major interstates (I-10, I-45 and I-69), and there is a system of toll roads managed by Harris County Toll Road Authority (HCTRA) and Fast Bend County Toll Road Authority (FBCTRA), including the Hardy Toll Road and the West Park Tollway.

There are major railroad marshalling yards to the north-east of the city and petro-chemical storage facilities. The Metropolitan Transit Authority of Harris County (METRO) has developed a light rail service since 2004, with three out of five lines now operational.

The Houston Airport System operates three airports, the major George Bush International Airport of Houston (IAH) with five runways and five terminals, with William P Hobby Airport for domestic and Ellington for military traffic. With the establishment of NASA's Johnson Space Center and Mission Control, this has also contributed to Houston becoming a modern centre of Space R&D around which supporting technologies have gravitated.

Transactional Forces and Typology

Houston, on the confluence of the Buffalo Bayou and White Oak Bayou inland from Galveston Bay, initially developed with its trade in the cultivation and export of cotton. With the discovery of oil and the development of railroads, this was to significantly change the local pattern and speed of development. The natural harbours of Galveston and Texas City by the coast were too small, and subsequent to the development and dredging of the Houston Ship Canal, the Port of Houston was expanded with the construction of the Turning Basin Terminal with its docks and later Barbours Cut at Morgan's Point and Bayport Terminal.

In parallel with this, new islands have been developed in Galveston Bay in the south-east benefitting from land reclamation, and oil refineries have been located around the San Jacinto River. All of this significant oil-based activity, however, has affected the marine development around the bay and the eastern areas of the city. With many areas of development in low-lying outer land, however, due to their proximity to the coastal zone, the impact of climate change and flooding remains an increasing risk in the region.

From the early Allen's Landing place, the city centre developed in the separate neighbourhoods of Uptown, Midtown and Downtown around the junction of the two bayous, with its NE-SW orthogonal grid street system. On either side of this are other, generally north-south orthogonal streets platted in previously marshy areas.

The developed central city is defined and now encircled by a network of major highways with the inner loop I-610 and outer Beltway SH8 and External Grand Parkway SH99 running through predominantly industrialised and suburban areas. This typifies Houston as a highly automobile-focused modern American commercial city with its pervasive petro-chemical industries.

With limited regulation or zoning restrictions, the city has evolved variously with its mixed residential and industrial communities. Along with specialist areas such as the TMC and the oil-based industrial areas in the Energy Corridor, peripheral edge cities of Greenspoint and Greenway Plaza have developed. The Woodlands area, which was developed separately to the north as a special purpose district, as a HUD new town, has also evolved into an economically dynamic area. Besides its proximity to the coast, the city of Houston has evolved in a concentric manner defined by its heavily dominant highway network.

4.4.6 SEATTLE, WASHINGTON STATE, USA

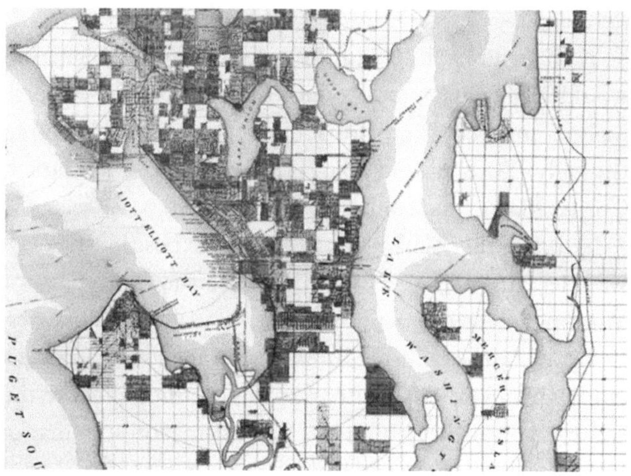

FIGURE 4.4.4 Seattle – early configuration, 1890.
Source: WMoL.

Background and Origins

Seattle is the largest city of Washington State, located in Elliott Bay on an isthmus in Puget Sound on the west coast of America in the Pacific Ocean, south-east of Victoria on Vancouver Island opposite. With evidence of early civilisation of indigenous peoples, the local tribes of Salish, Suquamish and Chinook lived in coastal areas, dependent on fishing.

During the 16th century CE, Europeans arrived seeking a sea route for the Northwest Passage, but it was not until 1792 that Captain George Vancouver from the British Navy mapped the long island, which came to be named Vancouver Island, Puget's Sound and the Salish Sea. British merchants set up the North West Company as the earlier Hudson's Bay Company with the fur trade in the region.

In 1851, a group of American pioneers led by Arthur Denny crossed the Rockies to reach Alki Point, now West Seattle, the Chinook naming it New York Aki, or Some-day New York. Others settled across the water around Pioneer Square, naming it Duwamps in the Lushootseed language. Subsequently, in 1869, the city was incorporated under the name of Seattle.

The first activities revolved around the lumber industry and the port and use of Yesler Way from Pioneer Square to the sawmill that came to be known as Skid Road. A setback for the city was the Great Fire in 1889 that destroyed most of the city. The original wooden houses were all rebuilt in brick, but also at a raised level due to the swampy ground conditions.

The next development resulted from the discovery of gold in the Sierra Nevada and the Klondike Gold Rush when the city became the jumping-off point for Alaska- and Yukon-bound miners. With the construction of the railroad, the port grew and shipbuilding was also further developed.

With significant agriculture in the hinterland, farmers would bring their produce into the city for sale. Accordingly, a number of individual markets were established in the early 1900s around Pike Place, with the Main Market, Corner Market and Sanitary Public Market.

In 1909, the Alaska Yukon Pacific Exposition (AYPE) was held in Seattle, now the site of the University of Washington, to promote the Pacific Northwest industries. Designed by the Olmsted Brothers, it was laid out around a circle with an axis orientated towards Mount Rainier in the south-east with key buildings emphasising forestry, manufacturing, machinery and agriculture.

The Lake Washington Ship Canal was realised in sections and eventually completed in 1934 to connect Puget Sound via a series of Chittenden Locks to Salmon Bay. Then it was connected via Fremont Cut to Lake Union and Portage Bay, and then via Montlake Cut to Union Bay to the fresh water Lake Washington. This was a significant, much-needed achievement after many alternative canal plans by the US Army Corps of Engineers.

Between the world wars the city suffered in the Great Depression. While shipbuilding was key in World War I, the development of aircraft manufacturing was a feature of World War II with the expansion of the Boeing aeroplane factory. It attracted other supporting technologies, and a period of considerable industrialisation and urbanisation ensued.

Modern City – Buildings and Arteries

The topography of the city of Seattle is varied, with seven hills that include Capitol Hill, First Hill, Beacon Hill, High Point in West Seattle and Denny Hill. It was reshaped partially, however, by major regrading between First and Beacon Hills and by the construction of a seawall and artificial Harbor Island at the mouth of the city's Duwamish Waterway with West Seattle Bridge. Discovery Park abuts Puget Sound in the Magnolia neighborhood, with several trails around lakesides and waterfronts.

Bellevue in the east is accessible via Lake Washington Evergreen Point floating bridge (I-90), which opened in 1940; this was expanded, dualled and rebuilt in 1993 and connects directly with the Mount Baker Tunnel into the city with its twin tunnels via Mercer Island.

The Port of Seattle is a major employer which merged with the nearby Port of Tacoma and is also the operator of the Seattle-Tacoma International Airport (SEA) known as Sea-Tac. A number of major industrial companies are headquartered in Seattle, such as Boeing with its manufacturing plants in Renton and Everett.

During the 1980s, a number of new technology companies established themselves in the area. These included the tech giants Microsoft, since moved to Redmond across Lake Washington, and Amazon, and also biotech companies and the Bill & Melinda Gates Foundation in different specialised commercial centres.

In 1962, the Century 21 Exposition World Fair was held in Seattle and the iconic Space Needle tower was unveiled. More recently the Seattle Great Wheel was erected in 2012 in Pier 57 downtown.

Previously the city had a system of streetcars, but they were removed in the late 1930s. With a network of buses and trolley buses, the Sound Transit Central Link LRT runs north-south between the University of Washington and Angle Lake (Sea-Tac) via downtown Pioneer Square with certain sections elevated. From King Street Station, the Sounder commuter rail is run by BNSF and freight by Amtrak, and there are several ferry companies operating in and around the Sound.

The central area has an orthogonal street grid system with the main I-5 and SR99 highways running north-south and I-90 east-west. The Alaskan Way Viaduct replacement tunnel following the earthquake collapse is a huge double deck bored-road tunnel 17.5 m. in diameter.

Downtown is built up with groups of major commercial office towers, such as the Amazon Building, Washington Mutual (WaMu), and the Columbia Centre. It is the fastest-growing US city with considerable in-migration from

South Asian countries – there is no income or corporate income tax in the city.

The population of the city is approximately 0.75 million, with 4 million in the metropolitan area. The city is, however, located in an active earthquake zone on the Pacific Rim and suffered earthquakes most recently in 1949, 1965 and 2001.

Transactional Forces and Typology

The region around Seattle has been shaped by the water from all sides located in the Central Basin of Puget Sound. In the south, unusually around 1910 the White and Green Rivers changed course to join and flow together into the Duwamish River into Elliott Bay and the man-made Harbor Island was created. The Cedar River was also diverted from the Black River to run into Lake Washington.

With these developments and after the Lake Washington Ship Canal was realised to connect the Sound with a route via the lakes to Lake Washington, this significantly reshaped maritime communications and transport links and strengthened the activities of the Port of Seattle. With its strategic location on the Pacific Coast, this consolidated its trading position. From its early logging and lumber activities to more recent manufacturing, with enhanced rail and road connections, this cemented its position as a regional trading centre on the west coast of North America. Now with its modern tech giant corporations, it is an active and progressive city with a strong medical sector also.

The city of Seattle has coastal, estuarine and riverine characteristics, with its unusual location on a narrow land isthmus with Puget Sound to the west and Lake Washington to the east. Westwards with its main harbour at Elliot Bay, this acts as an oceanic port for the Pacific. Southwards is the Duwamish Waterway that connects with the Green River, and to the east is fresh water Lake Washington. With its unique natural geography, the city has been laid out and configured around these unusual features.

It is a hybrid city, having not only a coastal perimeter but internal lakes and adjoining hills. This mixed topography creates a series of quite different districts, with many having marine, lake or river outlooks. These assets have been generally recognised in the built development of the city. With its north-south axis and mainly orthogonal street pattern, its land and water edges and forested areas demand careful preservation in this environmentally attractive region that has become increasingly metropolitanised in the Puget Sound region.

4.4.7 VANCOUVER, BRITISH COLUMBIA, CANADA

FIGURE 4.4.5 Vancouver – early configuration, 1907.
Source: DMG.

Background and Origins

With its location on the west coast of the North American seaboard, **Vancouver** as the capital of British Columbia in Canada is in a striking position at the Burrard Inlet where the Fraser River runs into the Georgia Strait between the mainland and Vancouver Island. In a predominantly forested region, it is cradled between the Pacific Ocean and the snow-capped peaks behind the Coast Mountains.

It is generally recognised that the first inhabitants were probably Asiatic peoples who crossed the land bridge across the Bering Strait and moved southwards to the Fraser River. Known as the 'Sto-lo' (People of the river), they developed artisanal skills and lived off the sea by fishing and hunting. Particular indigenous groups were the Tsawwassen and Squamish who lived in villages around the coastal areas.

The first Europeans to arrive were the British with James Cook, arriving on Vancouver Island in 1778, naming it after George Vancouver, a surveyor who was assisting in mapping the North American Pacific Coast. With the search for the Northwest Sea Passage, the French and Spanish were also prospecting.

The Canadian interior was being surveyed with Simon Fraser's exploration of the Fraser River from the Burrard Inlet. With fish and fur trading, a major transformation resulted from the later Gold Rush that developed inland in 1858. Although short lived, with attraction for prospectors and immigrants, new roads were laid down, such as the North Road, and a fort was built at New Westminster on the Fraser River, which was declared the capital of British Columbia.

In 1867, the Dominion of Canada was created and transferred into the Confederation in 1871. New areas were developed at Port Moody and at Granville, which was to become the Terminus of the Canadian Pacific Railway (CPR), with these areas subsequently incorporated into the city of Vancouver in 1886.

But later that year a great fire occurred, destroying many of the wooden buildings in the city. The city was rebuilt and expanded with new streets and areas such as Granville and Hastings, and new residential areas, such as West End and Kitsilano. The CPR transformed access to the city, the port expanded as timber and trade developed and more immigrants arrived at the turn of the 20th century.

Modern City – Buildings and Arteries

Lion's Gate Bridge was built across the First Narrows to North Vancouver as early as 1938. The Second Narrows Bridge was built subsequently near Lynn Canyon, which connects the main route of Highway 1 to the city centre and Burnaby to the east. At the western end of the highway is Horseshoe Bay with the ferry terminal to connect with Vancouver Island and farther north towards Squamish and Whistler.

As an active port, ships pass through the narrows under Lion's Gate Bride upon entry to Vancouver, passing Coal Harbour, Beach Crescent and the waterfront, next to Downtown and Kitsilano on the southern side.

The city contains some fine older Art Deco buildings such as the Marine Building, the CP Railway Station, Harbour Centre Tower and the Cathedral.

The individual Yaletown, Gastown and Chinatown are distinctive districts. The promontory of the downtown peninsula is Stanley Park, a quiet green, wooded area with trails for walking or cycling across or around the Seawall Promenade with sea views from Coal Harbour to Lion's Gate Bridge.

South of the centre is False Creek and Granville Island, reached by the Granville Street Bridge. Originally a settlement for Squamish fishermen, then a light industrial area, False Creek South has been redeveloped as a residential area with a market, brewery and bars and restaurants as an attractive social location.

In contrast, False Creek North, which was formerly the site of railway yards, has been developed in a different manner with several groups of high-rise towers as new real estate on the site for Expo '86.

The whole of South Vancouver is a large area, mainly residential, with Kitsilano and its beach to the north and the University of British Columbia to the west. Here are the botanical gardens and Pacific Spirit Regional Park, and the fine Museum of Anthropology with its history of the local First Nations Haida and their culture.

Across English Bay is North and West Vancouver. Separated by the Capilano River with its salmon ladders and the pedestrian suspension bridge, the northern area is wooded with former lumber mills and is now mainly residential. Grouse Mountain is behind, with Provincial Park and its conservation reserve.

Expo '86 was held to commemorate Vancouver's Centenary around Canada Place in 1986. The site of Expo on False Creek is now a cruise terminal in the active harbour where freight and commercial maritime traffic mix, along with helicopters and seaplanes that ply the harbour.

The main metro system in the city, Skytrain, traverses the city and connects downtown with Vancouver International Airport (YVR) at Sea Island in Richmond to the south.

The modern Vancouver Public Library (designed by Safdie), Living Shangri-La (designed by Cheng) and One Wall Centre (designed by Perkins and Will) are notable new buildings.

The city is a very liveable place, due to its natural setting and surrounding hills, forests and islands. With frequent precipitation, it remains green and accommodates a mix of different and varied cultures, many from Asia, being the gateway to the Far East.

In 2010 the Winter Olympic Games were held in Vancouver and up-country at Whistler. The current population of the city is approximately 0.7 million and 2.6 million in the metropolitanised area known as Metro Vancouver Regional District comprising over 20 local municipalities.

Transactional Forces and Typology

This region of British Columbia being mainly forested, the initial activities to be developed around Vancouver were based on logging. With the abundance of conifers and maple trees and with the installation of sawmills, the lumber business was to develop around the local rivers used for transport, and these developments gravitated around Hastings Mill on Burrard Inlet. This industry remained as the principal activity, along with fishing and fur trading in the area.

Whilst the Gold Rush period was transitory, one of the most significant developments was the construction of the trans-continental Canadian Pacific Railway (CPR) with its terminus in the city. This consolidated communications with the interior and beyond, whilst the development of the port was to facilitate outward connections to Vancouver Island and along the Pacific coastline and to Polynesian Islands and places in Asia beyond.

As a coastal city on the peninsula between the Burrard Inlet and the Fraser River, protected by the outer Vancouver Island, Vancouver is ideally sited for maritime communications. Laid out in a rectangular grid pattern, the city benefits from a long coastal shore and river frontages and the oceanic West Coast climate. With a compact urban core, recent development of the city has been concentrated on mixed-use and high-rise residential developments in the West End, resulting in quite dense built-up areas. However, with its coastline and parks, such as Stanley Park, and its interconnected transport system, the city benefits from a number of attractive yet different residential neighbourhoods with distinctive characteristics, such as Granville Island and Gastown, Kitsilano and Coal Harbour, and North and South Vancouver. The following figures illustrate the modern configuration of the American cities as described and the evolution of their modern population.

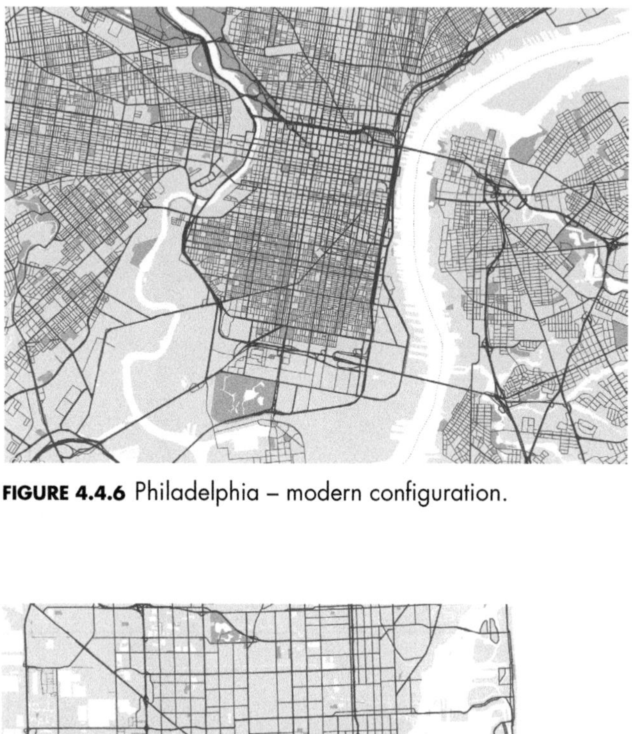

FIGURE 4.4.6 Philadelphia – modern configuration.

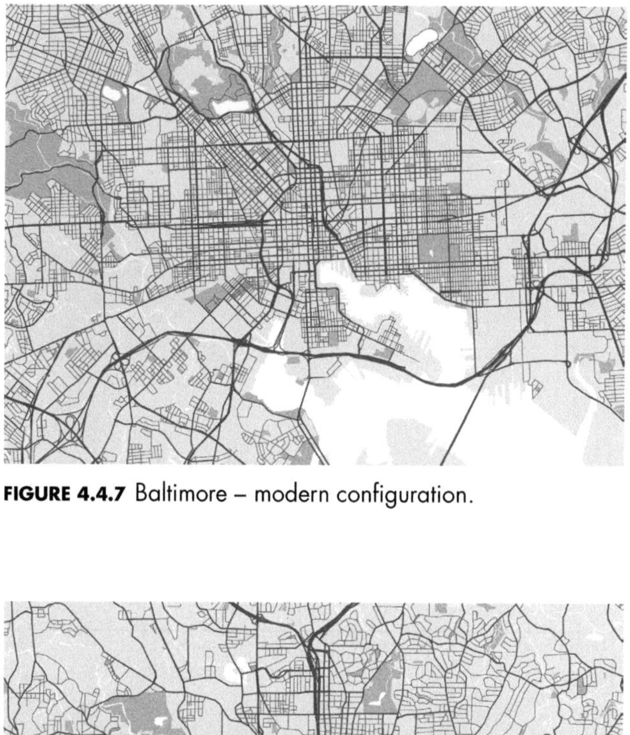

FIGURE 4.4.7 Baltimore – modern configuration.

FIGURE 4.4.8 Miami – modern configuration.

FIGURE 4.4.9 Atlanta – modern configuration.

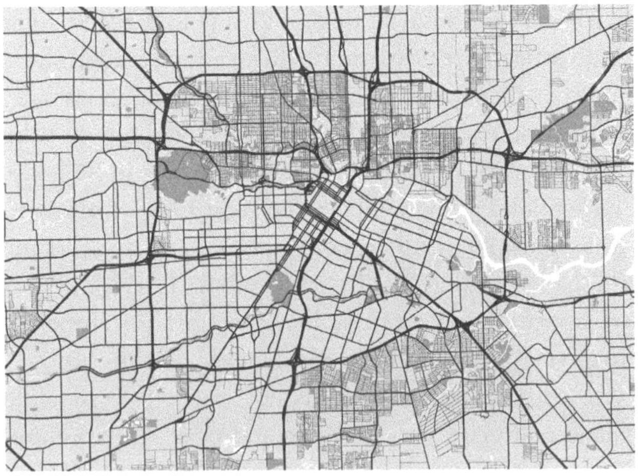

FIGURE 4.4.10 Houston – modern configuration.

FIGURE 4.4.11 Seattle – modern configuration.

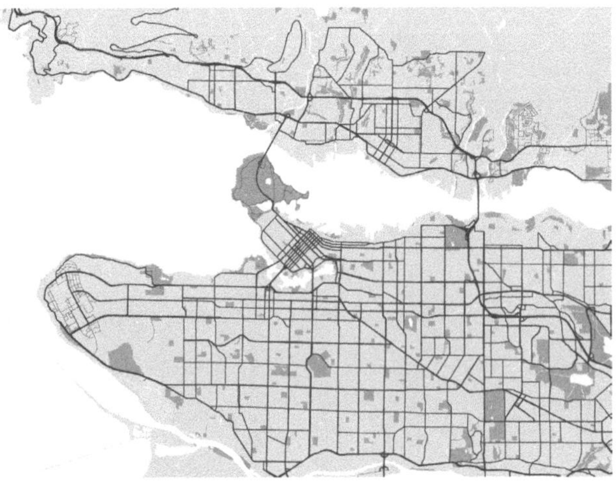

FIGURE 4.4.12 Vancouver – modern configuration.

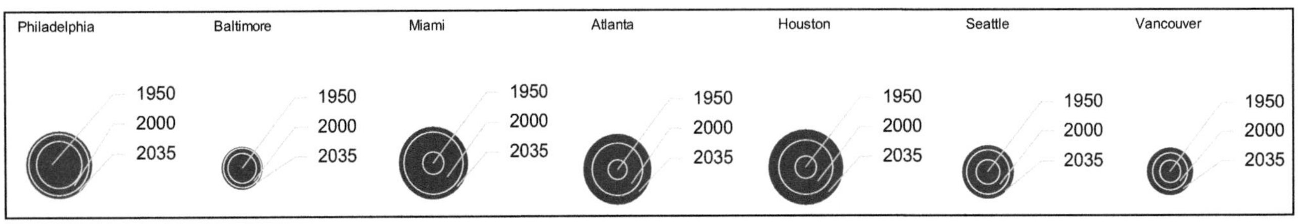

FIGURE 4.4.13 Modern American city populations.

Data source: Global Database of Metropolises 2020 – UN-Habitat (2020), UN DESA Population Division.

Further Reading

PHILADELPHIA

Kahan, Paul – *Philadelphia, a Narrative History*, University of Pennsylvania Press, 2024.
Kyriakodis, Harry – *Philadelphia's Lost Waterfront*, The History Press, 2011.
Minardi, Joseph – *City of Neighborhoods, Philadelphia, 1890–1910*, Schiffer, 2020.
Simon, Roger – *Philadelphia, a Brief History*, Penn. Historical Association, 2017.
Warner, Sam Bass – *The Private City*, University of Pennsylvania Press, 1971.

BALTIMORE

Crenson, Matthew – *Baltimore, a Political History*, Johns Hopkins Press, 2017.
Philipsen, Klaus – *Baltimore, Reinventing an Industrial Legacy City*, Routledge, 2018.

MIAMI

Ariza, Mario – *Disposable City, Miami's Future on the Shores of Climate Catastrophe*, Bold Type Books, 2020.
Bramson, Seth – *Miami, the Magic City*, Arcadia Publishing, 2007.
Weitz, Seth – *City of Hope, City of Rage, Miami 1968–94*, University of Alabama Press, 2024.

ATLANTA

Allen, Frederick – *Atlanta Rising*, Longstreet Press, 1996.
Ogbar, Jeffrey – *America's Black Capital*, Basic Books, 2023.
Pendergast, Mark – *City in the Verge*, Basic Books, 2017.

HOUSTON

Bradley, Barrie – *Improbable Metropolis, Houston's Architectural and Urban History*, University of Texas Press, 2020.
McComb, David – *Houston, A City*, University of Texas Press, 1981.
Sibley, Marilyn – *The Port of Houston, a History*, University of Texas Press, 2013.

SEATTLE

Dierwechter, Yonn – *Climate Change and the Future of Seattle*, Anthem Press, 2021.
Klingle, Matthew – *Emerald City, an Environmental History of Seattle*, Yale University Press, 2008.
Shobe, Hunter & Banis, David – *Upper Left Cities, a Cultural Atlas of San Francisco, Portland & Seattle*, Sasquatch Books, 2021.
Williams, David – *Too High and Too Steep, Reshaping Seattle's Topography*, University of Washington Press, 2017.

VANCOUVER

Berelowitz, Lance – *Dream City, Vancouver and the Global Imagination*, Greystone Books, 2010.
Francis, Daniel – *Becoming Vancouver*, Harbour Publishing, 2022.
Punter, John – *The Vancouver Achievement, Urban Planning and Design*, UBC Press, 2004.
Sommer, Warren – *The Ambitious City, a History of the City of North Vancouver*, Harbour Publishing, 2007.

4.5 New Capital Cities

American – Indian – Australasian – South American – African

Washington DC, New Delhi, Chandigarh, Islamabad, Canberra, Brasília and Abuja,
Naypyidaw, Astana, Nusantara and Putrajaya

4.5.0 INTRODUCTION – PLANNED CAPITALS

Over the centuries from the creation of early city-states and nation-states, capital cities have developed, often as the primate city in a country. Historically many of these have continued to serve for long periods as the national capital.

However, as economic and demographic circumstances have evolved in countries, these have led to movements for change and the development of alternative new capital cities. After the end of former empires and major world wars, territories and boundaries have changed and new independent States have formed in different regions. This has frequently led to aspirations for a State's own new or symbolic capital centre. In addition, as colonial empires have ended, new nations have found the need to move their national capital to a more strategic or central location for administrative, political, economic or symbolic purposes.

This has been the situation for several countries in Africa, notably Nigeria and Tanzania. Other significant instances have been for some of the largest and most populous countries, namely Brazil, India, Pakistan, Canada and Australia, each having only relatively recently created their own highly distinctive solution for a new capital that is more responsive to their own specific cultural, political and administrative needs.

In each of the most recent centuries key new capitals have been founded, beginning for example with Washington in the District of Columbia after the foundation of the United States of America in the 18th century.

In the 19th century, the capitals of New Delhi in India, Ottawa in Canada and Pretoria in South Africa were developed under British colonial occupation and supervision.

In the 20th century, the capitals of newly independent nations have been developed in Asia, at Islamabad in Pakistan in addition to new State capitals across India. Both Abuja in Nigeria and Dodoma in Tanzania are new capitals in Africa located at more geographically centric locations. In addition, transfers have been made to significant new modern capitals constructed at Brasília in Brazil and Canberra in Australia.

More recently, in the 21st century, new capitals have been developed at Naypyidaw in Myanmar and Astana in Kazakhstan in Asia, and others are continuing to emerge in growing nations, such as New Cairo in Egypt, Nusantara in East Kalimantan, Indonesia and Putrajaya in Malaysia.

4.5.1 WASHINGTON, DISTRICT OF COLUMBIA, USA

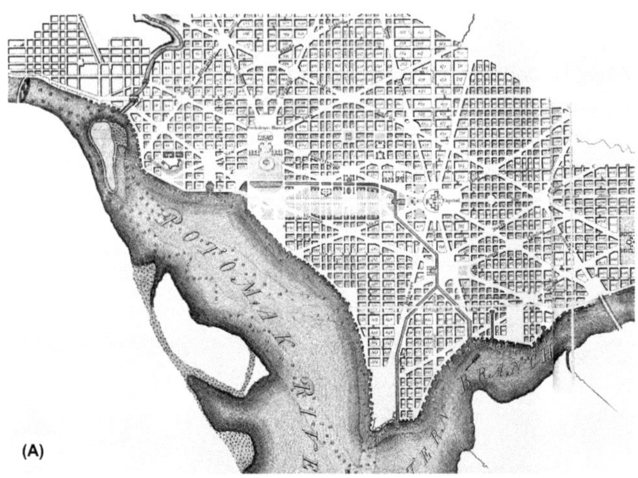

(A)

FIGURE 4.5.1A Washington DC – early configuration, 1792 (Ellicot).

Source: British Library.

DOI: 10.4324/9781003353348-23

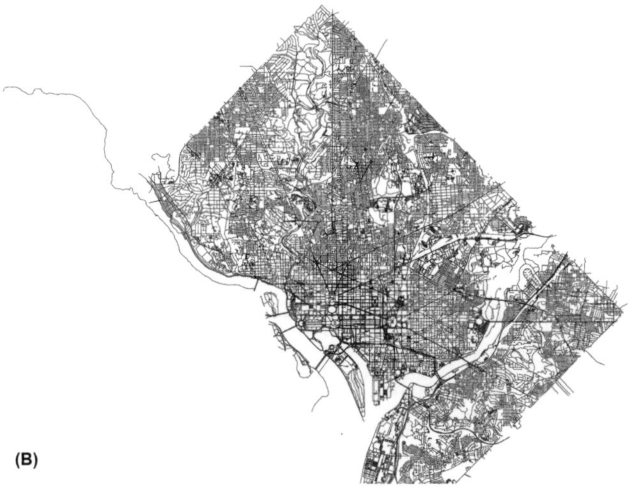
(B)

FIGURE 4.5.1B Washington DC – quadrilateral.

Background and Origins

Washington in the District of Columbia (DC), as the federal capital of the United States of America, is located on the Potomac River by the branch Anacostia River adjoining Maryland and Virginia. It was a swampy environment when first settled by George Washington at the time of the American Revolution.

After independence in 1776, for several years during the 1780s, there was no permanent home for the US Congress, which was moved around between Philadelphia, Princeton, Annapolis and New York. There was a plan for a dual capital in Georgetown, Virginia, and Delaware, and several other sites were suggested before an agreement was reached after a constitutional convention between the Northern and Southern states to site the capital at a virgin area of ten miles square, laid out as a diamond, by the Potomac River. The exact site was to be determined by three commissioners who narrowed the search to an area stretching between Georgetown and Alexandria in Virginia and Anacostia along the eastern branch of the Potomac.

Andrew Ellicott was commissioned to undertake field surveys and Pierre l'Enfant to identify areas for development that included the small towns of Hamburg and Carrollsburg below Georgetown, already platted for development. Accordingly, in 1791, President Washington confirmed to Jefferson the proposed siting of the national Capitol along Rock Creek and directed l'Enfant to prepare a plan for its development. He identified Jenkins Hill as the proposed location for the capital due to its suitable topography.

From this he laid out a grid of streets overlaid with key diagonal avenues, having a 'reciprocity of sight' between key landmarks, principally the Presidential Palace and Federal House, with the main avenue running on a north-west axis to connect the White House with the Capitol. The other feature was to be a canal developed between the Potomac and East Branch, around which development would occur, first heading east-west towards the Capitol and then turning south-north towards East Branch.

The other features were to be 15 main squares, designated for each state of the union, embellished with an obelisk at key nodes where the diagonal avenues intersect the streets, seemingly using the application of the golden section, or golden ratio, principle. It was divided essentially into sectors with a regular north-south and east-west grid of streets (numbered numerically and named alphabetically respectively) crossed by a number of grand diagonal avenues named after individual states.

The principal grand avenue of the National Mall was created with parallel gardens linked by key monuments to the Capitol. Much of this was influenced by l'Enfant's knowledge of the Gardens of Versailles created by Le Notre and the Avenue des Champs Elysées and Place de l'Etoile in Paris. Finally dedicated as the City of Washington in the Territory of Columbia, this came to be formalised in the Ellicott Plan published in 1792.

Accordingly, the prime axes became Pennsylvania Avenue that links the grand neoclassic US National Capitol running north-west with the President's House, the smaller White House by Lafayette Square, and Constitution Avenue running parallel to the National Mall. The main state institutions, the House of Representatives and the Senate, and the Supreme Court were grouped around Capitol Hill where the Capitol Dome was not completed until 1863, with the later Library of Congress and main east-west axis of the National Mall.

These public buildings were grouped in three areas in the capital triangle between the National Mall and Pennsylvania Avenue, the White House and Lafayette Square. All of this generated the formal plan that defined the subsequent development of the federal capital city.

In 1900, the centennial was celebrated of the founding of Washington as the seat of the national government. Plans were made for enhancing the central park area around the public buildings and a Senate Park Commission was appointed, overseen by Senator McMillan with Burnham, Olmsted, McKim and Moore. They reported with a plan to refashion the National Mall with a strengthened axis to the Capitol and featuring the Washington Monument. To the west it was proposed to extend access of the Mall to a new Lincoln Memorial along with a shallow reflecting pool.

The Reflecting Pool and the Lincoln Memorial were built and now connect with subsequently erected war memorials. This results in a fine and rather triumphal space. On the boulevard alongside the Mall are a number of national museums that have been built and institutions such as the Smithsonian.

The other axes comprise the diagonal avenues of Massachusetts and Connecticut Avenues, and New York and Rhode Island Avenues that create intersections at the traffic circles of Washington, Dupont and Logan Circles. These cross different districts that include Penn Quarter and the central Federal Triangle, and the cosmopolitan Adams Morgan district.

Modern City – Buildings and Arteries

In the commercial centre of Washington besides the federal government administration, there are now a number of major international institutional headquarters, such as the World Bank, the IMF and other financial institutions. A large number of foreign embassies are also located off Massachusetts Avenue.

Towards the west is Foggy Bottom where George Washington University is located, and the Kennedy Centre for the Performing Arts and the infamous Watergate Complex are nearby. Across Rock Creek sits the old city of Georgetown, which still retains its original character on the edge of the city, with its small, colourful row houses and dwellings, the old Chesapeake and Ohio Canal and with Georgetown University and Dunbarton Oaks Park nearby.

Externally the city is surrounded by the Capital Beltway, a major highway that runs around the city and connects with the interstate highways. A new modern metro has been built deep under the centre that also runs out to Dulles International Airport (IAD), and the main railway at Union Station is the terminus for lines that connect with the suburbs and other cities along the Eastern Seaboard with higher-speed trains since the original 'Baltimore & Ohio Railroad'.

Ronald Reagan Washington National Airport (DCA) is located nearby on the other side of the Potomac River and just beyond is the Pentagon, the symbolic home of the US Department of Defense and Arlington National Cemetery, besides a network of interstate highways.

The city is well laid out, spacious and expansive, in keeping with its symbol as the national capital of a major country. As the seat of government, it has situated here all of the great institutions of state, the executive, legislature and judiciary, and also offers many cultural additions of a modern federal capital, such as the National Air and Space Museum, the National Art Museum and the Natural History Museum.

Whilst the external veneer to visitors is commanding, there are several inner areas downtown with an underclass that has a high percentage of poorer black and immigrant populations. For them the central city is less accommodating, as are the fringes of many large modern American metropolises, in contrast with the mainly wealthier outer suburbs.

Greater Washington

Overall, Washington has become more decentralised beyond the National Capitol and the central area which is circumscribed by the outer Beltway. The Pentagon, national institutions and governmental lobbying lobbies are generally found within the Beltway, but private independent contractors, defence manufacturers and IT companies have gravitated outside and around the capital in neighbouring Fairfax County, with huge data centre capacity in northern Virginia.

These zones generally stretch out in two directions, west along the Dulles corridor towards Dulles International Airport and north-east along the Baltimore–Washington Parkway that ultimately connects with Baltimore Washington International Airport (BWI). This region, with its tendency towards becoming an 'aerotropolis', is developing into a significant metropolitan region with a large population and which is acting as a major zone of employment. It is now reaching out also towards connecting with New Jersey and New York and the extended North-Eastern 'Megalopolitan' Seaboard.

The population of the capital city is approximately 0.7 million and 6.4 million in the metropolitan area that stretches out into the Greater Washington Area.

Transactional Forces and Typology

The early selection of the site of the federal capital was adopted for administrative purposes from a virgin site on the Eastern Seaboard that would be well sited for the formation of the newly 'united' states. With its estuarine location and planned geometric layout, it was to serve as an accessible and symbolic centre for the newly constituted federation.

Since its realisation, Washington DC has consolidated and grown as the executive, administrative and governmental centre with the Capitol, which in turn has attracted both domestic and international institutions to gravitate around it so as to have representation in this increasingly powerful capital.

Most recently development has continued outside the DC capital and in neighbouring Fairfax County with the formation of an 'edge city' at Tysons Corner developing along the circumferential Beltway. Here with new office parks and supporting businesses gravitating to this area, this has transformed the pattern of development of the peripheral outer suburbs.

As a new planned city, the location of the capital was shaped by the topography in its original siting along the Potomac River and its Eastern Branch. Whilst maritime access was important, the planned development being both strategic and symbolic, a base orthogonal grid of streets was laid out, overlaid with a series of diagonal avenues. Within this, and its key features of a series of main squares, this determined the essential layout of the city. Upon this were sited the principal buildings of the Capitol, Federal House and President's House in an inter-related manner.

The key features of the streets and the main avenues were aligned according to planned axes that were laid out across the existing topography in a classical manner. This typology was generally European in manner and reflected the background of its French designer l'Enfant. With subsequent adjustments and refinements, the insertion of symbolic monuments and the creation of the National Mall and accompanying lake, the original layout has been substantially retained and elaborated upon.

Despite its estuarine origins, the planned city of Washington is strictly geometric with a north-south orthogonal grid of streets and buildings, overlaid with distinctive diagonal avenues that are inter-connected at traffic circles so as to create different quadrants radiating out from the Capitol. Now encircled by the circumferential Beltway, growth has led to the formation of a number of 'edge cities' externally, creating new poly-nodal structures on the periphery outside the original centre.

4.5.2 NEW DELHI, INDIA

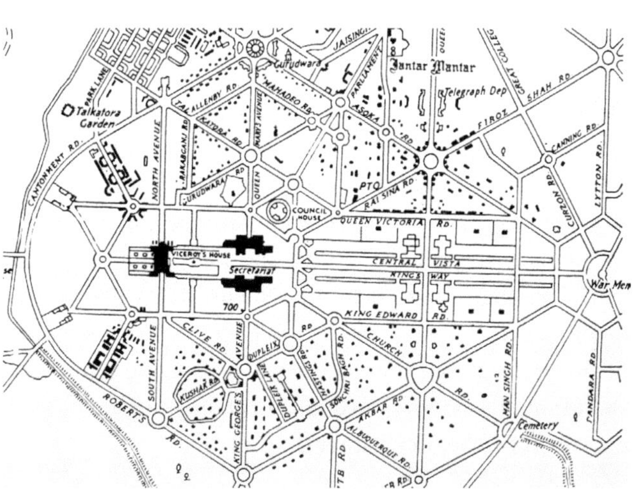

FIGURE 4.5.2 New Delhi – early configuration, 1916.
Source: British Library.

Background and Origins

At the turn of the 20th century, India was under the control of the British and had Calcutta as its capital. With difficulties in Bengal, Delhi was selected as being a better location to administer northern India as the new capital of India. Accordingly, a site on Delhi Ridge was selected by the British and intended for developing the city of New Delhi from where its empire could be managed more effectively. With the implantation of a ceremonial government house on Raisina Hill, from here the principal existing monuments and the River Yamuna of Old Delhi could be seen, which formerly had been the centre in the days of the Delhi Sultanate, or as Shahjahanabad with Chandni Chowk and Lal Qila (Red Fort).

New City – Plan

The British architect Edwin Lutyens was appointed, supported by landscape architect Gertrude Jekyll, to develop the plan, with Herbert Baker to design the Central Secretariat buildings, for a new capital that was to be 'imperial', with a blend of 'both east and west', or some form of 'Indo-Saracenic' architectural style. The foundation stone for New Delhi was laid in 1911 during the Delhi Durbar by King George VI.

As the seat of the government of India, Lutyens proposed a layout of hexagonal road patterns, with linking tree-lined avenues and boulevards. Kingsway, since Rajpath, was designed to connect the Viceroy's House to the west at Raisina Hill, through the middle of the North and South Block Secretariats, down to the war memorial India Gate. This was intersected at right angles by Queensway, now Janpath. Under these broad principles, the new capital was laid out and was inaugurated by the royal viceroy in 1931 with India Gate at the eastern end astride the main axis of the ceremonial Rajpath. Monuments were used to terminate the axes of imperial Delhi.

Besides monumentality, the entire layout had the feel of the 'garden city' planning movement, with diagonals and added climatic requirements for shade and ventilation and protection from the monsoon rains, for which particular design solutions were adopted on buildings such as verandas and window grills ('jaali'), with selected roof patterns ('chattri') and cornices ('chujja').

With its different sectors, administrative, legislative and residential, varied forms were proposed for the distinctive secretariat buildings, a separate residential Bungalow Zone, and for general housing, which was allocated to persons according to a 'warrant of precedence' where gradually government staff were moved from the Old Secretariat in Old Delhi.

The most imposing new edifices built were the Council House for Parliament and the Viceroy's House with their iconic roof styles in red and buff sandstone colour. The interiors were decorated with different coloured marble variously from Jodhpur (white), Baroda (green), Gaza (black), Alwa (pink) and Jaisalmer (yellow) that Mughal emperors formerly had employed.

Given the intentions for the new city, some significant developments were, however, to impact the intended governmental organisation. In particular, in 1935, the Government of India Act proposed an amended assembly to introduce representatives of both provincial regions and princely states.

With World War II and its enormous sacrifices, the world had evolved and religious tensions came to a head and were exacerbated between Muslims, Hindus and Sikhs. The imperial strategy was no longer tenable, and after British withdrawal was to lead to drastic partition of the country. In August 1947, India attained its independence to become the Republic of India as a self-governing independent country.

Partition had led to the creation of the separate Dominion of Pakistan (derived from the regions of Punjab, Afghan, Kashmir, Sindh and Baluchistan), and the Indian Punjab was divided in the west and Bengal was divided in the east. All of this was to result in significant movements of populations of differing religious faiths to different areas.

Modern City – Buildings and Arteries

From its colonial layout and beginnings, the city of New Delhi has grown extensively in all directions. The main axes remain as the original 'Rajpath' stretching from 'Rashtrapati Bhawan' (President's Palace) on Raisina Hill to India Gate with its perpendicular 'Janpath' leading to 'Indira Chowk' (Connaught Circus). It retains the administrative institutions in the central area and national institutions in the capital area. Beyond the President's Office and National Assembly, there are the historic Lodi Gardens and Buddha Jayanti Park and the Ridge beyond.

The first extension of New Delhi was developed in 1950 when the Central Public Works Department (CPWD) developed a new diplomatic enclave of Chanakyapuri along a new boulevard named Shanti Path (Path of Peace).

Administratively, in 1966 Delhi was converted into a union territory with a chief commissioner under the union government. In 1991, the National Capital Territory (NCT) of Delhi was formed to comprise the original central New Delhi and Delhi Cantonment areas plus the new extended municipal districts of North Delhi, South Delhi and East Delhi. These districts were further redrawn into 11 geographical districts in 2012.

The city of Delhi was a major junction on the established railway network with the Northern Railway and Delhi Southern Railway. It has five main railway stations at Old Delhi, New Delhi, Nizamuddin, Anand Vihar and Sarai Rohilla and an extensive network of commuter railway lines.

In terms of road transport, Delhi now has an Inner Ring Road and and Outer Ring Road and a dense network of connecting streets and roads with interconnecting national highways. The new metro was commenced in 1998 and now comprises an integrated system of ten lines and multiple stations with both elevated and underground lines, standard and broad-gauge track.

The modern city is growing into a highly dense populated city with the inter-related challenges of congestion, pollution and sanitation. With continuing in-migration there are large areas of informal settlements, and many micro-enterprises have developed, for example at the Mandoli suburb that contains an entire workforce focusing on the recycling of computer parts. New Delhi has now grown to comprise the enormous megalopolis that is considered the National Capital Territory (NCT) and the National Capital Region (NCR) which contains large urbanised parts of Delhi and also the surrounding metropolitan area that extends into the neighbouring Haryana and Uttar Pradesh states.

New Delhi itself has approximately 2 million inhabitants in its central area, but the population of all of Delhi is approximately 11 million and of the NCT is approximately 17 million. In considering the wider NCR it is estimated that this now contains up to 40 million people in Greater Delhi. This population growth and increase in vehicular traffic has had severe environmental impacts with poor air quality exacerbated by rural crop burning. On many days of the year air pollution far exceeds acceptable levels. The evolution of modern Greater Delhi and its surroundings is described further in Chapter 6.4.1.

Transactional Forces and Typology

With the decision to move the capital city of British India from Calcutta to Delhi in 1911, the initial mandate for New Delhi was to site and develop an imperial city that would symbolise its significance and monumental scale. To be located in an historic city which was formerly head of the Delhi Sultanate, New Delhi was to be sited to the south of Old Delhi with its fortified walls by the Yamuna River with an axial layout quite separate from the native city and the civil lines and cantonment for military use.

With its main governmental centre planned to be situated on the slightly rising land of Raisina Hill, the main axis was to run eastwards towards the ancient capital of Indraprastha along a central vista, with a perpendicular north-south axis and radiating roads on a hexagonal basis. On a virgin site, this infrastructure plan was complemented by a new central railway station and electric power lines and requisite water supply and sanitation routes. Monumental and distinctive in style, with its pink sandstone material and Indo-Saracenic features, this was to be a substantial and symbolic imperial capital city for administrative and control purposes.

As a planned new administrative centre following 'city beautiful' principles, New Delhi was to stand in stark contrast to the organic, densely populated Old Delhi. With the Legislative Chamber, Central Secretariat and Viceroy's House, this was to be the new imperial governmental centre with its supporting facilities. With some similarities with l'Enfant's plan for Washington DC, it would house the full complement of Indian civil service employees.

Whilst completed in the 1930s, it was unable to fulfil its imperial role for long, with World War II intervening and subsequent independence for India achieved in 1947. Redefined and restructured to reflect its new federal role, these governmental buildings have been transformed, streets have been renamed, functions have been reorganised and personnel have been reassigned. With the recent modern construction of a new triangular parliament building and projects for the redevelopment of the Central Vista and Kartavya Path, this area has evolved from its original purpose and yet still acts as the central governmental centre now for the extended federal government of India with Lok Sabha and Rajya Sabha, Lower and Upper Parliament Houses respectively.

4.5.3 CHANDIGARH, PUNJAB AND HARYANA STATES, INDIA

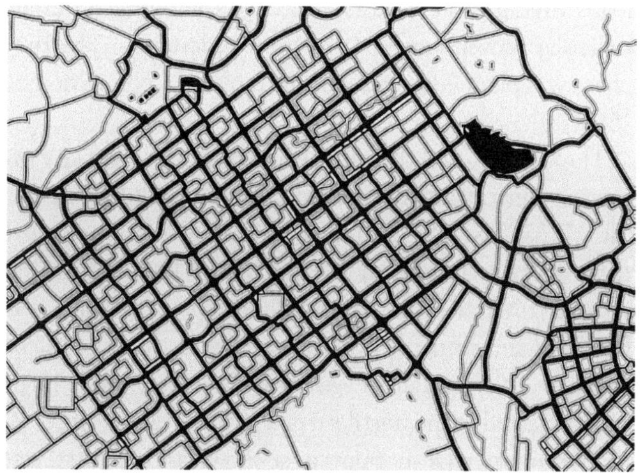

FIGURE 4.5.3 Chandigarh – early configuration, 1960.

Source: Le Corbusier.

Background and Origins

At the time of India's independence in August 1947 and the partition between the newly formed Pakistan and India along the Radcliffe line, the former Indian state of Punjab became divided with the previous state capital, Lahore, that fell into West Punjab, which was to become part of Pakistan. This was considered a considerable loss by India and left the need for a new capital for the state of East Punjab. As a result, Prime Minister Nehru made a political decision and commitment that an entirely new state capital should be created in the region that would act as a capital for the restructured state of Punjab in north-west India.

New City – Plan

The layout of a new state capital to be named **Chandigarh**, meaning 'Chandi fortress', was tasked to American planner Albert Mayer in 1949, on a site around a group of existing Punjabi villages. Mayer and his partner Nowicki developed a master plan based upon traditional 'city beautiful' principles with cellular neighbourhoods.

This work was assigned subsequently to Le Corbusier and his cousin Pierre Jeanneret, who with Jane Drew and Maxwell Fry continued to elaborate this as a new capital for the state of Punjab. Whilst Fry and Drew focused on the housing units and social facilities such as schools and hospitals, educational and recreational areas, Le Corbusier planned the Capitol complex. A highway grid was overlaid across the whole scheme, with a hierarchy of straightened roads dividing the sectors aimed at separating different vehicular movements (7 'V's) from pedestrians. The first phase was planned over 3,250 hectares for a population of 150,000.

The concept was the establishment of this new city on the northern plains not far from the Shivalik foothills of the Himalayas. The grid was divided into 56 principal rectangular sectors, each 1.2 km by 0.8 km, and having separate internal service roads. The Capitol complex was for the main government buildings comprising the major Executive Secretariat, Legislative Assembly and High Court located at the northern end as the 'head' of the city.

The commercial shopping area known as the 'heart' was planned around a central plaza, and adjoining this were green areas, the 'lungs', with gardens, streams and individual monuments surrounded by residential areas; the 'torso' with housing units; and green 'fingers' and highways as 'circulatory systems'.

As Le Corbusier focused on the design of the Capitol, he adopted trademark illustrations such as the 'open hand' welcoming sculpture that reflected his new 'machine age' approach, and the use of 'le Modulor' according to the adopted scale of a 'modular man' and golden section proportions, and in constructing the Tower of Shadows, Martyrs Memorial and Geometric Hill.

Modern City – Buildings and Arteries

With the foundation stone laid in 1952 and the essential initial phase completed by 1960, the population of Chandigarh began to increase, with the government acting as a major employer. This expanded with the creation of the new modern state of Haryana in 1966, when it was determined that Chandigarh should act as the combined capital for both of these adjoining states of the Punjab and Haryana administrations on common union territory. The most striking buildings in the Capitol complex are Le Corbusier's brutalist concrete Legislative Assembly and High Court buildings and the huge eight-storey Secretariat, 250 m. in length now divided in two for each state, with patterned grill facades (brise-soleils) and occasional walls of colour.

Whilst initially slow to develop, the new city was considered at the outset to have had excess circulation space. However, the increase in car ownership over the years has gradually filled the hierarchy of streets that serve each sector, and this layout works well with main boulevards for through traffic and separate internal service roads.

The whole site has matured after 50 years with established landscaping and a mix of ground coverage and grown trees that afford climatic protection in the summer heat. Its own University of Punjab has also been established in Sector 14 with its core buildings and Gandhi Bhawan designed by Jeanneret. With a population now of approximately 1.2 million people in this union territory, it has fulfilled its proposed role as a new joint state capital, with its own identity and reasonable proximity to Delhi, and within a natural environment which makes it a liveable place, despite its harsh brutalist design.

With the neighbouring areas of Mohali and Panchkula on the periphery, Mohali grew as an industrial estate and the grid pattern of development by sectors across Chandigarh has been extended here. With growing local markets, there are many areas of informal housing also that are present around the city. The current challenge now is to be able to manage additional low-income settlers that are continuing to migrate into the city. Overall, the larger metropolitan area now accommodates approximately 2 million people.

Transactional Forces and Typology

The driving force in the establishment of the new city was political, so as to fulfil the governmental functions in the provision and administration of a newly established Indian state and to be able to provide an identity and facilities for displaced persons. This had its own challenges with the migration of populations of mainly Hindu and Sikh religions who had to be accommodated.

With its structured infrastructure this has created its own attractiveness and pull forces with continued ongoing migration into the city. Peripheral informal unplanned areas have come under pressure in comparison with the original formally planned areas in the centre.

With its generally flat topography, the area in the northern plains has harsh winter and summer periods with extreme seasonal temperature differences. As a wholly planned new city, with its central administrative Capitol and orthogonal layout and individual industrial sectors, it has a highly structured layout and zoning plan with residential areas separated from industry which is being maintained. At the periphery, however, this contains unplanned areas that are evolving organically with large informal settlements and denser populations in the greater metropolitanised area.

The governmental centre or Capitol complex is unique in having been designed in a brutalist modern monumentalist style. Despite challenging the local Indian culture, it does afford the symbolic stature of a new state capital and provides modern residential and commercial facilities within a road-based transportation network.

Other New Indian State Capital Cities

In 1956, the States Reorganisation Act was passed in India, which led to the rationalisation of several States by language, in particular with the creation of Punjab and Haryana, Gujarat and Maharashtra, Andhra Pradesh and Telangana, Nagaland and Arunachal Pradesh in north-east Assam, each to have its own capital city.

For example, **Gandhinagar**, designed by PM Apte and HK Mewnda, was named after Mahatma Gandhi who was born there, just north of Ahmedabad on the Sabarmati River, as the new capital of the state of Gujarat, replacing Ahmedabad.

With decolonisation, new administrative facilities were developed with designs by Otto Konigsberger in **Bhubaneshwar**, with its ancient temples and Kalingan architecture, as the capital of the renamed state of Odisha.

The new state of Andhra Pradesh, with **Amaravati** as its capital, is also under construction on the banks of the Krishna River in the Guntur district. This was conceived with the assistance of the government of Singapore as master planner in order to be a sustainable city, but it ran into opposition from local landholders.

4.5.4 ISLAMABAD, PAKISTAN

Background and Origins

The Indus Valley civilisation flourished in the alluvial plains of the region around Harappa when the earliest ancient settlements were made around the period of the 20th century BCE. At this time, groups of baked-brick buildings were realised with relatively sophisticated supporting infrastructure. Later, from the 11th century CE onwards, there were periods of different Mongol tribes marauding in the region, including Chinggis Khan and Tamburlaine.

On the Pothohar plateau of the Punjab region there is an area of early Asian settlements near the surrounding Margalla Hills and Pass not far from the hill station of Murree. With high patterns of annual rainfall in the region, the groundwater accumulation in the area has been regulated by a series of artificially constructed reservoirs.

Much later in the 20th century, after the creation and independence of Pakistan in 1947, the new **Islamabad** Capital Territory of Pakistan was created and founded in 1960. With the plan to replace the port city of Karachi as the nation's federal capital, and with a location nearer to Kashmir and around the Indus River Valley, the capital was transferred to an area adjoining Rawalpindi near where the Pakistani Army is headquartered.

The newly developed area of Islamabad adjoins the older former cantonment area of Rawalpindi just to the south in Punjab Province. Together they form a larger metropolitan area whose activities are interconnected although remaining separate from Islamabad Capital Territory.

New City – Buildings and Arteries

The new city of Islamabad was designed by Constantinos Doxiadis as a planned 'dynapolis' with the city divided into five zones, two principally for urban development. Here within a larger area with different alphabetically named sectors in a grid pattern numbered from E7 to I18 comprise residential areas, the university, software technology parks and commercial, health and education sectors. The main axis is Constitution Avenue, which contains the National Assembly and most of the governmental institutions in the Secretariat complex.

The modern Pakistan Monument landmark, representing the four Punjabi, Pakhtun, Sindhi and Balochi cultures, was inaugurated in 2007, and the large modern striking Faisal Mosque was built. With its residential, commercial and industrial sectors, Islamabad is a contrasting modern city in the region connected with new major highways to Peshawar and Rawalpindi.

Now quite mature, the Islamabad Capital Territory although suffering a major earthquake in 2005, is well established as the new metropolitan capital in the rapidly urbanising country of Pakistan. With the establishment of the capital territory, it has become a place of attraction for many migrants from different regions across the country and of different cultural groups.

With its traditional businesses, crafts and textiles, light manufacturing has become established, along with new commercial areas and business parks with developing IT companies. Besides the National University of Science and Technology (NUST), there are a large number of new educational institutes and research institutes in order to train the growing workforce.

Transport is provided by the Rawalpindi Islamabad Metrobus on dedicated bus lines that run along the Islamabad Expressway and Srinagar Highway with Islamabad Railway

Station in Sector I9. Municipal services are provided and managed by the Islamabad Metropolitan Corporation. Islamabad International Airport (ISB) serves the capital and Rawalpindi.

The population of Islamabad is approximately 1 million with approximately 2 million and growing in the greater federal capital territory.

Transactional Forces and Typology

With its location beyond the Indus River between Multan and the hill station of Murree and towards the northern mountains of Kashmir, Kabul and Peshawar, Islamabad lies on old trading and mountain-crossing routes. Selected as a geographically suitable location for the capital territory of the emerging country of Pakistan, as a new city its modern layout and configuration stands out in otherwise traditional village settings of this northern region adjoining the Himalayas.

As a new formal city, sited and created in the north of the country, its modern characteristics and gridded sectoral layout differs from natural organic Punjabi towns and cities. Coupled next to traditional Rawalpindi, its role as an administrative centre with institutions and diplomatic quarter, residential and commercial areas are laid out in distinctive sectors. With its pull for new migrants, further adaptability is required from a rigid plan for supporting migrant workers and their activities along with the challenges of housing and informal settlements. This reflects similar pressures seen in the new state capital of Chandigarh in neighbouring Punjabi India.

4.5.5 CANBERRA, AUSTRALIA

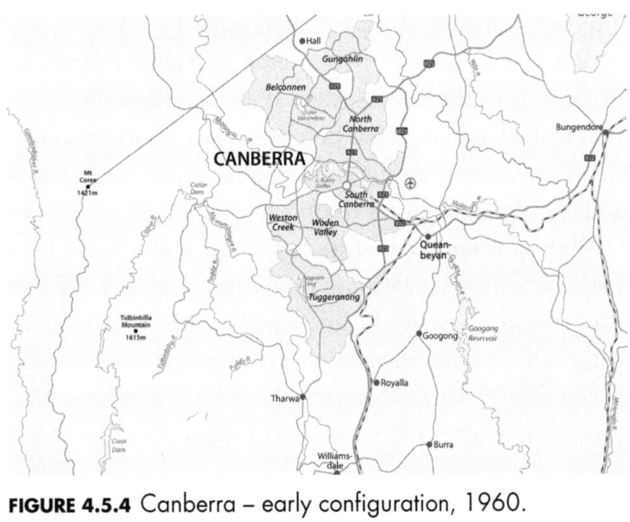

FIGURE 4.5.4 Canberra – early configuration, 1960.

Source: iStock.

Background and Origins

The original planning of a capital for the new Australian Capital Territory (ACT) began following the agreement in 1908 to establish a new seat of government in the Yass-Canberra region to be developed within an enclave of New South Wales between Sydney and Melbourne. This area is bounded by the watersheds of Nass Creek in the south and Cotter River in the west.

As a result of a competition, the design by Walter and Mahony Griffin in 1912 for a new capital on a virgin site at **Canberra** was selected. With the indigenous Ngunnawal in the area, 'Kamberra' is the Aboriginal name for meeting place – this was also an area of early English settlers and homesteads around 1850.

It was planned as a garden city to be laid out in a series of polycentric areas, to house the new federal government of Australia within the National Triangle to contain the main parliament buildings. The Provisional Parliament House was opened here in 1927 and with the damming of the Molonglo River, this building was used for the next 60 years.

Further expansion of the capital was overseen by the National Capital Development Committee, who reached an agreement in 1957 for the creation of artificial Lake Burley Griffin, and also the main principles of development that were to follow with a plan of linked urban centres connected by parkways with North and South Canberra.

The principal layout comprised a land axis aligned with Mount Ainslie running north-east with parliament and Capital Hill and a crossing water axis with the lake created by damming the Molonglo River. The central district was arranged around an equilateral triangle with three nodes of the federal government centre, the civic centre and the commercial centre connected by Constitution, Commonwealth and Kings avenues.

Modern City – Buildings and Arteries

The dominant Capitol was sited on a hill with the central executive, legislature and judiciary building either side. The main layout was planned as a series of polygonal areas on a radio-concentric basis linked by axial parkways to a municipal and market centre. In addition, there were to be separate suburban and residential districts planned, although not all were fully realised.

The basic layout was arranged on either side of the lake with Parliament House sited on one side on Capital Hill, in front of Red Hill, with a long axis of Commonwealth Avenue leading from Capital Circle to the city hall on the opposite City Hill aligned with the Ngunnawal hills in the distance.

Also, Anzac Park, flanked by eucalyptus trees, is aligned diagonally opposite, leading to Mount Ainslie, which forms the perpendicular of the triangle with the symbolic Australian War Memorial. The other axis, Kings Avenue, links with Market Centre across the lake.

But it was not until much more recently that a New Parliament House with its distinctive 80-m.-high flagpole was constructed on Capital Hill (designed by Mitchell Giurgola); it was opened by Queen Elizabeth II in 1988 and contains the main government executive buildings.

This building is surrounded by a series of circular and radial roads where most of the foreign embassies are also located. As a modern and a finely designed building, with furnishings made from local materials and having Aboriginal designs in the Great Hall, it contains both the House of Representatives and the Senate. The colour scheme is green in the House and red in the Senate, following the tradition of the British Houses of Parliament.

The grounds are finely landscaped, and the New Parliament House is elevated, overlooking the lower Old Parliament House. With the buildings and infrastructure now well established, and government institutions expanded, it has grown into a well-developed and resourced federal capital on a dedicated site with gradually expanding suburbs, such as in Molonglo Valley.

With its own Canberra Airport (CBR) and inner suburban railway station at Kingston, plans for high-speed rail connections to the national capital from Melbourne and Sydney have not yet been formalised. The population now is approximately 0.45 million people.

Transactional Forces and Typology

On a virgin site of formerly First Nations Ngunnawal inhabitants, the location of Canberra was selected principally to be approximately midway between the established competing conurbations of Sydney and Melbourne. In the hilly area and surrounded by natural parks, creeks and forests, it is attractively sited and formally planned, well connected by new highways in the south-east region of Australia. As a formal national capital, its built areas are essentially residential and institutional with some commercial but no industrial activity.

As a modern city, focused on governmental administration, it has matured following the original aspirations for the layout of the capital city, in a similar vein to those formal geometric principles as aspired to and established in Washington DC, New Delhi and also more recently in Brasília.

As an inland, formally planned, governmental territory in a natural landscape, the layout of Canberra was initially planned on 'garden city' principles around an artificial lake. Nestling among the hills, its role and design has provided the pull of development for governmental and diplomatic residents. As a modern national capital, with its administrative and symbolic role, it fulfils its role in a well-planned exemplary manner as a liveable green city with a marine climate, unaffected by typical modern population pressures of over-urbanisation and congestion.

4.5.6 BRASÍLIA, BRAZIL

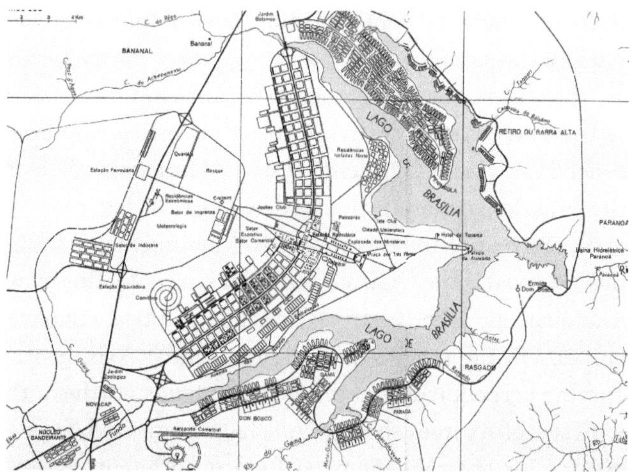

FIGURE 4.5.5 Brasília – early configuration, 1960.
Source: British Library.

Background and Origins

Brazil is an enormous tropical country which encompasses the entire Amazon River region. Whilst its coastal and inland characteristics differ greatly, as early as Brazil's independence, a new interior capital was planned in the country so as to move farther away from the concentrated south-east region. Initial surveys were undertaken and a new government agency, NOVACAP, was created to oversee its realisation.

With the goal of locating the capital away from Rio de Janeiro and nearer to the geographic centre of the country towards the highlands, once the political decision had been taken by President Kubitschek to move the federal capital to the centre of the country, design commissions were awarded and works commenced for the new capital of **Brasília** in 1956 on a selected virgin site. Whilst essentially designed as a governmental centre with an administrative role, economic activity developed in the area, not just with agriculture and food-processing activities, but also light industry.

New City – Buildings and Arteries

After a design competition, the intention was to build the infrastructure for the new city according to the cross-axial Pilot Plan of Lucio Costa. With its planned layout symbolically shaped like a cruciform, or aeroplane shaped in plan, the two main axes of the capital were laid out around artificial Lake Paranoá where the administrative institutions and buildings were to be located, designed by Oscar Niemeyer and with landscaping by Roberto B Marx.

The focal point for the government is defined by the twin vertical towers of the administration and with both the white dome and complementary bowl of the Congresso Nacional by its landscaped lake. At the end of the main axis are the striking modern concrete buildings in the Square of the Three Powers, with the Palace of Justice, Supreme Federal Court and Presidential Palace (Alvorada), much of the built design being influenced by Le Corbusier.

At one end of the east-west Eixo Monumental (Monumental Axis) is the Quartel General do Exercito, the army headquarters, with its strong lines and elliptical concrete curves. On the other axis is the Residential Axis, gently curving perpendicular to this which extends north-south with sets of six storey housing blocks or numbered 'Superquadra' for the resident population in different sectors. One other iconic building is the separate Metropolitan Cathedral, with its curved concrete ribs shaped as a crown of thorns and with sunken floor.

Built in the 'cerrado' with teams of 'candangos', workers from the north-east of Brazil, the earliest built areas such as Taguatinga that were started as a construction camp became popular settlements which have since grown and become established around the city, despite efforts by the authorities to remove them.

Whilst it has taken over 40 years to become fully established as a city, with all of the federal government machinery, international embassies and supporting population, it is now a fully functional, modern, formally planned centre, although very different from other denser, unplanned areas in other regional Brazilian metropolises.

Transportation is focused essentially on roads and cars, with bus transport along the two main axes which have the Rodoviaria Central Bus Station at their intersection. The multi-span arched Juscelino Kubitshek Bridge crosses the artificial Lake Paranoá ceremoniously here. The Federal District also has a modern metro system with two lines and Brasília Juscelino Kubitshek International Airport (BSB) is located to the south of the city.

The population of the main Federal District is approximately 0.5 million, and the overall population of the city is approximately 2.8 million, an attraction for new migrants.

Transactional Forces and Typology

As a formally planned new capital city, Brasília was implanted purposely in a more central location of the country, with the aim of rebalancing existing growth that has concentrated in the south-east region around São Paulo. As a governmental centre, it attracts administrative supporting services and personnel into the Federal District. As the city has matured and varying supplemental activities have also developed, the pull for migrants has broadened the economy into wider sectors and varied services as its population has increased. Whilst taking several decades to become fully established, it is now the political hub of the federal states of Brazil.

With its axial layout and gridded sectors, the new planned city is implanted around its originally defined sectors and organised Federal District. Whilst quite rigid in plan, with purposefully zoned sectors, in contrast at the periphery newer settlements have emerged in an unplanned manner with a developing informal sector. As a distinctive new Federal Capital, it follows the pedigree of Washington DC and Canberra with their axial arrangements on new virgin territory, but with more severe Le Corbusian modernistic design aspirations under a challenging tropical savannah climate.

4.5.7 ABUJA, NIGERIA

Background and Origins

With increasing congestion and population growth in the coastal city of Lagos and with regional ethnic tensions, the policy was adopted of developing an entirely new federal capital in the geographic centre of the country of Nigeria in an area formerly occupied by the Gbari people.

After detailed research, the master plan for the new city of Abuja was published in 1979 to comprise a new federal capital of Nigeria near Suleja. The plan was developed by Planning Research Corp (PRC); Wallace, McHarg, Roberts and Todd; and Archisystems, with more detailed design of the core by Kenzo Tange and Urtec.

The city was developed in phases, with Phases 1 and 2 in the central core with the Central Business District (CBD) stretching from Aso Rock in the east to the Old City Gate in the west. The Central District contains the main federal government institutions, the Senate and Eagle Square. Phase 1 development included ten cadastral zones, such as the Garki, Wuse, Asokoro, Maitama and Guzape districts. Phase 2 included another 16 cadastral zones and further 13 cadastral zones in Phase 3, sub-divided into areas all with a grid pattern of roads and streets. Beyond, the city has extended outwards, in clusters of satellite settlements that comprise commercial and industrial zones.

Overall, the Federal Capital Territory comprises 7,300 sq. km. and comprises six local government areas managed by the Federal Capital Development Authority. From its dense core, Abuja has been sprawling outwards since the rival of migrant peoples, with new settlements developing in the peri-urban areas.

New City – Buildings and Arteries

The newly evolving built city of **Abuja**, located in the centre and tropical rainforested plateau of Nigeria, was finally incorporated as the new Federal Capital Territory (FCT) of Nigeria in 1991. As a totally new planned city, its metropolitan area has become established and grown very quickly.

The city has been divided into distinct areas with several satellite districts in different cadastral zones and developed in phases. The main landmarks are the Presidential Complex, Senate, Central Bank, Eagle Square, National Mosque and National Ecumenical Centre located in an area between the stunning natural Aso and Zuma Rocks. The federal ministries are located together along with the park and several new universities, and the Millenium Tower (designed by Nicoletti).

Already well established, there have been many people migrating to the new capital in order to establish commercial and business activities. The population of the city now is approximately 1.25 million and the metropolitan area has already reached 6 million. With new highways and connections, it has relatively quickly established itself as the centralised capital of this large and fast-growing country.

In terms of transport, the city is surrounded by the Outer Northern and Outer Southern Expressways (ONEX and OSEX) and connecting highways with Kaduna and Lokoja. A new long railway line is planned between Lagos and Ibadan to be extended north to Kaduna (standard gauge) to replace the former old Cape gauge railway to Kaduna.

A new section has been built to connect Abuja with Kaduna under Chinese finance since it is not located on the main rail network. Other main lines connect with Port Harcourt and Warri and the iron mines around Itakpe.

Abuja International Airport Nnamdi Azikiwe (ABV) is at the edge of the city, whilst the light rail connection has been terminated there. A new district, Centenary City, is under development at the edge of the capital territory as a free trade zone that celebrated the centenary of Nigeria in 2014 with a signature tower.

The region is tropical, with agricultural activities generally in raising maize, sorghum and groundnuts. It is also the source of several mineral-rich materials. With its fast growing population as a Central African city, it has many informal 'Bush bars' that are a feature of the new city.

Transactional Forces and Typology

In a primarily forested tropical region, the FCT area lies to the east of the River Niger and does not benefit from immediate natural transport routes. Subsequent to its planning, installations and development, new infrastructure connections have had to be constructed to provide links both inland and to the coast. Being outside a region of rich natural resources, the city has been established for political purposes as an administrative capital of the federal states. Since its recent realisation, it has prospered by attracting new migrants to the area, which is providing its own challenges with the pull of work prospects leading to satellite development and informal settlements.

As a new planned inland capital territory, Abuja has been developed in accordance with a defined master plan in a similar manner to Canberra and Brasília, although with less striking landmark buildings or building types. As a gridded city, with a central spine of Independence and Constitution avenues leading to the Three Arms Zone of the executive, legislature and judiciary, its districts are now becoming overdeveloped and in turn surrounded by informal settlements creating satellite areas. From its original urban core, new poly-nodal and peri-urban developments are surrounding the city whose infrastructure is being challenged by the attraction of new residents.

4.5.8 OTHER NEW ASIAN CAPITALS

Naypyidaw, Myanmar (Burma)

The site of **Naypyidaw** is located roughly midway between the former capital of Yangon and Mandalay in the centre of Myanmar near the existing city of Pyinmana. As a former Kingdom of Pyu culture, it has a long history of Buddhist Burmese culture that suffered Qing invasions during the Sino-Burmese War and was involved in the Anglo-Burmese War during the 19th century. After occupation by the British, it was also overrun by the Japanese in World War II, becoming independent post-war in 1948 as the Union of Burma. However, its more recent history has been shaped by a series of ongoing civil wars.

In a flat area between mountain ranges, it was chosen as being more in the geographic centre of the country. Just to the east of the north-south-running Yangon-Mandalay Highway, the new National Assembly area is located on an axis running south-west to north-east. The Naypyidaw Union Territory comprises Ottera and Dekkhina districts and a number of existing townships.

New Capital Territory

Naypyidaw as the new capital of Myanmar is located centrally as the Naypyidaw Union Territory. It replaced Yangon (Rangoon) as the administrative capital in 2005 and acts as the seat of the government with parliament and its ministries. As a greenfield site, it was built up over the decade 2002–12, more centrally located within the country than the old capital and less exposed to cyclones than its former coastal location. The city is divided into different zones:

- Presidential Palace, parliamentary complex, ministries and city hall, military base and zone, diplomatic zone and embassies, residential zones for government employees, hotel zone and villas, commercial shopping zone and recreational zone, lake and park.

Naypyidaw has a central area surrounded by a number of townships. With a new railway station at Naypyidaw, it supplements Pyinmana on the main Yangon–Mandalay railway line parallel to the expressway on the same route. With its International Airport, Nay Pyi Taw (NYT), it has improved connections, although limited public transport and social infrastructure within the city. Whilst still a new and young city, the current population of approximately 1 million people is forecast to grow significantly.

Transactional Forces and Typology

At a central location by the Shan, Kayah and Kayin states, the motivation was entirely political for the relocation of the capital. In a hot and hilly inland area, north of Yangon and the Irrawaddy River, it is gradually becoming established as the administrative capital of the country, with its institutions and supporting agencies.

As a new planned capital, Naypyidaw is divided into different functional areas with a gridded central core and surrounding nodal centres. Its union parliament buildings serve as a focal core of the new city along with its ministries and the diplomatic zone. Surrounded by new residential and commercial areas, it is seeking also to retain its historic culture.

Astana, Kazakhstan

The city named **Astana** (formerly Akmoly and Nur-Sultan) is located in the centre of Kazakhstan on the Ishim River in a semi-arid steppe region. As a location on the great Silk Route, it was first settled around 1830 by Siberian Cossacks where a fortress was built by the river. In 1869, it became the centre of Akmolinsk Oblast and gradually developed on local trading routes. During the Soviet era, it came under the Kazakh Soviet Socialist Republic and armament industries were developed.

After World War II, its surrounding territory was encompassed in the Virgin Lands Campaign to increase grain and agricultural production and was renamed **Tselinograd**. Further development took place along with Soviet housing schemes until the collapse of the Soviet empire, and the city was renamed **Akmola**.

In 1998, the city was declared as the new capital of the Republic of Kazakhstan in the centre of the country to replace Almaty, which is located off-centre in the mountainous south-east. It was renamed **Astana** with extensive new developments planned under the design of Japanese architect Kurokawa. In 2019, when the president resigned, it was briefly renamed

Nur-Sultan after the former president Nazarbayev, but the city name was subsequently restored to **Astana**.

The economy is based on trade and light industry with State-owned companies, and since the city's nomination as the capital, it has acted as a catalyst for new development. With new investors in a newly created special economic zone and Astana International Financial Centre, small and medium businesses have increased and there has been additional in-migration to the capital.

The city comprises four districts that were created by presidential decree – Almaty, Saiyarka, Yesil, on the left bank of the River Ishim, and Baykonyr. Kurokawa's plan was both to enhance the existing city and also to expand to the south and east sides of the Ishim River. With a new dedicated governmental headquarters and diplomatic quarter, the Palace of Peace and Reconciliation Congress Hall was built (designed by Foster) in 2006 for mixed religions and new universities and colleges.

Several headquarters of State-run corporations have been built in the city and special new buildings were also built for Expo 2017, which focused on future energy, with the spherical National Space Centre and Central Concert Hall. With a grand new railway station, Nurly Zhol, it has a high-speed rail connection to Almaty. The new Nursultan Nazarbayev International Airport designed by Kurokawa is located to the south of the city. The population of Astana has now reached approximately 1.3 million people and is still growing with the Astana Grand Mosque built in the city centre.

Transactional Forces and Typology

With its location in the centre of the steppes of Kazakhstan, the region is rich in natural resources, minerals and oil and gas. From its former trading origins, the country is now focused on the exploitation of its rich petroleum and gas reserves and exportation via pipeline for refining. This has led to the development of major State industries responsible for their management and development.

The rationale of the decision was to relocate the capital as a city of 'State importance' in the geographical centre of the country. As a central, landlocked, planned city, Astana is laid out with a central core and radiating road axes in all directions. With a gridded layout on either side of the Ishim River, it is divided into four districts: Almaty, Yesil, Saryarka and Baykonr. Its symbolic centre has been created with a new governmental headquarters alongside the river and with Independence Square and its tall, symbolic column. It is also serving as a key location under the Chinese Belt and Road Initiative.

Nusantara, East Kalimantan, Indonesia

With its many varied islands, Indonesia is a highly populated country, especially on the islands of Java and Sumatra. With less population on the other islands of the Indonesian archipelago, the decision was taken in 2019 to build a new, more centrally located capital in East Kalimantan on the island of Borneo and to relocate the government institutions there, to be developed in phases under a ten-year plan. A region of tropical rainforests, it was formerly the home of ancient Hindu kingdoms and later became part of the Dutch East Indies in the 18th century.

New Capital Territory

Due to the congestion and over-population on the main island of Java and at Jakarta, the government has planned a new capital city being built at **Nusantara** (Ibu Kota Nusantara – IKN) in East Kalimantan on the border of the Kutai Kartanegara and Penjaman North Paser regions, between the cities of Samarinda and Balikpapan on the island of Borneo in the Makassar Strait of Indonesia.

Its name embodies the national vision known as Wawasan Nusantara, or 'Vision of the Indonesian Archipelago for 2045', as foreseen by the Ministry of National Development Planning of Indonesia (Bappenas) in order to reduce the burden on Jakarta as a national hub.

With the creation of a Capital City Authority, a design contest was held for the master planning and design of the key governmental administration buildings. The city is to be developed in phases, with the first phase focusing on the provision of basic infrastructure and utilities, followed by the development of the core centre with the central government area and ministries and the president's palace.

As a coastal location with a Central Capital Region and Extended Capital Region, new infrastructure will be built that will include port and harbour facilities, a new International Airport at Penagam, and transport interconnections with the Trans-Borneo Railway and Balikpapan Highway.

Transactional Forces and Typology

The creation of this new capital city in Kalimantan is expressly designed as a policy measure to rebalance the development forces applying on the existing capital of Jakarta in Java which is suffering from a high rate of growth in population and also having significant environmental challenges with sinking ground levels and rising sea levels.

As a major development more central within the Indonesian archipelago, this seeks to counteract excessive growth forces with new development in the less populated East Kalimantan region in the island of Borneo.

As a coastal city, planned on a virgin site, the Central Capital Region is to contain the Governmental and Administrative Centre. With a port area adjacent to the Teleri River Delta, the Extended Capital Region is planned to accommodate the currently resident population.

Putrajaya, Malaysia

Putrajaya was established as the administrative and judicial capital of Malaysia in 1999 as an addition to the legislative capital, Kuala Lumpur, which remains the national capital of Malaysia. As the Federal Territory of Putrajaya, it is on the land of a former rubber plantation in the state of Selangor, designed with 20 precincts around a series of lakes with its main axis of Persiaran Perdana. Most of the government ministries are now located here, with the Federal Court and the Ministry of Finance.

Closely connected by road and rail with the adjacent capital of Kuala Lumpur, it has grown to approximately 0.1 million people, adjoining the expanding area of Cyberjaya to the west and neighbouring Kampungs.

New Administrative Capital

With the significant pressures and existing congestion in the area around Kuala Lumpur, Putrajaya was selected to act as a new administrative centre of government with the main ministries relocated here. Sited along with MSC Malaysia, a special economic zone that covers much of the Klang River Valley, new infrastructure was developed for institutional precincts and new governmental facilities.

Carefully planned around a series of lakes, Precinct 1 contains the prime minister's office and Putra Square and mosque. Other precincts are linked by bridges towards the Ministry of Finance and the Ministry of Foreign Affairs, leading towards the International Conference Centre (ICC). Putra Sentral is the main transport hub, with the mass transit MRT to the west of the main axis and connected by a surrounding network of expressways.

Transactional Forces and Typology

Although adjacent to the existing city of Kuala Lumpur, the relocation of the administrative centre to a purpose-built site around a series of lakes in the Klang Valley provides relief to the very congested urban centre.

Laid out as a landscaped series of precincts around the river and lakes, this area provides a dedicated site for the Governmental Administrative Centre, with Kuala Lumpur remaining the national capital of Malaysia. With modern infrastructure and facilities, it is designed with 'intelligent city' features in the midst of an already existing major conurbation. The following figures illustrate the modern configuration of the new capital cities as described and the evolution of their modern population.

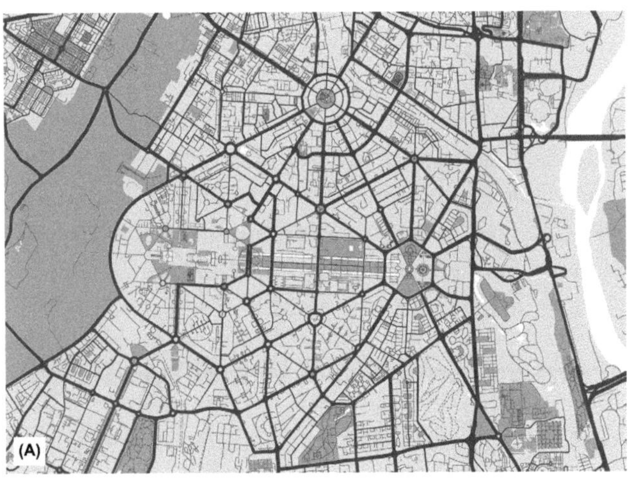

FIGURE 4.5.6A New Delhi – modern configuration.

FIGURE 4.5.6B Greater Delhi – modern configuration.

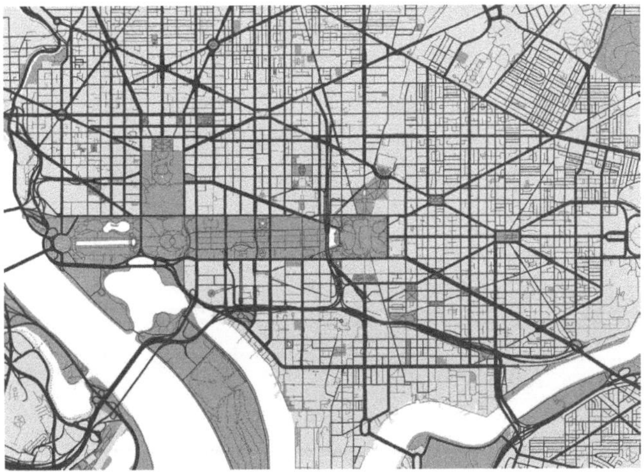
FIGURE 4.5.7 Washington DC – modern configuration.

FIGURE 4.5.8 Chandigarh – modern configuration.

FIGURE 4.5.9 Islamabad – modern configuration.

FIGURE 4.5.10 Canberra – modern configuration.

FIGURE 4.5.11 Brasília – modern configuration.

FIGURE 4.5.12 Abuja – modern configuration.

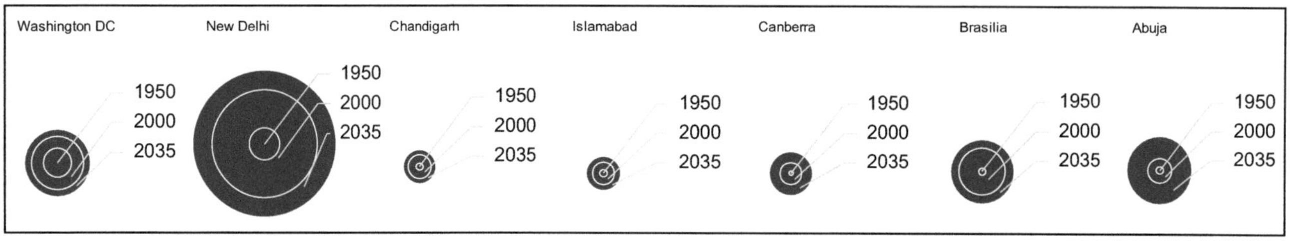

FIGURE 4.5.13 New capital city populations.

Data source: Global Database of Metropolises 2020 – UN-Habitat (2020), UN DESA Population Division.

Further Reading

WASHINGTON DC

Allen, William C – *History of the United States Capitol*, US Government Printing Office, 2001.

Herron, Paul – *The Story of Capitol Hill*, Van Rees Press, 1963.

Mann, Nicholas – *The Sacred Geometry of Washington DC*, Barnes & Noble, 2007.

Reps, John W – *Monumental Washington: The Planning and Development of the Capital Center*, Princeton University Press, 1967.

Seale, William – *The White House, the History of an American Idea*, American Institute of Architects Press, 1992.

DELHI

Collins, Larry & Lapierre, Dominique – *Freedom at Midnight*, Vikas Publishing House, 1997.

Irving, Robert – *Indian Summer, Lutyens, Baker and Imperial Delhi*, Yale University Press, 1983.

Michell, George – *The Royal Palaces of India*, Thames & Hudson, 1999.

Tillotson, GHR – *Tradition of Indian Architecture*, Yale University Press, 1989.

CHANDIGARH

Bahga, Sarbjit & Mirchandani, Arun – *Making of Chandigarh, a Vintage History*, White Falcon Publishing, 2023.

Joshi, Kiran, ed. – *Documenting Chandigarh*, Grantha Corporation, 1999.

Prakash, Vikramaditya – *Le Corbusier's Chandigarh Revisited*, Routledge, 2023.

Sharma, Sanjeet – *The Corb's Capital*, A3 Foundation, 2008.

ISLAMABAD

Moatasim, Faiza – *Master Plans and Encroachment, the Architecture and Informality in Islamabad*, University of Pennsylvania Press, 2023.

Yakes, Orestes – *The Birth of a Capital*, Oxford University Press, 2001.

CANBERRA

Brown, Nicholas – *A History of Canberra*, Cambridge University Press, 2014.

Statham, Pamela – *The Origins of Australia's Capital Cities*, Cambridge University Press, 1989.

BRASÍLIA

Gautherot, Marcel & Frampton, Kenneth – *Building Brasilia*, Thames & Hudson, 2010.

Williams, Richard – *Modern Architectures in History*, Reaktion Books, 2009.

ABUJA

Elleh, Nnamdi – *Architecture and Politics in Nigeria*, Routledge, 2016.

Keeton, Rachel & Provost, Michelle, eds. – *To Build a City in Africa, a History and a Manual*, nai010 Publishers, 2019.

Ofoegbu, L – *The Story of Abuja, Nigeria's New Capital*, Createspace, 2016.

NAYPYIDAW

Charney, Michael – *A History of Modern Burma,* Cambridge University Press, 2009.

ASTANA

Menser, Philipp – *Architectural Guide,* DOM Publishers, 2015.

NUSANTARA

Hudalah, Delik – *Planning Indonesia's New Capital City, behind Nusantara,* Routledge 2025.

Widana, DKK, Damayanti, D & Andriyani, E., eds. – *Management Development of the Capital City Nusantara,* Lambert Academic Publishing, 2023.

PUTRAJAYA

King, Ross – *Kuala Lumpur and Putrajaya, Negotiating Urban Space in Malaysia,* University of Hawaii Press, 2008.

Part V

Emerging Global Cities

5.0 Introduction – Emerging Global Cities

From the historically established and more recently planned and capital cities, the subsequent focus is on emerging global cities and the transactional forces which have led to their creation during the 21st century.

Industrialisation led to urbanisation, and urbanisation has resulted in the agglomeration of built-up areas as a result of both push and full factors. Yet 'de-industrialisation' has also occurred in modern evolving economies. No longer just core areas with a periphery, cities have become multi-centred, and with globalisation transnational organisations have expanded by siting themselves beneficially so as to secure the best comparative advantage that different urban regions afford around the world.

Some industries and services are very dependent on the presence of proximity, whilst other businesses are less 'place-bound' such that multinational and transnational organisations have developed in the digital world led by dominant American 'tech giant' corporations.

Historical European cities and modern North American cities have had their growing challenges and pressure between city and suburbs. South American and African cities have had the contemporary challenges of greater population growth, in-migration and import substitution. South Asia and South-East Asia have similar population pressures and contain the largest number of mega cities in the still-developing world.

Besides the already huge expanding metropolises of around 10–20 million people, the new megalopolitan agglomerations of China are now emerging as enormous clusters of even more than 100 million inhabitants and workers. Whilst the extent of urbanised land areas are growing, the population numbers are growing accordingly, but not at increasing density.

Here lie the contemporary challenges of these mega cities and agglomerations – not just in resources and housing, but in basic services, sanitation and supporting infrastructure, with the problems of congestion, pollution, health and the risks of contagion and ongoing sustainability for an extremely large and concentrated volume of people.

This **Part V** accordingly addresses the emergent mega cities that are both innovative and trans-national and also selected mega cities in Africa, Asia and Latin America with their challenges of scale, resources and services in the Global South (see Table 5.01).

DOI: 10.4324/9781003353348-25

Table 5.0.1
Emerging Global City Types

	City	Topography	Plan	Trans. Forces	Type
5.01 EMERGING GLOBAL CITIES					
5.1 *Innovative*					
	Singapore	Island	Zonal/organic	Capital	Transnational
	Hong Kong	Island/coastal	Linear/organic	Maritime	Transnational
	Seoul	Riverine	Organic	Spec. Capital	Transnational
	Taipei	Estuarial	Recti-linear	Capital	Transnational
	San Francisco	Coastal/bay	Grid	Technology	Transnational
	Tel Aviv	Coastal	Linear/zonal	Technology	Transnational
5.2 *Transnational Global*					
	London	Estuarine	Zonal/organic	Capital	Global
	New York	Coastal	Ortho-grid	Services	Transnational
	Los Angeles	Coastal	Grid/poly-centric	Services	Global
	Dubai	Coastal/desert	Linear	Services	Transnational
	Shanghai	Coastal/riverine	Organic/zonal	Technology	Megalopolitan
	Sydney	Bay/estuary	Poly-centric	Sta. Capital	Transnational
	Melbourne	Bay	Recti-linear	Sta. Capital	Modern
5.3 *Mega African*					
	Cairo	Riverine	Poly-centric/organic	Capital	Agglomeration
	Khartoum	Riverine/desert	Ortho-grid	Capital	Agglomeration
	Lagos	Coastal/lagoon	Organic/zonal	Sta. Capital	Trop. Afr-informal
	Kinshasa	Riverine/pool	Grid/organic	Capital-minerals	Trop. Afr-informal
	Nairobi	River valleys	Organic	Capital	Agglomeration
5.4 *Mega South American*					
	São Paulo	Coastal range	Poly-centric/organic	Sta. Capital	Agglomeration
	Rio de Janeiro	Coastal/bay	Organic	Sta. Capital	Lat. American
	Buenos Aires	Coastal	Grid/axial	Capital	Autonomous
	Santiago	Hilly basin	Radio-centric	Capital	Lat. American
	Lima	Coastal/desert	Axial	Capital	Port
5.5 *Mega Asian*					
	Mumbai	Coastal	Peninsula/linear	Maritime	Transnational
	Hyderabad	Plateau	Radial	Services	Indian
	Karachi	Coastal/estuarial	Sectoral/organic	Sta. Capital	Agglom-informal
	Dhaka	Delta	Organic	Capital	Informal
	Bangkok	Riverine	Axial/organic	Capital	Thai
	Manila	Coastal/bay	Radial/organic	Capital	Port
	Jakarta	Coastal	Sectoral/organic	Sta. Capital	Desa-kota
	Kuala Lumpur	River valley	Zonal/organic	Fed. Capital	Malaysian
	Ho Chi Minh City	Delta/riverine	Grid/organic	Agro-industrial	Vietnamese

Table 5.0.2
Emerging Global Cities Population Growth (thousands)

City	Country	1950	2000	2020	*2035**	Area
Singapore	Singapore	1,016	3,914	5,935	*6,480*	Urban agglomeration
Hong Kong	China, SAR	1,682	6,664	7,548	*8,127*	Urban agglomeration
Seoul	South Korea	1,021	9,879	9,963	*10,286*	Urban agglomeration
Taipei	Taiwan	551	2,642	2,721	*2,890*	Urban agglomeration
San Francisco-Oakland	USA	1,855	3,230	3,314	*3,652*	Urban agglomeration
Tel Aviv-Jaffa	Israel	418	2,739	4,181	*5,262*	Metropolitan area
London	UK	8,361	7,273	9,304	*10,556*	Urban agglomeration
New York-Newark	USA	12,338	17,813	18,804	*20,817*	Urban agglomeration
Los Angeles	USA	4,046	11,798	12,447	*13,778*	Urban agglomeration
Dubai	UAE	20	907	2,878	*3,526*	Urban agglomeration
Shanghai	China	4,288	14,247	27,058	*34,341*	City proper
Sydney	Australia	1,690	3,780	4,926	*5,875*	Metropolitan area
Melbourne	Australia	1,332	3,361	4,968	*6,056*	Metropolitan area
Cairo	Egypt	2,494	13,626	20,901	*28,504*	Metropolitan area
Khartoum	Sudan	183	3,505	5,829	*9,608*	Urban agglomeration
Lagos	Nigeria	325	7,281	14,368	*24,419*	Urban agglomeration
Kinshasa	Dem. Rep. Congo	202	6,140	14,342	*26,682*	Urban agglomeration
Nairobi	Kenya	137	2,214	4,735	*8,499*	City Proper
São Paulo	Brazil	2,334	17,014	22,043	*24,490*	Metropolitan area
Rio de Janeiro	Brazil	3,026	11,307	13,458	*14,810*	Metropolitan area
Buenos Aires	Argentina	5,166	12,504	15,154	*17,128*	Urban agglomeration
Santiago	Chile	1,322	5,658	6,767	*7,461*	Urban agglomeration
Lima	Peru	1,066	7,294	10,719	*12,972*	Metropolitan area
Mumbai (Bombay)	India	3,089	16,147	20,411	*27,343*	Metropolitan area
Hyderabad	India	1,096	5,650	10,004	*14,152*	Urban agglomeration
Karachi	Pakistan	1,055	9,825	16,094	*23,128*	Urban agglomeration
Dhaka	Bangladesh	336	10,285	21,006	*31,234*	Metropolitan area
Bangkok	Thailand	1,360	6,395	10,539	*12,680*	Urban agglomeration
Manila	Philippines	1,544	9,958	13,923	*18,649*	Metropolitan area
Jakarta	Indonesia	1,452	8,390	10,770	*13,688*	Metropolitan area
Kuala Lumpur	Malaysia	262	4,176	7,997	*10,467*	Metropolitan area
Ho Chi Minh City	Viet Nam	1,213	4,389	8,602	*12,236*	Urban agglomeration

Note: * *Forecast.*
Source: Global Database of Metropolises 2020 – UN-Habitat (2020), UN DESA Population Division.

5.1 Innovative Cities

Asian 'Tiger' Cities – American 'Tech Giant' Corporations

Singapore, Hong Kong, Seoul, Taipei, San Francisco and Tel Aviv

5.1.0 INTRODUCTION – INNOVATIVE CITIES

Whilst the 19th century was dominated by industrialisation, subsequent to the conflicts of the two world wars, the 20th century has been a period of advancement and innovation. In particular, during the last half of that century, under a broadly more peaceful era, employment evolved in the tertiary sector with the expansion of services and into digital technology and with continued concentration and urbanisation of peoples on each continent into the 21st century.

Today, with an increased emphasis on direct transactional services between people, the provision of more responsive and immediate personalised services has grown, aided by the digital transformation and abetted by electronics. This has not just focused on communications, but also on production through robotisation and the more recent introduction of artificial intelligence into mainstream analytical processes.

This digital economy has rapidly outgrown the former traditional economies, and new centres of employment have grown around information services, finance and media, with the dominance of the American 'tech giant' companies which has directly impacted both the location and the forms of built development around urban areas, especially in Asian 'tiger' cities.

These new forces have in many areas led to increased de-centralisation and the development of 'edge' cities on the urban fringe. In addition, the requirements of 'proximity' have changed, enabling many less place-bound companies to develop as emerging global companies are established around the world in different zones having the requisite land, resources, skills and labour to remotely manufacture electronic components and systems to support the world-wide digital economy. These forces have increasingly led to the development of poly-nodal centres in cities.

5.1.1 SINGAPORE, SINGAPORE REPUBLIC

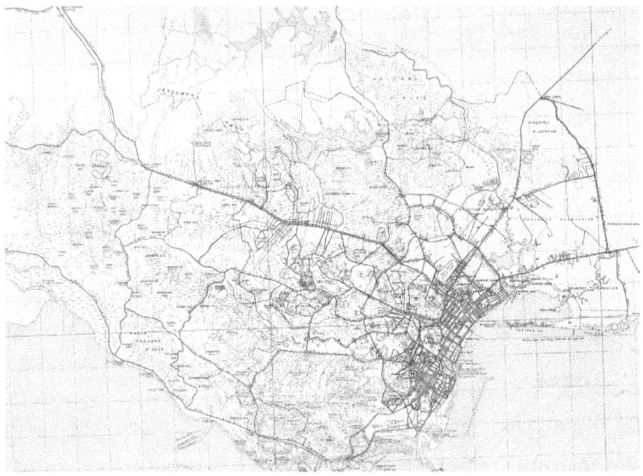

FIGURE 5.1.1 Singapore – early configuration, 1914.
Source: British Library.

Background and Origins

Originally known as the island Temasek, then as Singapura during the 14th century CE, **Singapore** was acquired from Johore in Malaya by Stamford Raffles in 1819. Under the Straits Settlement in 1824, the British agreed on territorial arrangements with the former Dutch occupiers around the Strait of Malacca and promoted its development as a free port zone. The heart of the original town was the Singapore River with its godowns, mills and tanneries, merchant houses and Fort Canning Hill. Ships were unloaded in the outer harbour and freight transported by lighters and junks. At the crossroads between Sumatra, Borneo and Malaya, this was where the sea lanes of the Strait of Malacca and the Sunda Straits met, reaching beyond connecting India with the South China Sea.

With the resources of tin and rubber in Malaya, spices in the Moluccas (Celebes), silk from Indochina and other commodities, it was ideally placed for both staple and regional

DOI: 10.4324/9781003353348-26

entrepot trade. The opening of the Suez Canal led to a significant change in shipping patterns with new steamer routes between continents and with Singapore acting as a coaling point. Immigration increased with Tamils from India and Chinese from Fujian and Guangdong, which resulted in a mixed multi-ethnic environment evolving from a Malay-dominant to a Chinese-dominant community. In particular, the Straits Chinese formed a most influential group.

The island city had originally been laid out in separate zones for different ethnic communities – the European town in the north, Chinese communities south of the river and the Chulia Company for the Indians. There were also areas for Arabs, Bugis and Malays. Elaborate temples and mosques were built for religious assembly.

At the turn of the 20th century, large offices and banks were built in the central area, and the residential suburb of Tanglin was developed. Other areas were less prosperous with their side streets, 'shop-houses' and often unsanitary and overcrowded areas. Land was at a premium across the island of Singapore. As an open city, both populations and commodities evolved, extending later in the 20th century to supporting the development of discovered oil fields in and around Sumatra and Borneo. In 1924, the Johore Causeway was built between the Malaysian peninsula and the island of Singapore, which provided an alternative to the ferries. The world's largest dry dock was also opened in 1938, named after King George VI.

With its colonial past and as a British fortress, Singapore's World War II surrender to the Japanese in 1942 was traumatic; the Japanese had renamed it Syonanto occupied until its defeat in 1945. Post-war Singapore had to re-establish itself. As a small island, it was challenged with the closure of the British military bases and Sembawang. After a period of self-government, under a referendum, the population of Singapore voted to join in a federation with Malaya to become Malaysia.

However, tensions arose with the Malays, who wanted the Bumiputra to control their institutions. This resulted in the eventual expulsion of Singapore from the federation upon which Singapore had established itself as an independent republic in 1965. As founding prime minister, Lee Kuan Yew set about establishing rigorous policies that would ensure that both Singapore's physical infrastructure and its institutions would be developed robustly. With sustained investment, this enabled Singapore to lay the ground for a highly competitive economic environment for both new and established industries.

Modern City – Buildings and Arteries

From the moment of arrival at Changi Airport and taking the MRT mass transit into the centre, it is apparent that Singapore is a very organised and special modern place. From its early settlement beside the Singapore River, it still retains some of its old charm, as evidenced by the classic restored Raffles Hotel and Padang Field in front of the colonial Supreme Court.

Its modern image is perceived from the enormous Marina Sands Bay resort, built on reclaimed land and designed by Moshe Safdie, with its triple towers, shopping facilities and the Sky Park garden deck of Gardens by the Bay. Between 1965 and 2020, Singapore grew from approximately 580 to 730 sq. km. through land reclamation. Along the Singapore River there are small 'godowns' (warehouses) on the boat quays, with bumboats and restaurants nestling alongside the riverbank at Clarke Quay, albeit at a much smaller scale than formerly. However, low-rise 'shop-houses' have been demolished to make way for larger commercial projects.

Downtown now has a raft of high-rise commercial buildings around the Financial District, such as One Raffles Place, beside the old Colonial Quarter and Old Parliament House, Empress Place Building, the Supreme Court, St Andrews Cathedral and city hall. There are the botanic gardens, Chinatown and Orchard Road where there used to be spice and fruit plantations, the shopping mall and Old Cavanagh Bridge and Clarke Quay. In addition, the elegant temples and mosques are special features, such as the onion-domed Sultan Mosque, and also Sentosa Island to escape to for recreation or gambling.

Public transport works most efficiently in Singapore, especially the MRT transit system, which is punctual and frequent. Due to its small land area, ownership of cars is discouraged (an expensive Certificate of Entitlement for ten years is required) and restrictions on use with the effective Electronic Road Pricing (ERP) system that controls this through congestion pricing arranged by the Land Transport Authority. All public transport modes are well integrated and inter-connected throughout the whole island. Small and compact, the 'Lion City' is an efficient, busy, dense and urbanised place. The Pan Island Expressway (PIE) also crosses the island.

In addition to its mix of languages such as English, Tamil, Mandarin and Malay, the colloquial 'Singlish' dialect still remains. Natural resources are a problem, with fresh water having to be imported from neighbouring Malaysia. Agricultural land is scarce, and intensive farming systems are used for growing fruits and vegetables, and with pork imported from nearby Pulau Bulan in Indonesia. Due to its location in a monsoon region and due to its island nature, it remains susceptible to the impact of climate change. As a small intensively populated manufacturing centre and hub, Singapore also tends to act as a 'heat sink', located as it is in the centre of a tropical rainforest region.

It is continuing to pursue different initiatives, such as the Centre for Liveable Cities and embracing new technologies as part of the Smart Nations Project. One North is a biomedical research centre in a dedicated district that is experimenting with driverless cars. Its home-grown high-tech industry has been promoted with the manufacture of semi-conductors and computers, such as those by Seagate, a manufacturer of disk drives. With its dirigiste economy, it has developed as a global financial centre in the region, assisted by the abolition of withholding tax.

As Singapore is a small, crowded island, the Housing Development Board (HDB), established in 1960 and charged with addressing the local housing challenges, needed to build new public housing to eliminate 'kampongs'. An Urban Renewal Unit was set up, restructured as the Urban Renewal Department and the Urban Renewal Authority in 1973, to focus on the planning of 'new towns'.

With new satellite towns such as Toa Payoh and Punggol in the north-east, the goal was to create new neighbourhood centres with a sense of place. A recent development, the Oasis Terraces, is an interesting example of a new community hub with a social mission to contribute to the liveability of an increasingly densely populated island city. As the city-state prospered, it has become an attractive destination for immigration.

The government maintained a strict policy of law and order, and the provision of public housing was a key plank of government policy that commenced in the 1960s. Examples include Sky Ville, Sky Terrace and other interesting examples, such as the Interlace, which combines housing and public spaces, designed by Paul Rudolf in 1987. The model has evolved into the provision of condominiums rather than rental units.

With its limited land area, the city maintains a policy of maintaining a hierarchy of urban parks that are distributed across the city, such as open space for recreational purposes. More green buildings are being built, for example Marina One, Oasis Hotel and the Tree House. A green approach is also being pursued in a biophilic manner, the Singapore Green Plan 2030, with vertical farming at Sky Greens.

The education sector is very strong in Singapore with the two universities, National University of Singapore (NUS) and Nanyang Technological University (NTU). The 'T-score' is a competitive measure for the future schooling selection of students. Despite its challenges with an ageing population, Singapore retains a well-educated population.

Climate change is a real challenge – the new planned Changi International Airport (SIN) Terminal 5 development design will be raised 5.5 m. above mean sea level. With shortages of natural materials, sand dredging has been required from neighbouring coastal areas, but this is being restricted by adjacent countries. Water is also a precious resource in Singapore, which has a '4 Tap' policy supply system – firstly from its own reservoirs, secondly from desalination, thirdly from recycled sewer water and finally using imported pipe water from neighbouring Malaysia.

Jurong Island to the west is a centre for chemical refineries, and underground caverns have been excavated for oil storage. Land reclamation is challenging, and refuse has been deposited offshore at Pulau Semaku landfill sites since Lorong Halus was closed in 1999. Since the 1970s, with the ongoing entrepot trade, the port has continued to be expanded under containerisation, as its major port has grown to become the most advanced containerised facility in the region.

Singapore continues to act as a trans-shipment hub with its trades in oil and commodities. As a high-tech centre, it has new underground sub-sea cables (Echo and Bifrost) connecting with the West Coast of the USA and has its own data centres. Its sovereign wealth fund, Temasek, actively promotes its own companies, such as Singapore Airlines and SingTel, in addition to offshore investments.

The population of Singapore is approximately 5.7 million with a high density of approximately 7,800 pers./sq. km.

After the end of Premier Lee Kuan Yew's era and Chinese challenges in the region, Singapore has developed its own Air Defence Force base. Generally, Singapore is a carefully controlled and structured society that guards its own prosperity as a small sovereign nation in a dynamic economic region. Under his leadership, with a strong line of 'dirigisme', he set about laying out a route of 'path dependence' that would enable this small country to both stand on its own feet and prosper in the region. With the establishment of industrial estates, such as at Jurong and elsewhere, along with favourable economic policies, foreign investors remain attracted to Singapore.

Transactional Forces and Typology

This small island of Singapore, located at the southern end of the Malay peninsula, has a strategic location in the Strait of Malacca. With its siting, it found itself on the maritime spice trade route with commodities being exchanged in each direction. As the city was settled, it grew around its developing coastal perimeter with wharves and godowns for the storage of materials.

After its more recent separation from Malaya, Singapore developed and extended its own entrepot trade which, with modern advanced port and airport facilities, has led to the trans-shipment, handling and re-export of many different commodities. With the development of oil and gas facilities offshore, this has also resulted in the provision of support services to the developing local petrochemical industry.

Under this modern form of guided democracy with some authoritarianism, labour laws were revised, education strengthened, a sovereign wealth fund created, tax incentives provided and businesses promoted. With the development of the port and shipyard, airport and metro infrastructure, and support for the offshore oil industry, pharmaceuticals and other industries were attracted to the country. With the attraction of foreign capital, this led to its evolution as a regional financial centre at a location having efficient communications.

From the 1980s, growth took off dramatically with the attraction of additional skilled and foreign labour. As a city-state, despite its small size, it has been able to shape its own strategy grounded on the basis of maritime and air commerce, construction, real estate and financial services. The future challenges are the form of new political leadership required and with its small land area, management of natural materials and water resources that need to be guarded preciously.

As an unusual island city-state, its development has resulted from favourable economic policies that have promoted its self-sufficiency and participation in international trading as an entrepot port. Singapore remains an independent sovereign state located on strategic maritime trade routes. From its central core, the perimeter has been expanded outwards with residential communities interspersed with commercial and industrial areas, with a dense pattern of radial and sectoral development.

Due to its physical limitations, land reclamation has been playing a pivotal role in aiming to provide additional land horizontally for industrial facilities, but cognisant of key climate change challenges given its island nature and coastal location. However, with built developments focused around the metro MRT lines and public transport system axes, the use of private cars and ownership is actively discouraged. As the rate of population growth continues, the outcome is towards further vertical development and increased densification of buildings across the limited urban area.

5.1.2 HONG KONG, SAR CHINA

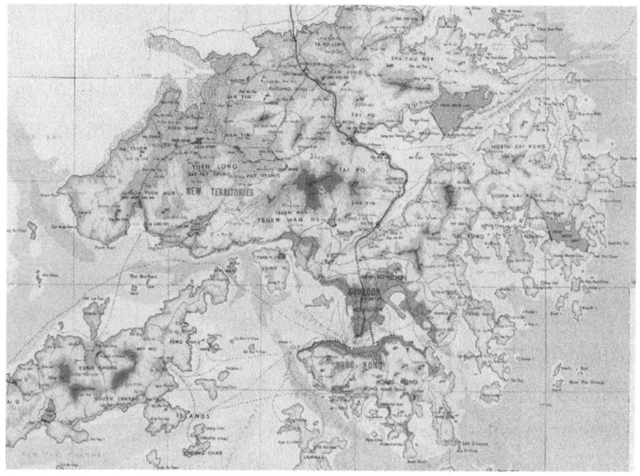

FIGURE 5.1.2 Hong Kong – early configuration, 1952 (GSGS).
Source: Maphouse.

Background and Origins

The British staked their flag on rocky Hong Kong Island, located on the South China Seas, near the Pearl River Delta and opposite Macau, in 1841. With the presence of some Hakka and Tanka people, Hong Kong's significance then and to the present was as a way station into China. When the Treaty of Nanking ended the Opium Wars in 1842 and China ceded **Hong Kong** Island to Britain, to be managed as a Crown Colony, it also opened up five Treaty Ports that included Canton and Shanghai for trading. In 1860 this ceding was extended to include Southern Kowloon and Stonecutters Island under the Treaty of Peking and was extended further in 1898 under a Second Convention to include leasing Northern Kowloon and the New Territories.

This was the basis upon which many mercantile entities, such as Jardine Matheson, developed locally and under its operation as a free port. With the development of steamships in the 19th century, Hong Kong became an important coaling station and a key maritime hub with a shipyard on the Chinese coast. Other rival ports were Canton inland on the Pearl River Delta, and farther north Shanghai and Tientsin. Hong Kong developed its service and trade entrepot economy over time with other 'hongs' and diversified from light manufacturing of textiles into other administrative activities, such as shipping, banking, insurance and real estate.

The new town of Victoria was developed in the Central District. The commanding point of Hong Kong Island remains the Peak that overlooks the South China Sea outwards and Stonecutters Island and Kowloon inwards. On steep terrain, this was accessed by the man-made meandering Peak Road or by the Peak Tram cable railway. With the small, rocky land area comprising Hong Kong Island, plus Kowloon, Lantau Island and some smaller fishing islands and the New Territories, it became a highly populated and densely urbanised area but with challenging geography and seasonal typhoons. The institutions and administration of the colony continued to be strengthened by Britain and in 1911 the University of Hong Kong was established.

Difficulties developed, however, in the region in the 1930s with the Sino-Japanese Wars. Then the invasion of Hong Kong in December 1941 during World War II resulted in the collapse of the garrison after the Pearl Harbor event. Post-war reforms were needed, and a new Hong Kong Planning Unit was established with the goal of establishing a new municipal council. This had to take into account significant change in circumstances with the newly formed People's Republic of China in 1949, with whom Hong Kong was linked with treaty obligations, under a wholly different regime.

Gradually, the transformation of Hong Kong evolved in the 1950s from being mainly an entrepot port to becoming an industrial colony. With a large increase in in-migration from China, manufacturing of textiles increased and next started to be focused on plastics, benefitting from the supply of cheap labour and giving a comparative advantage. It was also able to benefit from its extensive trading networks. With Hong Kong's limited land area and natural resources and constraints on development, new beneficial infrastructure was realised, for example construction of the Cross-Harbour Tunnel. During the 1970s efforts were made at developing and further extending the economy, and in the financial sector new capital markets were developed as the financial centre expanded.

A further change that was to have a major impact was the PRC's 1990 creation of the special economic zone (SEZ) at Shenzhen, right on the border with Hong Kong in the north. With significantly increased manufacturing resources, this had the impact of enhancing the attraction of the area to multinationals and strengthening this industrialising area.

Modern City – Buildings and Arteries

As Hong Kong prospered post-war during the 1960s, population growth and in-migration put pressure on housing requirements. Accordingly, a number of new towns were planned across Kowloon and the New Territories in order to try to better accommodate this growth. With satellite towns of Kwun Tong and Tsuen Wan, new towns were developed at Sha Tin and Tuen Mun, and later at Tseung Kwan O and Tin Shui Wai, with industrial estates at Tai Po and Yuen Long, and eventual North Lantau New Town (renamed Tung Chung New Town in 1996).

The waterfront behind the marine piers on Hong Kong Island comprises the main commercial area. Here are the towering new buildings around Central Square, such as the HSBC Building (designed by Foster) with its precise feng shui, Bank of China Tower (designed by Pei) and the IFC Tower (designed by KPF) with 100 storeys. In the Central District a number of elevated pedestrian walkways were built to connect buildings, along with a series of escalators up the hillside.

Eastwards along the island is Victoria Harbour and the Convention and Exhibition Centre and Causeway Bay with its houseboats and typhoon shelter, and tucked away is the Happy Valley (horse) Racecourse. The south side of the island is more rugged and wooded past the Aberdeen fishing harbour, with Stanley and its old fort and rocky bays beyond.

Across from the island is the mainland of Kowloon, between which the Star Ferry plies regularly. With the tolled road tunnels, the busy Western and Eastern Harbour Crossing and the central Cross-Harbour Tunnel carry very high volumes of traffic. Tsim Sha Tsui (TST) is the main commercial district with the Golden Mile up Nathan Road and the old Peninsula Hotel nearby. The tall International Commerce Centre (ICC) (designed by KPF) dominates the landscape here, and New Kowloon continues to be redeveloped around Union Square, with a major rail and metro terminal developed at Kowloon Station and the M+ Building (designed by Herzog de Meuron).

To the west, the Tsing Ma double deck road and rail bridge spans across to Lantau Island to connect with the growing modern Hong Kong International Airport (HKG) at Chek Lap Kok, built upon reclaimed land and which is also connected with the Tuen Mun–Chep Lap Kok link road and tunnel and across Tsing Yi and Stonecutters Island.

Beyond Kowloon, the New Territories are more suburban after the intensity of the central areas, that stretch out northwards to the Chinese border, and Shenzhen with its SEZ and the rapidly expanding neighbouring Guangdong province and Guangzhou (Canton) beyond.

Public transport is excellent with mass transit, ferries, historic trams, bus and rail connections. The long-distance high-speed train connection with China commences at the new station in West Kowloon. The MTR is a highly efficient transit system, carefully designed with its broad gauge of five metro lines, four crossing between Hong Kong Island and Kowloon in a tunnel, and the others running out to the airport and Lantau Island via the Tsing Ma Bridge and to Tuen Mun and the north-west.

Using development rights over its depot, MTR has developed a huge commercial area at Tai Wai Station. The development of other land areas continues apace with redevelopment of the old Hong Kong airport of Kai Tak Airside as a cruise terminal and continuing other areas of land reclamation.

Most of the undeveloped land in Hong Kong is owned by the government, which only releases land for special developments such as MTR expansion, with very high premiums for commercial property development. Some commercial malls are even being developed in a vertical manner due to the limited land area. Much of the residential population is housed in individual public rental housing estates.

In 1989, a major infrastructure development plan, the Port and Airport Development Strategy (PADS), was prescribed for the realisation of both a new port and and international airport (HKG) at Chek Lap Kok on reclaimed land beside Lantau Island, with its first phase to be built by Handback.

With a population of 6 million in 1990, the overall population has now grown to approximately 7.5 million, with over 90% Han Chinese. Kowloon once had one of the world's densest urban neighbourhoods, the old Walled City having extremely high density before its eventual demolition in 1992.

The circle was finally closed with the return of the former colony of Hong Kong to China in June 1997, intended to remain for a further 50 years under the mantra of 'one country, two systems'. Whilst under challenge as the pace of change is accelerating, Hong Kong is fully subsumed into the People's Republic of China (PRC) as a Special Administrative Region (SAR), in a similar vein to Macau.

At the time of Handback, concerns over ongoing democracy in the future arose with changes to the Legislative Council and the movement of people and businesses out of Hong Kong. Formerly its commercial business, community and real estate business were dominated by just a few large families. Hong Kong is striving to retain its economic dynamism and remains a major financial centre in the region despite some setbacks in its evolving relationship with mainland China, with a decreasing number of years of the dual system model remaining.

The most important recent major infrastructure feature is the Hong Kong–Zhuhai–Macau (HKZM) Bridge and Tunnel crossing, a significant engineering achievement of a 55-km. fixed-link connection, mainly viaduct, across the Pearl estuary from Hong Kong to Macau that now connects these three major cities in China.

Transactional Forces and Typology

As an island territory, Hong Kong's existence and evolution developed on the basis of its trading role, located strategically in the South China Sea. Foreign trade with China was originally regulated through monopoly of the 'cohongs' in Canton, with the sale of the principal commodities of porcelain, silk and tea. It was as a result of the Treaty of Nanking after the Opium Wars that the colony was ceded to Britain for 99 years, and the creation of the five Treaty Ports, including Canton and Shanghai, enabled trade to be developed.

With increasing cotton and textile trade in particular, more Chinese labour was attracted to the area, and commerce grew along with its expanding entrepot trade in the region. The infrastructure on the small island and adjoining New Territories was improved, further merchants and entrepreneurs were attracted, and the population and its income grew. Despite occupation in World War II, post-war industrialisation developed and along with Macau, Hong Kong has acted as a gateway into China and especially Guangzhou (Canton) and as it developed as a regional financial centre.

With neighbouring Shenzhen since its SEZ and subsequent expansion, Hong Kong's location at the entrance of the Pearl River estuary was to be a pivotal advantage. Now with the end of the colony's lease after Handback, the economies of the separate poles of development of Hong Kong, Macau, Shenzhen, Zhuhai and Guangzhou have become inextricably linked, which is modifying the balance of forces between them.

Unique as a small tropical island in a strategic location in South-East Asia, but with limited buildable land area, the growth of Hong Kong came from trading. With annexation of adjoining territory and reclamation of coastal land areas, it had to build vertically in order to accommodate populations at very high densities. With corresponding commercial developments for business, the city has grown into a compact financial centre for real estate and business services and by harnessing the burgeoning Chinese and South-East Asia markets.

With construction of extensive new modern infrastructure, rail and metro, port and airport, and with tunnelled road and rail connections, this coastal island city attains exceptional high-density patterns of development, clearly evident in the vertical dimension. As an island type, its development is being maximised in using all available land and reclaimed areas in order to support its growing population and business activities.

Since Handback of both Macau and Hong Kong to China, the area has yet wider economic significance, as the whole Greater Bay Area of these cities plus Guangzhou has resulted in its evolution as a megalopolitan area around the Pearl River estuary.

In consequence, new infrastructure links have been and are continuing to be extended with mainland China on the western side beyond Lantau of its major airport and port, with new land areas continuing to be reclaimed, which is resulting in an extended region with three major urbanised nodal centres in the Bay Area. From being an inward-looking protected island, Hong Kong is now becoming a pivotal cog in an even wider region reaching right into Guangdong province and the urbanised areas on the Pearl River beyond developing around Guangzhou in China.

5.1.3 SEOUL, SOUTH KOREA

Background and Origins

'Seoul Special City', the modern named capital of South Korea, is located on the Han River that flows to the western Yellow Sea, surrounded by a hilly landscape with the Bukhan Mountain on the northern periphery of the city. Previously named 'Hanju', 'Wiryeseong' and 'Namgyeong' as the formal capital during the Three Kingdoms, Jamham period, that is, of Goguryeo (or Goryeo), Baekje and Silla. In 1394 CE King Taejo made the city (Hanseong) the first capital of Joseon, a long dynasty that was to last for around 500 years.

With massively constructed fortress walls, much of these no longer existing, two of the eight great gates remaining around the compass of the city are the Sungnyemun (South Gate – Namdaemun) and ornamental Heunginjimun (East Gate – Dongdaemun) that are still standing. There was a series of five palaces that were constructed, initially Gyeongbokgung in 1395, the 'Palace of Shining Happiness', now rebuilt with its walls, courtyards and pavilions in the centre. The nearby palaces of Changdeokgung and Changgyeonggung to the east are in Changdeok Park, completed in 1412 and 1483. The other two palaces were Deoksugung and Gyeonghuigung.

This dynasty remained until 1910 when Japan annexed Korea, renaming the capital Gyeongseong, or Keijo in Japanese, when the city walls were further demolished and streets rebuilt. After World War II, in 1948, the first Republic of Korea was founded as the country was divided between North and South under separate zones of occupation of Soviet and American influence respectively. The charter of the Special City of Seoul was created in 1949.

With the subsequent catastrophic Korean War that was to develop when the capital was transferred temporarily to Busan, the outcome at armistice was the permanent division of the country in 1953 between South and North Korea separated by a demilitarised zone (DMZ). With large numbers of refugees gravitating to Seoul, there started a period of reconstruction and modernisation in the south. With increased urbanisation, both population and density of the city of Seoul increased significantly.

Modern City – Buildings and Arteries

Seoul is a polycentric city with a number of districts, the river running east-west across the city and crossed by a large number of road bridges. Besides the old historical Palace Quarter in the Jongno and Jung districts, and central Insadong, the modern financial centre is in Yeouido Island on the Han River along with the National Assembly. The new economic district of Gangnam to the south was annexed, and Olympic Park was developed in Songpa to the southeast for the 1988 Olympic Games. The Gwangjin district has rivers on each side.

The 25 districts ('gu') are further subdivided into different neighbourhoods ('dong'). The shopping area of Insadong is made up of several winding alleys ('golmok') with wooden ('hanok') buildings. The area alongside the Cheonggyecheon stream has been cleverly redeveloped and revitalised after an elevated freeway constructed above the channelled river was removed. Now there are pedestrianised pathways on either side of the riverbanks near the Dongdaemun market and Jongno with the Bosingak Bell Pavilion.

In the middle of the city is Namsan park and hill, with the tall N Seoul Tower on its summit, and to the north of the palaces is the outer Bukhansan National Park with its ridge trails and valleys.

At the end of the Korean War with the DMZ created between the two Korean countries, a green belt was established around the city to also serve as some form of security buffer. New buildings were generally of brutalist concrete designs, with some of these having been replaced recently with a number of more modernist buildings. These include, for example:

- 63 City Skyscraper, Yeouido Island 1985, Jongro Tower (designed by Vinoly) 1999, the Seoul Museum of Art (designed by SeMA) 2002, the Korea National Museum 2005, Samsung Town (designed by KPF) 2008, Floating Islands, Han River 2011, New City Hall 2012, Glazed City Hall Jung District (designed by iArc) 2012, Dongdaemun Design Plaza and Metonymic Park (designed by Hadid) 2014, Seoullo 7017 Skypark (designed by MVRDV), Seoul Station 2017 and Lotte World Tower (designed by KPF) 2017.

From its earliest streetcars, the extensive Seoul Metropolitan Subway now has a dense network of rail and metro lines both across the central area and the greater metropolitan region to the Incheon metropolis and the satellite cities in Gyeonggi province.

The nine metro lines and two subway lines run out to Incheon with the main railway lines of KORAIL and the commuter and KTX high-speed rail lines from the main Seoul Osong Station to Busan and Honam, with the AREX line between the Gimpo (GMP) and Incheon (ICN) International Airports, and suburban rail to nearby Gwacheon. The new Incheon International Airport, constructed on reclaimed land between two existing islands, was initially served by a dedicated Maglev line. A number of other new lines under realisation are operated by the Seoul Metropolitan Subway Corporation (SMSC) and the Seoul Metropolitan Rapid Transit Corporation (SMRTC).

In terms of highways, modern expressways in Seoul have been built on each side of the Han River across the city, including the Gangbyean Expressway, Olympic Expressway and Dangbu Expressway, each of which connects with the main ring roads and Incheon Airport. There are now many road bridges that span across the river, such as Banpo Bridge, Hannan Bridge, and the Seongsan multi-span bridge. With a distributor road network, Korean expressways operate the main toll roads, and the Gyeongbu Expressway AH1 runs south to Busan.

With an ethnically homogeneous population, Seoul has been one of the fastest-growing regions economically since the end of the Korean War as one of the Asian 'tiger' economies. With its modern manufacturing and IT sector, it has been driven by fast-growing automotive and aerospace sectors, and computer and biotech manufacturers. Large, newly industrialised companies developed, initially under family-controlled corporations, or 'chaebol', that had government support, commencing with import substitution activities, such as Hyundai, Samsung, LG and SK groups, specialising in particular sectors.

Much of the housing in the city has been procured from new vertical apartment blocks, such as in the Gangnam district of southern Seoul, having significant repetitiveness. Some new parks have been developed and with the conversion of the former Yongsan Garrison, a further urban park is being created. A greenbelt policy for the city was established in 1971. With limited developable land area and very high population growth pressures, the net impact of this containment has been to lead to significant increases in land prices. As redevelopment occurred, this led to a large increase in high-rise apartment dwelling units but a considerable deficit in affordable low-income housing.

The new Special Self-Governing City of **Sejong** was created 100 km. south of Seoul in 2012 with the goal of creating a new smart city to which many of the government departments could be relocated. With a greenbelt policy and regional economic development plan, the intention was that it should become the new governmental administrative capital by the year 2030 and have a population of half a million people from both the North and South Chungcheong provinces.

Seoul Metropolitan Government (SMG) is outward looking and supports City Net and other international collaborative initiatives. As defined, Seoul Special City has a population of approximately 10 million people within the Seoul National Capital Area, whereas the metropolitan area of Greater Seoul has a population of around 25 million people. The population density of the whole city region is approximately 19,800 pers./sq. km.

Greater Seoul has expanded as the Seoul Capital Area (SCA), which constitutes the wider Seoul Special City and the Districts of Jongno, Yeouido and Gangnam, the Metropolitan City of Incheon and also the province of Gyeonggi. These are all tightly interconnected by road and rail, with the Seoul Ring Expressway connecting the peripheral and satellite cities.

Transactional Forces and Typology

With its coastal plains and internal forests, Korea is endowed with natural resources. Its trading had evolved around the region but changed in the 20th century following Japanese occupation, World War II and subsequent political division after the Korean War. The forced outcome was to lead to a period of significant reconstruction, with a focus on education and training which led to rapid industrialisation.

With its investment in human resources, significant development was focused around the Han River in Seoul, with the creation of new industries and manufacturing, from automobiles to electronics and IT, with major manufacturers headquartered there. These extended to neighbouring areas and also Incheon. With specialisation in electronics and digital media, this led to rapid growth across South Korea with it becoming one of the Asian 'tiger' economies which is continuing with its specialist aerospace, bio-technology and robotics industries.

As a riverine city, Seoul has developed from its original Palace Quarter and administrative Yeouido Island around the Han River on either side, with added districts of Gangnam, Songpa, Gwangjin and Yongsan. As this growth extends to the outer areas of Incheon and Gwacheon, the city is evolving with ongoing poly-nodal developments as the population increases and industrialisation continues.

With Seoul located just to the south of DMZ, the trend for expansion is inherently southwards, although constrained by the greenbelt and as the neighbouring political threat to stability from Pyongyang in the North remains. With this continuing and rapid growth, Greater Seoul is now showing the characteristics of an emerging megalopolitan region.

5.1.4 TAIPEI, TAIWAN, RoC

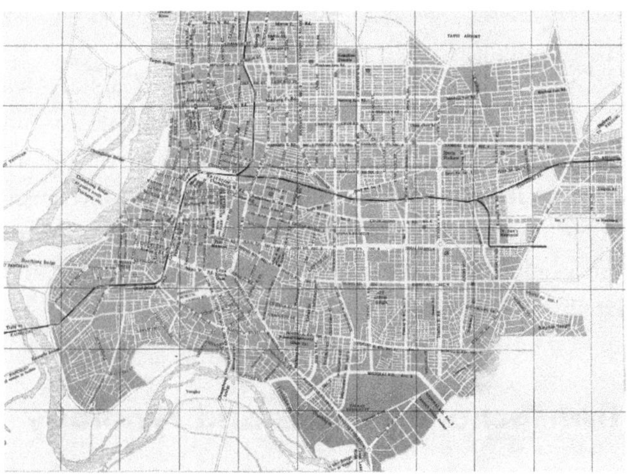

FIGURE 5.1.3 Taipei – early configuration, 1970.
Source: Maphouse.

Background and Origins

Taipei is the capital of Taiwan, the Republic of China, formerly known as Formosa, an island in the East China Sea. In the north of the island, in the Taipei Basin, it is bounded by the Keelung and Xindian rivers that join into the Tamsui River. It was initially inhabited by Ketagalan aborigines and is bounded by volcanic mountains.

It was taken over by Han Chinese migrants from the Yellow River Basin region. With periods of Dutch and Spanish occupation of Formosa, under the United East India Company (VOC), the Dutch set up a fortress at Fort Zeelandia at Anping in the south-west of the island, and the Spanish built Fort San Domingo at Tamsui. Under the Chinese Qing dynasty, Taiwan was annexed under the jurisdiction of Fujian province for 2,000 years and the Han people developed the agricultural sector on the island. After the Opium Wars and the Sino-French War, land reforms were introduced and infrastructure developed under Chinese Governor Liu.

With the Sino-Japanese War in 1895, after the Treaty of Shimonoseki, Taiwan came under control of the Japanese empire, with Taipei being renamed Taihoku. During the subsequent period, cash crops of rice and sugar were increased, industrialisation commenced and infrastructure developed. There followed a process of assimilation, aiming to bind the population more fully into Japanese culture.

In 1911, the Kuomintang Nationalist Party (KMT) was founded by Sang Jiaoren and Sun Yat-sen after the Xinhai Revolution and Chiang Kai-shek was deputed to form a National Revolutionary Army aimed at unifying mainland China based in Nanking. However, this was challenged by newly formed Communist groups that had established control in the rural areas and was unsuccessful.

After the long Sino-Japanese War and the Pacific theatre of World War II came the eventual defeat and surrender of Japan. In succession the forces of the Chinese Communist Party (CCP) coalesced to create the formation of the People's Republic of China (PRC) under Chairman Mao in 1949 with its capital in Beijing. The Kuomintang had to retreat to the island of Taiwan and focus on stabilising it as the Republic of China (ROC), the capital city being Taipei since this time.

Modern City – Buildings and Arteries

After a period of martial law, and defensive acts as construction of the Central Cross-Island Highway, the country began to find its way and prosper around the 'Taiwan miracle' which evolved with rapid industrialisation and economic growth. Declared a special municipality in 1967, Taipei annexed outlying areas and continued to grow. With its expanding economy based on modern technology and electronics, its main emphasis has continued to focus on information and communications technology, with innovations and start-up companies that have attracted global manufacturers into the city, becoming now the world's major producer of semi-conductors.

The centre of the city of Taipei comprises the districts of Da'an, Datong and Songshan, which are densely populated. The Zhongzheng district, named after Chiang Kai-shek, is the central district that contains the National Chiang Kai-shek Memorial Hall alongside the twin National Concert Hall and National Theatre in the great Liberty Square. Nearby is the Presidential Office Building and Peace Memorial Park with the National Taiwan Museum. Farther east, in the Xinyi district, is the national Sun Yat-sen Memorial Hall in honour of the founding father and the tall Taipei World Financial Centre, Taipei 101, with its distinctive stepped pagoda silhouette of 101 floors, 500 m. high (designed by CY Lee), opened in 2004, with internal mass dampers against typhoon-induced sway and earthquakes.

The Datong district is one of the oldest districts and contains the Dalang Dong Bao'an Temple, the Xiahai City God Temple and the Taipei Confucius Temple. The Ba'an district is the commercial area and the Sangshan district has Taipei Main Station with an underground market and the main hub for both the rail and metro stations. The central Taipei Songshan Airport (TSA) and airbase is nearby. Farther out is the Shilin district with the National Palace Museum and the Jiantan Mountains overlooking the Keelung River. The Wanhua district contains the Longshan, Qingshan and Tianhou temples.

Administratively, Taipei is a special municipality directly under the Executive Yuan central government and has 12 separate districts. The population of the central city is approximately 2.7 million. The area now known as the Taipei-Keelung Metropolitan Area, or Greater Taipei, that includes coastal Keelung City to the east, is a highly urbanised developing area stretching from Taoyuan in the west and includes New Taipei City as a new district of the city. All of this area is integrated with new rail and metro systems. In including neighbouring Taoyuan City to the west, where the Taipei-Taiwan Taoyuan International Airport (TPE) is located, the total population is approximately 9 million people.

Urban public transport is very good, with the modern metro of five underground lines, a circular line and additional VAL elevated light rail lines and a new line out to Taoyuan International Airport, along with dedicated fast bus lanes. All main rail lines now run underground as part of the Taipei Railway Underground Project. Across the country Taiwan High Speed Railway (THSR) lines connect with Taichung and Zuoying near Kaohsiung in the south-west of the island.

Transactional Forces and Typology

With its location in the Taipei Basin, original activities of trading tea and camphor developed around the Tamsui River and its tributaries in coastal areas. With new mining areas being developed in the north-east at the time of the Japanese occupation, it was not until after World War II and the creation of the Republic of China (RoC) with Taipei as its independent capital that the economic miracle of Taiwan developed as one of the fastest-growing Asian 'tiger' economies.

With its industrial and commercial activities and substantial foreign direct investment, the focus on information and communications technology developed, with its attraction for foreign US 'tech giants', and the IT sector took off with new technology and software parks. This led to strengthening of the financial sector also and subsequent focus acting as incubators for new R&D companies.

The riverine city of Taipei has grown around the central districts of Datong, Zhongzheng, Zhongshan and Songshan. In the central areas with a grid configuration, since the blocks are 500 m. in dimension and larger than in New York, they are infilled with crossing lanes and alleys. Development is extending outwards from the historic centre, especially eastwards, overlooked by the tall, iconic Taipei 101 Tower. With its extenuating underground metro network, its urban built form assimilates with comparable modern Asian cities across China, Japan and Korea along with its high-speed rail connections across Taiwan to Taichung and Kaohsiung in the south. The significant growing conurbation of Greater Taipei is being created in the corridor from Taoyuan in the south-west via New Taipei City through the main Taipei urban core to Keelung in the north-east.

5.1.5 SAN FRANCISCO AND BAY AREA, CALIFORNIA, USA

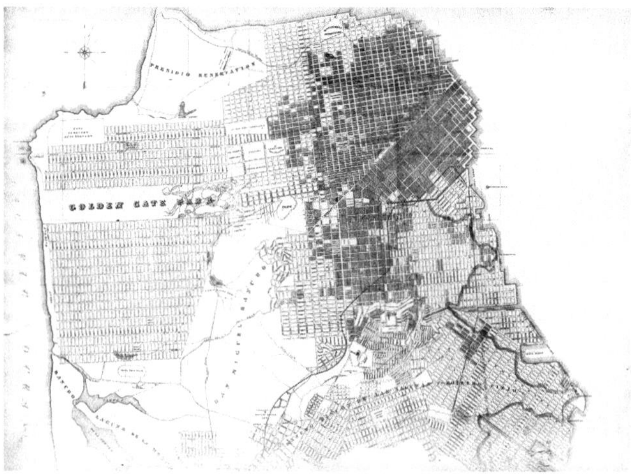

FIGURE 5.1.4 San Francisco – early configuration, 1873.
Source: iStock.

Background and Origins

The state of California is situated in a geologically unstable region with a number of faults, namely San Andreas, Hayward, Garlock, San Jacinto and Nascimento, between the tectonic plates. In recent times there have been recurrent earthquakes of varying scale, in 1872 Owen's Valley, 1906 San Francisco (8.3), 1933 Long Beach, 1971 San Fernando, 1989 San Francisco (7.1) and 1994 San Fernando (6.7), the greatest destruction being of the city of **San Francisco** in 1906.

The area was inhabited originally by Yelamu local tribes, of Ohlone-speaking people, who had developed activities of fishing along the coast and subsistence farming inland. The first records of external visitation followed the arrival of the Spanish (Cabrillo) at Point Reyes Promontory outside San Francisco Bay in 1542 and the Portuguese (Cermeno) in San Francisco Bay in 1595. In between, the British, under Francis Drake, anchored at Point Reyes in 1579, naming it Drake's Bay and the region 'Nova Albion'.

With first periods of occupation, the Spanish established early Jesuit missions around 1680, which were settled under the Laws of the Indies. When in 1775 the Spanish entered the mouth (La Boca) of San Francisco Bay, they established a 'presidio' and 'mission' at Mountain Lake. Subsequent Franciscan missions in San Francisco were established down the coast, at Monterey and San Diego, and at intermediate locations. By 1835, 21 missions had been secularised, including Santa Barbara, San Luis Obispo and San Diego.

With the California Republic declared in 1846, it was not until 1848 that all the Mexican lands north of the Rio Grande were ceded under the Treaty of Guadalupe Hidalgo to the USA. In 1850 California became the 31st state, having formerly been administered under the Spanish alcalde system of law. The state capital was moved variously between San Jose, Vallejo, Beneficia and Sacramento, where it finally settled in 1856.

Whilst principally an agricultural region, in the 1850s minerals were discovered in the Sierra Nevada foothills, which led to the Gold Rush and westward migration of peoples seeking gold and silver in the region, such as 'Okies' from Oklahoma and Arkansas.

The city of San Francisco sits at the tip of a peninsula between the Pacific Ocean and San Francisco Bay on the west coast of North America. With the Diablo Coastal Range mountains beyond, it is a striking setting with the smaller cities of Berkeley and Oakland in East Bay, and Palo Alto and San Jose in San Francisco Bay, and with the Marin Headlands and Point Reyes peninsula to the north. In the Bay Area, the cities of both Vallejo and Beneficia were platted during the 1850s, the oldest areas before the newer areas of Oakland and San Jose were developed.

The original settlement was a Spanish mission settled around Yerba Buena on a small grid in 1776. The Americans took possession of the city in 1846, and a further area was developed on a separate grid that runs at 45 degrees to this, which has resulted in angled street junctions, Market Street being the dividing line. The original plaza, bounded by several large streets, is now Union Square.

At the end of the American–Mexican Wars in 1848, the state of California became part of the USA. It grew very fast to become a significant trading centre, especially with the Gold Rush in the region at that time. New piers and warehouses were developed for shipping and the waterfronts were developed.

Infrastructure started to be developed with the Union Pacific Railroad westwards from Omaha and the Central Pacific eastwards from Sacramento meeting at Promontory Summit in 1869, thus completing this route into California across the mountains. Other railroads were also developed, such as the Atchison, Topeka and Santa Fe Railroad in 1880. In 1871, the development of Golden Gate Park was started, a large rectangular area of the Outside Lands of the city, formerly sand dunes. Now a heavily planted and wooded area, it is a finely landscaped recreational area to the west of the city.

In 1905, Daniel Burnham prepared a master plan for the development of San Francisco with a new Civic Center and land set aside for parks and recreation. Whilst striving to attain a new formality of layout and with diagonal boulevards, it was not progressed, partly due to its timing. Water was also essential and with the establishment of the Board of Water Commissioners under Michael O'Shaughnessy, a number of dams, reservoirs, pipelines and hydro-electric schemes were established in the Tuolumne Valley in 1923 that serviced the increasingly urbanised city.

In the bay, shipbuilding was developed in East Bay around Hunters Point with Kaiser Shipbuilders, which built many ships during World War II, and there are naval facilities at Mare Island. Post-war contractor Kaiser assisted with the development of Kaiser Community Homes that were developed across the region. Social infrastructure was developed also with new schools and educational establishments and new universities at Stanford, Palo Alto and Berkeley, and the beginning of the University of California multi-campus system.

With its downtown and hilly neighbourhoods and street grid pattern over-riding the topography, and quaint cable cars, San Francisco has distinctive districts of quite differing character from Pacific Heights and Nob Hill to the Golden Gate Park and the Presidio. South of the CBD is Mission District with its Latino culture and distinctive street art and murals and grand Victorian houses from the time of the Gold Rush.

Downtown contains the Financial District near the marine wharves and the piers, with the Ferry Building to provide access to Oakland, and Union Square and Market Street. Nearby is Chinatown with a significant Chinese population along Grant Street and Nob Hill looking out over Fisherman's Wharf and North Beach. Also along Jefferson Street is Little Italy with its characteristic seafood restaurants.

The Powell to Hyde and Mason Street cable car passes here and provides access east to the piers. From these piers it is possible to see Alcatraz Island across the bay, which served as a high-security federal penitentiary from 1934 to 1963.

The zig-zagging Lombard Street climbs to Russian Hill, and Coit Tower can be reached from Telegraph Hill Park where semaphore was used to signal the arrival of merchant ships. Pacific Heights and the Marina District are residential areas with large mansions from Alta Plaza to Lafayette Park.

To the north by Marina Drive is Fort Mason, where the army had barracks during the American Civil War. Farther south is the area of Haight Ashbury, made famous in the progressive 1960s, Buena Vista and Corona Heights Parks, and the winding streets of Twin Peaks.

Modern City – Buildings and Arteries

On the promontory and bounded by scenic Lincoln Boulevard is the historic Presidio, a camp of originally adobe buildings founded by the early Spanish settlers. Mission Dolores, the original Spanish colonial mission, is nearby. Behind Fort Point is a large natural landscaped recreational area of over 1,400 acres with eucalyptus and cypress trees.

Here is the southern abutment of the Golden Gate Bridge, which was completed in 1937 (designed by Ellis) and managed by the California Toll Bridge Authority. Iconic, with its twin steel, golden-coloured tapering towers and central span of 1,280 m., it carries around 150,000 vehicles per day, with trains on the lower deck.

The other major crossing is the San Francisco–Oakland Bay Bridge, opened in 1936 and which crosses the bay to Berkeley and Oakland with its multiple spans. This also carries the Bay Area Rapid Transit System (BART), which is the public transport system in the Bay Area, opened in 1972, linking San Francisco with Oakland. Each structure is now upgraded according to seismic requirements to countervail the underlying threat from the San Andreas tectonic fault on the connecting bridges.

Golden Gate Park is one of the largest urban parks within which is situated the established de Young Museum and the California Academy of Sciences (designed by Piano). With its landscaped roof and separate planetarium and aquarium for exploring the lands, oceans and skies, it is a modern educational place.

The redeveloped Yerba Buena Centre has its gardens, and the Embarcadero Centre has the distinctive tall triangular Transamerica pyramid (designed by Pereira). At the junction of Columbus Avenue and Broadway is the City Lights Bookstore, one of the centres during the 1960s Beat Generation. The whole Bay Area of San Francisco is renowned for its local activism. Since this period, different activist movements have arisen, from peace protests to civil rights and environmental protests. The city consequently has orchestrated several revolts, notably against construction of the Embarcadero Freeway in the 1970s and the Yerba Buena urban redevelopment. Institutions developed to support these movements, and whether downtown or in the peninsula, in East Bay in Berkeley, or in North or South Bay, around these areas citizens have prevailed in grassroots activism.

In terms of industry and technology, the city developed an emphasis on defence systems, which led post-war to the development of research universities and expansion of both Stanford and Palo Alto. Within Stanford Industrial Park came the invention of the vacuum tube, the audio oscillator and the first transistors and semi-conductors, such that the region was to expand to become the Silicon Valley of today, now renowned for its IT and technological developments and home to single incubator industries such as Hewlett Packard, Texas Instruments and Apple. This activity has since extended into bio-tech industries with significant facilities around the Bay Area.

As a vibrant multi-cultural city, San Francisco oozes variety and with its imposing sights, modern history, music and culture, it is one of the most iconic modern cities of America. San Francisco International Airport (SFO) is located to the south of the city in San Mateo County. The population of the city is approximately 0.9 million, with 4.5 million in the metropolitan area.

Transactional Forces and Typology

As a coastal city located in San Francisco Bay, early activities were marine based with fishing and whaling in the vicinity. Trading developed up and down the Pacific Coast and was impacted by the Gold Rush in the region in the mid-19th century. Due to the setback of repeated earthquakes, especially in 1906, significant rebuilding and upgrading of the urban infrastructure have been required. Finance and commerce developed in the city, along with industrial activities such as shipbuilding at Hunters Point.

With its hilly and varying topography on a peninsula enclosing the bay, different neighbourhoods developed, on opposing headlands and within the bay, which with the benefit of connecting infrastructure has led to the urbanisation of the metropolitanised area in nucleated settlements. With its universities and research centres, this has led to the development of several of the American 'tech giant' corporations in the surrounding area of 'Silicon Valley'.

As an undulating city on a peninsula with an extended shoreline, its uniqueness has been secured with patterns of orthogonal street grids where the geography permits with its iconic cable cars running up- and downhill. Its marine surroundings have also defined the environment, where road and rail infrastructure links have been designed to afford accessibility in the metropolitan area.

As the city has developed and with the realisation of the wider BART public transport system, this has integrated access around the Greater Bay Area, stretching beyond the urban peninsula to other urban centres in the East Bay of Oakland and San Jose. As a unique coastal city surrounded by water, its infrastructure has been expanded to connect with the peripheral poly-nodal urban areas developed around the bay and with significant international airports at Oakland (OAK) and San Jose (SJC).

Bay Area, California

As the city of San Francisco sits astride the southern peninsula adjoining the Pacific Ocean, to the north after the Golden Gate Bridge are the rugged Marin Headlands and Point Reyes Peninsula and San Pablo Bay crossed by Richmond Bridge towards Berkeley. Crossing the Oakland Bay Bridge to the East Bay, there is the backdrop of the Diablo Coast Range inland to the east on the other side of the bay. To the south, the San Bruno Mountains enclose the area beyond Palo Alto and San Jose. Santa Clara Valley originally was an alluvial plain that drains into the north San Francisco Bay and to the south into Monterey Bay. With its favourable climate and soil, it was ideal for fruit farming, as developed by the Del Monte food company.

This southern part of the **Bay Area** of San Francisco which corresponds to the Santa Clara Valley is the area that has become known as Silicon Valley. Post–World War II, speculative building of suburban industrial parks commenced across the San Francisco Peninsula following US Route 101, when in 1948 Menlo Park developed a new zoning ordinance for 'administrative, executive, professional and research institutions'. This encouraged development all along Sand Hill Road that was to facilitate the beginnings of new incubator start-up businesses. With an airbase also at Moffett Field in Sunnyvale, this was used as a base for aeronautical research around which aerospace companies began to gravitate. The name emanated from the attraction of a large number of innovators into the area of high-tech corporations that were built upon the invention and development of silicon-based microchips.

This came as a result of having available land, materials, educated personnel and resources of venture capital in the region. It became an attractive environment with excellent communications and now largely urbanised as a result of annexations, and due to its attractiveness to entrepreneurial technologists. As a metonym, it is symptomatic of the IT high-tech sector with its rapidly expanding growth in northern California. The population of the Bay Area region is approximately 9 million, including San Francisco, Oakland and San Jose.

With Stanford University that was founded in 1891, under Prof. Frederick Terman, a technology park known as Stanford Industrial Park, later Stanford Research Park, was set up in 1951 at Palo Alto that served to attract many new companies. They were established initially around the development of solid-state technology, such as Shockley Laboratories, Fairchild Semiconductors and PARC, each developing their own inventions and the integrated circuit (IC) chip. With both the forces of military research and the need for venture capital for civilian technology, subsequent waves of innovation led to the establishment of more hardware companies that were to grow into major international corporations, such as Hewlett-Packard Computers and Intel, in undertaking the development of the metal oxide silicone (MOS) transistor, and corporations such as CISCO Systems.

The physical area of Silicon Valley comprises the centres of Redwood, Palo Alto, Menlo Park, Mountain View, Sunnyvale, Cupertino, Santa Clara and older San Jose localities, in the nine-county Bay Area stretching from the Bay to the foothills of the Santa Cruz Mountains. With its attractive climate and resources, and access to an expanding and educated workforce, it has become one of the largest high-tech manufacturing regions in the USA. The subsequent development of the Internet led to the creation of numerous software companies, such as SGI and Netscape, and the establishment of the headquarters of high-tech emerging companies in the area, such as Apple's Infinite Loop at Cupertino, Facebook in Menlo Park and Google's Googleplex at Mountain View. These have continued to be expanded with the construction of the Apple 2 Campus (designed by Foster/Olin) and Facebook West (designed by Gehry/CMG) and has been complemented more recently by a large number of media companies.

Regional Transactional Forces

Despite its location in a region of inherent geophysical instability, from its original plentiful agricultural resources surrounding the Bay Area, communication infrastructure was developed as mineral exploitation was sought in waves, from gold and silver, to oil and gas exploitation in the region. But it was research and the invention of the micro-chip that was to trigger entrepreneurial technologists and venture capitalists to settle in the region around which skilled resources gravitated that were to assist in developing the computer industry. With its base of hardware systems development, this evolved to software development, media and Internet applications.

The characteristic urban typology that has resulted around the metropolitan Bay Area is more of suburbanised industrial clusters and modern interspersed office parks interconnected by highways. With localised hubs and clusters of privately owned high-tech industries, research institutions on large lots with extensive parking areas, this is the typical development pattern along newly developed and landscaped corridors. Nestling around the bay with a backdrop of mountains, prime sites have been developed as campuses and corporate headquarters of the American 'tech giant' companies. With extended low-rise areas of modern research facilities, it relies on its automobile dependency for inter-connections between these high-value clusters of IT companies.

5.1.6 TEL AVIV, ISRAEL

FIGURE 5.1.5 Tel Aviv – early configuration, 1944.
Source: Maphouse.

Background and Origins

Tel Aviv is located directly on the Mediterranean coast of Israel. Without a real port, the original place was founded as a housing settlement on virgin sand dunes in 1909 under a plan by Austrian Wilhelm Stiassny, not far from the old walled city of Jaffa, which has its own small natural harbour and was the site of a very ancient settlement as early as 1800 BCE.

The original town of Jaffa has a rich history with its own special characteristics, like an amphitheatre on the side of a hill. Around Clock Tower Square, with its mosques and minarets and small trading port, it was surrounded by orchards where oranges were grown, so successful that 'Jaffa' oranges are traded worldwide. With extensive in-migration and pressures on housing, the surrounding city of Tel Aviv gradually annexed the area of Jaffa. Since the 1920s with protests of the local resident Arabs, urbanites and Fellahs who were dispossessed of their land, with further tensions in 1936, the orchards were taken over and built upon with the Jewish 'settlement offensive' and 'wall and tower' land division by the kibbutzen.

Under the British Mandate for Palestine and the High Commissioner Herbert Samuel in the 1920s, a master plan for Tel Aviv was prepared by Patrick Geddes in 1925 based on 'garden city' principles. As demand grew, new settlements took place, with one street network interestingly laid out in the shape of a menorah. A number of modern commercial and residential buildings were constructed in a defined area in the Bauhaus style, a few storeys tall, concrete with distinctive rounded corners. Whilst localised, these still remain around Dizengoff Square (Tel Aviv's first mayor), which has led to Tel Aviv having the label of the 'White City'. The main streets of Rothschild Boulevard, Herzl Street and Allenby Road with the Great Synagogue were realised at this time.

After its annexation, the old city of Jaffa was essentially destroyed. In the 1950s, the old quarter within, renamed Tel Aviv-Yafo, was redeveloped and the antique city lost most of its previous historical characteristics. A further extensive process of urban renewal was undertaken and ruins were destroyed, which included removal of the track and old train station where an early railway line had been built between Jaffa and Jerusalem after promotion by Montefiori under a concession with the former Ottomans and French funding and put into operation in 1892.

Modern City – Buildings and Arteries

After foundation of the State of Israel in 1948, with new immigration especially from Eastern Europe, the population of Tel Aviv grew more rapidly. As the State's commercial and financial centre, with its businesses and markets, more and more people gravitated to the city, which grew linearly from north to south, especially along the Ayalon Corridor. As a natural river valley, development of the corresponding infrastructure of both road and rail axes on either side is very constrained.

The first skyscraper of Shalom Meir tower was built in 1965. New commercial districts developed in the centre of the city with the tall Azrieli Towers, Dizengoff Centre, Sarona District, the university and the Financial Centre. With a new wave of immigration from the former Soviet Union in the 1990s, pressure continued on the housing sector. Other developments have been constructed farther out at Petah Tikva and Rehovot, but this did not ease the housing situation, with more recent immigration continuing, for example from Sudan. Due to water shortages, agriculture was enhanced with hydroponic irrigation. A most significant event was in November 1995 when Prime Minister Yitzhak Rabin was assassinated in the city centre.

But the city, seeking its own independence and need for survival, sought ways to continually reinvent itself as a 'high-tech' development centre specialising in software development and also in defence and security systems. It achieved this successfully through training and education, research and development with an increasingly specialist workforce.

A new direct fast rail line was built from Tel Aviv to Jerusalem to replace the old winding uphill railway route which opened in 2020. With improved connections also to Haifa and Beersheva from Central, HaShalom and HaHagana stations, and a new metro system with three new lines under development, these each assisted in alleviating severe traffic congestion in and around the city. The main Ben Gurion International Airport (TLV) is located at nearby Lod. A further economic factor is the recent exploitation of a major oil and gas field, Leviathan, just offshore in the Eastern Mediterranean Basin.

One of the older parts of the city is the Yemeni Quarter, with its bustling market at the end of Ha-Carmel Street and its mix of restaurants. With Mediterranean beaches and fine weather, new facilities and a diverse young population, it has become a vibrant place. With expanding built-up areas in and around Tel Aviv, it has grown beyond the Yarkon River, the original northern edge of the city, and is a place of multi-culturalism with a continuing flow of new immigrants. The current population of the city is approximately 0.5 million with up to 4 million in the metropolitan area.

Transactional Forces and Typology

With its desert origins, on this coastal region of the eastern Mediterranean, the original settlers' objective for Tel Aviv was envisaged to be a future modern city having European aspirations. Adjoining the old port of Jaffa around which coastal trade took place, it was planned as a Hebrew centre to co-exist also with surrounding Arab settlements. This was outlined under new plans in the 1920s, with both water and power being pivotal for its existence.

As a series of waves of refugees from European pogroms began in the 1930s, building within the municipality focused on establishing a new home for Jewish settlers. With the creation of the State of Israel in 1948, surrounding areas of land were annexed and a unified city of Tel Aviv–Yafo was created. From this, with its outer irrigated agricultural areas, commercial and industrial activities began to develop in new neighbourhoods. This was to lead to the growth of a highly developed, modern, urbanised centre that was to focus on high-tech and specialist industries.

As a coastal Mediterranean city, the city has developed linearly from its central core, north and south around the Ayalon River. With geographic constraints, this is now the main north-south communication artery, a narrow river valley that also acts as a main corridor for both road and rail transport. Whilst localised centres have developed, these poly-nodal centres are growing within the overall Gush Dan region or Greater Tel Aviv, with new growing centres radiating eastwards at Petah Tikva, south at Ashdod and north at Netanya, all interconnected by road and rail infrastructure with the centre. With available land at a premium, as population density is increasing at the centre and with new commercial and residential developments, it is dependent on outer growth poles and satellite areas to accommodate new development. The following figures illustrate the modern configuration of the innovative cities as described and the evolution of their modern population.

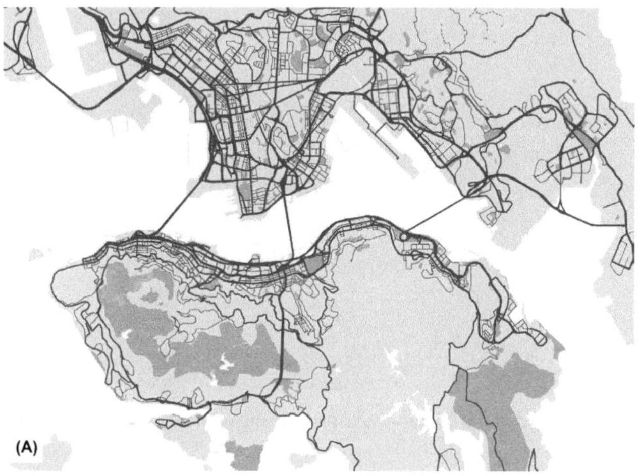

FIGURE 5.1.6A AND B Hong Kong Island and New Territories – modern configuration.

FIGURE 5.1.6A AND B *(Continued)*

FIGURE 5.1.7 Singapore – modern configuration.

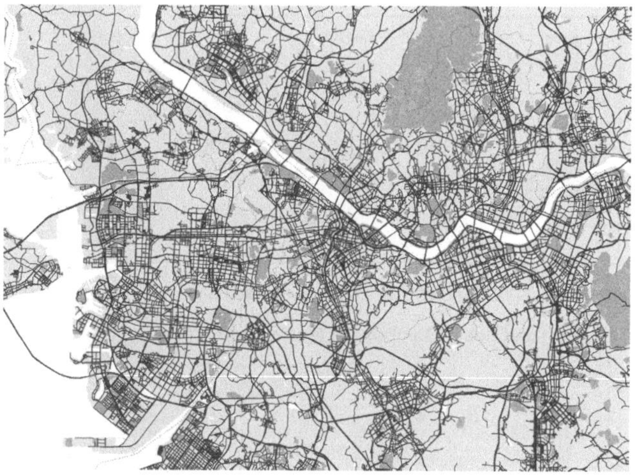

FIGURE 5.1.8 Seoul – modern configuration.

FIGURE 5.1.9A San Francisco – modern configuration.

FIGURE 5.1.9B San Francisco Bay Area – modern configuration.

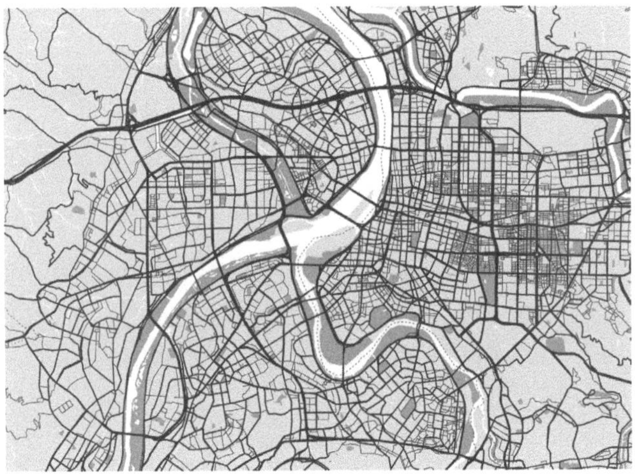

FIGURE 5.1.10 Taipei – modern configuration.

FIGURE 5.1.11 Tel Aviv – modern configuration.

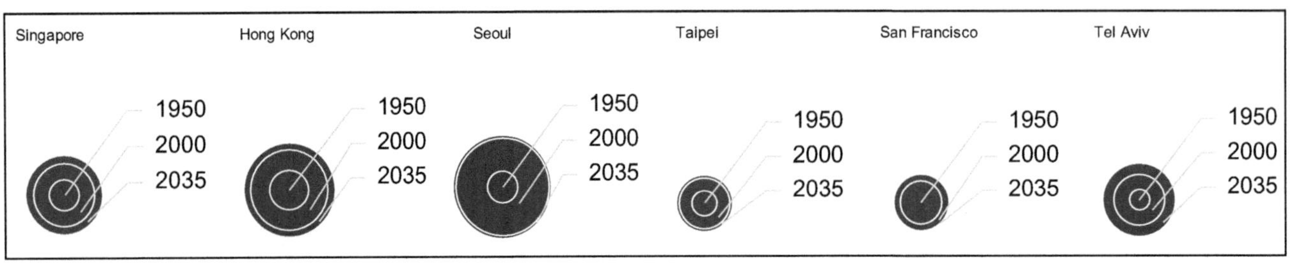

FIGURE 5.1.12 Innovative city populations.

Data source: Global Database of Metropolises 2020 – UN-Habitat (2020), UN DESA Population Division.

Further Reading

SINGAPORE

Barr, Michael – *Singapore, a Modern History*, Bloomsbury Academic, 2020.
Hamnett, Stephen & Yuen, Belinda – *Planning Singapore, the Experimental City*, Routledge, 2019.
Perry, John – *Singapore, Unlikely Power*, Oxford University Press, 2017.
Rowe, Peter & Hee, Limin, eds. – *A City in Blue and Green, the Singapore Story*, Springer, 2019.
Tan, Kenneth Paul – *Governing Global City Singapore*, Routledge, 2017.
Turnbull, Mary – *A History of Modern Singapore 1819 – 2005*, NUS Press, 2020.
Vasagar, Jeevan – *Lion City – Singapore and the Invention of Modern Asia*, Little Brown, 2021.

HONG KONG

Cameron, Nigel – *An Illustrated History of Hong Kong*, Oxford University Press, 1991.
Morris, Jan – *Hong Kong, Epilogue to an Empire*, Penguin, 1993.
Shelton, Barrie – *The making of Hong Kong from Vertical to Volumetric*, Routledge, 2010.
Tsang, Steve – *A Modern History of Hong Kong*, Hong Kong University Press, 2004.

SEOUL

Busquets, Joan – *Deconstruction/Construction Cheonggyecheon Restoration Project in Seoul*, Harvard University Press, 2011.
Joe, Wanne – *Traditional Korea, a Cultural History*, Chung'ang University Press, 1972.
Joo, Yu-Min – *Megacity Seoul, Urbanization and the Development of Modern South Korea*, Routledge, 2019.
Kim, J & Choe, S-C – *The Making of a Metropolis*, John Wiley, 1997.
Kyoung-Ho, Shin & Timberlake, Michael – *Seoul, Korea's Global City*, Routledge, 2022.
Yun, Jieheerah – *Globalizing Seoul, the City's Cultural and Urban Change*, Routledge, 2017.

TAIPEI

Allen, Joseph – *Taipei, City of Displacements*, University of Washington Press, 2011.
Kwok, Reginald – *Globalizing Taipei, the Political Economy of Spatial Development*, Routledge, 2006.
Marsh, Robert – *The Great Transformation, Social Change in Taipei, Taiwan Since the 1960's*, Routledge, 1996.
Rubinstein, Murray – *Taiwan, a New History*, Routledge, 2018.

SAN FRANCISCO & BAY AREA

Brook, J, Carlsson, C & Peters, N – *Reclaiming San Francisco, History, Politics, Culture*, City Lights Publisher, 1998.
Cole, Tom – *A Short History of San Francisco*, Heyday, 2015.
Golden, R & Unterberger, R – *San Francisco, Portrait of a City*, Taschen, 2022.
Richards, Rand – *Historic San Francisco*, Heritage House Publisher, 2011.
Saxenian, Annalee – *Regional Advantage, Culture and Competition in Silicon Valley and Route 128*, Harvard University Press, 1994.
Shaw, Randy – *The Tenderloin*, Urban Reality Press, 2015.
Sloan, Davies – *Geology of San Francisco Bay Region*, University of California Press, 2006.

TEL AVIV

Rotbard, Sharon – *Black City, White City*, Pluto Press, 2015.
Schlor, Joachim – *Tel Aviv, from Dream to City*, Reaktion Books, 1999.
Visscher, Jochen – *The White City*, Jovis Verlag, 2019.

5.2 Transnational Global Cities

Europe – America – Australasia

London, New York, Los Angeles, Dubai, Shanghai, Sydney and Melbourne

5.2.0 INTRODUCTION

Modern global cities are those cities that are involved in the global economic network of trade. Not necessarily capital or primate cities, they have grown and been able to benefit from both their particular geographical location and their combined human, physical and financial resources, which through having an inherent comparative advantage, have grown and developed to a considerable size and of dominating scale.

Characteristically, these global cities are recognised through their physical size and scale, having headquarters of multinational corporations, with energy companies, manufacturing giants and logistics centres acting as hubs. With extensive financial service companies in finance, insurance, real estate and research institutions, they provide comprehensive integrated services that afford a breadth of employment opportunities in supporting transnational corporations.

Different indicators or yardsticks can be used to develop and assess the comparative rankings of these 'global cities', which include parameters such as:

- Total output, size of the financial services sector, investment in research and development, wage levels, transport infrastructure, housing affordability, green credentials, smart facilities, lifestyle, etc.

These global cities have often emerged from either established or planned cities, as new 'transnational' world cities on different continents, which attain high scores on factors such as quality of life, innovation, research and development, skilled labour, green and smart credentials.

Whilst European cities were previously at the forefront during the Industrial Age, these have now been overtaken by cities of the South-East Asian 'tiger economies' and within China and with the American 'tech giant' corporations in cities on each coast of the USA in the modern digital age.

5.2.1 LONDON, UNITED KINGDOM

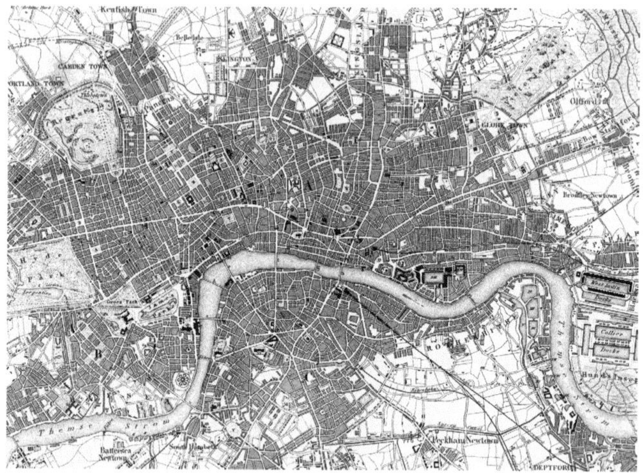

FIGURE 5.2.1 London – early configuration, 1851 (Heck).
Source: iStock.

The foundation, origin and earlier development of **London** and its historic districts have been described in Chapter 2.4.1. Significant changes have resulted in its growth over the years and its evolution into a modern city since the end of the last world war which are addressed in this section.

Modern City – Buildings and Arteries

Extensive bomb damage was inflicted across London during World War II, and an extended period of consolidation and rebuilding was required. Fortunately, some serious governmental reflection had been going on even during the war concerning reconstruction, with policies concerning land development, betterment and distribution of the industrial population being addressed under the Barlow Commission, the Abercrombie Plan for the County of London in 1943 and the Greater London Plan in 1944.

* Basildon, Bracknell, Crawley, Harlow, Hatfield, Hemel Hempstead, Stevenage and Welwyn Garden.

DOI: 10.4324/9781003353348-27

These had proposed the adoption of a surrounding Green Belt and green wedges, which set the scene for the containment of future development around the city and for managing population overspill with planned satellite suburbs and new towns farther out on the periphery. Eight* new towns were planned around London with four nearby in the county of Hertfordshire in the spirit of the first planned garden city at Letchworth in 1907 and Welwyn Garden City in 1920, following the aspirations of Ebenezer Howard.

The Festival of Britain was held on the South Bank in 1951, during a period for new hope and aspirations. But it was not until the 1960s, when the economy fully revived, that considerable numbers of new housing schemes were required for the growing population and new growth areas and public housing schemes were developed. Huge vacant sites, for example the Barbican in the City of London, an area that had been heavily bombed during World War II, were comprehensively redeveloped in the 1970s.

Docklands

As a mercantile city that had developed around its port, with its colonial empire and global reach, subsequent to the Industrial Revolution it was the institutions and financial infrastructure that supported the importance of the city in financing, insuring and trading commodities and products from around the world. This connectivity amongst the markets was emphasised eventually with Prime Minister Thatcher's 'Big Bang' in 1986 which with deregulation, secured its ongoing role as a global city.

During the 1970s the economy had stagnated and in the 1980s, with the decline in shipping in the Port of London and new methods of trade using containerisation, the old docks were abandoned and taken over by the newly formed London Docklands Development Corporation (LDDC) as a dedicated enterprise zone, and plans for new commercial and residential areas were developed. There were several other areas of urban renewal undertaken also on the South Bank and towards the East End. The old working markets, as Covent Garden (fruits and vegetables), Smithfield (meat) and Billingsgate (fish) were all relocated, leaving a battle to preserve the original historic buildings such as at Covent Garden.

The new concept of enterprise zones had been introduced into the Isle of Dogs with the creation of the Docklands Corporation. The intention was to give tax breaks to attract private investment and to incentivise redevelopment and increase employment in deprived areas. With the closure of the East and West India Docks in the 1960s and the Royal Docks in 1981, there were large areas ripe for redevelopment with shipping having been relocated to estuary ports with the coming of containerisation. By 1990, the LDDC had acquired large areas of land, much of which was sold to the private sector for development.

Whilst requiring better infrastructure and transport, the first line of the Docklands Light Railway (DLR) was opened in 1987 and was subsequently extended to Canary Wharf, where a major commercial development was being undertaken by Olympia & York. With a new road access tunnel, the Limehouse Link, this was to enhance access and communications. As the public infrastructure was improved, private commercial and residential developments were taking off, large in scale, but not always of the best quality. A new compact London City Airport (LCY) was also created at the site of the former Royal Docks.

Despite the financial crash in the early 1990s, the area weathered the storm and re-emerged to become a new centre for financial services to the east. The City of London itself, which has its own planning authority, also managed to attract a large number of property developers and by relaxing some of its previous height and bulk constraints in the vicinity of St Paul's Cathedral, gained major redevelopments, for example Broadgate and Ropemaker Place. The city grew in the provision of financial services and a new commercial centre was developed around the old docks, in Canary Wharf, that became a significant new centre of employment to the east of the original city (see Support Material: Annexe 5.2A – London Significant Modern Buildings).

Housing

With over-crowding across London, housing has always been a demanding sector with the need for public housing schemes across the capital and in different boroughs. After World War I, in 1918 Tudor Waters published a significant report on housing that was to set the standards for the period to come. It encouraged local authorities to provide new housing on undeveloped land in the city outskirts in concert with tramway extensions. Standards were to be more generous (viz. 'Nothing gained by overcrowding' Report by Unwin), as defined in the new housing manual. Whilst increasing provision, the eventual outcome was to lead to a pattern of 'ribbon development' along main transport arteries, which was emulated across the urbanised region.

Pressures on land development continued to occur, contained by the 'green belt' and then by leap frogging outwards in a movement of suburbanisation around London, with occasional growth poles as envisaged under the

Greater London Plan. Whilst managed and contained in some respects, suburban sprawl ensued.

After World War II, the London County Council (LCC) had developed a formidable architectural design department. Many skilled designers under the leadership of Leslie Martin, who had contributed to the Festival of Britain site on the South Bank in 1951, developed creative public housing schemes across different boroughs, notably at Alexandra Road, Brunswick Centre, Churchill Gardens, Golden Lane, Lillington Gardens and Roehampton Estate.

Much later, with the Summer Olympics in 2012, regeneration was further reinforced on a new site at Stratford, where an Olympic Village was constructed to be converted for subsequent housing and the Jubilee Underground line was extended, leading to the promotion of this former brownfield area of east London.

Administration

Post-war, with the growth of the population, existing administrative arrangements were found to be inadequate to handle the growing demands being placed upon local government, and a Royal Commission Enquiry on Local Government was set up in 1960, which recommended replacement with a Greater London Council (GLC) and to be increased in size to cover an extended area of 32 new London boroughs with functions divided between the council and the boroughs. With the additional creation of the Inner London Education Authority (ILEA), it operated since 1965 with power oscillating between the two national parties, Tory and Labour. However, by 1980 different rival factions developed and in order to rein in the power of opposing socialist leaders, it was abolished eventually by PM Thatcher.

Subsequently, more powers were devolved to the local authorities and a new Greater London Authority (GLA) was established in the year 2000 having a mayor and a 25-member local assembly located in a newly built city hall. Mostly funded by local government, it also acts as a 'percepting' authority and raises money from local council taxes. It also introduced an area-based zone of congestion charges on vehicles in the central area in 2003 in order to reduce congestion and pollution, more recently extended to include an Ultra Low Emission Zone (ULEZ) for road vehicles.

The DLR light railway had been pivotal in enhancing accessibility that supported the redevelopment of the London Docklands. The extension of the Jubilee Line 'Underground' eastwards had also been an essential component along with upgrading the 'Overground' Railway. With the new fast 'Crossrail' across town (Queen Elizabeth Line) to link Paddington in the west with Stratford in the east, via the central stations of Bond Street and Farringdon, this was to strengthen rail passenger movements, in terms of speed and access across London. Across the river, the Thames Barrier was built at Woolwich and opened in 1982, which due to the general tilting of the whole London Basin plus climate change, aims to protect the city from increasingly frequent and intense storm surge events, which now requires the flood gates to be raised more often.

Transactional Forces and Typology

As a city dependent on its maritime trade since the Industrial Revolution, the Port of London has played a pivotal role in its development. Whilst this has shaped the narrowing estuary with its wharves and the East End, the transformation of the former docklands is now complete, as redevelopment has occurred. Containerisation has moved handling downstream, and new businesses have evolved instead in the service sector, with financial services and the digital economy coming to the fore.

With de-industrialisation and changes in the financial and service institutions with computerisation, changes have developed at the workplace with fewer floor area requirements. Formal corporate ties are being loosened and with the gig economy, there is more flexibility in the workforce.

Demand for good homes still remains and with a deficit in affordable homes, the housing sector is unbalanced. But across the whole urban area, accessibility and good transport facilities are still vital.

From its origins on both banks of the River Thames, London has developed with its separate cities of Westminster and Southwark and the City of London itself. Now more amorphous growth has taken place with infills, renewals and replacements, as the built-up area of Greater London is constrained within an encircling green belt. With its radial and circumferential infrastructure axes, of both road and rail, satellite images demonstrate the ribbon development in the built-up areas of this region.

With its differing multicultural residential neighbourhoods, London and its boroughs has a poly-nodal structure which is self-supporting at a local scale. Less dependent on its maritime and industrial role than previously, now the whole city and urbanised region is embedded commercially in the global network of services and media, transformed and driven by the immediacy and speed of communications yet still concerned with proximity in the functioning of the modern digital economy.

5.2.2 NEW YORK, NY, USA

Background and Origins

With the narrows at the base of the peninsula first discovered by the Italian Verrazano, it was not until Henry Hudson in 1609 voyaged south from Newfoundland, acting on behalf of the Dutch, that the Inner Bay and the land of the Lerape were discovered. Hence the first settlers in Manhattan Island were Protestant Walloons from Holland, whom under trading rights of the Dutch West India Company began the development of New Amsterdam in New Netherland in 1625. With a fortress and surrounding plots and dwellings at the tip of the island built behind a palisade, from here a small network of streets was laid out across the peninsula with wharves facing east. The adopted name 'Manhatta', coming from the Munsee Algonquian Native language, means 'small island'.

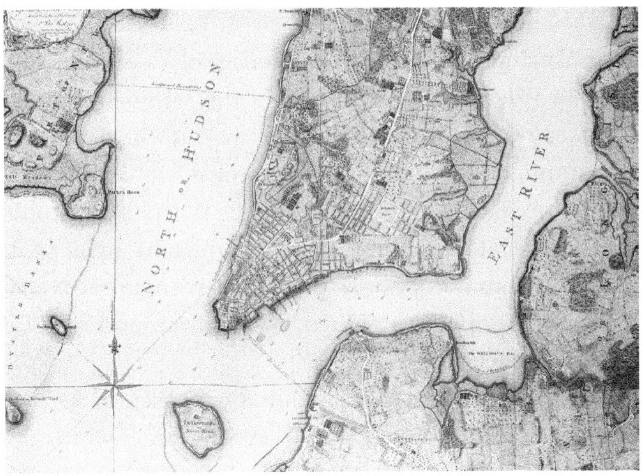

FIGURE 5.2.2 New York – early configuration, 1770 (Ratzer).
Source: British Library.

Streets were developed with a main street called Broadway and a small canal beside Water Street and fortified with a northern wall along Wall Street. Along the waterside, developments began to extend up the shoreline along both the broad western Hudson and the East Rivers, with ferry crossings installed across to the Jersey and Long Island shores. Other smaller Dutch settlements were established, such as Bowery Village and Harlem, and farther up the Hudson River the town of Albany developed as a centre of trade, especially fur, in this emerging region.

In the mid-17th century, after the power of the English Crown strengthened, in 1675 the Dutch had lost control to the English. King Charles II secured and bestowed the area of New Netherland as a royal province to his brother, the Duke of York, which led to its renaming as the Province of New York, and who started planning new streets and building sites within individual wards. This extended north beyond the border of Wall Street, with further recti-linear platting of new sub-divisions having principal north-south avenues and crossing east-west streets.

At this time, the streets of **New York** broadly followed two orientations, hugging the rivers, with a grid from the Battery parallel to the Hudson River and Hudson Street up to Christopher Street, and another grid parallel to the East River and Water Street, up to Grand Street. In the middle a third orientation developed broadly north-south on either side from the Bowery towards Union Square.

The surrounding area was subsequently divided into counties, namely Queens, Kings (Brooklyn), Richmond (Staten Island) and New York (Manhattan), with the eventual later addition of Bronck (Bronx). During the 17th century, with religious persecutions across Europe, under Dutch Calvinism, the area tolerated religious pluralism as the first permanent immigrants began to arrive of French Huguenots and Sephardic Jews. Enslaved African Americans were also being brought in as slaves.

But with the Revolutionary War, the British retained occupation of the city with the Battles of Brooklyn Heights and Harlem Heights in 1776. It was not until 1783 that General Washington eventually entered the city, causing the defeat and evacuation of the British troops. New York was finally to become not only the state capital but also briefly the national capital managed from city hall. Subsequently, as late as 1790, it was determined that one in five households retained at least one slave. There was only a gradual process of emancipation, with eventual full abolition not to be attained until 1827.

As recovery from the war began, a charter was granted in 1797 to the Manhattan Company to develop utilities and infrastructure in the city, and a federal Naval Yard was developed on the East River to support maritime trading. From a survey made in 1796 of the land north of Hudson Square, a Commission was formed to assess the layout of land north of a line running east-west across Washington Square.

The Commissioners Report in 1811 confirmed the basis of the planned gridiron system, which was to be adopted with these 12 avenues (100 ft. wide) and 155 cross streets (60 ft. wide) running across Manhattan to be paved, effectively confirming the earlier survey between the northern 23rd and 93rd streets. This was to form the basis of the familiar resulting New York planned grid (of 260 by 700 ft.) with around 2,000 blocks, typically of 200 by 600 ft. with the longer sides facing north and south. The only exception to this was the diagonal running Broadway and occasional open public spaces such as Washington and Madison squares.

During 1832, a significant cholera epidemic broke out. Public health was generally poor and only improved when the construction of the Croton Aqueduct was realised with the High Bridge bringing fresh water to the city from Westchester County and the realisation of a proper sewerage network. The need for this was further accelerated after the Great Fire of 1835.

After the Civil War in the 1860s, the City of New York benefitted from its strategic location for trading, with new infrastructure and railroads as finance, banking and manufacturing facilities grew, enabling it to develop into the first built American metropolis. As immigration continued, and with increased affluence, new commercial and residential buildings were developed, from fine brownstone buildings along 5th Avenue and with major retailers to emerging major high-rise commercial buildings downtown.

With the growing population in the city, during the mid-19th century, the need for open spaces for recreation purposes became evident. With the appointment of commissioners in 1853 to assess siting, a competition was held for converting the land between 5th and 8th Avenues and 59th Street and 110th Street into a grand park. The winning scheme by Olmsted & Vaux, 'Greensward', divided this into two parts, with the southern area having varied topography and winding paths and the northern area having a lake and water reservoir connected to the aqueduct to serve the city. This Central Park was also to have a set of transverse sunken crossroads to enable crossing by horse carriages to separate pedestrian and wheeled transport, intended to become a large area for recreation, a green 'lung' for the residents of central New York.

The city benefitted from its natural location having a deep harbour protected by outer islands, with channels leading from the main ship channel into the narrows between Brooklyn and Staten Island. With a long line of wharves constructed along Manhattan from the Battery to South Street on the East River, these were also extended on the western side up the Hudson River able to wharf the first trans-Atlantic 'clipper ships'.

With its river connections upstate and subsequent rail connections west to Chicago and south to Baltimore and beyond, its communications reinforced the city's position as a manufacturing and processing hub, with its attractive location leading to it becoming the headquarters of many large corporations. With its maritime connections and benefitting from its man-made infrastructure, as the Erie Canal opened in 1825 to connect Albany and Buffalo across New York State and with railroad lines and highways into Long Island, the city was the first in North America to develop its full industrial capability.

Further waves of immigration occurred with Irish, German, Italian, Polish and Russian migrants. In 1921, the National Origins Quota Act aimed to balance the racial and religious mix of immigrants; however, the wide ethnic diversity of New York continued with localised enclaves around the city, especially of African Americans in Harlem.

During the 19th century, much of the development of New York evolved under the patronage of Tammany Hall, built in 1812 as a meeting place for the Tammany Society that was founded for Native Americans but who welcomed new immigrants. This came to exert considerable influence under the control of different 'bosses'. With Ellis Island as the first point of arrival for many immigrants arriving in the USA from Europe or farther afield, many areas of poverty and poor housing existed, especially around the Five Points area near the Bowery.

As the population grew, with industrialisation the city also developed its own financial services, upon which the early industrial barons relied. In particular, investment banks gravitated around Wall Street and emerging trading markets, and with other institutions and law firms these were to provide the depth of resources used to finance, promote and realise major domestic and international projects.

Districts

New York City is typically identified by its core of Manhattan Island, with its dense vertical buildings in the central core rising from the bedrock upon which they are founded. It was not until 1898 that the five boroughs in New York – Manhattan, Brooklyn, the Bronx, Queens, and Staten Island – were consolidated into a whole to act as New York City (NYC) under one mayor. Whilst having different characteristics and individual 'presidents', as an ensemble they house 10 million people and also provide daily employment for around an equal number. During the 20th century a strong mayoral–council system developed that aimed to promote the interests of all boroughs and as a major international city the mayor now carries considerable political weight.

With its prevailing regular grid system, there are different neighbourhoods across the city. The largest open area, Central Park, remains the green oasis in the midst of the daily thrust of constant traffic. Adjoining this to the north is Morningside Heights and the district around Columbia University with its rocky granite outcrops and rising elevation up to the district of Harlem.

Southwards down Manhattan both the Upper West Side and Upper East Side have distinctive features of mainly residential areas. Midtown developed as the commercial area with the Theatre District around 42nd Street and with major department stores. The large Rockefeller Center complex, the major transport hub of Grand Central Station, the great Empire State Building and the classic Chrysler Building each stand out as dominant iconic landmarks in Midtown surrounded by a large number of commercial skyscrapers that are continuing to be built around them.

Other distinctive districts have also evolved in different neighbourhoods, with Chelsea and the Garment District (now the Fashion District), the Meatpacking District and more recently the rehabilitated old railroad Hudson Yards.

Farther south the regular grid is broken up by the occasional squares of Madison Square, Union and Stuyvesant squares, then Gramercy Park, around Greenwich Village and Washington Square Park where residents helped to pass the city's Landmark Preservation Law. Broadway cuts across the rectilinear grid with the triangular Flatiron Building and farther south there are the SOHO (South of Houston Street) and TriBeCa districts, Chinatown and Little Italy, with Canal Street and the Lower East Side. The Financial District and the Old Stock Exchange remain around Wall Street and in Lower Manhattan, Battery Park City adjoins the site of the World Trade Center and its rebuilt replacement.

Although the city is regulated by its gridiron layout, on each corner there are extensive views in each direction. With the waterside never far away of either the Hudson River or the East River, with old wharves and dominant Brooklyn, Manhattan and Williamsburg bridges. All of this encapsulates Manhattan before venturing out to the other districts of Brooklyn, the Bronx, Queens, and Staten Island.

Buildings

Since the original Commission on the Height of Buildings and the Zoning Commission, the size and form of the streetscape in New York has been driven by the prevailing planning and zoning laws. The first zoning law in 1916 determined the footprint, lot, floor area, number of floors and setback restrictions in order to preserve daylight, which led to a ziggurat, or stepped, profile of the early tall buildings. Development was permitted to a certain height, above which the floors were to be cut back at an appropriate angle so as to admit light to the streets, but with 25% of the plot area allowed to be continued to unlimited height. These laws evolved from initial consideration of just building bulk to more modern modelling of light angles and sky exposure planes.

As the state laws were varied from time to time, this led to evolution in the permitted height, shape and configuration of the tallest commercial buildings in the city and their relationship with adjoining streets. Although developed in prescribing the private realm, they also framed the public realm.*

There was a revision to the city zoning law in 1961 which altered the emphasis to consideration of floor area ratios (FAR) and building bulk using setbacks, and also, in exchange for private developers providing public spaces, additional storeys would be allowed. Also transferable development rights (TDR) from adjacent sites was allowable for additional permitted development.

During the 21st century, designs have evolved from postmodernist styles at the end of the 20th century to parametric styles and beyond. With recent advancements in technologies and structural design, super-tall or very tall super-thin pencil buildings have been designed on narrow plots able to be founded on the NY bedrock, which may have either perforations at periodic levels or dynamic mass dampers that have enabled wind forces to be attenuated on these very high, slim buildings.

With the transport hub, the Oculus (designed by Calatrava) at the rebuilt World Trade Center, major new iconic buildings are Hearst Tower (designed by Foster), Spruce Street (designed by Gehry) and super-tall buildings such as the slender 432 Park Avenue (designed by Vinoly) with

---

* Whilst the grid has divided the whole of Manhattan into potential rectangular building blocks of 200 by 600 ft., aligned with the long axis east-west, there is potential for unusual development. For example, the 600 ft. length may be subdivided into three, resulting in square plot areas of 200 by 200 ft. This has led to tall towers with lower podiums, stepped developments or even super-tall developments above 1,200 ft. (or 400 m.). The Rockefeller Center is an example that profited whereby a major group of buildings was constructed between 5th and 6th avenues and 48th and 51st streets, comprising one tower, the RCA (GE) Building, a private street and a sunken plaza with several other lower-rise buildings.

further iconic buildings continuing to be added to the central cityscape (see Support Material: Annexe 5.2B – New York Significant Buildings).

In terms of **Housing**, since the mid-19th century, with the growth in population, there were a large number of tenement houses built, such as the notorious 'dumb-bell' tenements that were cramped, almost lightless and had severe overcrowding. With the Tenement House Acts passed in 1867 and 1879 that required fire escapes and windows, and a final Tenement Act in 1901, the new law aimed at providing courtyards with improved light and air access. As a result, the Lower East Side became home to many of these 'new law' tenements, especially with back tenements. Whilst physical regulation of private developers was introduced, the challenges especially for housing poor immigrants remained.

The introduction of zoning and building regulations was to have an effect. After the Great Depression in the 1930s, emphasis was placed on particular new public or social housing developments. These tended to be developed in cycles according to the emerging housing models that prevailed across the city in different periods. Some of these, as cited hereafter, had their own character in different boroughs, for example:

Forest Hill Gardens	1909	Queens	Olmsted
Sunnyside Gardens	1923	Queens	Stein and Wright
Hillside Homes	1933	Bronx	Stein
Harlem River Houses	1937	Manhattan	Brown, Ginsbern, Wilson
Williamsburg Houses	1938	Brooklyn	Shreve, Lescaze
Red Hook Houses	1938	Brooklyn	
Queensbridge Houses	1939	Brooklyn	
Stuyvesant Town – Peter Cooper Village	1947	Manhattan	Clavan

During the 1930s, under the New Deal, the Federal Housing Administration (FHA) introduced federal mortgage insurance in 1934, which stimulated the construction industry. This began to change the shape of social housing with, for example, the Harlem River Houses and the Williamsburg Houses in Brooklyn, funded by the Public Works Administration (PWA). The New York Housing Authority was also created in 1934.

Later, during the 1960s, with a period of significant urban unrest in America, President Johnson launched the Model Cities Program in 1964, aimed at targeting remaining slums, through the creation of community development agencies. Despite several demonstration projects, this exercise never really took off, with a strong reaction against sweeping urban renewal programmes, such as at Harlem and Cooper Square in New York introducing the practice of 'incentive zoning'.

After the end of World War II, with the creation of the United Nations, a site was sought for a new world headquarters for the UN, to host the General Assembly and Secretariat Building. In 1946, JD Rockefeller donated a sizeable site on the east side of Manhattan, adjoining the East River that was subsequently designed by Le Corbusier and associates and constructed. With a World Fair in New York, the 'City of Light' in 1939, New York again hosted the World's Fair in 1964 under Robert Moses (with Walt Disney presentations), who had been responsible for realising several major highways across the city.

Transport Arteries

With Manhattan virtually an elongated island, transport in and out of the city has been determined by its crossings, bridges and tunnels. With the surrounding Hudson River and the East and Harlem Rivers, these have been challenges in order to satisfy the daily flow of people and goods in and out of the city. The first major bridge across the East River, the Brooklyn Bridge, was commissioned in 1869.

Farther beyond, an outer network of arterial roads and parkways was developed around the city. In a regional context, routes from Long Island, New Jersey and upper New York State each lead into the city. In 1929, a Regional Plan for New York and its environs was developed by Clarence Perry with a proposed highway loop around Manhattan that would include the George Washington Bridge (designed by Ammann) in 1931, the Lincoln Twin Tunnels and the Holland Tunnel built in 1927 under the Hudson River into New Jersey. The major infrastructure of these tolled bridge and tunnel crossings was developed by the Port Authority of New York and New Jersey created in 1921 that was to benefit from tax exemptions (see Support Material: Annexe 5.2B – New York Significant Bridges and Tunnels).

Under the efforts of Robert Moses in developing parkways during the 1930s across Long Island, a number of new bridges were also built to supplement the original

Brooklyn Bridge across the East River that included the Williamsburg, Queensboro and Manhattan bridges, the Triborough Bridge complex (RF Kennedy) and Hell Gate Bridge, Queens' Midtown Tunnel and the Hugh L Carey Tunnel into Brooklyn. The third tube in the Lincoln Tunnel was built in 1957, the second deck of the George Washington Bridge was built in 1962 and the Verrazano Bridge was built later in 1964.

For aviation, LaGuardia Airport (LGA) was opened in 1939 and post-war, in 1948, a new airport called Idlewild was opened in Queens, on the border of Jamaica Bay. Idlewild subsequently developed into a major international airport and was renamed John F Kennedy International Airport (JFK) in 1963. It is now leased and operated by the Port Authority of New York and New Jersey and has multiple international terminals.

In terms of rail, both the East and the Hudson rivers had to be crossed in order to gain access to Manhattan. The New York and Harlem Railroad was opened in 1831, with its extensive goods yards, that have recently been converted around Madison Square Garden. The New York and New Haven and the Hudson railroads were developed during the 1840s, as was New York Central and, in turn, the Long Island Railroad (LIRR).

Within Manhattan, new elevated rail lines were built on steel structures up the east side on 2nd and 3rd avenues, and the west side on 8th and 9th avenues. These replaced early omnibuses and horse-drawn streetcars and were the beginning of the mass transportation system now embedded in the city. With the continuing growth in population in densely built and populated Manhattan, new underground lines were next developed with the first subways of the Interborough Rapid Transit Company (IRT) and the Brooklyn Manhattan Transit Company (BMT) as public-private companies. New lines soon followed in creating the dense linear subway network that exists today.

New railroad terminals were required downtown – with the financing of Cornelius Vanderbilt, Grand Central Terminal was developed on land acquired between Madison and Lexington avenues in Midtown, with a major covered double-deck terminus and classical facade, which opened in 1871. With its large footprint, this led to the subsequent property development using 'air-rights' over the terminal, with the MetLife (PanAm) Building. The other major terminal realised was the Pennsylvania (Penn) Station for the Pennsylvania Railroad, recently modernised.

Administration

In the early 1970s, the city of New York was under great duress financially, the Urban Development Corporation (UDC) collapsed in 1975 and New York City Securities were downgraded. Close to default and without federal support, in October 1975 eventually a rescue package with a programme of federal loan guarantees had to be instituted, and it was not until the 1980s that financial health returned.

Significant upgrading and investment had become a necessity in the city's infrastructure that was a challenge for subsequent mayors. In 2007, Mayor Bloomberg set about developing a strategic plan for the future development of the city. PlaNYC was developed with a set of initiatives which defined measurable objectives for enhancing specific sectors and infrastructure of the city.

The subways and bus network are run by the Metropolitan Transit Authority (MTA), each having their own extensive network. Given the traffic demand, levels of pollution and traffic congestion in the city, new methods of 'congestion charging' have been implemented in order to control the traffic volumes entering Manhattan, especially at peak periods. Accordingly, for the area south below 60th Street, a new cordon has been instated in order to charge vehicles driving into the city on weekdays, and parking charges were raised.

As the modern city evolved from manufacturing and warehousing to finance and services, now in the digital era the emphasis is on logistics and 'just-in-time' deliveries ready for collection and 'last-mile' deliveries to final customers and consumers. The central street networks accordingly are having to be adapted to neighbourhood loading zones (NLZ) for truck parking and goods transfers which are transforming movements in central neighbourhoods. The population of New York City is approximately 8.5 million at an average density of 10,500 pers./sq. km. and having 20 million in the metropolitan area.

Transactional Forces and Typology

Whilst not the political capital, by many measures the city of New York is perceived as the effective business and commercial capital of America. Since its early settlement and trading, its natural features with a large, protected harbour have been instrumental in the building of perimeter wharves along Manhattan such that its river and marine connections have been pivotal for the movement of goods and people, later enhanced by the addition of railroads.

The other fundamental but less evident feature is the bedrock of the city, upon which numerous large and subsequent skyscraper buildings were able to be founded and with considerable density with the development of technology. This has served to create a pole of attraction where the density and proximity of businesses has afforded work opportunities for the continually evolving labour force. Different types of businesses have coalesced around different neighbourhoods, according to the grouping of differing cultures, traditions and races. Hence in a cyclical manner, with its phases and waves of varying migrants, this has served as the business engine for New York. As buyers and suppliers to and from each other, these continued interactions of inputs and outputs, materials, labour and technological skills have generated the growth of intensive commercial and business activities across the city.

With its familiar central grid of cross streets and avenues creating regular-sized blocks in Manhattan, the built form of the city has been regulated by evolving zoning laws in each district, enabling developers to benefit from afforded opportunities. This has given the city such distinctive patterns of individual or stepped skyscrapers in different blocks. With its street transport axes, strengthened by rail arteries built both above and below ground, these have led to extended development to the outer districts.

The typology associated with constructed Manhattan and its skyline is defined by the planned horizontal rectangular grid and having distinctive vertical 3D built forms of varying bulk and height, elevated or set back, as determined by available ground footprints and associated air-rights for development.

With New York comprising not just central Manhattan, but the individual boroughs of the Bronx, Brooklyn, Staten Island and Queens, each has its own distinctive physical and cultural characteristics linked by their bridge crossings into each interconnecting area. Combined, these form an integrated metropolitan area under one sole administrative jurisdiction of the New York City Authority. However, Greater New York extends outwards along with the adjoining New Jersey region to constitute a burgeoning greater 'megalopolitan agglomeration' of the North-Eastern Seaboard of the USA, reaching both northwards to Boston and southwards towards and encompassing the cities of Philadelphia, Baltimore and Washington DC. This is addressed further in Chapter 6.2.1.

5.2.3 LOS ANGELES, CALIFORNIA, USA

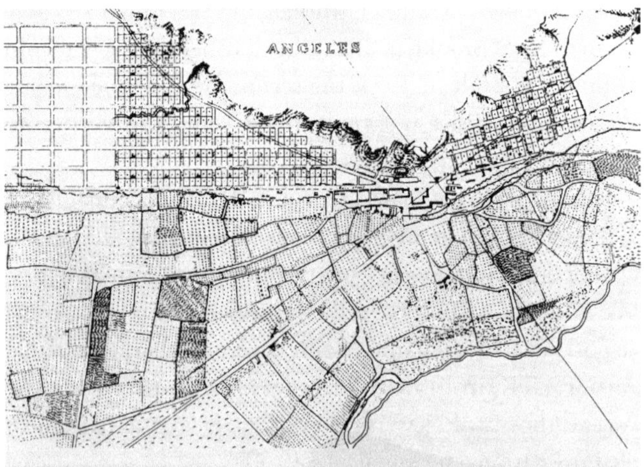

FIGURE 5.2.3 Los Angeles – early configuration, 1849.
Source: Historic Urban Plans.

Background and Origins

Los Angeles lies in a basin of Southern California alongside the Pacific Ocean, with a backdrop of the San Gabriel Mountains farther inland. It was originally home of the Chumash and Tongva tribes in the coastal region and lies on the San Andreas fault, the geophysical boundary between the Pacific and North American tectonic plates.

The area was first discovered by Europeans with the Spanish arriving from Central America and by João Cabrilho, Portuguese, who reached Santa Catarina Island in 1539 and Santa Monica Bay by sea. But it was not until 1769 that the Franciscan missionary Juan Crespi and a group of Los Pobladores who reached the area established a settlement that came to be known in 1781 as El Pueblo de la Nuestra Señora la Reina de los Ángeles. Not far from the Los Angeles River, settlements gradually grew around this plaza, which exists as an historic district in the centre of Los Angeles today.

As a 'pueblo' (civil) setting rather than a 'presidio' (military) or 'mission' (religious) in Alta California, a settlement of four square leagues was established, with one league in each cardinal direction, but orientated at 45 degrees from the north, which remains the basis of central downtown today. With New Spain attaining independence from the Spanish Empire, and as part of Mexico until 1846, the settlement was taken over by the Americans at the end of the Mexican–American War.

Whilst having both flat and hilly characteristics, the land had plenty of potential, yet water was critical. Despite lying on the geophysical fault, petroleum was discovered in the area at Inglewood in 1892, a significant find which was to generate much potential. With the platting of new land, the construction of new railroads also began with the Southern Pacific and the Santa Fe and new Union Station. The infrastructure of the city was also developed, with major aqueducts realised by the LA Water Department of the Board of Public Works under the leadership of Mulholland, to bring drinking water from Owens Valley in Sierra Nevada via aqueduct to a storage reservoir in the San Fernando Valley and also via the Colorado River aqueduct. Here large agricultural areas were developed for citrus fruit production.

At the same time, significant development occurred with one of the largest electric transport systems, the Pacific Electric Railway. Developed by Huntingdon, its red tram cars carried 100 million passengers annually to Long Beach, Santa Monica and Pasadena where Huntingdon was able to develop residential land tracts supported by this new infrastructure.

During the 1920s, interstices between these rail tracks began to be filled in and developed as car ownership increased. In a Report for the Traffic Commission of the City and County of LA, a major traffic street plan was outlined in 1924 by Olmsted, Bartholomew and Cheney. This led to decentralisation and further dispersal into outer areas. As new origins and destinations developed, new boulevards were built, such as Wilshire Boulevard connecting downtown with the coast which was the beginning of the realisation of a number of major highways after the Arroyo Seco Parkway in 1938.

The first municipal zoning ordinance had been introduced in 1908 for residential and industrial land use. New built districts continued to be created and annexed and Los Angeles hosted the Summer Olympics in 1932. By now rail traffic was declining as road traffic was starting to grow exponentially. But the next year there was the Long Beach earthquake.

The first 'Air Meet' was held in Los Angeles in 1910 and was the beginning of a period of aviation development and subsequent production of propellor-driven aircraft. In the 1920s, there were already over 50 flying fields around the area of Los Angeles. During World War II, the region became a centre of manufacturing, especially of aircraft, with Douglas, Hughes, Lockheed and Northrup each contributing to the war effort from Hughes Airfield, Lockheed Air Terminal in Burbank (BUR), and Westchester Airport, which eventually became Los Angeles International Airport (LAX). This continued after the war with the development of jet aircraft from the Jet Propulsion Laboratory with CalTech at Pasadena, and new residential communities were built at Westchester, Westside Village and Toluca to accommodate the aircraft workers. Other airports in the area now include Long Beach (LGB) and Santa Ana (John Wayne) Airport (SNA) in Orange County.

During World War II, the Kaiser Company built a large number of ships at Richmond and San Pedro, and a new steelworks to support this was built at Fontana, formerly an agricultural area with coking coal imported from Utah. In addition, Kaiser subsequently set up a method of building prefabricated houses, 'Kaiser Homes', followed by Kaiser-Frazer Motors automobiles. The steel plant was to remain in operation until 1983, when it was finally closed due to cheap competition and as an anti-pollution measure since it was a key contributor to smog in the area. Whilst planned to be rebuilt with new housing developments, these never succeeded and the area remains a blighted neighbourhood east of Pomona.

The core areas of Los Angeles continued to be developed, radiating out from the central area with Main Street in both directions along the coast, constrained between the canyons of the Santa Monica Mountains in the north-west, via Long Beach and the port towards Orange County and south of the Santa Ana Mountains. Inland development continued up to the San Gabriel Mountains with denser areas east of Hollywood at Burbank, Glendale, Pasadena and east to El Monte, Pomona and beyond, housing new groups of Hispanic and Chinese residents. The new film and media industry also established itself and began to develop around Hollywood.

With continued growth and the need for more homes, further annexation and land platting occurred, with more highways, new communities and low-density 'edge cities' developing in every direction. Much of this was developed with the support of the Federal Housing Authority (FHA) that after World War II guaranteed mortgages on new properties.

In 1939, the Los Angeles Freeway Plan was developed, converging downtown with the Pasadena, Hollywood and Harbor freeways. In 1956, after the Interstate Highway Law, the wider interstate highway system was developed across California with the Santa Ana and San Bernardino freeways. Knitted together by intervening streets and highways, the region of California developed to become wholly auto dependent, as an archetypal grid city, its distinctive neighbourhoods all interconnected.

From the earliest communities, such as Lakewood Park, developed by Kaiser, and with a regional shopping centre in 1950 near Long Beach, dozens of incorporated municipalities continued to emerge farther out in Orange, Santa Ana and Anaheim, covering former countryside with roads and tract housing. There were setbacks, however, in the development of the city itself and inadequate provision of housing that resulted in the Watts riots in 1967, subsequently documented by the Kerner Commission, and further riots across the city in 1992.

Accordingly, the modern city has become a collection of multiple suburban developments and the accretion of many new neighbourhoods as Greater Los Angeles now extends over many square miles of southern California. Highways, malls, parking lots, warehouses, industrial parks and horizontally built stretched developments are penetrated by localised nodal cores of commercial high-rise vertical developments near highway intersections (see eResource: Annexe 5.2C – Los Angeles Significant Buildings and Freeways). And in 1971 another major earthquake occurred at San Fernando.

Modern City – Buildings and Arteries

The modern 'City of Angels', a sprawling multi-centred place of numerous towns of nearly 500 square miles that makes up the city of Los Angeles, with residences averaging only 1.4 dwelling units per acre, stretches from Long Beach in the south to Santa Monica, Beverly Hills and Hollywood, all merging and knitted together by a network of wide highways. As Dorothy Parker notoriously noted, there are '72 suburbs in search of a city'.

Districts

The original downtown grew around Central Avenue, the 'stem' of Los Angeles. Here was the original commercial centre with city hall, commercial offices, hotels, Union Station and Market Street. The Financial District developed around it, which grew extensively at the end of the 20th century with Pershing Square and Bunker Hill as key features. The civic areas are city hall, with Civic Centre in Grand Central Square. With an increasing population from Latin America it has become a Catholic city, exemplified by the modern cathedral La Reina de Los Angeles (La Placita).

Leading westwards towards the coast is Sunset Boulevard, a contrasting winding scenic road past Beverly Hills where the rich and famous reside in their mansions. The road drops down through the canyon towards Santa Monica, an elegant resort with its palisades, pier and promenade, where the beach stretches westwards and on towards Malibu.

Hollywood has its own flavour, as a centre of the entertainment industry, with the circular Capital Records Building, its boulevard, the strip and 'Walk of Fame', studios and theatres. Hollywood is home to Universal Studios, the Hollywood Bowl and the Hollywood Sign up in the hills near the Griffith Observatory. In the 1920s, the film industry became established in Hollywood, Burbank and Culver City with MGM, Warner Bros and Paramount as providers of film and music. 'Film noir' and 'TV flix' are new creators in Hollywood with Universal City and the Smart Strip.

In 1955, Disney created Disneyland in Anaheim, a whole new environment to cater for such entertainment desires in the post-war era of growth, with extended aspirations of the American Dream citing 'Frontierlands, Autopia and Tomorrow's Land'. Not just an amusement park, with hotels, it became a dedicated destination for the whole family.

Farther beyond, the new town of Irvine was built, the site now of USC Irvine. On the other side of the canyon is Venice Beach, a rougher place with its Muscle Beach, hippy homes and the boardwalk. Tucked way are some gems such as the Getty Center (designed by Meier), an extravagantly endowed museum and arts complex. Nearby is LACMA, the LA County Museum of Art, with its eclectic collection and the La Brea tar pits where oil continues to ooze from the ground, with the Automotive Museum and the new Academy Museum of Motion Pictures (designed by Piano).

As a financial and commercial centre, Los Angeles does not rival Wall Street in New York but has commercial towers such as the Wells Fargo Building, Wilshire Grand and the First Interstate World Center (designed by Pei) in the financial centre. Nearby is the parametric Walt Disney Concert Hall (designed by Gehry) and the Broad Museum (designed by Diller Scofidio + Renfro) bringing geometric shapes to the centre alongside the modern Catholic Cathedral of Our Lady of the Angels (designed by Moneo). Nearby also is Caltrans District 7 HQ Building in Main Street designed by Morphosis.

But the megalopolitan city of Los Angeles now stretches east to west from the border of Ventura County, from Santa Monica, Beverly Hills, Hollywood and Culver City to Downtown and East Los Angeles, and north-south encompassing Burbank, Glendale and Pasadena through the central districts of Inglewood and Southside towards Long Beach. It is an enormously developed but low-density region – the population density in the whole city region is only approximately 4,600 pers./sq. km.

Altogether, the whole of the dispersed sprawling city of Los Angeles has generated another style of life, from motor cars to movies, always changing and continuing to re-invent itself, unsurprisingly nicknamed 'La-la Land'. Blessed with over 330 days of sunshine each year, with a long coastline, inland hills and canyons, it has become a magnate for the music industry and media despite its increasing pollution, congestion and paucity of water. The alchemy of Los Angeles as described by P Lunenfield is based on 'earth', 'water', 'air' (aviation) and 'fire' (oil). In re-imagining LA he added 'entertainment' – the other basic desire of human motivation in America, along with its new worlds of the 'metaverse' with activities of 'gamification' and 'Disneyfication'.

Transport

With half a dozen major airports, a myriad of parking lots and drive-ins, and hundreds of miles of freeways, it is the land of the automobile, 'autopia' indeed – 'without one, you are no-one'. A 'multitude of multitudes', a city that created the 'Autopocene' that has attained global significance.

Transport infrastructure is one of the main economic essentials, and without having a natural port an international one needed to be constructed at Long Beach. Here the connecting multi-modal Alameda corridor was realised to carry much of the country's freight in or out across the Pacific. This vital infrastructure was developed with the twin ports of Los Angeles and Long Beach (PoLA and PoLB).

With their twin sites, these two ports were extended extensively, in particular with containerisation, where all of the Pacific world trade of the USA passes in and out of the West Coast. With its supporting connections, both rail and road, the corridor was developed to supply the ports of Terminal Island and the Port of Los Angeles and Long Beach Harbor in San Pedro Bay with new basins developed.

In terms of aviation, Los Angeles International Airport (LAX) with its four main parallel runways 3,000 m. in length, handles around 90 million passengers each year and a very large volume of freight. The interstate highways of I-5, I-405 and I-110 carry millions of vehicles each day in and around each part of the city. Downtown is ringed by the Golden State Freeway, the Santa Monica Freeway and the Pasadena Freeway (see Support Material: Annexe 5.2C – Los Angeles Significant Buildings and Freeways).

The railroad supports long-distance freight trains that carry containers, bulk and fuel on the BNSF Railway and Union Pacific Railroad rights of way. Metrolink provides commuter rail services stretching from Lancaster to Oceanside and San Bernardino via Union Station.

More recently there has been a focus on urban public transport with the opening of the Metro Blue Line to Long Beach in 1990, followed by the Red Line and the Green Line. This is the beginning of the development of a more comprehensive metro system, with Lines A–E across the sprawling metropolitan area of Los Angeles and a new subway planned under Wilshire Boulevard. The central population is approximately 4 million, with 13.2 million in the metropolitan area and 19 million in the CSA extending out to Orange County with centres at Santa Ana, Irvine with its UCLA campus and Anaheim.

City of Quartz

Sprawling city that it has become, Los Angeles has distinctly different neighbourhoods and population groups from urban Downtown, Southside, East and Westside, with black communities and Latinos that differ from the whiter suburbs of Santa Monica, San Fernando Valley and Orange County and beyond. The city is fractured and retains many extremes of poverty and gang-related homicides and memories of the Watt's rebellion in 1967 where the Watts Towers still prevail.

Major developments changed the old downtown as Bunker Hill lost its former residential nature and along the expanding Figueroa corridor with Hill Street separating the different worlds of gentrified and corporate Bunker Hill and Broadway with the adjoining Skid Row, as the old Angels Flight funicular was removed and eventually replaced. The Community Redevelopment Agency took over and the urban village of South Park developed. Several deprived poor areas in and around MacArthur Park exist where gang culture (like the CRIPS and Bloods) were prevalent.

Boyle Heights, for example, an area of east Los Angeles, was built in the original 1880s property boom. Over time, under trickle-down effects, new owners took over, filtering down to lower-income owners. As freeways were built around the area, it became isolated, was red-lined and deteriorated as the I-5 freeway was built over Hollenbeck Park. With mainly Hispanic and Latin residents, the population changed and was politicised and with organised protests, this challenged the status quo and a gradual process of upgrading and gentrification re-emerged with pedestrianised areas and better access through construction of the new metro Gold line.

With its highways and freeways for cars and trucks, public transit has not yet been fully developed for the carless urban underclass without requiring high operating subsidies. With de-industrialisation of manufacturing and import substitution from Mexico and China, the economy is now heavily service orientated and fewer major corporations are headquartered downtown. New areas and tracts that are being developed and platted for low-density, single-family housing, are mainly in outer Orange County and beyond.

Due to its pivotal location with the Pacific Rim, under Mayor Bradley after the 1984 Olympics, the 'LA 2000 – a city for the future' plan was produced in 1988, as a manifesto of new regionalism targeted on new growth management measures. Although challenged by UCLA's LA School of Planners, it generated discussion on the re-industrialisation and internationalisation of capital, under increasing competition from South-East Asia and China.

Excessive growth led to increased pollution, and smog became a problem in the 1990s, with a number of air-quality management districts (AQMD) introduced. There was also a movement against non-incorporation and the creation of large, low-density residential developments, for example under the Lakewood Plan that by contracting out directly vital services, was able to avoid property taxes.

One of the distinctive developments post-war has been the development of the regional shopping mall. As an alternative to congested downtown, the growth of suburbanisation and the use of the automobile has enabled distinctive new zones of attraction of consumerism to be developed as large anchor department stores act as a magnet for consumers. Along with specialty shops, they attract families to shop and to be entertained at all times of the day. This has led to the creation of extensive paved parking lots in order to accommodate the transitory visitors.

Transactional Forces and Typology

With its backdrop of mountains, the agricultural land leading down to the coastal plain around Los Angeles was transformed by the discovery of underground oil fields in Southern California. As land was platted and railroads were built, this was heavily exploited as Los Angeles developed at a low density with mixed development sprawling across large land areas. From the San Fernando Valley, its central area and basin was developed and extended through annexation of the adjoining territory, most recently around the Alameda Corridor down to Long Beach. Here both the ports of Los Angeles and Long Beach now serve as the prime trans-shipment point for all incoming and outgoing landed goods on the West Coast of America, supported by the major international airport of LAX.

With the overbuilt network of freeways that dominate the surroundings, its relentlessly built orthogonal grids and streets cover most of the built-up area, where automobile accessibility is the essence of existence. The separate residential districts and enclaves of Burbank, Pasadena, Santa Monica and Malibu exist with many developed commercial areas around the main downtown financial district. Whilst oil is the lifeblood of the city, water is the scarce resource for survival upon which the metropolitan area is wholly dependent in pumping water in from Owen's and Mons Lake in outlying Sierra Nevada and from the Colorado River. The Pacific Palisades fire in early 2025 has also had a devastating impact on central residential areas.

Los Angeles is unique as a modern American automobile-orientated city. It is a city type of extended longitudinal highways, orthogonal grids, vertical towers, 'edge cities', suburban sprawl, built-up strips, with signs and billboards – truly an amalgam of 'suburbs in search of a city'.

With its auto dominance, paved parking areas in many districts out-number natural ground areas. As a planned, built-up, emerging city that continues to annex and plat more land, it is an expanding low-density megalopolitan region without equal, as aspired to by Walt Disney as 'Autopia'. Given its resource requirements and energy consumption and pollution, its sustainability across the ever-widening Greater Los Angeles region poses significant ongoing challenges under the presence of geological faults and with climate change leading to the propensity for more frequent serious localised fires.

5.2.4 DUBAI, UNITED ARAB EMIRATES

Background and Origins

Dubai was originally a small fishing and pearling village on the creek in the region of the neighbouring Trucial States in the Arabian Gulf. With white adobe houses and distinctive wind towers, and small alleys and the souk of Deira, the buildings nestled comfortably on either side of the creek.

In a strategic location in the Gulf for trade, in the early 19th century arrangements were entered into with the British as protectorate over local shipping, which served their self-interest with the East India Company. As a small sheikhdom, Dubai has been ruled closely by the al Maktoum family that has retained tight control of the State over the years.

The early rulers had broad mercantile objectives as traders and established Dubai as a free port in 1901, removing trade barriers to take advantage of steamers travelling to India and to trade with the East India Company. This trade continued and expanded to include the re-export of commodities, particularly gold, when the pearl market declined.

Often in competition with Abu Dhabi, and having little oil of its own, it set about developing its own markets and infrastructure. With a new land law, a master plan was prepared for the Dubai Emirate around the historic creek in 1959 as Sheikh Rashid Al Maktoum borrowed funds from Qatar, with which it was aligned through a common currency union using the Qatar and Dubai riyal, to construct a toll bridge across the creek to the Deira side in 1963.

In the post–World War II period, when Britain eventually withdrew its troops East of Aden in 1968 whereupon Dubai entered into a union through the creation of the United Arab Emirates in 1971 with the former seven Trucial States, with the adoption of a new currency, the UAE dirham to replace the riyal. With these changes, recognising that it could not compete with the extensive wealth of the capital Abu Dhabi City, it set about seeking aggressively to diversify its economy.

Modern City – Buildings and Arteries

In order to improve communications, the new Dubai International Airport (DXB) runway and terminal was constructed with the first duty-free shops in 1970. This was followed by the development of a deep-water port, Port Rashid (designed by Halcrow). This attracted considerable marine traffic and was expanded to 35 berths in 1975. Further to its success, a yet larger and deeper port Jebel Ali was constructed and opened in 1979.

With its ability to capture regional maritime traffic, after Dubai Ports was created it developed to become DP World and with the largest Gulf port for container traffic in addition to its growing overseas interests.

The Al Maktoum family rulers continued with emphasis on the entrepot trade with the development of new facilities such as the Jebel Ali free trade zone. With further real estate developments based upon land reclamation, as the artificial Palm Islands, at Jumeirah and Deira, it began to become a more significant regional financial centre.

The population of Dubai grew with the migration of people from both India, Banians and Baluchis, as shopkeepers and workers, and Persia, as merchants and craftspeople, who settled in the area of al Bastakiya along with others from Abu Dhabi and Sharjah.

Whilst neighbouring countries with a wealth of oil wells attracted a large number of migrant technical workers to support the oil boom, in Dubai with its more diversified economy, different skills were required. With only a small local population, this resulted in a large number of migrants, classified as 'permanently transient' that constitute now approximately 90% of the population. This reliance on foreign labour leads to a very transient labour force that rises and falls dependent on demand. In consequence, and with the concentration on tourism and hotels, retail, finance and real estate, it became over-extended and suffered significantly in the global financial crisis of 2008, although since recovered.

With its deliberate policy of diversification and deployment of migrant labour, Dubai has become a real regional economic hub for people, goods and capital, attracted by its free trade zones. The impact of political conflicts in the region has also served to make it a place of refuge for both labour and capital. Whilst guided by generally 'dirigiste' policies, major state-owned enterprises have developed such as Emirates Airlines, Etisalat, DP World, Dubai International Airport (DXB) and Dubai Drydock and major local real estate players such as Nakheel (Palm Island), Emaar and Sama Dubai. Whilst the free trade zones served to attract and anchor new industries, this has also been extended to comprise Media City and Internet City.

The Al Maktoum International Airport (DWC) is being expanded at Jebel Ali in the south where an entire new free economic zone is being developed for both passenger and freight.

In terms of built form, it has developed a significant skyline with major achievements such as Burj Khalifa (designed by SOM), the world's tallest building 830 m. high with its necessarily deep foundations in the sand, and tower blocks and hotels such as Burj Al Arab (designed by Atkins) and others in the Business Bay District and Dubai

Marina. With its new extended airport and metro lines, tram and monorail, constructed for Expo 2020, it is also developing a high-speed rail to connect across the UAE. Sheikh Zayed Road acts as the central main urban artery.

With the new Dubai Opera and modern museums, it has continued also to extend its cultural and sports facilities. Whilst of only a small land area, with its very high percentage of expatriate workers, mainly Asian, it maintains its position as a significant business and transit hub that links the continents of Asia and Europe through the Gulf and the Middle East.

Its longer-term strategy to 2030 is to extend diversification but with sustainable industries focusing on knowledge-based and innovative industries. Susceptible to external economic shocks, it is a creative but very small country with outward-looking policies, needing to continue to seek areas of economic advantage given its location in the challenging desert climate region of the Arabian peninsula.

The population of the whole of Dubai is approximately 3.3 million with only approximately 10% of the population being UAE nationals.

Transactional Forces and Typology

At a strategic location on the Gulf Coast, Dubai was well located on trading routes between the east and west that suited its role as exporter or re-exporter of commodities in its entrepot trade. The expansion of the port and surrounding facilities with its harbour and dry dock attracted maritime activity which along with expansion of the airport has led to significant handling capacity of re-exports and logistics capability. In addition to its natural materials, this has reinforced its role as a finance centre in the region, as well as for insurance, real estate and business services. Through the supply of such services it has established itself as a leader in the region where, due to other conflicts in the Middle East, the Emirate of Dubai has been able to step up as an emerging and stable financial centre and a magnet for tourism.

From its original activities as a small fishing harbour on the creek and with the ownership of the majority of the land remaining in the hands of the ruler, infrastructure was progressively developed for both a major new port and airport with related highways and bridge across the creek. Beyond this, with the shapely areas of land formation and reclamation, new communities such as the artificial Palm Island and The World have been realised in the Gulf coastal water which has enabled new residential areas to be accommodated.

As a perimeter coastal yet desert environment, the city has been able to grow with the formation of new areas and the annexation of land in the development of different industrial and commercial sectors. This has resulted in the realisation of a more extended and specialised polynodal development pattern of the land areas surrounding Dubai linearly along the Gulf coast between Abu Dhabi and Ajman, from the sabka plain reaching to the very arid interior. However, in terms of sustainability, Dubai is very dependent on the growing demand for both fresh water and generated electricity for cooling.

5.2.5 SHANGHAI, PEOPLE'S REPUBLIC OF CHINA

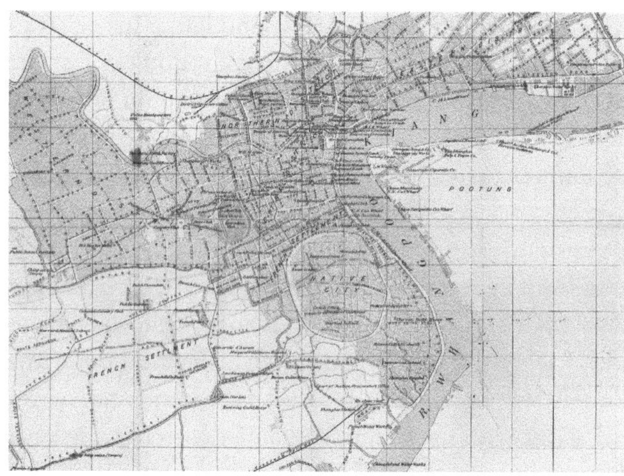

FIGURE 5.2.4 Shanghai – early configuration, 1923.
Source: Maphouse.

Background and Origins

From its original walled Chinese town in an area of tidal flats, beyond its fishing traditions **Shanghai** emerged as an early trading port on China's eastern seaboard on the southern estuarial bank at the mouth of the River Yangtze.

The earliest settlements of Shanghai were recorded during the Warring States period in China (475–221 BCE) between the Yue and the Wu kingdoms. Close to the intersection of the Huangpu and Woosung rivers, a circular walled city was built. Later in the Three Kingdoms period (220–80 CE), the primary settlement was at Qinlong, up-river from the present-day city, that began to serve as a gateway for the transport of goods, and under the Eastern Jin dynasty, another settlement developed at Hu Tu Lei.

But it was under the Song dynasty that it was deemed to have become more formally established, around 1074 CE, and in the centuries thereafter that Shanghai gained its trading and cultural ascendancy. The old Chinese city developed behind an annular wall with arrow towers, and land and water gates. With its creeks, paths followed the lines of these waterways, which were tidal and needed protection and crossing points.

During the Ming dynasty five main streets were developed running east-west around the core of the city Xian Shu, and commercial activities developed along the main riverbanks. As trading developed, more wharves were built, and guilds were established as the number of boats mooring grew significantly. The Woosung River frequently flooded until it was later controlled by dykes and narrowed into the smaller tributary named Suzhou Creek as it is known today. Other creeks such as the Yang Jin Bang were also channelled and eventually built over.

Outside the Walled City, areas were built up and developed organically over many years. However, all of this was to change radically when as a result of the Opium Wars with China and seeking access to Canton, the principle of extra-territoriality was established in the Treaty of Nanking in 1843. Further elaborated upon in the Treaty of Wanghia the next year, this was to designate a number of treaty ports as open for trade for foreigners and agreement was reached with the city's representative, Taotai, over a number of land regulations that formalised areas of settlement for foreigners.

Under these treaty port 'concessions' that were agreed with Britain and the USA, which were eventually combined into the International Settlement, and with a separate French concession area also agreed, the Bund became lined with dedicated buildings where maritime trade was focused along the banks of the Huangpu River.

Between them, once the settlement areas were agreed, a commercial area developed along the Nanking Road, which became the earliest centre of retail activities. At the sharp turn of the river, a waterside boulevard or 'Waitan' was built up as individual wharves were developed along the riverbank along with a Customs House and a series of neoclassical buildings that housed new banking institutions. The Shanghai Municipal Council, established in 1854, then set about developing the infrastructure for the city along with the provision of public service facilities.

The settlement became the favourite location for foreign businesses and significant trade developed with commodities stored in the 'godowns' of tea, porcelain and silk. Simultaneously the sale of imported opium developed, in the potent narcotic opium trade that had become highly profitable. Different currencies were used, including the silver 'tael' in different weights. The number of merchants increased significantly along with small businesses in 'shop-houses' where 'shed people' had to live employed by the few wealthier 'taipans'.

A mass housing type, 'Li-long', emerged, a type of terraced housing style having small internal courtyards on narrow gated lanes and alleyways across Shanghai. Whilst short of natural light, these became the main housing stock prevailing at that time for the employed mill and factory staff in the port area. They evolved in design and although of high density, gradually gained improved facilities with different styles of apartment and garden Li-long.

Several areas were defined by the boundary creeks, such as Suzhou Creek in the north and Defence Creek in the west, and along roads parallel to Nanking Road and out to 'Bubbling Wall'. The American Settlement was in Hongkou, north of Suzhou Creek, in a nascent industrial area with its main street known as Broadway, and the French Concession was smaller, with its missionary foundations, bordered by the Old Chinese city to the south.

In 1900, the Concession areas were expanded with the enlargement of the Western and Eastern Districts of the International Settlement and also the enlargement of the French Concession. Population increased with in-migration, and by 1910 a small network of urban tramways had been built, extra-settlement roads were increasing and the western suburbs were growing. By 1916, the railways were developed with connections built between the North and South stations. Department stores and municipal buildings were introduced, and hotels, clubs and recreational facilities were built in the soft silty soil of Shanghai. The Bund was reclaimed and widened, and Public Gardens was created on the foreshore by the Creek where several consulates were housed.

With the eventual end of the treaty ports era, a significant increase in traffic resulted, especially from newer ocean steamships and with increased inland river traffic upstream connecting with the Grand Canal, and a customs service was developed with a low tariff regime. The coastal regions developed extensively with their outward-looking port cities more than the other less developed interior areas of China.

New Chinese settlements developed around the concession areas in Shanghai where ancient walls were eventually pulled down to be replaced with roads, and built-up areas were consolidated. In 1919, Sun Yat-sen had a plan for developing the Great Port of Pudong across the Huangpu

River, but this was unrealised. In 1927, a city government of Greater Shanghai was created and it was declared a 'special administrative city' responsible to the central government.

Further, in 1929, a new City Planning Commission was formed and developed a Greater Shanghai Plan that was intended to create a new focus for the city around a civic centre at Jiangwan, cruciform in plan, with axial boulevards and malls capped by semi-circular roads around a grid of streets. Whilst the Mayor's Building was built, this 'city beautiful' plan was halted as the Sino-Japanese War intervened during the 1930s.

Much was to change during the next two eventful decades, firstly with World War II, when a small area of Hongkou was used as an area of confinement, 'shitei chiku', for Jewish refugees, until eventual liberation. Next, the subsequent conflict between Nationalists and Communists then intervened until the creation of the new People's Republic of China in 1949 when an entirely new socialist planned administration was to emerge.

After a short period of Sino-Soviet collaboration, when new industrial suburbs were developed and residential block apartments were built, with the urgent need for the development of new infrastructure, a period followed of significant migration from rural areas to the city. As economic growth eventually took off after the Great Proletarian Cultural Revolution, so did the increase in population in the urbanised areas.

Modern City – Buildings and Arteries

It was not until 1988, as a result of the 'Comprehensive Plan for Shanghai', which focused on developing the area on the east side of the Huangpu River at Pudong connected by a series of bridges and tunnels, was this commenced. It included a series of preferential policies for an export processing Zone at Jonquiao, a free trade zone in Waijacquio, a deep-water port and the proposed development of the Lujiazui Financial District.

In 1991, the Shanghai Development Corporation (SDC) invited five international design firms to prepare a master plan for the development of the area of Pudong, but these were not pursued. Rather, with support from the French planning agency IAURIF from Paris, the Lujiazui District development began with the realisation of the Pearl of the Orient Tower as a landmark with its distinctive tall tripod and twin orbs, around which the financial area was to be developed.

Following an updated master plan in 1999, and a series of 5 Year Economic Plans, the World Expo was held in Shanghai in 2010 and a pilot free trade zone for Linyang was established in 2013. Along with a new offshore deep-water port in Yangshan connected by the Dongshai Bridge, with a new airport in Pudong, connected by Maglev, and new tunnels and bridges across the Huangpu River, this modern infrastructure led to exceptional development and expansion of the city.

Infrastructure was further developed with the construction of an Inner Ring Road and elevated highways surrounding the entire city and linking the two halves of Shanghai, Pudong and Puxi, east and west with the Nanpu Bridge, with its coiled access ramps and also Yangpu Bridge across the Huangpu River. The Outer Ring Road was built with crossing tunnels under the river and further bridges were realised, such as the Xupu, Lupu and Minpu bridges.

Tree-lined Century Avenue was developed as a main axis across Pudong, and other streets were laid out as major buildings were developed in this area. With a series of very tall buildings, this has resulted in a Manhattan-type profile to this area, with Shanghai Tower (designed by Gensler), the Jin Mao Pagoda Tower (designed by SOM) and the World Financial Centre (designed by KPF) that have totally transformed the skyline. Modern Shanghai is unrecognisable from old Shanghai, except for some of the neoclassical bank and hotel buildings remaining along the Bund. With its intense development of infrastructure and major commercial buildings developed in Pudong on the opposite eastern side of the Huangpu River, the central area has been transformed (see Support Material: Annexe 5.2D – Shanghai Significant Modern Buildings).

Since development of the special economic zone (SEZ) in 1993, and the World Expo held in 2010, the view from the Bund across the river to Pudong has been transformed, with the TV tower surrounded by a series of modern skyscrapers. Some areas of heritage have been retained, and the limited open spaces that remain are the People's Park, now People's Square, and the rebuilt Yu Gardens area in the old city.

As one of the largest cities in China, having a key role as financial centre, and with extensive manufacturing and production facilities, it is one of the major regional drivers in wealth creation in central China. With new infrastructure arteries, major tunnels and bridges, airport and major container port and high-speed rail connections, 14 subway lines and the iconic Shanghai Maglev Train (SMT-Trans Rapid) to Pudong International Airport (PVG), it has become the centre of a powerful economic region. This area has even been connected by the longest sea structure, Donhai Bridge, to link with the distant Yangshan Islands.

The city population is approximately 26.3 million and approximately 34 million in the metropolitan area. It is administered by a Municipal People's Congress under the direct control of the State Council of China, with the city of Shanghai having special status located in between the provinces of Jiangsu and Zhejiang. The density of the whole city region is approximately 27,900 pers./sq. km.

Transactional Forces and Typology

Situated at the intersection of the major Yangtze River and the Yellow Sea, upon a triangular land mass shaped by river deposits, Shanghai and its surrounding region have developed as a dominant economic centre. As a coastal city on north-south maritime routes, the great inland river provides a riverine route that penetrates westwards deep into central China.

From its earliest trading and fishing activities as a coastal and estuarial city, and from the production of silk, located not far from the Grand Canal, traditional industries flourished. With its settlement on the Huangpu River by the lower Great Yangtze River, customs duties were collected on behalf of Jiangsu province after the Treaty of Nanking opened up Shanghai to international trade as one of the treaty ports. With the British and American International Settlement and the French Concession, these were to change the predominant pattern of development as international trade grew irrevocably in the city around the riverside. Downtown developed around the Bund where commercial activities were centralised as the headquarters of banking and trading institutions became established.

With defined territorial boundaries by the Bund within the city, these settlements radically modified the urban development around the Huangpu River with its jetties, wharves and handling equipment for the exchange of raw commodities and supporting the manufacture of new fabrics. Despite periods of political turbulence, the thrust of economic development continued as infrastructure was developed and built settlements grew around the centre and Puxi District. The most recent developments have extended farther outwards on the eastern side of the river in Pudong, which has now been developed into a major new commercial, financial and technological district.

With its coastal location by the Yangtze River estuary, the old city developed on the western side of the divided Huangpu River from Dianshan Lake. The old city was settled at the location where Suzhou Creek joins the Huangpu River at a sharp right angular bend in the river as a strictly vernacular original Chinese city. Its organic origins were to change significantly in the 19th century with the planned International Settlement and French Concession, enabling foreign managed zones to be laid out and developed beside the river. Accordingly, these zones were developed having very different characteristics alongside the circular Old Chinese City. With the riverside Bund and long East and West Nanjing roads, Beijing and Huaihai roads, these axes then shaped development around the old city and with People's Park.

The characteristics of modern Shanghai now are of a dense, modern, urban core around the river surrounded by a ring of new infrastructure, roads and railways, crossed by a central elevated axis of highways, with a number of underground river crossings, and also an extensive metro network. With only a limited Chinese old city remaining, growth has been essentially vertical, constituting a dense central pattern of modern commercial and residential blocks.

With the new Huangpu River bridge crossings and tunnels, the opposing area of Pudong has been developed as a special economic zone since 1990 as a counter-vailing pole of commercial development. Now spreading outwards radially from the urban core, the region of Greater Shanghai has developed as one of the densest metropolitan areas in China, having a number of poly-nodal satellite centres.

With a central core, main axis and external 'butterfly wings', the 2035 Vision for Shanghai is planned to contain 25 million people in 3,200 sq. km. developed in multiple cores and corridors, a developed region with poly-nodal centres. With its vision as a 'socialist international metropolis', this will be consolidated on the mainland with three offshore islands, including Chongming in the Yangtze River delta, which will provide some more open relief from the dense urbanised agglomeration that has developed.

5.2.6 SYDNEY, NEW SOUTH WALES, AUSTRALIA

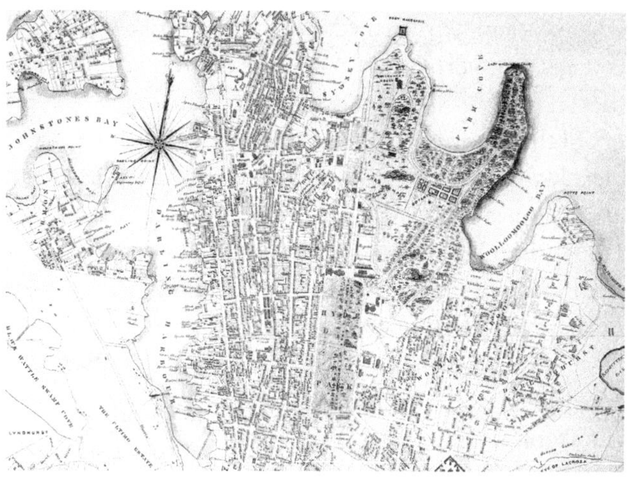

FIGURE 5.2.5 Sydney – early configuration, 1854 (Woolcott & Clarke).

Source: British Library.

Background and Origins

The location of **Sydney** harbour is one of the most dramatic in the world with its nestling in the estuary of the Parramatta River around Sydney Cove in Port Jackson just north of Botany Bay where early convicts were settled arriving from Britain. With its curves and inlets originally inhabited by Aboriginal peoples, it is a stunning location around this natural harbour, named after Viscount Sydney.

Discovered and colonised at the end of the 18th century, these foreigners were used to create the first settlement around Sydney Cove and Paramatta. By 1830, the outlines of a new settlement were in place, around which new immigrants were attracted to settle, and in 1850 the first railway was built from Sydney to Parramatta. The discovery of gold in the Blue Mountains encouraged further influx of people with segregated reserves allocated for Aboriginal peoples.

In 1856, Sydney became capital of the colony of New South Wales where agriculture became established with extensive pastoral activities of sheep and cattle. In 1901, the Commonwealth of Australia was founded with NSW as one of the six states of the federation. With the production of wool and the discovery of coal as well as gold, these became primary exports of the region.

With different districts and reserves, the city began to be consolidated as infrastructure was realised. With construction of the Sydney Harbour Bridge, this enabled the central city area to develop further by being connected to the North Shore with the city railway and highway. The Rocks and the Circular Quay were key areas adjacent to Sydney Cove which act as the focal point of the city at the site of the southern abutment of the Harbour Bridge. This giant, hinged, steel bridge was opened in 1932, known as the 'coat-hanger', and the Argyll Cut through the rocks led down to the quays where ocean liners berthed. With the old storehouses and cottages, its original wharves were the centre of marine activities and a place of ferry crossings in the harbour.

On the eastern side are the Royal Botanic Gardens, previously the site of the 1880 Garden Palace Exhibition and also the former NSW governor's residence at Government House. Lachlan Macquarie, governor between 1810 and 1821, was responsible for much of the early development of the colony, leading to its city status in 1842. Past Farm Cove is the winding road towards the location known as Mrs Macquarie's Chair, where there are fine gardens and the Tropical Centre.

Adjoining this is The Domain, a fine park and public space, and along Macquarie Street are a number of other early buildings, old State Parliament House, the State Library of NSW, the Mint with its two-storey veranda, and Hyde Park Barracks. Farther around is Kings Cross and Darlinghurst with its Victorian houses and verandas in Potts Point and the Naval dockyard. Paddington has neat, restored, terraced houses and leafy Paddington Street with local weekly markets and Centennial Park nearby. Oxford Street heads out past the Sydney Cricket Ground eastwards to the popular Bondi Beach.

Modern City – Buildings and Arteries

Beyond the harbour is Bennelong Point named after an Eora elder, where in 1957 a competition was won by Jorn Utzon to design a new concert hall and opera house on this stunning harbour promontory. Its design like soaring sails that could be seen by land, air or sea, was to provide some controversy, not least initially due to its un-buildability. With the engineers of Arup, his team found a solution that enabled the structure to be modelled as circular arcs. Once this was achieved, the design was turned into a construction plan and the segmental shell roof could be built. Tiled with arrow-shaped panels and with a wide glazed foyer, the whole came to realise Utzon's ambition of appearing as three groups of giant sails in the harbour. Inside, the concert hall and opera house, opened in 1973, have their own independent layouts for acoustic purposes and form a fine venue for musical arts.

Farther south is the city centre and the CBD with the Strand shopping arcade where the main commercial buildings are accommodated. The Sydney Tower, 300 m. high, is located here also with its observation deck and twisted steel supporting cables, completed in 1981, and the other tall octagonal concrete building is 25 Martin Place (MLC Centre – designed by Seidler) and Aurora Place (designed by Piano). Adjoining this is Hyde Park, a green oasis with the ANZAC Memorial and Pool of Remembrance.

To the west lies Darling Harbour and Cockle Bay with the docks and former railway yard. As part of the 1988 Bicentenary, an urban redevelopment plan was undertaken around the harbour and Pyrmont Bridge, and with the National Maritime Museum and Aquarium.

With construction of the immersed tube Sydney Harbour Tunnel in 1992, development has extended extensively in North Sydney, Mosman and the suburbs and the northern shores beyond Kirribilli Point. This now forms part of the Sydney Orbital network around the city. The journey previously to Manly was known as the 'seven-mile journey', with the ferry ride from the Circular Quay, a fine trip to Manly Cove and North Head.

The Cross-Harbour City Tunnel and Lane Cove along with the M3 and M5 tunnels are now tolled highways. The Sydney Harbour Bridge, which also has twin railway tracks, carries eight lanes of traffic managed under tidal flow arrangements.

With the Summer Olympics in Sydney in the year 2000, much new development was undertaken, especially in the transport sector. With the light rail system from Central Station and the Monorail around Darling Harbour, public transport was improved. A new Central City Precinct Development is growing around Central Station. The Sydney metro continues to be extended, with a new Cross-Harbour line to link the North West line from the north-west suburbs to Chatswood, with the planned South West line from the south-west suburbs to Sydenham, through the central business district.

To the south of the city is the original Sydney Kingsford Smith Airport (SYD), which was developed on residential land as runways were extended out southwards into Botany Bay. A new airport, Western Sydney International (Nancy-Bird Walton) Airport (WSI), is under construction at Badgerys Creek in order to enhance capacity as part of the Western Sydney Aerotropolis plan.

Sydney, as capital of the state of New South Wales, is an exceptional multi-centred world city, where resident 'Sydneysiders' are able to enjoy the natural setting around the harbour with its jagged inlets and fine climate. One downside, however, as a result of climate change, is the increasing prevalence of bush-fires that have been occurring in recent years, and higher frequency of intense storms and flooding in the region.

With the shape of the harbour and inlets, the city is divided geographically into the Inner Area and Inner West, the Eastern and Western suburbs and Northern and Southern Sydney. The Greater Sydney Commission in 2022 categorised the wider region as the growing Six Cities Region of:

- Eastern Harbour City, including Sydney CBD and North Shore,
- Central River City around Parramatta,
- Western Parkland City and Central Coast City and
- Outer Newcastle City and Illawarra-Shoalhaven City.

In terms of administration these areas subdivided into districts, and 33 local government areas (LGA) participate in a Regional Organisation of Councils (ROC). With the established southern suburbs, the North Shore includes the northern suburbs and beaches stretching out beyond Middle Harbour to North Harbour and Manly Peninsula. The population of the city is approximately 5.5 million but has a generally low population density beyond the inner area.

Transactional Forces and Typology

Having a natural harbour formed at the mouth of the Parramatta River, with its coves and inlets, this location was an ideal place for the Eora Aboriginal inhabitants for fishing, hunting and animals. With foreign settlers, new districts developed around the coastal city as it grew up from Darling Harbour and its docks to the Circular Quay past the botanic gardens to Kings Cross and Darlinghurst. The city centre was consolidated and with the inner suburban residential area of Paddington and the established built-up area stretched between Darling Harbour and Darling Point.

Besides the Sydney Harbour Bridge it was the construction of the Sydney Harbour Tunnel in 1992 that opened up subsequently the north side and shores from Mosman and the outer suburbs beyond, shortening the travel time into the city and transforming the reach and extent of the broader metropolitanised area.

The city of Sydney is characterised by its extraordinary natural geography, being situated in an enormous natural harbour with multiple bays and inlets. Having been established initially on the southern side of the Parramatta River, with its main bridge and tunnel crossings, development has now extended along the opposite northern shores. Stretching north, south and westwards from its platted central area, it is radiating outwards and coalescing around poly-nodal centres which are interconnected with the central core. Surrounded by natural parks and environmentally sensitive wetlands, the scope for future expansion in the centre is limited, which is leading to more distinct developments in outer suburban areas around newly developing localised cores.

5.2.7 MELBOURNE, VICTORIA, AUSTRALIA

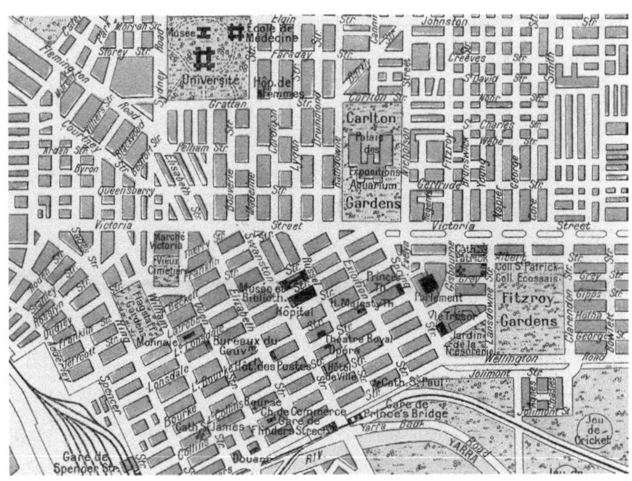

FIGURE 5.2.6 Melbourne – early configuration, 1911 (Baudelot).

Source: Maphouse.

Background and Origins

On the traditional land of the Aboriginal Kulin Nation in Southern Australia, the first settlements were established after the 1802 arrival of Captain Phillip in the Rip entrance from the Bass Strait at the bay, now named Port Phillip, as a penal colony with expelled foreign persons from Britain. Free settlers arrived later from Tasmania, and others followed as a crown settlement was made at Birra Runga, 'the place by the Yarra River'. The Colony of Victoria was legally established in 1851 with the city named after a British prime minister, Viscount Melbourne, and a subsequent gradual process of disposition of land from Aboriginal groups by European settlers.

At the north of the bay, the first central layout was made to a gridiron plan with 48 rectangular blocks, with north-south avenues and east-west streets, such as Collins and Flinders streets. It soon became well populated and with the discovery of gold, and the Gold Rush, people gravitated to the area, leading to its gradual transformation into a significant metropolitan area.

After the Federation of Australia was created in 1901, **Melbourne** served as the interim seat of government before the new location of Canberra became the permanent capital in 1927. Melbourne became a central place of business and trading around sheep and wool, and cattle meat and dairy, and its agricultural products became key exports, leading to the expansion of the port. The city's civic buildings were developed with the town hall, old treasury building and gold vaults, law courts, parliament house, government house, state library, National Gallery of Victoria and gothic St Paul's Cathedral (designed by Butterfield) and clubs.

The domed Royal Exhibition Building was built for the 1880 International Exhibition by the botanic gardens, subsequent to which a land boom occurred across the city, with new buildings and shopping areas such as the Block Arcade. Utilities were improved and tramways extended across the city and the Centennial Exhibition was held in 1888, although a period of recession was to follow.

Having adopted many attributes of its original occupiers with addiction to sport, following the Australians' overseas cricket success in 1882 from which 'the Ashes' emanated, the Melbourne Cricket Ground (MCG) became a key centre that has grown in size to its present high capacity. The other notable landmark is the Shrine of Remembrance, the fine World War I memorial to those inhabitants of Melbourne that served in the Great War, erected in 1934.

With the early development of the Railway and Flinders Street Station, the city also had a dense array of cable cars and tramways that criss-crossed the streets, as the city spread out into the suburbs to St Kilda, Fitzroy, Kew and Richmond, with individual neighbourhoods each having their own identity and characteristics, many with large, detached houses. The city was also laid out with several attractive urban parks, such as Carlton and Fitzroy Gardens and plentiful open space beyond, and with diverse ecology there are fine beaches at St Kilda, Brighton and Sandringham to the east of the city, out to Mornington Peninsula National Park beyond.

Modern City – Buildings and Arteries

Concerning building control in the city, in 1958 the height limit was removed for building construction downtown with the construction of Orica (ICI) House, and in 1969 the Melbourne Transportation Plan was developed, which focused on the construction of a series of new highways. New sections were eventually built, such as the City Link, East Link and Peninsula Link, but they took time to be realised.

In terms of infrastructure, highway development followed the railway development initially and its construction of the underground City Loop. The Central Business District has been extensively developed with a comprehensive modern street tram network as well as the heritage City Circle and other routes. Recent focus has been on development of the Suburban Rail Loop (East) in twin tunnels underground and the airport link.

Major highways have been built, such as the West Gate Freeway and the bridge over the Yarra River, the West Gate Bridge and the City Link Tollway with Bolte Bridge that closed access to Victoria Dock. Consequently, new built developments emerged around the areas of Southbank and Docklands by the Southern Cross Station. In the outer area there is the partial M80 ring that connects with the M8 highway in the west and M1 and M3 in the east.

The Port of Melbourne is a major commercial port with significant container cargo located on the Yarra with the separate Port of Geelong for general cargo and oil in the south-west of the bay. Melbourne Airport (MEL) was opened in 1970 at Tullamarine to the north-west of the city.

The inner-city areas and terraces of Collingwood and Flemington have become overdeveloped, having attracted many recent migrants from South-East Asia. In consequence the Yarra River has become a dividing line between the lower-income north and more affluent south Melbourne. With a large Greek population and numbers of Vietnamese, and its own Chinatown, it has become a very multi-cultural society. Since the turn of the century, growth has developed outwards around green wedges beyond the city's defined urban growth boundaries. In more inner areas there have been areas of urban renewal around suburban Port Melbourne and Melbourne Docklands.

New buildings have developed, for example Australia 108, a super-tall tower in Southbank, and nearby Eureka Tower, modelled to look like an enormous surveyor's staff. Federation Square was redeveloped and opened in 2002 after building over the railways at Swanston and Flinders streets with modernist designs of lanes and vertical 'shards'.

The economy has developed to comprise modern biotech companies, and the pharmaceutical industry has grown along with the financial, real estate and media sectors. With a large Convention and Exhibition Centre, it has a diversified modern service economy and strong educational sector with its universities.

As a dynamic modern multi-cultural city, Australia's second-largest city in the Pacific with good infrastructure, fine coastline and environs, it is one of the most liveable cities. The population of the city is approximately 5 million and growth has been targeted at 'central activities districts'. As a modern, outward-looking city, it looks to South-East Asia with its extensive trading links.

Transactional Forces and Typology

With its Bay Area and large natural harbour, as well as its hinterland of grasslands and bush, the settlement of the city of Melbourne at the head of the protected bay by the curving Yarra River grew around trading and agricultural and farming activities. As it has developed, manufacturing activities increased, and more recently it has become the centre for financial services. Whilst not a dense city, the rate of in-migration has grown recently and its size is now vying directly with Sydney in scale and level of activity as a competitive and well-resourced modern coastal city in Victoria with scope for expansion.

As Melbourne is a coastal city that sits at the north of Port Phillip Bay, with its urban core, the city is split by the River Yarra. Development has occurred on either side of the river, extending radially outwards in each direction and circumferentially around the bay. The central area, defined by the Hoddle grid that runs almost north-south, there are a number of old terraced areas where the CBD fronts onto the Yarra.

The newer suburbs were laid out generally on a one-mile grid system with radial boulevards. As a coastal and riverine city situated at the head of a large, enclosed bay, it has developed into an active city divided into sectors with its downtown docks and railyard, inner commercial and outer suburban areas, interspersed with an abundance of fine parks and gardens. The following figures illustrate the modern configuration of the trans-national cities as described and the evolution of their modern population.

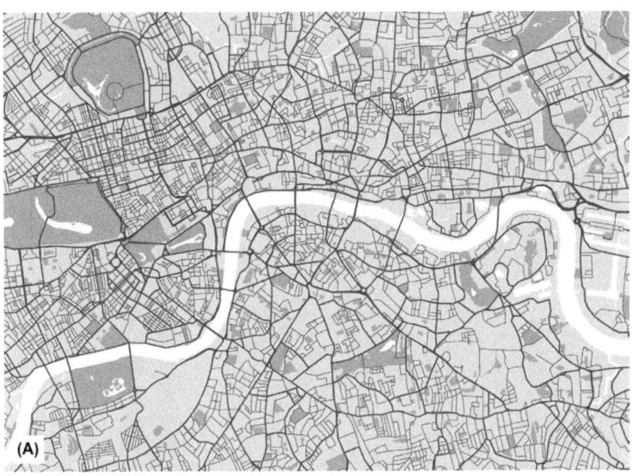

FIGURE 5.2.7A London – modern configuration.

FIGURE 5.2.7B Greater London – modern configuration. *(Continued)*

FIGURE 5.2.8A New York – modern configuration.

FIGURE 5.2.8B Greater New York – modern configuration.

FIGURE 5.2.9 Los Angeles – modern configuration.

FIGURE 5.2.10 Dubai – modern configuration.

FIGURE 5.2.11 Shanghai – modern configuration.

FIGURE 5.2.12 Sydney – modern configuration.

FIGURE 5.2.13A Melbourne, modern configuration.

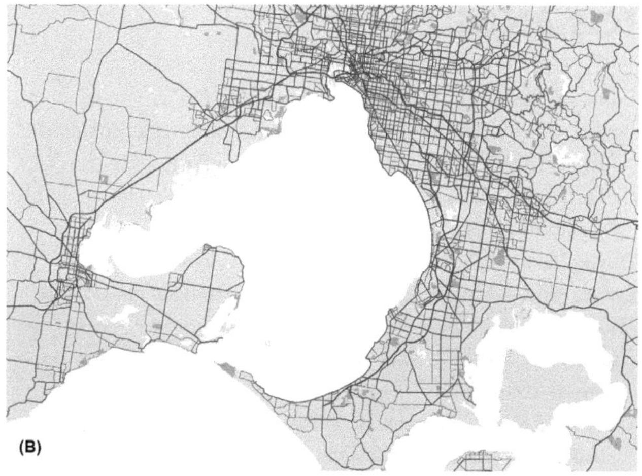

FIGURE 5.2.13B Greater Melbourne Bay – modern configuration.

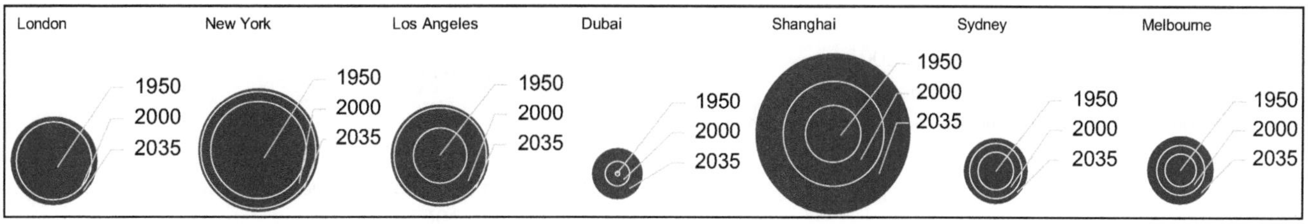

FIGURE 5.2.14 Transnational global city populations.

Data source: Global Database of Metropolises 2020 – UN-Habitat (2020), UN DESA Population Division.

Further Reading

LONDON

Cheshire, James & Uberti, Oliver – *London, the Information Capital*, Penguin, 2016.

Hall, Peter – *London 2000*, Faber & Faber, 1971.

Jenkins, Simon – *A Short History of London*, Viking, 2019.

Kynaston, David – *City of London 1815–2000*, Chatto & Windus, 2011.

Marshall, Bruce – *Building London*, Mainstream Publisher, 2007.

Sutcliffe, Anthony – *London*, Yale University Press, 2006.

NEW YORK

Busquets, Joan & Katsikis, Nikos – *Manhattan, Regular Grid for Ordering an Island*, Applied Research & Design, 2017.

Caro, Robert – *The Power Broker, New York*, Vintage Books, 2008.

Henry, Sarah, ed. – *New York at Its Core*, Museum of the City of New York, 2017.

Jordy, William – *American Buildings and Their Architects*, Doubleday, 1972.

Koolhas, Rem – *Delirious New York*, The Monacelli Press, 1994.

Lankevich, George – *New York City*, New York University Press, 1998.

Wolfe, Gerard – *New York, a Guide to the Metropolis*, McGraw-Hill, 1994.

SHANGHAI

Denison, Edward & Yu Ren, Guang – *Building Shanghai*, John Wiley, 2006.

Hsu, Immanuel – *The Rise of Modern China*, Oxford University Press, 2000.

Jackson, Isabella – *Shaping Modern Shanghai, Colonialism in China's Global City*, Cambridge University Press, 2019.

Keay, John – *China, a History*, Harper Press, 2008.

Wasserstrom, Jeffrey – *Global Shanghai 1850 – 2010, a History in Fragments*, Routledge, 2008.

LOS ANGELES

Banham, Reyner – *Los Angeles*, Allen Lane, 1971.
Bottles, Scott – *Los Angeles and the Automobile, the Making of the Modern City*, University of California Press, 1991.
Carle, David – *Introduction to Water in California*, University of California Press, 2016.
Davis, Mike – *City of Quartz*, Verso Books, 2006.
Davis, Mike – *Magical Urbanism, Latinos Reinvent the US City*, Verso Books, 2024.
Esperdy, Gabrielle – *American Autopia*, University of Virginia Press, 2019.
Gottlieb, Robert – *Reinventing Los Angeles*, Massachusetts Institute of Technology, 2007.
Haddad, Paul – *Freewaytopia, How freeways Shaped Los Angeles*, Santa Monica Press, 2021.
Hise, Greg – *Magnetic Los Angeles*, Baltimore, 1997.
Lunenfeld, Peter – *City at the Edge of Forever*, Penguin, 2020.
Patsouras, Nick – *The Making of Modern Los Angeles*, ORO Editions, 2024.
Starr, Kevin – *California*, Modern Library, 2005.
Starr, Kevin – *Coast of Dreams*, Allen Lane, 2005.
Wilkman, Jon & Nancy – *Picturing Los Angeles*, Gibbs Smith, 2006.

DUBAI

Davidson, Christopher – *Dubai, the Vulnerability of Success*, Oxford University Press, 2008.
Elsheshtawy, Yasser – *Dubai, Behind an Urban Spectacle*, Routledge, 2009.
Hashim, Alamira – *Planning Abu Dhabi, an Urban History*, Routledge, 2018.
Krane, Jim – *City of Gold, Dubai and the Dream of Capitalism*, Picador, 2018.
Reisz, Todd – *Showpiece City, How Architecture Made Dubai*, Stanford University Press, 2020.

SYDNEY

Connell, John, ed. – *Sydney, the Emergence of a World City*, Oxford University Press, 2000.
Moorhouse, Geoffrey – *Sydney, the Story of a City*, Houghton Mifflin Harcourt, 1999.
Morris, Jan – *Sydney*, Faber & Faber, 2010.
Shafir, Yvonne – *Sydney*, Parkstone Press, 2000.
Turnbull, Lucy – *Sydney, Biography of a City*, Random House, 1999.

MELBOURNE

Baine, Jim – *A Financial Tale of Two Cities, Sydney and Melbourne's Remarkable Contest for Commercial Supremacy*, UNSW Press, 2007.
Briggs, Asa – *Victorian Cities*, Penguin, 1990.
Buxton, Michael & Goodman, Robin – *Planning Melbourne, Lessons for a Sustainable City*, CSIRO Publications, 2016.
Sipe, Neil & Vella, Karen – *The Routledge Handbook of Australia Urban and Regional Planning*, Routledge, 2017.
Wilson, Granville – *Building a City, 100 Years of Melbourne Architecture*, Oxford University Press, 1981.

5.3 Mega African Cities
North – Central – East Africa

Cairo, Khartoum, Lagos, Kinshasa and Nairobi

5.3.0 INTRODUCTION – AFRICAN CITIES

Other than Asia, Africa has a number of countries that have some of the highest birth rates in the world. With less developed areas especially in sub-Saharan Africa, there is a strong centripetal force of attraction for in-migration from rural to more urbanised areas.

Despite poor levels of infrastructure and sanitation in particular, large informal settlements have gravitated at the fringe of many African capital cities today. Due to the lack of rural opportunity, this process is self-sustaining for migrants due to the higher probability of their finding some forms of employment in the urbanised areas than elsewhere.

However, due to the lack of formal structures, such as land registration and administration, municipal tax take is low, leading to a vicious spiral of low wages and poor services and amenities. As farther land areas are annexed, there is an ever-increasing expansion and growth of peripheral urbanised areas, both in terms of built area and population, especially the young. In addition, localised areas of regional or tribal conflict have exacerbated this migration trend in several countries in Africa, putting further pressure on the need for the provision of basic urban services.

As an example, in the Global South both Cairo and Kinshasa in particular are African mega cities having in excess of 15 million people and that are still growing fast, followed by Lagos (around 10 million) and Khartoum.

In West Africa, across Nigeria there are mixed tribes of different linguistic groups that include Yoruba, Edo, Ijaw and Igbo in coastal areas and inland Hausa, Fulani and Chamba, with major centres of population above 1 million in Kano, Ibadan, Benin City and Port Harcourt.

In East Africa, Nairobi and Dar es Salaam are the largest cities, having similar challenges of in-migration, urbanisation and management of informal developments in peripheral urban areas and the growing need for provision of basic services for these rural migrants.

The late 20th century was characterised across Africa as a period of post-colonisation, when after a period of continuing growth with de-agrarianisation, global forces resulted in 'push' from the countryside and 'pull' from the cities that generated significantly high rates of urbanisation.

With high birth rates and inadequate provision of housing, this has led to dependence on the informal sector as the only way to procure shelter for many families. This prevailing vicious situation of inadequate housing and sanitation, with over-crowding, under-employment and poverty, has become the greatest challenge currently in the growing mega cities of Africa and the Global South.

5.3.1 CAIRO, EGYPT

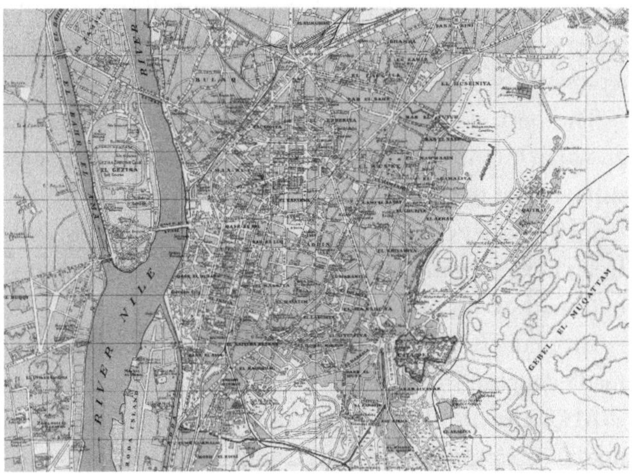

FIGURE 5.3.1 Cairo – early configuration, 1920 (Nicohosoff).
Source: Maphouse.

Background and Origins

In dynastic history the earliest capital of Egypt around 3000 BCE was at **Thinis**, originally founded by King Menes in upper Egypt which was made capital of ancient Egypt during the First Dynasty. The capital was moved to **Memphis**, just south of the Nile delta where the lands of Upper and Lower Egypt meet, by King Djoser during the Third Dynasty of the Old Kingdom. Djoser had commissioned the earlier construction of the stepped pyramid at Saqqara, around 2600 BCE.

DOI: 10.4324/9781003353348-28

Nearby at **Giza**, the Great Pyramids of the Kings (one of the wonders of the ancient world) and the Great Sphinx were subsequently built of sandstone covered in limestone under the Pharaoh Khufu in the Fourth Dynasty around 2500 BCE and the city, with its landmark temples, monuments and tombs, thrived with traders.

Farther south, **Thebes** in the Upper Nile during the period of the Twelfth Dynasty of the Middle Kingdom around 2000 BCE had become the most significant city. Upstream here on the eastern and western banks of the River Nile are the places of Karnak and **Luxor**, with their temples and the ancient necropolis on the west bank of the famous Valley of the Kings and Queens. During the Eighteenth Dynasty of the New Kingdom, the Theban Kings ruled after the conquest of Nubia around 1500 BCE and the capital returned to Memphis until the decline of Ancient Egypt.

Heliopolis had been one of the oldest cities in the pre-dynastic period of ancient Egypt, located where the Nile widens to form a delta. Now known as Ain Shams, the 'Eye of the Sun', it is sited in the north-east of modern Cairo. With a temple dedicated to the sun god Atum, it flourished as a seat of learning in the Thirty-first Dynasty at the time of Alexander the Great, around 300 BCE, until it lost significance in the following era when Egypt became a province of the Roman Empire in 30 BCE.

The port at **Alexandria**, named after Alexander, had grown in importance in the Mediterranean under the later sultanates at the head of the delta of one of the most fertile long river valleys. The site of **El Fustat** just south of the Nile Delta next grew in significance. It was expanded and grew to form the basis of a new Arab capital.

Around 970 CE the Fatimids, descendants of Muhammad's daughter, took control and created a new city known as **Al Qahira**. Whilst El Fustat was burnt such that it could not be occupied by invading Crusaders, Saladin took control with extensive building and construction of the citadel in 1176 CE. With its defensive walls and Fatimid gateways, this unified the whole of the city as **Misr**.

The next period was under Mamluk rule during the 13th and 14th centuries, when the city prospered as capital of the caliphate. With the fortifications and a number of mosques with striking minarets, its markets grew into a thriving centre on the eastern side of the River Nile, with its rich floodplain suitable for agriculture. With its maze of lanes dividing up different quarters, it became a bustling trading place. Sadly, not many remnants of this old city still exist.

In the following Ottoman period from 1516 CE it was taken over by Ottoman Turks, with **Cairo** as the capital of a province that was ruled from Constantinople. Trading grew as new markets opened up in and around the Mediterranean. After a brief, unsuccessful intervention by Napoleon at the Battle of the Pyramids, Egypt reverted to Ottoman rule.

The foundations of modern Egypt were then laid effectively by Muhammad Ali Pasha in the mid-19th century, and the central area of Cairo was developed further by his grandson the Khedive Ismail. New buildings and suburbs were developed and with new developments at Gezira Island, along with commencement of the Mahmudiyah Canal to link Alexandria with the Nile.

The most significant project then to have a major influence in the region was the opening of the Suez Canal in 1869, developed by Ferdinand de Lesseps, without the need for any locks. Around this time three new cities were also built along the Suez Canal, Port Said in the north, Ismailia and Port Tewfik in the south, as the canal finally linked the Red Sea and the Mediterranean via the Great Bitter Lake.

Modern City – Buildings and Arteries

After World War II and the Egyptian Revolution of 1952, the rate of urbanisation increased significantly with continuing growth in and around Cairo and the strategic development of satellite centres. Now the central area formed around Tahrir Square is very congested, and there are continuing pressures on urban services. On the west bank of the River Nile is Giza, which is also a very large, urbanised area.

With its location on the fertile banks of the River Nile, Cairo (**Al-Qahirah**) has continued its expansion on either side of the river and spreading out in the delta to the north. With a government decree in 1966 to protect the agricultural lands, subsequently amended several times, these restrictions led to the development of satellite centres being developed as new towns in desert areas. However, as the authorities did not police the development of agricultural land, informal building has developed, with the result that over half of Greater Cairo is now covered in informal settlements.

The pattern of this unplanned urbanised development has been of extended sprawl as the formerly agricultural land has been subdivided under 'gamaya' schemes, resulting in dense development of built types of reinforced concrete blocks with brick infill that have become predominant across suburbanised areas. As these have increased in coverage horizontally, they are now also increasing in scale vertically.

There is also evidence of extensive informal settlements all around the city, for example at Boulaq of 'ashwaiyyat', or haphazard places, of approximately 0.5 million, cut off from the city by railway lines and a canal. In many areas of the informal sector, these are characterised by Zabbaleen waste pickers collecting waste for recycling. In recent decades, extraordinarily the urbanised area has continued to double in size, every ten to 15 years. The current population of the city is approximately 10 million, with 20 million in the metropolitan area. Greater Cairo, however, extends yet farther beyond in all directions, with amorphous and informal settlements, under poly-nodal settlements which in total reach around 100 million people. The population density of the whole city region is approximately 44,900 pers./sq. km. – one of the densest world cities.

Of particular significance, as has been demonstrated by Shlomo Angel in *Planet of Cities*, between the years 1940 and 2000 whilst the population of the city of Cairo increased by a factor of six, the urbanised area increased by a factor of 16. This has put huge pressure on infrastructure and services as the urban peripheral area has been extensively developed and with deficient sanitation.

In terms of transport, having one of the earliest tram networks in the city at the beginning of the last century, this has been complemented with a modern new metro in this century. With new highways and a ring road, the transport network remains congested around the city, which is also very polluted.

Cairo International Airport (CAI) is located to the north of the city at Heliopolis and has been expanded incrementally. A number of new highways are being built in the Memphis Necropolis region. The Al-Kablat highway is planned to cross just north of the Great Pyramids in a north-south direction and new highways are also being built east-west just to the south of the pyramids; a further east-west highway crosses the Saqqara Necropolis between the Djoser Pyramid and the Red Pyramid in Dahsur. A new museum, the Grand Egyptian Museum, has been built at Giza in the vicinity of the Great Pyramids.

New Administrative Capital Cairo

As a result of inherent congestion, an additional area of New Cairo is now planned and is under development with the New Administrative Capital (NAC) as a new centre for the government of Egypt. Under design by Mohammed Talaat Architects (MTA), there is a plan for a mixed-use Pyramids Centre with a tower and mall, with three diamond-shaped commercial buildings and a new Masged Masr Mosque. An entirely new business district is due to be developed, planned to accommodate approximately 7 million people by 2050 and linked to the existing city of Cairo in the Greater Cairo region.

This NAC, named 'Wedian', being built east of Cairo, half-way towards Suez, is outside the Second Greater Cairo ring road and due to be connected by Cairo Metro, as a place of government and relief for congested Cairo. Still under realisation, being built in phases, there is a new presidential palace and Ministry of Defence headquarters, the Octagon. With a green mall and public park and perpendicular axes modelled on the layout of Washington DC, the Administrative Capital Development Authority (ACDA) manages the land, planning and infrastructure.

Transactional Forces and Typology

With its location at the junction of the coastal Nile and the Nile River Delta, Cairo has a strategic location on the eastern Mediterranean of North Africa where once on the Asian spice routes it afforded access for the transport of goods of Indian textiles and Yemeni coffee, and people on annual haj to Makkah. From the Mamluk period through to the Ottoman period, with the fertility of the surrounding land dependent on the flooding reach of the adjoining River Nile, it has acted as a trading centre at a pivotal crossroads between the Mediterranean, Africa and Arabia.

From its very ancient foundations and riverine settlements around an island on the Nile, recent modernisation of Islamic Cairo began in the 19th century with Europeanisation under the Pashas, with new buildings and infrastructure that set the framework for the future. With Gezira Island just to the west are the central Great Pyramids of Giza, with the downtown to the east, beyond which is the area for the modern planned development of New Cairo and its administrative centre.

Interspersed in many areas are either formal 'gamaya' sub-divisions for housing or many spontaneous informal areas populated by migrants searching accommodation in ever-urbanising areas around the perimeter of expanding Greater Cairo. With this comes the increasingly poly-nodal development of newly urbanising areas, stretching into marginalised land with inadequate service provision. While this land take is extending relentlessly, it is not leading to a corresponding increase in density.

5.3.2 KHARTOUM, SUDAN

Background and Origins

Khartoum (Al-Khartum), the capital of the Sudan, is located at the confluence of the White and Blue Nile Rivers. Geographically it is divided into a tripartite city with the separate areas of Khartoum, Khartoum North and Omdurman to the west. As a result of the conquest of Mamluk Egypt by the Ottomans, the city was founded in 1821 by Ismail Pasha north of the old city of Soba.

Sudan has a very ancient history that goes back to the time of the pharaohs in ancient Egypt around 1500 BCE. At that time the Kingdom of Kush existed in the Nile Valley with its capital at Kerma and with Nubian pyramids at Meroe. The Nubians subsequently created three kingdoms around the River Nile, named Nobatia, Mukuria and Alodria, around 400 CE. The capital of Alodria was Soba, which was located close to the confluence of the two Nile rivers.

With most of this Sahel region subsequently overrun by Bedouin nomads, the Funj Sultanate formed the Kingdom of Sennar in the 16th century. But they were to come into conflict with the Ottomans who had established themselves in Lower Nubia as the Ottoman Ruler of Egypt and invaded the north of Sudan. However, as Muhamed Ali Pasha took control of the Ottomans and became governor as khedive of Egypt, he extended this territory to include Sudan for his successor, Ismail Pasha.

With its location on the confluence of the Nile Rivers, Khartoum had grown into a centre of regional trade. But it was also a focus of the slave trade from Southern Africa to the northern Ottoman region where caravan trade routes existed across Africa and intersected at Khartoum. A new Islamic movement was established by Muhammed Ahmad, al-Mahdi, to expel foreign rulers.

As the British held power over Egypt, the region had become effectively an outpost of the Egyptian Army, but the Anglo-Egyptian garrison fell to the Mahdists with the defeat of General Gordon in the siege of Khartoum in 1884, which led to the subsequent formation of a new national government located at Omdurman.

But later in 1898, General Kitchener overthrew the Mahdists in the reconquest at the Battle of Omdurman. Since then, the Anglo-Egyptian government of Sudan prevailed for the next 50 years, with its capital rebuilt at the opposite side at Khartoum, parts of it laid out based on a group of Union Flags with a series of diagonal streets overlaid on a rectangular grid. As an Anglo-Egyptian condominium, it remained under joint rule until the Egyptian Revolution in 1919, when the independent Kingdom of Egypt was formed in 1922.

Between the two world wars, the British administered Sudan as two effectively separate territories, North and South. After World War II and with the Egyptian Revolution in 1952 when Nasser took power in Egypt, Sudan was granted self-government and attained its independence in January 1956, as the Republic of the Sudan with Khartoum as its capital.

After further periods of ongoing conflict, it was not until later in 2011, after a referendum, that South Sudan was to formally secede to become the separate entity of the Republic of South Sudan, having its own capital in the south at Juba. With its oil resources, it is near a mineral-rich region, but it has suffered extensively with the civil war in 2023 and conflict with the south.

Modern City – Buildings and Arteries

Omdurman initially outgrew Khartoum and was more of a cultural nexus for Sudan. Khartoum North was divided into blocks and connected with Khartoum proper with its bridges across the Blue Nile. The newer Al Mak Nimr Bridge and the Kobar Bridge opened in 2014 across the Blue Nile and complement the original Blue Nile Road and Railway Bridge built in 1909. Across the White Nile is the Omdurman Victory steel truss bridge that connects Khartoum to Omdurman, built in 1926.

Khartoum has its formal zones with embassies and a university and with a mix of commercial and residential areas and some modern buildings and the international airport right in the centre. With Nile Street alongside the riverbank, some main streets are tree lined, the central souq and Grand Mosque are in the centre of the city of Khartoum. The Al-Mogran Project was developed in the triangular junction of the two rivers, but the city has suffered much decay and destruction.

Across the river, by Tuti Island, is Omdurman with its mixed residential and cultural areas, markets and settlements that expand westwards towards Um Bada in increasingly arid and unplanned areas. In 1991, a major relocation of squatters was undertaken with 0.5 million people relocated to peace camps.

The current population of the city is approximately 5 million, with more in the wider metropolitan area. There are, however, very large areas of informal settlements which, with formerly very high rates of migration from rural areas, are distorted now by very large volumes of displaced persons.

Transactional Forces and Typology

At the confluence of both the Blue and White Nile Rivers, the earliest settlements occurred at Khartoum, which was a natural centre for regional trade. As a hot desert location, caravan trade routes intersected with the merging rivers in the city. With the region rich in minerals, more recently oil has been exploited, especially in the south, with a pipeline built across the country to Port Sudan, that was previously connected by rail.

The economy of the city, however, has suffered drastically from the regional disputes in Darfur and also the recent political separation and conflict with South Sudan. In addition, desertification and soil erosion remain further problems for the region.

As a riverine but desert city in a Sahelian country, the city of Khartoum has developed in the geographic triangular areas created naturally by the junctions of the two Nile rivers, of the centre of Khartoum, North Khartoum and Omdurman across the river to the west. As development has extended out in all directions, beyond the bridge crossings over the narrower Blue Nile, there have been peripheral planned developments interspersed with many very large areas of informal, unplanned development. The challenges of war, displaced refugees, increased urbanisation, climate effects and need for essential services are very demanding challenges prevailing both in the centre and at the periphery.

5.3.3 LAGOS, NIGERIA

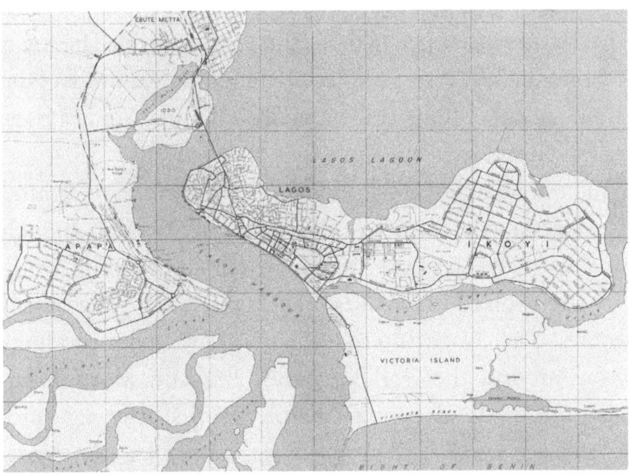

FIGURE 5.3.2 Lagos – early configuration, 1960 (Mouton).

Source: Maphouse.

Background and Origins

Lagos was mainly populated originally by Yoruba people, with the Edo people emanating from the Kingdom of Benin in southern Nigeria. Here the earliest trading and centre of crafts developed around Benin City, which was the capital of the Benin Oba. With its thick city walls, it was laid out with a central king's palace and with a grid of main and narrower streets dividing different neighbourhoods and wards. Occupied by the Portuguese in the 15th century, its trading continued until it was gradually reduced to its Edo heartland. All remnants of this city were destroyed at the end of the 19th century by the colonial occupiers.

Lagos (land of the lakes), with original islands of Iddo, Idoyi and Lagos, was occupied and named by Portuguese settlers at the same period and connected to the west of Benin City by a system of inland waterways. The Kingdom of Lagos first came under the influence of Benin but subsequently within the bights of Benin and Biafra it had become a key slave-trading place, such as at Badagry near neighbouring Porto-Novo. With the eventual treaty between Britain and Lagos that abolished slavery there in 1852, Lagos was declared a British colony in 1862 under the Gold Coast, and then as part of Southern Nigeria, it later fell within the British colony and protectorate of Nigeria that was declared in 1914.

In 1922, under Lord Luggard as governor of Nigeria, a policy statement, 'Dual Mandate for Tropical Africa', was published, which set out a framework for spatial and racial separation of communities that was also to apply to British colonies. Infrastructure was developed with new streets, utilities and tramways under the auspices of the Lagos Executive Development Board. Railway lines were also built linked to the port and northern Kano, which contributed to enabling the export of raw materials. The Cape gauge Lagos Government Railway line to Ibadan was developed in 1896.

Lagos remained the capital of Nigeria until the country became an independent federation in 1960. Lagos State was created in 1967 with other towns in the western region and acted as both state and federal capital up until 1976. The capital of Lagos State was moved to Ikeja in 1976, and a new Federal Capital Territory (FCT) was created for Nigeria at Abuja, in the centre of the country, in 1991.

The geographical coastal setting of Lagos is unusual with its marine entrance, Commodore Channel, that provides water access to the creeks and Lagos Lagoon, which is a major inland lake. On the western side of the entrance is Apapa and Lagos Harbour, with the main Eti-Osa Island and Lekki Peninsula to the east.

The present city is formed from the Ikoyi and Victoria islands that are connected to the mainland by the Carter and Eko bridges and now the long Third Mainline Bridge across the lagoon, which is framed by Lagos Harbour. With the main formal centre, development has occurred around the lagoon, with informal areas such as the Makoko informal floating settlements by the lagoon's edge.

With the principal product of the country being mainly agricultural produce, of cocoa and palm oil, cassava and rubber, along with leather and textiles, this was to change significantly with the discovery of oil and gas fields around the Niger River Delta during the 1970s.

Modern City – Buildings and Arteries

The Central Business District is on Lagos Island with the major commercial developments where most industries and commercial activities are now located, which is separate from the mainland. Massive urbanisation and migration to the city have ensued in recent so that it is now highly populous with a very large informal sector around the perimeter.

The port has grown with servicing the local oil industry, and infrastructure has developed around it with both road and rail connections. Now large ships also queue at Apapa Port and Tin Pan Island to collect containers, with another terminal planned at Lekki. New highways include the Lagos-Ibadan Expressway that connects with the Trans-Saharan and Trans-West African coastal highways. Lagos Murtala Muhammad International Airport (LOS) is located to the north-west of the city at Ikeja.

Other industries have developed outside the oil and gas industry with the production of cement and steel and some automobile assembly plants, along with the pharmaceutical industry. In addition, the financial services sector has grown with insurance and real estate, and new tech start-ups are thriving in the Yaba suburb. Also, the 'Afropolitan' film industry is newly entrenched, called 'Nollywood'.

With increasing migration from rural areas, the pressure on land and new settlements has become demanding. As a consequence, significant development has been undertaken to the north-west of the city on the mainland, considerably expanding the coverage of the metropolitan area, with other local governmental areas being annexed to the city.

With its low-lying land and mangrove swamps by the coast and lack of easily developable land, the focus has been on land reclamation and development of new areas. This is being pursued in two ways, either in the lagoon, or outside in the ocean proper. Within the lagoon, large areas of land are being reclaimed and shaped into new residential communities, such as at Banana Island and Epe Marina. A new viaduct has been built across the bay, Third Mainland Bridge connecting with Lagos Island.

Outside the harbour, with the construction of a new deep-sea retaining wall, whole new areas are being developed from the ocean at an area named Eko Atlantic. Here huge volumes of material have been moved in order to create a new man-made area, an extensive area of new real estate beside the ocean. In addition, a new deep-sea port is being developed at Lekki, with its free trade zone where also a large area of land reclamation is underway so as to create a privately financed port with an oil refinery at Dangote and fertiliser complex.

Within the city, with growing population pressures, attempts at urban upgrading have been undertaken, with development of the Marina District and due to increased demand for new commercial and residential areas. As elsewhere, private individual gated communities are also being constructed for more wealthy patrons, which is segregating communities and leading to social fragmentation.

Eko Atlantic City is also a planned commercial district on reclaimed land on a new peninsula adjoining Victoria Island of Lagos City and Lekki to the east. Developed in partnership with the Lagos State government, it is dependent on a new seawall constructed in the Atlantic Ocean with revetment, protection and dredging to create reclaimed land. With commercial and residential areas, a group of high-rise towers is under construction with new roads and infrastructure. Flooding, erosion and coastal protection here remain particularly challenging.

Separately the area of Makoko has developed as a huge settlement of hundreds of thousands of people on wooden shacks floating or perched on stilts in the lagoon. The informal sector is all pervasive, estimated at about two-thirds of the employment of the overall economy, with micro-enterprises, salesmen and hawkers across the congested city.

These large areas of informal settlements are around the lagoon, where micro-enterprises have developed, such as the Otigba computer village where thousands of workers are involved in IT and computer systems, gadgets and games. The Alaba International Market trades products from around the world. At the Olusosun waste dump also hundreds of people, mainly youths, scavenge piles of garbage for discarded materials, copper wires and batteries for recycling. Since the city has a huge challenge in dealing with municipal waste, it has set up an innovative 'We-cycling' programme to encourage collection and sale to material consolidators.

The current population of the city of Lagos is approximately 8 million and is approaching 15 million people in the metropolitan area, as one of the largest and fastest-growing cities in Africa. With its constant congestion, it has a new light rail rapid transit system, besides many informal taxis and jitneys and also water ferries. With its shadow entrepreneurs and informal economy, the yellow 'danfo' minibuses abound. The whole of the Lagos metropolitan area has considerable mixed challenges of growth, congestion, health and sanitation to cope with under the continuing forces of migration and urbanisation in this part of Nigeria, with the local government struggling to provide basic services.

Transactional Forces and Typology

As a major centre on the coastal Bay of Benin, much has changed from its former trade and exploitation of African labour to the sugar plantations in the Caribbean. As a harbour around an internal lagoon, it has evolved from being a regional source of food, palm oil, leather and textiles, to becoming a major supplier of oil and gas from the area.

With its maritime location in a zone of natural mangrove swamps, waves of rural migration from the interior have put significant pressure on the urbanising area. Whilst seeking methods of developing new areas to accommodate population through land reclamation, socially there is significant segregation with the separate development of formal and informal areas, where the latter are increasingly marginalised. Physically with the risk of further coastal zone degradation and erosion due to rising sea levels, flooding also remains an increasing risk.

As a coastal city facing both outer ocean and inner lake, the city of Lagos has dual characteristics as a rapidly expanding African city. On the one hand, surrounded by tropical mangroves and wetlands, there are marginal areas with extensive informal settlements. On the other hand, large new areas of land are being reclaimed off the Lekki Peninsula for commercial and industrial purposes.

This dichotomy is exacerbating tensions in a globalising city with rapidly expanding urbanised areas between aspiring migrants gravitating around the greater informal sector that operates under its own system of rules, in comparison with the established formal sector operating in an increasingly globalised external market. The impact of the new deep-sea port at Lekki is tending to encourage bi-polar development at the eastern end of Lagos as distinct from the established western extremes of the city.

5.3.4 KINSHASA, DEMOCRATIC REPUBLIC OF THE CONGO

Background and Origins

Kinshasa is the capital of the Democratic Republic of the Congo (DRC), located on the Congo River in the western part of central Africa. On the other side of the river is the city of **Brazzaville**, which is the capital of the Republic of the Congo, formerly part of French Equatorial Africa, making these adjoining twin cities. They have a wet tropical climate.

With the original Bantu tribes and indigenous population of the Humba and Teke tribes, it was colonised during the 19th century by the Belgians and controlled as the Congo Free State and named Léopoldville after King Leopold II of Belgium.

Discovered and developed by the explorer Henry Stanley, it acted as a port upstream of the Livingston Falls, a series of rapids downstream of the Malebo Pool (Lake Nkunda) and towards Matadi on the Lower Congo River. In order to avoid these rapids, a railway line was built between Matadi and Léopoldville so as to act as a portage railway to carry goods by detouring around these pools. A pipeline was also constructed, and the city developed and grew, acting as the capital of the Belgian Congo.

The old city ('ville base') is on a flat alluvial plain near the river, whilst newer developments were undertaken on eroding soil of the hills. During the 1920s, the city was developed on a geometric pattern with segregation of African and European neighbourhoods via a cordon sanitaire. The local resources and rubber industry were also developed, and the capital was moved to Léopoldville from Boma near the coast in 1923.

In 1960, Congo gained its independence but the south-eastern state of Katanga announced its secession from the country. With its abundant mineral resources of copper and cobalt mines in Shaba Province, it is a most valuable region, having one-third of the world's cobalt reserves. In 1965, under a coup d'état, General Mobuto seized power and renamed the country the Republic of Zaire, the capital Kinshasa and himself Mobutu Sese Seko.

However, during the 1990s, a long period of destabilisation occurred with the demise of Mobutu, the Rwandan genocide, two Congo wars and deaths of thousands and subsequent migration of millions of people in the region. This devastated the DRC, and with ongoing conflicts in the region, the situation remains very unstable.

China became involved in the region, procuring half of the metal exports from the area. The infrastructure is generally inadequate, both roads and railways run by SNCC and poorly maintained. This is now a highly populated and urbanised region, not just in Kinshasa but also in neighbouring Brazzaville. Due to the political instability, this has resulted in very poor conditions of health and deteriorating nutrition and overall human welfare.

Modern City – Buildings and Arteries

Kinshasa is the seat of government with the individual buildings of the Palais de la Nation, Palais du Peuple and Palais de Justice. Downtown commercial areas are developed to the south around the Boulevard du 30 Juin in Gombe. The port extends along the riverside and Kinshasa International Airport (FIH) is at N'Djili to the south-east. A new development of La Cité du Fleuve was planned on land reclaimed from the Congo River.

Most recently, with extensive in-migration, the city has grown significantly but in a much more haphazard fashion with significant informal settlements around the city. The current population of the city of Kinshasa, which is also one of the DRC's provinces, comprising the districts of Lukunga, Funa, Mount Amba and Tshangu, is approximately 8 million, with 17.5 million in the surrounding metropolitan area.

Kinshasa is primate city with very challenging social conditions and a population many times larger than the second most populated city of Lubumbashi. As the rate of in-migration continues, more land is being appropriated for informal settlements and is becoming increasingly urbanised. Impoverished and homeless 'street children' (shegués) are also most prevalent, with the lack of permanent accommodation and need to resort to street vending and begging.

Transactional Forces and Typology

With its location on the southern bank of the Congo River, which is a source of significant hydro-electric power, Kinshasa is a mineral-rich region. With its cobalt and copper reserves, it is also a source of rare earth elements that are a vital source for the modern digital economy, especially for lithium batteries. Whilst the mines are located in the south-east of the country, these are a vital source of income for the DRC once commercialised.

The river acts as a strategic transport facility, and the city extends linearly along the riverbank with its quays and jetties and ferry connections with neighbouring Brazzaville, where imports and exports travel via rail to the coastal port of Pointe Noire. With its economic potential, political instability has prevented resources from being properly focused on the needs of the growing number of urbanised poor that are continuing to settle in the capital.

As a riverine city on a major central African river, the city of Kinshasa has grown unusually on the south bank immediately opposite an equally urbanised zone of Brazzaville on the other bank of the river around l'Ile M'Bamou despite being in separate countries. Whilst there is no physical fixed link, planning of a bridge to connect the two potential twin cities remains under discussion. With an orthogonal gridded central area of avenues and rues, Kinshasa is now sprawling linearly south and eastwards as informal settlements gravitate off and around the main highways. Whilst low density, the numbers and sizes of settlements continue to grow with the ongoing arrival of new rural migrants, refugees and displaced persons.

5.3.5 NAIROBI, KENYA

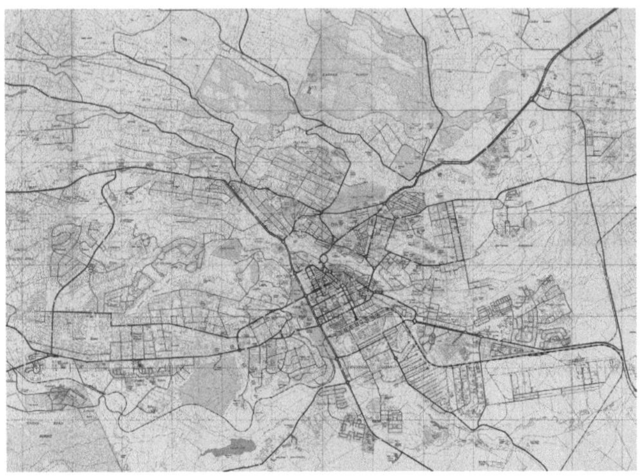

FIGURE 5.3.3 Nairobi – early configuration, 1968.

Source: Maphouse.

Background and Origins

Close to the equator and nearly 2,000 m. in elevation, **Nairobi**, as the capital of Kenya, is at the centre of East Africa. Formerly occupied by the pastoralist Maasai tribe, with a comfortable climate and good natural resources yet large population, it is located near the African Rift Valley. Heavily dependent on seasonal rains, it can suffer from long periods of drought, making the eastern areas yet more arid.

To the west and the interior Aberdare highlands, the soil is richer, cash crops and tea are grown on the hillslopes, and cattle graze before the hills that rise towards Mount Kenya, and the Nairobi River flows across the city.

Activities differed by areas having varying tribal practices – more nomadic to the east, sedentary to the west and agricultural to the north. The Maasai are a colourful tribe in the plains, with groups tending to gravitate around water sources and trading routes. With wonderful wildlife, the modern focus is on new conservancies aimed at balancing the use of land in a sustainable manner for animals, wildlife and biodiversity.

Under a British colonial administration, the city grew due to the realisation of the railway into Uganda. Considered to be a suitable encampment site during construction in 1899, it grew to replace Machakos as provincial headquarters. As a depot, streets were laid out around it and with a railway station, this became the initial centre of the settlement. From the 1920s it started to become better established with a mix of local Africans, Indian labourers and British administrators.

In 1948, a master plan for the city of Nairobi was developed by a group of professionals – White, Silberman and Anderson, South African planners. Probably influenced by their own context, where segregation was prevailing, they developed a plan for different commercial and residential areas, and zoning proposals for housing that suggested one house/acre in European areas, two houses/acre for Asians and ten houses/acre for Africans.

Open spaces and parks were designed with broad thoroughfares planned for traffic. Whilst this plan for the city and its suburbs was broadly pursued, it was soon overwhelmed due to the rapid increase in population and with in-migration from rural areas. Unusually also, Nairobi has a national park just on the outskirts of the city where wildlife roam alongside the Athi River.

Post-war and after the Mau-Mau rebellion, the city consolidated, and following independence in 1963 Nairobi became the capital of the Republic of Kenya; developed its new parliament building, city hall and Kenyatta International Convention Centre in the centre; and added a new university. The original residential areas in the north-west, towards Spring Valley, had higher income areas that retained their English place names, whereas the more populous areas developed to the south, such as at Embakasi, Umoja and Mathare.

Modern City – Buildings and Arteries

The city covers an ever-expanding area with the government district located in the centre surrounded by the commercial and business areas. Farther out, development sprawls in a haphazard manner with a high rate of urbanisation as more and more people are migrating to the urban centre. As crime increases and security deteriorates, the better off have retreated into gated residential communities with larger estates remaining in the outer areas, such as at Karen and Muthaiga.

The commercial centre of the city has developed with a number of new commercial buildings, such as the Britam Tower, Times Tower and Telposta, but more recent developments have gravitated farther out around Upper Hill and the suburban Westlands, such as the UAP Tower along with newer shopping malls. Nairobi is also the headquarters of two important UN institutions, the UN Environmental Programme and UN Habitat, founded here in 1972 and 1977 respectively.

Besides the formal planned sector, there has been a huge increase in the informal sector with often illegal settlements. This is putting more pressure on services and resources and in particular in the sprawling areas on the periphery as Kibera, not far from the city centre, where conditions are very crowded and unsanitary. With constant pressure from the growing population in the informal sector, the city is unable to provide adequate essential services. In several areas different housing models have been tested, for example at Dandora, where new lots have been provided on a serviced grid to house a large number of residents, but this informal sector is growing very fast as inward migration continues into the urbanised area.

In terms of roads, the new Uhuru and Thika highways have been built along with the outer Eastern and Northern bypasses, and Airport Road has been upgraded. The Nairobi Expressway from Jomo Kenyatta International Airport (NBO) to Westlands is a part-elevated toll road built with Chinese finance. However, many more rural roads are in very poor condition. Within the city there is considerable congestion with urban traffic, matatus, taxis and buses prevalent.

From the coastal port of Mombasa, the original metre-gauge railway meandered uphill to Nairobi where the city developed around the railway terminus. A modern, straighter, standard-gauge railway has been built by Chinese from Mombasa to Nairobi with many viaducts. This route, financed by China, was opened in 2017 and is being extended towards Uganda via Naivasha and Kisumu to the border and on to Kampala on a direct route across the Rift Valley. Outside the city a new terminus of Syokimau Station has been built nearer the airport with a connection to the older central station. The population of the city is approximately 4.5 million and 9.5 million in the metropolitan area.

Transactional Forces and Typology

As a colonial settlement, the city of Nairobi was located in a more inhospitable area of land midway between the coast and Lake Victoria, on a railway route. With an equatorial climate and near the Rift Valley, the small Nairobi River acted as a water source before reaching the nearby escarpments. With land more suited for cattle grazing, agricultural farmers sought more favourable areas up-country for cash crop production. As it grew as a centre of communications, the city developed as a growing municipality in the early 20th century and now houses the regional headquarters of international businesses.

Laid out by colonial settlers in an undulating but essentially arid area, a network of roads criss-crossed the valleys interspersed with trees and shrubs. Zoned for different populations of African, Asian and European settlers, the city's implantation follows the geophysical features at the foot of the Ngong Hills. As the population has grown and with in-migration from rural areas, new areas have been developed around the commercial centre with light industry around the railway sidings.

As urbanisation has increased, the number of informal settlements has grown intensively in any intervening spaces. Now a poly-nodal structure is developing outside the original core and administrative centre around city hall, with formal residential areas at Upper Hill, Westlands, Langata and Dagoretti. However, in contrast there are growing extreme areas of poverty in the absence of adequate serviced facilities, in the Kibera and Mathare districts, which are continuing to extend yet farther outwards. The following figures illustrate the modern configuration of the mega African cities as described and the evolution of their modern population.

FIGURE 5.3.4A Cairo, modern configuration.

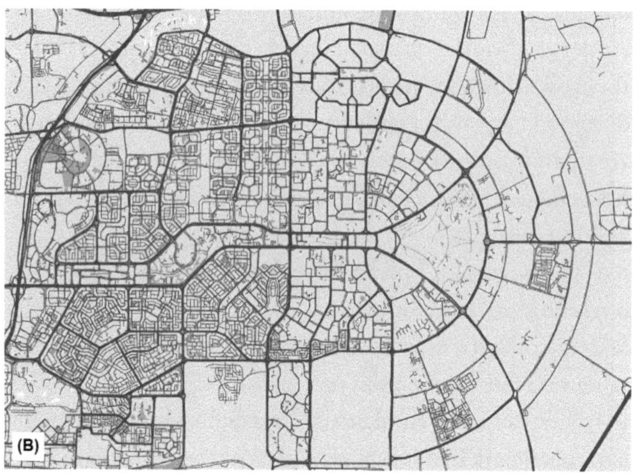

FIGURE 5.3.4B New Cairo, modern configuration.

FIGURE 5.3.5 Khartoum – modern configuration.

FIGURE 5.3.6 Lagos – modern configuration.

FIGURE 5.3.7 Kinshasa – modern configuration.

FIGURE 5.3.8 Nairobi – modern configuration.

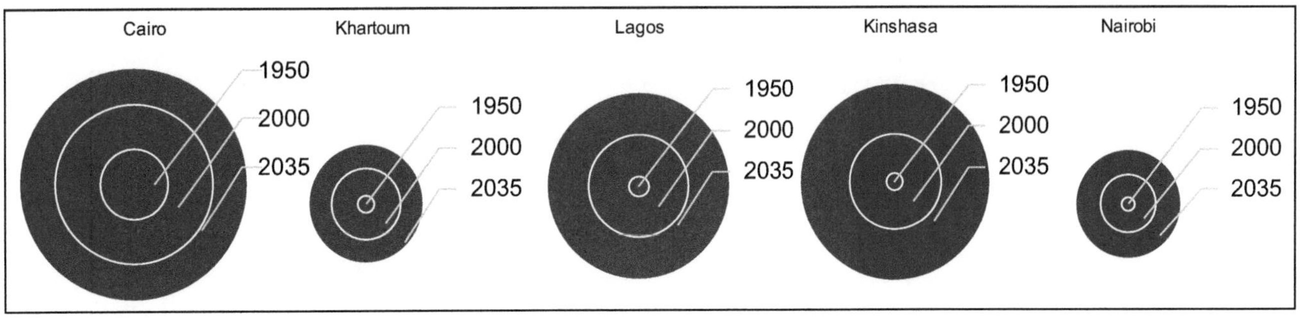

FIGURE 5.3.9 Mega African city populations.

Data source: Global Database of Metropolises 2020 – UN-Habitat (2020), UN DESA Population Division.

Further Reading

CAIRO

Alsayyad, Nezar, ed. – *Routledge Handbook on Cairo*, Routledge, 2023.

Angel, Shlomo – *Planet of Cities*, Lincoln Institute of Land Policy, 2012.

Clayton, Peter – *Chronicle of the Pharaohs*, Thames & Hudson, 1994.

Nims, CF – *Thebes of the Pharaohs, Pattern for Every City*, Elek Books, 1965.

Ruthven, Malise – *The Great Cities – Cairo*, Time Life Int., 1980.

KHARTOUM

Mahjoub, Jamal – *A Line in the River, Khartoum, City of Memory*, Bloomsbury Publisher, 2018.

Melly, George – *Khartoum and the Blue and White Nile*, Legare Ind. 2022.

LAGOS

Adelusi-Adeluyi, Ademide – *Lagos, Mapping History, Place and Politics*, Ohio University Press, 2023.

Mann, Kristin – *Slavery and the Birth of an African City, Lagos, 1760 – 1900*, Indiana University Press, 2007.

KINSHASA

Malu-Malu, Arthur – *Le Congo, Kinshasa*, Karthala, 2002.
Shapiro, David & Tambashe, Oleko – *Kinshasa in Transition*, University of Chicago Press, 2003.

NAIROBI

Charton-Bigot, H & Rodriguez-Torres, Deyssi, eds. – *Nairobi, Today the Paradox of a Fragmented City*, Mkuki Na Nyota Publisher, 2010.
Ese, Anders & Ese, Kristin – *The City Makers of Nairobi, an African Urban History*, Routledge, 2020.
Smith, Constance – *Nairobi in the Making*, Boydell & Brewer, 2019.

5.4 Mega South American Cities

Atlantic Coastal and Amazon Brazil – Southern Argentina – Pacific Coastal Chile

São Paulo, Rio de Janeiro, Buenos Aires, Santiago and Lima

5.4.0 INTRODUCTION – LATIN AMERICAN MEGA CITIES

The South American continent as a resource-rich region was populated by early agrarian communities that gravitated around riverine regions, for example the extensive Amazon basin and the Parana River. With its original coastal fishing and farming communities, and with foresters and animal husbandry groups, settlements gravitated around such strategic locations in both the lowlands and highlands.

As European explorers ventured around the continent, reconnoitring groupings of indigenous peoples, settlements developed along the coast at natural harbours, from which external trade routes became established. Along the South Atlantic, on the Brazilian coastline, significant ports were developed at Recife, Salvador, Rio de Janeiro, Santos and Rio Grande. On the Pacific coast, along the Chilean coastline, similarly ports were established at Antofagasta, Valparaiso and Concepción. It was around these places and their hinterlands that significant growth was centred.

With the division into Portuguese and Spanish regional spheres of influence, initially '370 Leagues west of the Cape Verde Islands', coastal regions were colonised and influenced first, before reach was extended inland to some of the established interior mountainous peoples and cultures, such as in the Andes. With mineral and agricultural resources, industrial and agro-processing developed especially around established coastal cities and ports. Several primate cities developed as centres of trade, for both export and import, laid out in accordance with the 'Laws of the Indies'.

During the late 20th century the established colonial coastal cities retained their attraction, and with rural migrants and higher birth rates, have experienced significant over-crowding at their cores. Lack of adequate services and housing has caused development, in particular of peripheral urban areas where populations have depended on the informal sector to establish settlements in marginal areas.

Often in hazardous or unstable areas on steep slopes, temporary shelter has been fabricated in peripheral areas, without access to proper utilities and sanitation services. Such 'barrios' and 'favelas' have become predominant semi-permanent features on the outskirts of many South American mega cities that currently act as homes to millions of young people.

In the 21st century, particular South American cities have entertained significant growth, for example both Mexico City and São Paulo, have now grown into mega-cities that have over 20 million inhabitants and are still growing fast. Several other cities have over 10 million in their metropolitan areas with significant continuing in-migration from rural areas.

Settlements that arose in the Caribbean and South America emanated as a result of actions of both Spanish and Portuguese explorers developed under the 'Laws of the Indies'. Both San Juan in Puerto Rico and Santo Domingo in Hispaniola are typical Spanish typologies found in the Caribbean, whilst Bahia, San Salvador is prototypical of Portuguese coastal development on the South Atlantic shoreline. On the Pacific coast, Spanish settlements had different characteristics, as exemplified by fortified Lima with its grid and coastal Callao, and colonial settlements, such as at mountainous Cusco.

5.4.1 SÃO PAULO, BRAZIL

Background and Origins

As the largest and most exotic South American country, Brazil stretches from the Amazon rainforest through the Cerrado, to the great escarpment of the coastal Atlantic region and Serra do Mar. The area around São Paulo by the Tiete River was originally occupied by Tupiniquim tribes organised in 'caciques', having agricultural and fishing practices.

DOI: 10.4324/9781003353348-29

The Portuguese settled and established a mission at São Paulo dos Campos de Pivatininga' in 1554 CE that depended on subsistence crops and local labour. Located in the south-east of the country, **São Paulo** is just inland from the port of São Vicente and became a stopover for earlier explorers heading into the interior, the 'bandeirantes' who were seeking materials for their own benefit.

It was not until around 1691, however, when gold was discovered in Minais Gerais, that it became more frequented. Agricultural practices also began to shift to the production and export of sugar cane with the requirement for significant labour and consequent enslavement of workers. Sugar remained the major cash crop until it was overtaken by its replacement, coffee, that led to further development and construction of railway lines for access.

In 1822, as the Empire of Brazil, it became independent from Portugal, under Emperor Pedro I, who declared São Paulo an imperial city. In 1869, São Paulo was connected to the Port of Santos by the Santos–Jurdiai railway, along with other lines which centred on São Paulo that supported coffee production from the plantations and also milk from Minais Gerais dairies. Private developers were entitled to ownership of bands of land 30 km. on either side of a railway, which enhanced land value and encouraged development.

Slavery was eventually abolished in 1888 and in the following year Brazil became a republic with a new federal constitution. With the economy focused essentially on the export of coffee, both road and rail infrastructure was developed and it grew as a regional centre with continuing immigration increasing especially from Europe, particularly from Spain, Portugal and Italy. Whilst planters were the elite, workers were housed in 'corticos', which were overcrowded tenements.

Particular garden neighbourhoods were developed in Higienopolis, Jardim Europa, Jardim America and Jardim Paulistano, and Haussmann-like boulevards were developed in and across the city.

Between the world wars there was an unsettled period during the 1930s during the Great Depression, before a period of industrialisation was to commence after the ending of World War II.

Modern City – Buildings and Arteries

São Paulo is the modern capital of the State of São Paulo, with a mayor and cabinet, and the city is divided into 32 sub-prefectures. As state capital, the conurbation, which includes the city of Guarulhos, has the densest built-up central commercial area and older 'Jardins' district.

The city has developed from the centre and has become increasingly dense with vertical structures all the way to the periphery, which is typified also by extensive, sprawling, informal settlements often on sloping land, or on any available vacant land area, whatever its suitability.

The main axes of the city, especially the Avenida Paulista, developed from the Plan of the Avenues in the 1920s and earlier plans for along the Tiete River. Later, since the Urban Development Plan of the 1960s, expansion has been undertaken aggressively in all directions with only very limited zoning control.

With high in-migration from the countryside, the city was subject to extreme growth of urban 'corticos' for workers, and at the periphery the centre has been surrounded by many informal settlements where 'favela' had been established on any available and often hilly vacant land under a process of self-building. One of the largest is 'Paraisopolis' in the Vila Andrade District – it is estimated that over 2 million people live in favelas there.

In contrast, the wealthy live in protected areas such as Faria Lima often as gated communities or 'condominios fechados' within the city, as defence from the rising levels of urban crime that are found elsewhere, exacerbating the gap in housing provision. During the 1960s, large-scale public housing products were undertaken with developments in São Paulo Cidade Itaquera and Cidade Tivadentes with multi-storey apartment blocks.

The original heart of the city was Praca da Sé with its cathedral, which is now surrounded by broad heavily trafficked avenues of bustling activity with a dense pattern of commercial skyscrapers and multiple lanes of traffic in each direction.

From the areas of Republica, Bras, Liberdade and Bela Vista, the Avenida Paulista typifies the sheer scale of development and on beyond to Jardim Paulista, America and Europa. With its different commercial neighbourhoods of Itaim and Brooklin to the south on the banks of the Rio Tiete, the built development extends relentlessly, reaching skywards in any available airspace.

Post-war industrialisation was to focus on a programme of import substitution with the emergence of a new automobile manufacturing industry and subsequent aero industry. The economy more recently has been transformed from a previously industrialised economy to an essentially service economy. The corporate headquarters of many companies are now located in downtown São Paulo and with the stock exchange, many science and technology companies, healthcare and new research institutes have been established, along with a concentration in finance, real estate and insurance services.

Outside São Paulo is São José dos Campos, an R&D innovation centre in the Paraibo Valley that is focused on the aerospace industry with the Instituto Aerologico de Aeronautica (IAA) and Tech Park. Another commercial centre is around Itaim Bibi in the suburb of Brooklin.

The modern metro transit lines help to transport hundreds of thousands of commuters each day across the city, with a major underground interchange between the Red and Blue Lines at the station of Sé. With Lines 1–5 plus Monorail Line 15, the metro has 89 stations.

There are also a number of Rapid Bus Lines (Passa Rapido) that interact with both the rail and the metro stations, the two main railway stations being Estacao de Luz and Julio Prestes. The commuter rail system of the Companhia Paulista de Treno Metropolitanos (CPTM) has seven lines and 94 stations. A new high-speed line is also planned to be built out to connect São Paulo to Rio de Janeiro.

São Paulo has two main airports – Guarulhos International (GRU) and also São Paulo–Congonhas (CGH) – that both carry significant passenger and cargo traffic.

The city has an extensive network of expressways that both enter and surround the city. These include the peripheral beltway Rodoanel Mario Covas and the main Rodovias BR-116, SP 270, 280 and 330. The main arteries run along either side of the Rio Pinheiros and the Tiete River – the Anchieta SP150 and Imigrantes SP160, which roll down alongside the steep cliffs of the Serra do Mar to the Port of Santos. In the inner city there are the central underground Tunel São Joao Paulo II and overhead viaducts mid-town.

With its old cathedral in the Praca da Sé, the Monastery São Bento, the elegant Jardins neighbourhood, many museums and theatres, São Paulo is both a commercial and cultural centre. Each of the individual neighbourhoods of Sé, Republica, Bela Vista, Liberdade and Jardim Paulista has its own unique and distinguishing attributes.

The Parque do Ibirapuera provides a small green lung of open space, as well as the Parque Burle Marx farther out, but the whole sense of the city is one of continuous intensity of all types of human activity of the resident Paulistanos, day and night.

The population of the city is approximately 12.5 million, with 22 million in the metropolitan area. Given the extensive outlying urbanised areas of Campinas, Santos, Sorocaba and São José dos Campos, the now greater metropolitan area named the São Paulo Macro-Metropolis has a population that exceeds 30 million inhabitants. The population density of the whole city region is approximately 14,500 pers./sq. km.

São Paulo is a significant South American state capital as an extremely dense, cosmopolitan, highly congested sprawling city with extensive informal settlements, that has many population and growth challenges. As the centre of a continually growing metropolitan area, it has developed into a truly megalopolitan region. There are clusters of new economic development around Faria Lima, Vila Olimpia and Berrini Nacoes Unidas along the Rio Pinheiros.

The city has been subject to significant growth pressures with urban agglomeration in its peripheral areas. New industries have grown in the São Paulo Metropolitan Region, particularly in an area around the municipalities of Santo André, São Bernardo do Campo and São Caetano do Sul, known as ABC Paulista in attracting new migrants.

In terms of sustainability, the city has a water crisis in trying to deal with its growing population. Both the Tiete and Pinheiros rivers are very polluted such that they cannot be cleaned to drinking-level standards, and the Cantareira water system has inadequate capacity. Pollution and air quality remain key issues.

Transactional Forces and Typology

With its location on the crest of the Serra do Mar, inland between the South Atlantic coast and the hinterlands, with its suitable agricultural land and location to channel the importation of labour, São Paulo was well suited for the production and exportation of sugar cane. As mining also developed in the region, the road to Santos was created along with its port, which with the realisation of new railway lines also enabled it to handle the main cash crop of sugar, followed by subsequent coffee development.

From sugar to the coffee barons and the dairy milk producers of Minos Gerais, new businesses developed in São Paulo that evolved from plantations to subsequent industrialisation, which led to the city's development as a manufacturing centre. More labour was required and attracted to the city, which, based upon a national policy of import substitution, led to growth both at its dense core but also around the periphery. With weak zoning controls initially, this led to the bi-polar development of the city with socio-spatial segregation with its elite central areas surrounded by many marginal areas of informal settlement of favelas, generally on demanding and unsuitable geographical terrain.

On the periphery, informal housing has centred on illegal allotments having inadequate sanitation facilities. The tension between the forces of private capital, real estate interests and informal settlement has developed. The national government initiated a programme of housing expansion in 2009, named Minha Casa, Minha Vida – MCMV, My House, My Life – which did lead to some increase in housing provision in parts of São Paulo, mainly in the outer areas.

In 1973, the government created new metropolitan regions, of which Regio Metropolitana – RM São Paulo and a corresponding city region, Macro-Metropole – MM Paulista, encompassed the city. But within this region there are mixed and differing areas of declining industry in ABC Paulista and marginalised areas, such as Guarulhos.

Agro-industry is centred in Sorocaba, and science parks have been built, such as Campinas and São José dos Campos, resulting in a polycentric structure of this larger urbanised region. Whilst all of this is subject to a Macro-Metropolitan Action Plan led by EMPLASA, the regional planning agency, it is evolving into a truly megalopolitan agglomeration.

Whilst now a very dense and highly developed city with virtually little remaining open space, with its high density its core is very compact. Extending out in all directions, the built density of commercial and residential buildings remains at a very high level, with its commercial core surrounded on the one hand with well-off gated residential communities, and on the other hand with many areas of dense, poor neighbourhoods, informal favelas with inadequate services.

A hilly inland city, not far from the coastal hinterland, despite its organised central core, São Paulo now sprawls relentlessly encompassing the poly-nodal centres of Guarulhos, Itaquaquecetuba, São Bernardo do Campo and Osasco, and is extending towards Campinas, Santos and São José dos Campos into one continually expanding 'megalopolitan agglomeration'.

5.4.2 RIO DE JANEIRO, BRAZIL

FIGURE 5.4.1 Rio de Janeiro – early configuration, 1830.

Source: iStock.

Background and Origins

In the south-east of Brazil, Rio, originally inhabited by indigenous Tupi and Guarai peoples, was first occupied at the mouth of the river in Guanabara Bay by Portuguese settlers under Cabral in 1502 CE, known as **Rio de Janeiro**, the 'River of January'. Swamps were drained and water supplied via aqueduct. With its imposing location and Pão de Açucar, Sugar Loaf Mountain, it has the most striking topography, of the Corcovado Mountain, as well as the adjoining long and distinctive Copacabana Beach and Niteroi on the other side of the bay.

After several centuries of colonial occupation, in 1763 the city became the capital of the Vice-Royalty of Brazil and with the arrival of the Portuguese Royalty and Dom João VI in 1808, it was made both the capital of Brazil and of the Portuguese Empire, which lasted until 1889. During this colonial period, the city had a large influx of African slaves, around 0.5 million, who were landed at Valongo Wharf during the early 19th century.

During summer periods the capital was relocated to a summer palace in Petropolis, the 'Imperial City', a location just to the north-east of Rio founded by Emperor Pedro. With its nucleus, this site is the second planned city in Brazil after Recife that was eventually linked by highway and railway both to Minais Gerais and Rio.

In 1902, Mayor Pereira Passos introduced a redevelopment plan for the city, with the creation of new formal avenues and streetscape and removal of the poorer 'cortices', forcing the poorer population to move out to informal favelas.

Rio was retained as the capital of the Republic of Brazil until 1960, when the capital was transferred to the newly realised and more central inland city of Brasília. Later, in 1975, with the fusion of Guanabara and Rio, Rio de Janeiro replaced Niteroi as the capital of the state of Rio de Janeiro and also the Rio de Janeiro Metropolitan Region was created.

Rio has several different districts, starting from Sugar Loaf Mountain, which guards the entrance of the bay, the promontory of which is actually two peaks of granite and quartz, comprising the lower flat summit of Morro da Urca and the peak of Pão de Açucar, which can both be reached by cable car.

The historic centre of the city (Centro) revolves around the old Royal Palace, Paco Imperial, in Praca XV de Novembro, formally Lago do Paco, with its tree-lined square. With other historic buildings such as the São Bento Monastery, Municipal Theatre, its museums and national library, the old docks are situated at the end of Avenida Rio Branco, formerly Avenida Central. Guanabara Palace houses the state government.

After the historic centre developed around Guanabara Bay, subsequent development gravitated outwards and farther south along the coast to Zona Sul, after a tunnel had been built south through the mountains from Botafogo to connect with Copacabana and Avenida Atlantica. Further developments stretch westwards into the largest Zona Oueste.

In contrast, adjoining this beyond is Leme Beach and Copacabana Beach, one of the most iconic beaches that sweeps around to Forte de Copacabana, with its geometric wave paving (pedras portuguesas) designed by Burle Marx and beach protection. Beyond is Ipanema ('rough water') Beach, an equally stunning mile-long sandy beach and Leblon, past the canal that connects the lagoon, Lagoa Rodrigo de Freitas, with the Atlantic Ocean and the botanic gardens nearby.

On the summit of Corcovado Mountain stands the majestic statue of Cristo Redentor (Christ the Redeemer), inaugurated in 1931 on the centenary of Brazil and which towers over the whole of Rio, with iconically outspread arms.

Modern City – Buildings and Arteries

With other areas, such as Maracana, there are several buildings of renown, for example the large Maracana Stadium, the Museum of Modern Art and the expressive saucer-like Museum of Contemporary Art (designed by Niemeyer) opposite in Niteroi. With his Sambodromo also this acts as a parade venue for the annual Carnival, and with the vibrant musical culture in the city, Rio is the most exotic Brazilian city that thrives to the Latin American music of the local Cariocas, whether at carnival time or at football games, in areas such as Lapa and Gavea, Ipanema and Copacabana.

In the 1960s, oil exploitation began after the discovery of oil in the offshore Campos Basin, which has led to industrialisation around the port. The construction, pharmaceutical and health sectors have also been developed, as well as the education sector with universities and research institutes. The summer Olympic Games were held in Barra da Tijuca Olympic Park in 2016 when a new central light rail system, VLT Carioca, was developed.

Large-scale formal social housing schemes were built in the 1960s with dense, multi-storey apartment blocks, and significant developments have been undertaken on the outskirts. However, in contrast many private, gated communities have been developed as security deteriorated in more affluent areas, which emphasises the dichotomy in different types of communities.

With population growth, the northern parts of the city have become the most populous, and those that contain the most extensive 'favelas' built on the hillsides, such as Complexo do Alemão. With dense areas of informal housing such as Cidade de Deus and Rocinha in São Conrado, these have large populations and contain vibrant, although poor, communities involved extensively in self-building and in peripheral areas.

A new metro system was developed in the city across its challenging geography that opened in 1979, with four underground lines. A new metro link is planned across Guanabara Bay to the city of Niteroi opposite with Fortaleza da Santa Cruz to complement the ferry services and the long Rio-Niteroi road viaduct that was opened in 1974. An extensive urban commuter rail system also radiates out to more distant suburbs, run by Super Via. Rio de Janeiro International Airport (GIG) is located on Galeão Island in the bay on the site of an earlier air base. The overall population of the city is approximately 6.5 million and double in the metropolitan area.

Transactional Forces and Typology

From the sugar and coffee plantations in the region, and the traditional Brazilian industries of textiles, these gradually evolved to newer industries. As the economy was modernised the commercial and financial sector expanded in Rio, with new corporations being headquartered there and service industries increased. They were also transformed with the offshore discovery and exploitation of oil.

As the rate of urbanisation increased, in-migration followed, leading to housing shortages. This led to the growth of a large number of informal 'favelas' across the city, mainly on challenging hilly terrain or on vacant land where built development lacked proper access. All of this has contributed to the expansion of the metropolitan area of Rio de Janeiro.

As a coastal city on hilly terrain around the bay, characterised by the iconic Sugar Loaf Mountain, first centred around the historic core, Rio de Janeiro now stretches north and westwards, radiating out from the central area. With its parks and long perimeter beaches of Ipanema and Copacabana, it is a city of contrasts, rich and poor, with extensive expanding areas of informal favelas reaching up the hillsides, generally on inhospitable terrain.

Beyond are the growing suburbs of Barra da Tijuca and coastal areas of lagoons hemmed in by surrounding mountain Serra. Whilst development is mainly radial, determined by hillside contours, the opposing Niteroi across the bay so far is less intensively developed. With increased population and migration pressures and improved infrastructure connections, this is likely to develop as another pivotal and heavily urbanised node on the other side of the bay.

5.4.3 BUENOS AIRES, ARGENTINA

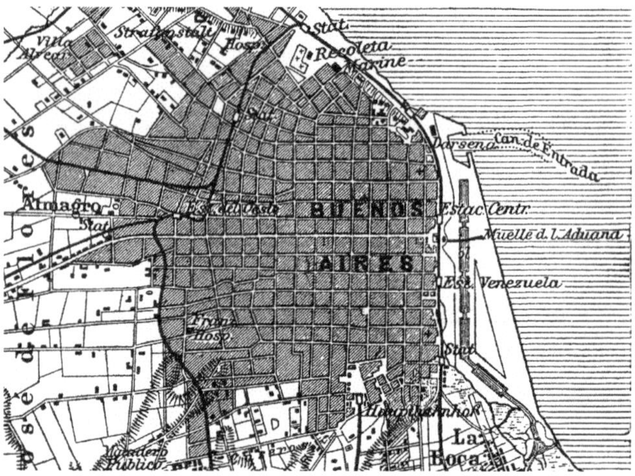

FIGURE 5.4.2 Buenos Aires – early configuration, 1897.
Source: iStock.

Background and Origins

As one of the most Europeanised cities in Latin America, **Buenos Aires** reflects much of this imported culture. Along with its indigenous population, with local culture from the rural pampas and puna, and the more distant mountainous regions of the Andes, cattle and sheep, grain and tobacco served as the basis of the national economy.

Located in the Rio de la Plata estuary, beyond La Plata and opposite smaller Colonia in Uruguay, into which flows the long Rio Parana, from the first settlements in the 16th century CE the city grew around the Microcentro banking area and old port of La Boca and the grid around the central areas of Plaza de Mayo, Plaza San Martin and Retiro. It was made capital of the Vice-Royalty of Rio de la Plata under the Spanish in 1776 and continued to grow until the end of the century.

After the Argentinian War of Independence, the independence of the United Provinces of Rio de la Plata was secured in 1816 at the Congress of Tucumen with Buenos Aires as its capital. A new Argentine constitution was formalised in 1853.

The heart of the city is Plaza Mayor with the Pyramid de Mayo dominated by the infamous Casa Rosada, the pink presidential palace where Eva Peron addressed her supporters. With its neighbouring administrative buildings – Cabildo (city hall), Ministry of Economy, Banco de la Nacion and Catedral Metropolitana – the city was well planned and laid out with the wide Avenida de Mayo that stretches out westwards, leading into the Avenida Rivadavia. The Federal Capital District area of the city was laid out in 1898.

At the other end of the avenue is Plaza del Congreso with the national Palacio de Congreso, the Senate and House of Representatives. Perpendicular to this, the wide Avenida 9 de Julio crosses Avenida Corrientes at the busily trafficked Plaza Republica, with its tall obelisk at the intersection. Here is Teatro Colón, the famous opera house with its main staircase and grand hall that has hosted many great international artists.

The San Telmo and La Boca districts are to the south where European settlers took up residence. The San Telmo district with its traditional houses and patios is best known as the home of Argentinian tango, and La Boca as home at La Bombonera of the Boca Juniors' football team. With colourful districts in El Caminito, nearby is the great iron Transporter Bridge Nicolas Avellaneda that used to carry steel gondolas across the Riachuelo.

Adjoining the extensive rail yards and cargo wharves are the commercial districts of Plaza San Martin and El Retiro. Here are the elegant Palacio San Martin and Haedo and the Circulo Militar, the Galerías Pacífico shopping mall and Edificio Kavanagh, once the tallest reinforced-concrete stepped building in South America when constructed in 1935. Opposite the main railway station, Estación Retiro, that connects with the subway system, is the Torre Monumental clock tower.

Recoleta nearby is an area of green plazas around the extensive Recoleta Cemetery that contains the tombs of former presidents and Eva Peron. Farther west are other coastal districts of Palermo and Belgrano with Parque 3 de Febrero and Bosques de Palermo along Avenida del Libertador.

With the Japanese Gardens, botanical gardens and Parque Thays nearby, these are favoured recreational areas for 'porteños'.

The city developed around the west bank of La Plata River, with the port area of Puerto Madero. Since the river estuary was shallow, an area for docking ships was required to avoid off-loading into barges. Accordingly, in 1897, a new dock area of Puerto Madero was constructed with four contiguous docks. However, due to its small size, it soon became obsolete and a new, much larger Puerto Nuevo was built with six docks and breakwater in the Retiro district which was to become the predominant port of the city.

In the mid-19th century, four new railway lines were built to connect with the interior, and at the end of the century streetcars were introduced on a radial network, with a new administrative area extended in 1888 beyond the historical centre bounded by Avenida General Paz. With electrification of the railways and the streetcars, the commuting distance increased and outer suburban residential municipalities were developed, 'alrededores', on a radial basis and with new highways.

Modern City – Buildings and Arteries

Besides its historic district, the Buenos Aires central business district, along with Microcentro, contains the original European areas of settlement. Close to Rio de la Plata estuary, having been protected by flood-control walls with channelled creeks, the area was contained and developed as a new business district. With the early stock exchange and post office, the first tall building post-war was the Alas Building.

In 1994, a revitalization project under a newly created Corporation for Old Puerto Madero began for the whole area aimed at renewing 170 hectares of the old port area. From its former warehouses, urban renewal included a new Puente de la Mujer (designed by Calatrava), the Argentine Catholic University, the Arts Pavilion and the Fortabart Art Collection. In addition, in the area of old Docks 1, 2 and 3, a whole series of new buildings and major towers have been built. These include Harbour Tower, Alvear, Renoir, Mulieris, El Faro and YPF (designed by Pelli) towers, among others. Further commercial areas have also been developed after the Madero Centre, such as Bouchard Plaza (designed by HOK) and other buildings at the new Catalinas Norte Office Park.

The Subterraneo de Buenos Aires (Subté) was the first underground network built in Latin America, with six lines that radiate out from the centre, except for a new Line H built parallel to Line C. There is a comprehensive commuter rail network now operated by Ferrovias and Metrovias under concession. With the active Port of Buenos Aires and Dock Sud, Ezeiza International Airport (EZE) is to the south-west of the city.

Beyond the central area, growth in population has led to increased urbanisation of the city and with inadequate housing provision, several informal areas of 'villas miseria' have developed in and around the periphery of the city. These, such as Villa 31, have accordingly put increasing pressure on the provision of basic services and sanitation.

Overall, Buenos Aires is a fine city with colour and Latin American character across the urbanised centre with its own mix of indigenous and European culture. The population of the city is approximately 3 million, with 16 million in the metropolitan area.

A fixed-link road connection across the wide estuary of Rio de la Plata to Colonia in neighbouring Uruguay remains only a remote scheme.

Transactional Forces and Typology

Located on the western shore of the Rio de la Plata, the early European migrants to Buenos Aires arrived thanks to 'fair winds' that had driven sailing ships across the Atlantic. As a suitable port at the confluence of the Uruguay and Parana rivers, with the pampas and extensive surrounding rich Argentinian lands, this was the basis for agricultural production of wheat and beef and was to become one of the world's major exporters at the turn of the 20th century.

With its waves of European migrants, other industries developed, mainly labour intensive, and with industrialisation, the network of railways was expanded across the country, radiating out from Buenos Aires. The port area and its wharves were developed to handle the processing of bulk commodities. Post–World War II, however, the economic environment changed with nationalisations and with political instability, and a pattern of economic crises ensued. Despite its rich natural resources that enabled the production of cotton, textiles and tobacco, soybean, wine and food processing, the industrial sector has suffered from periods of very high inflation and has declined.

As a coastal port city on the wide Plata estuary, with a planned central historical core, the city has grown in a radial manner. As new axes have been extended outwards, new suburban areas were created that have grown together in a more aggregated manner. With its central grid area, blocks ('manzanos') and main avenues beyond the city radiate outwards in a generally radio-centric manner, with localised centres of development creating an extended poly-nodal structure. As the old inner dock area has been regenerated, there are many outer areas of poorer inhabitants and increasingly larger areas of informal settlements and villas miseria in the metropolitanised region.

5.4.4 SANTIAGO, CHILE

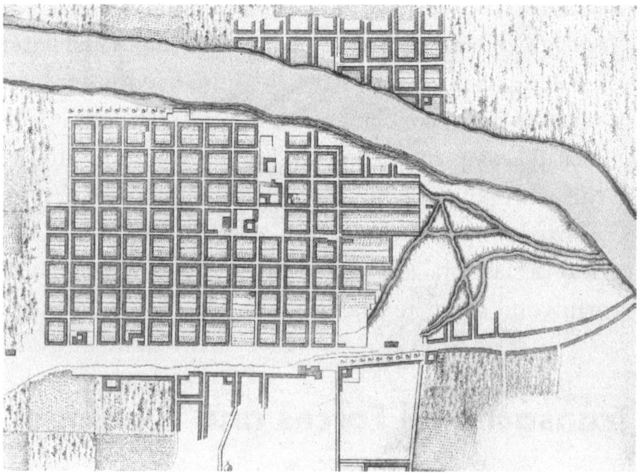

FIGURE 5.4.3 Santiago – early configuration, 1764 (Bellin).
Source: Maphouse.

Background and Origins

The country of Chile is of extraordinary geographical biodiversity, stretching as it does from the regions of the north and the arid Norte Grande and Norto Chico with the Atacama Desert, through the lush Central Valley, the Valdivian rainforest, the Lake District and Chiloe to Northern and Southern Patagonia and the colder Tierra del Fuego. With its long coastline, these regions range from forests and lakes through deserts to the towering Andes mountains, with volcanoes, fjords and ice fields in the south. Its coastal region, containing the northward-flowing Humboldt Current, also is the area where the continental plates of South America and Nazca collide, creating the Peru–Chile trench, which risks creating strong earthquakes, as has been evidenced in recent times.

From its early population of Picunche, it came under the influence of the Incas from their central Andean empire in the 15th century CE. The site of Santiago del Nuevo Extremo, known as Nueva Extremadura, was settled by the Spanish conquistador Pedro de Valdivia, who founded the city in 1541. Seeking wealth from 'encomiendas', the search for the best agricultural land 'haciendas' caused tension with the local Mapuche.

The site of **Santiago** just inland is dominated by Cerro San Cristobal in the foothills of the Andes, in the centre of the country and bisected by the Rio Mapocho. With its planned gridded core and blocks around Plaza de Armas, which was used formerly as a parade ground, this was developed as the civic centre of the city. Here the main institutional buildings, such as the Palacio de la Real Audiencia and the Metropolitan Cathedral were built.

With a major earthquake in 1647, the area was subject to frequent flooding. Around 1770, the embankments along the Rio Mapocho were stabilised and an arched stone bridge was built to connect each bank of the river, the Calicanto Bridge. On its declaration of independence, Santiago was proclaimed the capital of Chile in 1818 under the leadership of Bernardo O'Higgins. In the following period of Patria Nueva a series of improvements was built, such as the San Carlos water canal, the La Canada neighbourhood and Avenida Libertador Bernardo O'Higgins (La Alameda).

Around the central Plaza de la Constitucion is the presidential Palacio de la Moneda, Bolsa de Comercio and Barrio París-Londres with its fine mansions. With a number of churches and the Central Market, the district retains one of the oldest adobe buildings, Posada del Corregidor. On nearby Cerro Santa Lucía sits Castillo Hidalgo at the summit. To the west of the centre are Barrio Brasil, Barrio Concha y Toro and Parque Quinta Normal. Farther northeast along the Rio Mapocho are Barrio Lastarria and Barrio Providencia opposite Parque Metropolitano de Santiago, a large hilly park with the Statue of the Virgin at the summit of Cerro San Cristobal.

In the mid-19th century, under Mayor MacKenna, the urban infrastructure of the city was enhanced with the remodelling of the Cerro Santa Lucía and Avenida Alameda and the construction of a ring road, the Camino de Cintura, and the realisation of O'Higgins Park. An international exposition was held in 1875 in Quinta Normal Park when the railway network also started to be developed from Estación Alameda, Central Station (designed by Eiffel) and then Mapocho Station.

In 1910 also, with the City Centennial, further urban development improvements were made. At the beginning of the 20th century the city began to expand to the suburbs with an influx of population from rural areas, and new municipalities were created that included Providencia and Barrancas. However, with the collapse of the nitrate industry and the Great Depression, it was not to recover until the late 1930s. Post–World War II, the situation was to change as copper became the main revenue earner, but this was interrupted by the period of the Pinochet regime. With resumption of democracy, economic reforms led to new modernisation.

Modern City – Buildings and Arteries

Considerable redevelopment was undertaken in the centre with large real estate developments, and off-centre in the

Barrio El Golf District with Titanium La Portada tower, and in the modern commercial area with the Gran Torre Costanera (designed by César Pelli). The telecommunications tower Torre Entel was constructed near La Moneda in 1975.

Networks of modern free-flow tolled roads were expanded across the city, such as Vespucio Norte and Sur, with Autopista Central (part of Panamericana Ruta 5) running north-south and Costanera Norte running east-west.

In terms of infrastructure and public transport, the main avenues carry several lanes of 'Red' buses, and six underground metro lines carry high volumes of daily passenger traffic. The first line, L1, was opened in 1975, followed by Lines L2–L6, with three more lines planned to follow. There is also the Metrotren commuter rail system and connection with the airport at Pudahuel, Arturo Merino Benitez International Airport (SCL) at the north-east edge of the city.

Santiago is a place of much character located in the centre of a very elongated country that has a wide diversity of geographical and regional landscapes. But lying in an active seismic zone it suffered in a major earthquake in 2010. For the bi-centenary of the country that year, the Plaza de la Ciudadaria was remodelled. As a fast-moving and energetic city, Santiago is connected to the main port of the historic hilly neighbouring city of Valparaiso with its many funiculars, where the country's legislature is retained.

Modern economic reforms have led to new areas of real estate development, especially on the urban fringes and the foothills of Barrio Alto. Around the area of Avenida Providencia, this has developed as a major commercial hub with new shopping malls and pedestrianised areas such as Paseo Ahumada.

The population of the city is approximately 6.3 million and 6.9 million in the Santiago Metropolitan Region. The area of La Florida to the south of the city is a fast-growing and populous area. The wider conurbation of Greater Santiago extends outwards and comprises over 40 different communes.

Transactional Forces and Typology

With its lengthy Pacific coastline, historically indigenous peoples inhabited both coastal areas and fertile valleys. Beyond these under the Spanish conquistadores, the city of Santiago was in the forefront of the Reconquista settlements. After the conquest of land, the city was shaped around the river valley, crossing from the mountainous Andes towards the coast and with the cultivation of staple crops and the subsequent exploitation of copper. From the original central layout and grid of the city, its development was challenged by periodic floods and earthquakes.

On becoming the state capital, its influence was to rise during the 19th century, as the centre was remodelled around Santa Lucía Hill and as Alameda and Providencia avenues became main axes alongside the Mapocho riverbed. Whilst closely linked with the ports of Valparaiso and Concepción, Santiago is also located nearer the central trans-Andean crossing to Mendoza, at Paso Los Libertadores, where the Trans-Andean Railway was developed around 1910, improving connectivity.

From its central core and river crossing surrounded by a hilly area, the city developed in different districts from its hub in a generally radio-centric manner where the valleys afforded space for development. As urbanisation increased, the surrounding communities of Conchali, La Cisterna, La Granja and Barrancas were developed in the north, south and west. In 1958, the Intercommunal Plan of Santiago was prepared and aimed at containing central development but creating industrial belts along new axes.

With a rapidly expanding population, the development of the metro was vital and although only commenced in 1969, has now an extended urban network with perpendicular axes crossing in the central area. This forms an essential part of the Transantiago Master Plan project launched in 2007, with new lines continuing to be built especially to the expanding south-eastern area. As a populous Andean capital city, it is continuing to extend radially outwards from its central basin.

5.4.5 LIMA, PERU

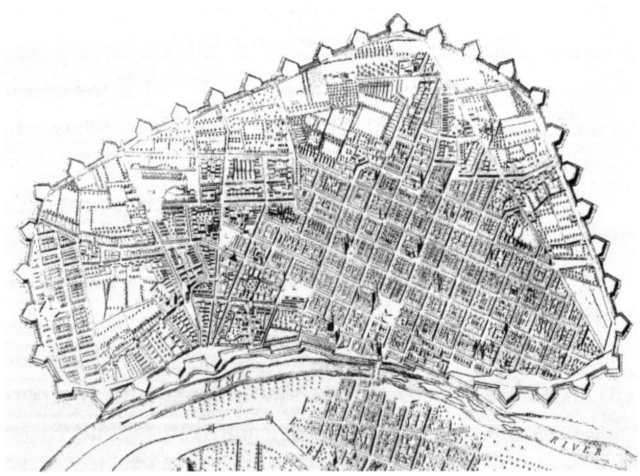

FIGURE 5.4.4 Lima – early configuration, 1755 (Rocque).

Source: Historic Urban Plans.

Background and Origins

Lima, the capital of Peru, is located on the Pacific coast looking westwards along the Rio Rimac from the valleys of the high Andean mountain range in the east. Known as Limaq in the Quechua language, with a history of ancient settlers, the earliest indigenous cultures were the Nazca and Wari during the Wari-Tiahuanco period. At the ancient site of Huaca Pucllana, the ruins of a fine adobe brick pyramid built around 40 CE are to be found near Miraflores and similarly, in adjoining San Isidro, Huaca Huallamarca, another ancient pyramid.

In the 13th century CE the indigenous Incas residing in the Cusco valley of the highlands were to have a significant impact on subsequent developments. Their influence was across a wide region of South America, the Tahauntinsuyu, 'Land of the Four Quarters', connected by a network of stone roads, often in very challenging terrain. Particular evidence of this is found at Machu Picchu near Cusco in the pre-Columbian era.

All was to change in the 16th century with the Spanish conquest of the region under Francisco Pizarro, advancing down the Pacific coast from Panama in search of El Dorado. Having taken over Cusco, he was to lay the foundations for the development of the central area of the city of Lima. As the 'City of Kings' (Ciudad de Los Reyes) under the viceroyalty of Peru, the city was laid out based on a grid, with fortified walls emanating from Plaza de Armas, with the Palacio de Gobierno (Casa de Pizarro) and Municipalidad where La Catedral and Palacio Arzobispal were located.

The colonial San Francisco complex was built comprising a church, convent and chapels of La Soledad and El Milagro with imposing facades of quincha (wood, rushes and adobe). It is also the site of the oldest university in the 'New World', the National University of San Marcos, founded in 1551, as well as other churches of San Pedro and Santo Domingo, palaces and mansions.

But due to its location on a geological fault, the city has suffered historically from a cycle of significant earthquakes in the years 1656, 1687, 1746 and 1940, with the result that most of the buildings standing today have been extensively rebuilt.

In the early 19th century, after campaigns by José de San Martin and Simón Bolívar, independence was achieved from Spain in 1821 and the Republic of Peru was created. In 1818, the original colonial walled fortifications and gates were demolished and new avenues were built across the city. A railway was built to connect Lima with the port of Callao and with an iron bridge across the Rimac River. Earlier fortifications were built here with the Fortaleza del Real Felipe.

Besides El Centro with Plaza Mayor and Plaza Bolívar, with the country's increased revenue from guano exports, the city developed in other individual areas, such as Miraflores along the coast and with its manicured park, the Malecon de Miraflores, on the cliff top and eastern border of the 'Paseo de la Republica. San Isidro is a green garden district adjacent to Miraflores with the coastal Circuito de Playas and where the restored Huaca Huallamarca adobe pyramid is sited.

Just to the south is the district of Barranco, with La Ermita church and the Puente de los Suspiros by the Parque Municipal, a mixed working area that had an early tram along Avenida Pedro de Osma.

At the turn of the 20th century, the grand avenues of Arequipa, Argentina and Brasil were further expanded, and in the 1930s the central administrative buildings and ministries were remodelled as the city was further consolidated.

Despite ongoing boundary disputes between Peru and Ecuador, after a period of military dictatorship, from an economy based on textiles and clothing, fish and agro-industry, industrialisation developed in and around the major port at Callao. With a growing economy in the 1960s, financial areas developed in the San Isidro District with commercial buildings such as the original Banco de la Nacional.

Modern City – Buildings and Arteries

The city has developed mainly in a triangular zone between the Old City, Cercado de Lima, Chorillos and Callao. The main modern districts comprise Centro, Miraflores, San Isidro and Barranco.

In recent times, periods of in-migration have ensued from the rural areas and the Andes, with people seeking better opportunities of work and education. This has led to greater pressures on the infrastructure and social facilities that had fallen behind as new areas were settled north of the Rimac River and also south and east along the Central Highway.

Such settlements have grown along the main river valley after migration from rural areas, leading to informal areas, or 'pueblos jovenes'. These 'barriadas' developed extensively during the 1960s, as well as other areas such as at 'lomas costeras', desert coastal ridges, where informal settlements have gradually accumulated on steep slopes. One particular area taken over and developed has been the Pampa de Comas settlement on the periphery of the city.

With the pressures of population on both land and housing, various experimental schemes have been undertaken, from basic sites and services to low-cost housing. One scheme, a demonstration project in northern Lima, funded by the state PREVI, was to test a series of self-build housing prototypes. Similarly, in Villa El Salvador District in the south, a basic grid with self-build clusters of housing has been developed for relocated inhabitants.

Most densely populated districts now are to the north and south edges of the city (Cono Norte and Cono Sur). Historically, this immigration resulted in a mixed variety of racial groups of Amer-Indians, Mestizos, Afro- and Euro-Peruvians. Different scattered communities have developed, with one of the poorest being Pamplona Alta, physically separated from the wealthier las Casuarinas, who are beginning to live in more secure, walled, and gated communities.

In terms of transport, the city is located adjoining the northern and southern branches of the coastal Pan American Highway (Route 1) that runs just inland. The development of the Metro de Lima y Callao has evolved slowly with north-south Line 1 opened in 2014 and underground Line 2 to Callao. In order to overcome urban traffic congestion, a number of dedicated bus BRT lines have been realised on the main avenues, such as Avenida Arequipa known as El Metropolitano. A future parallel regional railway, Treno de la Costa, is planned also along the coast from Lima south to Isa.

The population of the city is approximately 9 million, and 11.5 million including Callao in the surrounding metropolitan area. Although one of the fastest-growing cities in South America, it has high levels of poverty. As a primate city, its population is nearly ten times greater than the second most populated city Arequipa. The main Jorge Chavez International Airport (LIM) is located at Callao. The Pan American Games were held in Lima in 2019.

Transactional Forces and Typology

Traditional industries that were established in the city revolved around fishing, textiles and fabrics, leather and craft industries. Post–World War II, under import substitution policies, this led to increased food processing and light manufacturing with the beginning of industrialisation. With increasing prosperity, Lima has developed as a regional centre with a number of financial institutions and is the seat with the headquarters of the regional Andean Pact Economic Community (CAN) of the four adjoining South American countries. Its external outlook is either coastal habitat to the west or forestry reserve inland beyond the mountains to the east.

Lying on the edge of the Pacific Ocean, the historic central core of Lima was laid out by the Spanish on a rectilinear grid with fortified walls. Now developed along the Rimac River on both sides, it extends outwards towards the active port of Callao. The city sits at the crossing of the east-west axis to the port and the north-south axis parallel to the eastern range of mountains. These corridors now act as the main zones of development towards Carabayllo and Comas to the north, Santa Anita and Ate in the east, and San Juan de Miraflores in the south, an expanding metropolitan region.

Beyond its central formal planned districts, areas are typically being infilled with interspersed and accumulated unplanned informal buildings towards the periphery and occupied by rural migrants. With their increasing rate of expansion, this is really challenging the city in the provision of adequate services and infrastructure for these poorer communities. In addition, as an exposed coastal city, the area is suffering from some of the recurrent, more extreme events of El Niño. The following figures illustrate the modern configuration of the mega South American cities as described and the evolution of their modern populations.

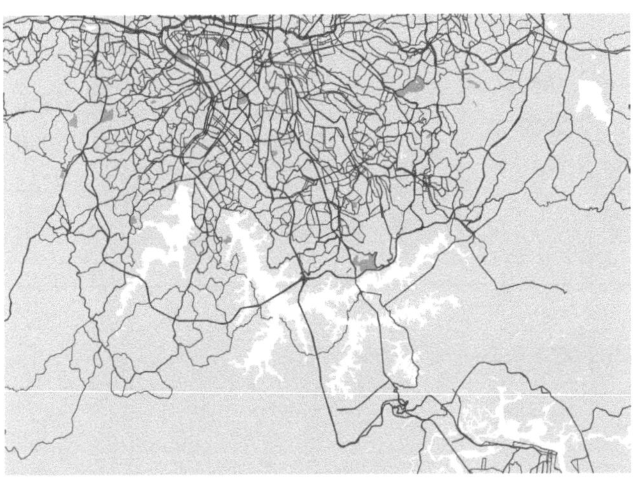

FIGURE 5.4.5 São Paulo – modern configuration.

FIGURE 5.4.6 Rio de Janeiro – modern configuration.

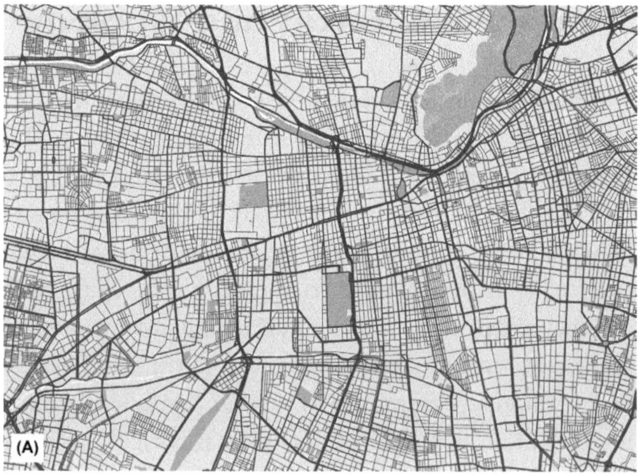

FIGURE 5.4.7A Santiago – modern configuration.

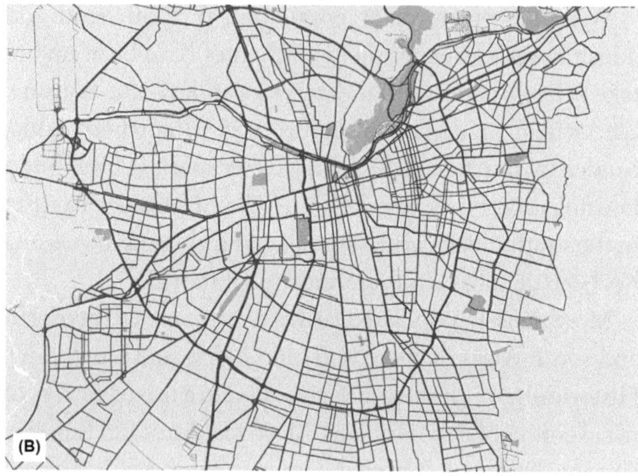

FIGURE 5.4.7B Greater Santiago, modern configuration.

FIGURE 5.4.8 Buenos Aires – modern configuration.

FIGURE 5.4.9 Lima – modern configuration.

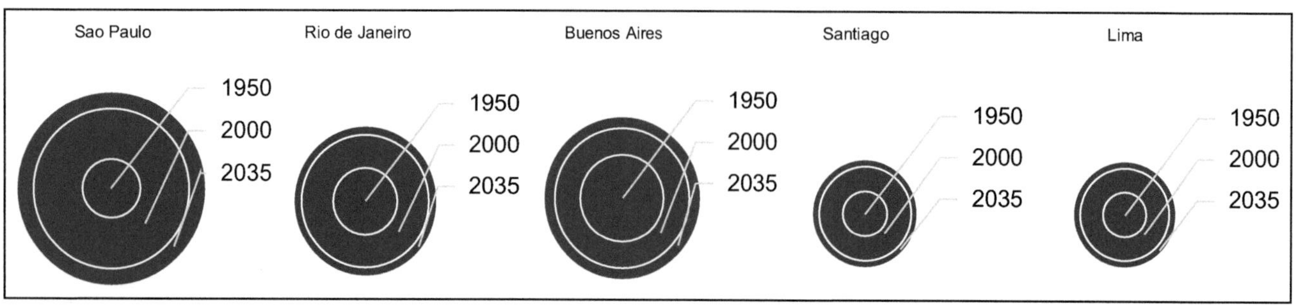

FIGURE 5.4.10 Mega South American city populations.

Data source: Global Database of Metropolises 2020 – UN-Habitat (2020), UN DESA Population Division.

Further Reading

SÃO PAULO

Correa, Felipe – *São Paulo, a Graphic Biography*, University of Texas Press, 2018.
Murray, M – *The Evolving Spatial Form of Cities in a Globalising World Economy, Johannesburg & São Paulo*, HSRC Publisher, 2004.
Sa, Lucia – *Life in the Megalopolis, Mexico City and São Paulo*, Routledge, 2007.

RIO DE JANEIRO

Carvalho, Bruno – *Porous City, a Cultural History of Rio de Janeiro*, Liverpool University Press, 2013.
Jaguaribe, Beatriz – *Rio de Janeiro, Urban Life Through the Eyes of the City*, Routledge, 2014.
Latotzki, Ingo – *Rio de Janeiro*, Parkstone Press, 2001.

BUENOS AIRES

Garner, James – *Buenos Aires, the Biography of a City*, St. Martin's Press, 2015.
Gorelik, Adrian – *The Grid and the Park, Public Space and Urban Culture in Buenos Aires 1887–1936*, Ubiquity Press, 2022.
Wilson, Jason – *Buenos Aires, a Cultural History*, Interhill Books, 2023.

SANTIAGO

Chastain, Andra – *Chile Underground, the Santiago Metro and the Struggle for a Rational City*, Yale University Press, 2025.
Walter, Richard – *Politics and Urban Growth in Santiago, Chile 1891–1941*, Stanford University Press, 2005.
Williamson, Edwin – *Penguin History of Latin America*, Penguin, 2009.

LIMA

Aguirre, Carlos, ed. – *The Lima Reader, History, Culture, Politics*, Duke University Press, 2017.
Higgins, James – *Lima, a Cultural History*, Oxford University Press, 2005.

5.5 Mega Asian Cities

South and South-East Asia – India, Pakistan and Bangladesh – Thailand and the Philippines – Indonesia, Malaysia and Vietnam

Mumbai, Hyderabad, Karachi, Dhaka, Bangkok, Manila, Jakarta, Kuala Lumpur and Ho Chi Minh City

5.5.0 INTRODUCTION

Over the past centuries different empires have ebbed and waned across regions of Asia, from the tribal, mediaeval and warring periods to the contemporary world. In different areas, emergent races have had prominent impacts in their varying spheres of continental influence, whether Mongol or Mughal, Khmer, Safavid or Ottoman.

In more recent years, after the turbulence of two extensive world wars, new areas of influence have come to the fore. The upheaval of these wars has created a new world order, the rise and fall of the Soviet Empire, the growing 'West' and a broader spread of development and growth. Now in the 21st century, China and South Asia has become the most populated and developed continental region. This expansion has been based on extensive urbanisation, which has exacerbated significant disparities between rural and urban areas, with the push and pull forces of attraction and in-migration to urban centres.

Southern Asia is the scene of major increases in population, with migration and urbanisation that has coalesced in several primate cities. The most significant countries and island areas around the South China Sea and the Pacific Ocean that have expanded besides China are Japan, Korea, Malaysia, Indonesia, Thailand, Vietnam and the Philippines, which between them now contain around 30% of the world's population.

More specifically, five of these countries contain a major metropolis, notably in Thailand, Indonesia, Malaysia, Korea and the Philippines, that each have a population in excess of 15 million people. Bangkok, Jakarta, Kuala Lumpur, Seoul and Manila accordingly can be considered primate capitals, with populations having five to ten times that of their second cities, namely Chiang Mai, Surabaya, Georgetown, Busan and Cebu respectively. With such a high rate of population growth and in-migration coming from surrounding areas, these metropolitan areas, or mega cities, will continue to dominate their neighbouring hinterlands.

Both China and India have two cities each in excess of 20 million people, namely Beijing and Shanghai, and Mumbai and Delhi. Similarly, Manila and Seoul in South-East Asia have over 20 million inhabitants. Other emerging South-East Asian primate cities that are the most significant mega cities, are Jakarta and Tokyo, which have metropolitan areas now in excess of 30 million people.

From this perspective, a mega city can be considered as any city that has a metropolitan area that comprises a range of 15–30 million people and beyond. With a developed central core, its neighbouring hinterland may be equally or more extensively populated as the actual core itself.

In India, the largest cities currently are Delhi, Mumbai (Bombay), Bengaluru, Hyderabad, Ahmedabad, Chennai (Madras) and Kolkata (Calcutta), all having distinctive and historic characteristics, along with other major cities such as Surat. Karachi and Lahore in Pakistan, and Dhaka and Chittagong in Bangladesh are the largest urbanised cities and are continuing to expand rapidly in their regions.

In the late 20th century, the established capitals across South and South-East Asia attracted migrants from both within and adjoining states, sometimes as a result of regional tensions, events or famines.

The largest metropolises are Delhi, Mumbai, Kolkata, Karachi and Dhaka, each having both high birth rates and rates of in-migration, and their cores have been overwhelmed with informal settlements extending out into peripheral areas, with dense concentrations of populations on marginal areas of land. Of 'shanty' construction, these temporary houses lack adequate sanitation services and scarcely offer adequate accommodation for squatting families. Of more concern, in many areas these peripheral urban settlements are at grave risk due to flooding and climate change, in particular Dhaka in Bangladesh.

DOI: 10.4324/9781003353348-30

India and its population remains constantly shackled by the religious 'caste' system with its rigid hierarchical grouping of people based on their hereditary occupations: viz. the Brahmins with an intellectual background, the Kshatriyas as warriors, the Vaishyas as traders and Shudras as workers. Outside this system are the Dalits, or untouchables, 'broken people' that undertake highly menial tasks and suffer significant discrimination. Despite some minor evolution, this system is deeply embedded in Indian culture and impervious to change.

Besides these mega cities there are also cities that are developing as megalopolitan agglomerations across the sub-continent that have evolved and expanded from their original core to an entirely broader structure of poly-nodal settlements. This evolution is occurring at an increasing rate as urbanisation continues, driven by high rates of migration into these areas.

5.5.1 MUMBAI (BOMBAY), INDIA

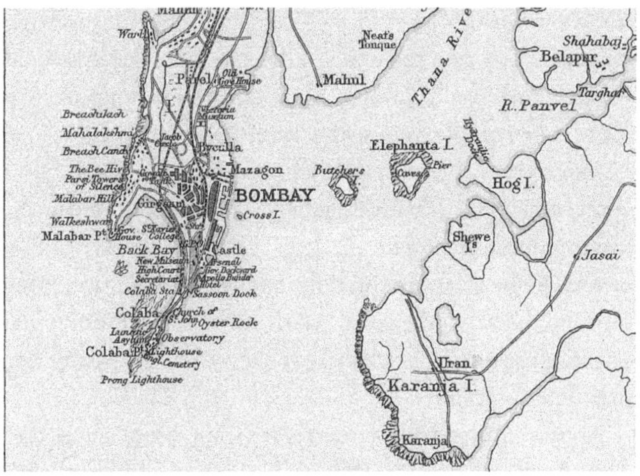

FIGURE 5.5.1 Bombay – early configuration, 1890 (Bartholomew).

Source: Maphouse.

Background and Origins

Located in the Konkan western coast of India, with its deep natural harbour, this place was first named Bon Baia (Good Bay) by the Portuguese, or Bombay. It became the capital of the modern state of Maharashtra in 1995 and was renamed **Mumbai** (or 'Mumba-Ai' – mother Mumba), and to become one of the most populous cities in the world.

On an archipelago, it was an original set of seven islands populated by Koli fishermen from Gujarat and actually adjoins the larger region of Salsette Island. Off the eastern shore are Butcher and Elephanta Islands, where there is evidence of early habitation and old caves that were carved in the 7th century CE with Brahmin images and a triple-headed Shiva statue. After the ancient dynasties and Mughal period, this area came to be occupied by early Portuguese settlers who built fortifications.

With its fort and city walls, it was taken over by the British Crown and leased to the East India Company. Here the land area was reshaped under a major project named Hornby Vellard with construction of the southern Colaba Causeway in the 18th century when new land was reclaimed to integrate the seven once separate islands; the causeway was completed in 1784.

This major project resulted in the city's consolidation and to it rising to prominence as a major Indian seaport across the Arabian Sea. Subsequent to the opening of the Suez Canal in 1869, and the introduction of steamships, this enhanced the attractiveness of the city and with trade to the east. With further consolidation, the surrounding infrastructure was developed along with construction of new roads and railways, and necessary improvements in sanitation.

With its Parsi community, the city attracted cotton traders, and the local population increased rapidly along with the construction of wharves and shipbuilding. With the esplanade overlooking Back Bay, under British rule an array of formal public buildings was realised that included the Government Secretariat, High Court and university library in a linear manner by the Maidan.

The first railway was built across the city from Tannah (Thane) to Bori Bunar Central. Its significance was portrayed by the grandest building of the Central Railway Station, with its enormous train shed, known as Victoria Terminus, now Chhatrapati Shivaji Terminal (CST), a grand mass of Indo-Saracenic domes that was opened in 1887 for Queen Victoria's Diamond Jubilee alongside the Municipal Corporation Building.

The railway infrastructure was developed further, with the Great Indian Peninsula Railway (GIPR) as the challenging line built through the western Ghat Mountains and incline to reach Poona in the Deccan Plateau. The Bombay, Baroda and Central Indian Railway (BBCIR) was also built northwards to connect with the cotton fields of Gujarat.

Hence the two main railway networks, the Central Railway and Western Railway, operate out of the Victoria Terminus (CST) and Churchgate respectively.

Due to the increasing levels of commercial activity, Bombay became known as the 'City of Gold' and 'the Gateway to India'. With the growth in demand, it became the foremost cotton-trading market for the textile industry. In addition to having neighbouring cotton-providing areas, it set up its own mills in the northern Byculla quarter to produce textiles, as at Girangaon, which were to grow in size to rival Manchester.

With the large docks, mills and railways, the city of Bombay was becoming increasingly industrialised and overcrowded with an increase in population and migration, as it was evolving from a trading centre into a growing industrial port town. The area around the fort was anachronistic and in the 1860s the gates to the walled area were pulled down and the water tanks sealed, and surrounding roads widened under the auspices of the Ramparts Removal Committee. Other urban street improvements were made, such as Elphinstone Old Cotton Green (now Horniman) Circle and around the town hall.

As a city of business and commerce, the city had to be made to work, hence civic improvements and public works were essential as parts of the fort quarter were dismantled. Land reclamation around Back Bay, Modi and Apollo Bay, all led to consequent real estate investments along Marine Drive. Whilst Malabar Hill with its Hanging Gardens became an exclusive suburb, in and around the centre public health and sanitation were often found to be wanting.

Proper water reservoirs eventually were built at Malabar Hill and Bhandawada, and water was diverted into Vihar Lake. Further capital projects, such as service and pumping stations, were also required and a municipal corporation was established in 1872 to manage these works, with an imposing headquarters building. Princes Dock was opened in 1880.

Civic buildings such as the Victoria and Albert Museum (now the Bhau Daji Lad Museum) in the botanical gardens were also developed, and other buildings in Bombay Gothic style, such as Gilbert Scott's University Library and the Rajabai clock tower near the Maidan, and the Venetian Gothic Government Secretariat and the CSMVS King Shivaji Museum. The symbolic landmark 'Gateway of India' arch monument by the ferry port and next to the Taj Mahal Hotel, was built in 1924 to commemorate the visit of King George V. The Ballard Estate was built in the vicinity of the Customs House, and the Kala Ghoda (Black Horse) District was developed around the central area to the east of the Oval Maidan Ground.

Modern City – Buildings and Arteries

The city was to continue to grow and consolidate its position as a significant regional business centre. With increasing wealth, financial institutions also grew. After India's independence in 1947, it became the capital of Bombay State. It was later subdivided in 1960 under linguistic lines, into separate Gujarati- and Marathi-speaking areas, now Maharashtra State, and renamed Mumbai.

With its burgeoning growth and in-migration, and despite huge congestion, Mumbai, the 'Gateway of India', has become the commercial capital of India, a global financial hub and one of South Asia's largest cities. It is also the centre of Bollywood productions with the Film City district. South of the Colabar causeway a new district with World Trade Centre exhibition centre has been developed.

Much of the city has remained unhealthy with its infamous 'chawls', narrow passages, poorly drained with dense pockets of population. With its mix of cultures and races, religions and caste identities, it is a challenging mix of peoples. In the centre of Mumbai is one of the largest informal areas in the city, called 'Dharavi', with over 1 million people. Despite its squalor and illegal settlements, there is extraordinary inventiveness with large numbers of children involved in recycling mountains of discarded materials. Many small workshops and micro-enterprises also flourish in this area to assist in self-build and rehabilitation and upcycling of waste materials. These repair tasks, known as 'jugaad', or frugal innovation, are targeted as supporting a more stable employment structure.

The current population of the city of Mumbai is approximately 12.5 million, with 24 million across the metropolitan area. The density of the whole city region is very high, at approximately 52,400 pers./sq. km., one of the densest world cities. With its configuration, it has ongoing pressures of urban heat exposure for its extensive urban population.

A new underground metro is being built across the central area of the city that will relieve congestion. With elevated Line 1 and Lines 2 and 7 already operational, Line 3 is under construction and another six lines are in planning and development.

In terms of highways, a coastal toll highway, the Bandra–Worli Sea Link (BWSL), has been built across the bay with a cable-stayed bridge that acts as a relief road towards Chhatrapati Shivaji International Airport (BOM). This will interact with the coastal road improvement scheme that is underway.

Navi Mumbai – New City

With large increases in population, massive growth has continued in recent years across the Metropolitan Region of Mumbai that needs to be accommodated. The first concept to support this was for a new city instigated by the

Municipal Corporation of Greater Bombay (MCGB) in the 1960s as a separate growth pole and industrial zone, with a new Port at Nhava Sheva and an industrial zone at Thane Belapur, with Taloja to be connected by the new Thane Creek Bridge. In addition, there was a potential plan for the state government to be relocated there.

Accordingly, the satellite town of **Navi Mumbai** – New Bombay – was planned to the east across the creek, with the aim of assisting in the dispersal of the growing population, along with a major port to be commissioned and with its own international airport.

Navi Mumbai was envisaged and sited on the eastern mainland across the creek from original Mumbai as a new growth pole planned to alleviate the existing highly congested city. Under the planning of the City and Industrial Development Corporation (CIDCO) of Maharashtra State it was commenced in 1971 in the Thane and Raigad districts. It has been connected to Mumbai by the enlarged Thane Creek Bridge, or Vashi Bridge, for both road and rail. The intention was to include the new Navi Mumbai International Airport (NMI), which due to its scale will have a significant influence over the whole area, along with the new major container port Jawaharlal Nehru at Nhava Sheva farther south.

New development has only been gradual and was sluggish into the 1980s, such that competition continued even with the extension of areas such as Varai Vivar in North Mumbai, seen as a more natural extension of the original existing city.

The new planned area over the creek encompasses a number of already existing multi-nucleated settlements, such as Panvel, which are to act as nodes of development. However, since development had only taken place gradually, in 1991 a new state-run Navi Mumbai Municipal Corporation (NMMC) was created and with new infrastructure connections, the pace started to improve. Besides such major infrastructure and industrial developments, extensive new areas of affordable housing were also planned.

This area is to comprise also the Navi Mumbai special economic zone (SEZ) for new industries. Whilst emphasis to date has been on major new infrastructure, future accommodation and sustainability and smart city components remain yet to be demonstrated. The current population of Navi Mumbai is approximately 1 million.

Transactional Forces and Typology

Located on the Arabian Sea, as a series of incorporated islands, Mumbai originated from its maritime trade. From early fisheries to trading raw materials with the Gulf and Africa, with more powerful sea-going vessels, longer-distance transport was able to be extended around the Cape to Europe, thus encouraging more traders.

With the opening of the Suez Canal, this significantly increased shipping links between Europe and the Far East, upon which Mumbai was to develop as a significant transit and coaling port for steamships. With the subsequent emergence of the oil-rich Gulf States, this created new links and trade between these countries too. From a landward perspective, Mumbai has been an entry port for peoples to Maharashtra State, the Deccan Plateau and beyond, and benefitted from this once its extensive road and rail transport routes had been developed.

As a maritime-facing elongated peninsula, formed from a series of islands surrounded by sea, Mumbai has been constrained by its limited land area. By undertaking areas of land reclamation around the bay from the sea, the metropolitan area of the city has gradually been consolidated, although scope for expansion only really remains to the north.

As contrast, a separate zone of expansion as a satellite city is being developed across Thane Creek to the east, so as to create a countervailing pole on low-lying land and mangrove swamps where Navi Mumbai is under development. Here there is significant space for expansion as extensive infrastructure for a new port and new airport is being planned under a special economic one and a new jurisdiction under the management of a special purpose company, NNMC. This is a significant development not yet fully materialised that will enable further industrial development of a bi-polar nature and should release growth pressures on the existing centre of Mumbai itself.

5.5.2 HYDERABAD, INDIA

Background and Origins

Hyderabad is located on the Deccan Plateau in the modern state of Andhra Pradesh on the banks of the River Musi around the artificial Hussain Sagar Lake that was created in 1563 CE. Originally named Baghnagar (City of Gardens), there is evidence of very early human settlement during the Iron Age.

Early dynasties were the Chalukya during the 6th–12th centuries CE, and the Kakatiya dynasty in the 12th–14th centuries who built the sprawling Golconda Fort lines of defence. This was followed by the Khalji dynasty and the Tughlaq dynasty of the Delhi Sultanate until being overcome by the Muslim Bahmani Sultanate in the Deccan Plateau and then by the Persian Qutb Shahi dynasty.

Muhammed Qutb Shah established the city on the banks of the Musi River in 1591 and had built the fine Charminar Monument with Four Minarets. It stands as a large archway at the crossroads of four major streets, and the large Makkah Masjid Mosque was built although not completed until 1693 to extend beyond the fort and artificial lakes and canals. The Qutb Shahi tombs contain the royal necropolis.

In 1687 the city was annexed by the Mughals, and the subsequent Muslim Asaf Jahi Nizam dynasty was founded with the following line of Nizams of Hyderabad. It became the capital of the Princely State of Hyderabad under the British Indian Empire when the British Residency was built in 1798. Under a subsidiary alliance system with the East India Company, in return for protecting the state, the British Indian Army was allowed to be stationed in the neighbouring Secunderabad Cantonment with its old Mud Fort and parade ground nearby to protect the capital.

During the 19th century, under the British Raj, railways were constructed in the region with Hyderabad Deccan, Secunderabad and Kacheguda stations and the general infrastructure and drainage was developed and enhanced, and as new factories came to be established there.

The historic area south of the Musi River forms the Old City, Purana Shahar, with the New City on the northern banks, connected by several bridges, the oldest being Purana Puh, Old Bridge built in 1578, and with older Secunderabad with its heart-shaped Hussain Sagar Lake, and the Tank Bund Necklace Road.

In the city, there are numerous buildings of character, having either Indo-Saracenic, Islamic or Mughal characteristics. These include the Chowmahalla Palace, Falaknuma Palace, Purani Haveli and King Kothi Palace. Institutional buildings include the High Court, Government City College and State Central Library, several museums and the University of Hyderabad and Osmania University.

Modern City – Buildings and Arteries

After the two world wars, and with the upheaval of independence and the creation of the Republic of India, a new constitution came into effect. The Princely State of Hyderabad was the last to join the Indian union. There were even some deliberations at the time of the preparation of the Indian Constitution, with the Chair of the Drafting Committee suggesting that due to its strategic location and amenities that Hyderabad might be considered as the second capital of India.

In fact, the State of Hyderabad was abolished in 1956, and reorganisation of the Indian states was made along language lines through recognising Telugu- and Urdu-speaking districts which resulted in Hyderabad State being divided into three parts which were merged with neighbouring states essentially to form the modern Maharashtra, Karnataka and Andhra Pradesh states.

For some time, this created significant tensions, and it was not until 2014 that it was agreed that part of Andhra Pradesh would be split off to form a new Telugu-speaking Telangana State, with the city of Hyderabad remaining the capital for both states until the new capital of Andhra Pradesh was completed at Amaravati after 2024. A new Telangana State executive building has been erected in Hyderabad in front of the lake.

Historically there were many bazaars with textiles and rugs, handicrafts, pearl and diamond jewellery and food products. However, there are also very large areas of informal settlements, especially around the core of the city, with a large number of suburban tenements, estimated at over 1.5 million informal inhabitants. The population the city is approximately 6.8 million and 10 million in the greater metropolitan area. A separate Greater Hyderabad Municipal Corporation (GHMC) was formed to manage all of the municipal infrastructure.

During the 1980s the industrial sector grew with large manufacturing companies, especially automotive. More recently in terms of the modern economy, there are several new research institutes, and biomedical and pharmaceutical facilities. Its newest industries have developed essentially from modern automotive and aeronautics, defence and electronics, to IT and financial services.

With industrial areas in the north, the north-west has more affluent residential areas. There has been a considerable construction boom, with the realisation of many multi-storey commercial and residential tower blocks across the city of 30–40 storeys which have been grouped especially around the IT corridor.

A special economic zone (SEZ), HITEC City – Hyderabad IT and Engineering Consultancy City – was developed by Larsen and Toubro with Andra Pradesh Industrial Infrastructure Corporation to the west of the city in 1998 and is now known as 'Cyberabad', focused on IT with many high-tech SME companies. It comprises Cyber Towers and Laxmi Cyber City, L&T Info City and CII-Godrej Green Business Centre, along with IT Park,

Cyber Pearl and several other research and campus developments. It has become a dynamic, entrepreneurial centre to the south-west of the city, and a further HITEC-2 City SEZ is under development.

In terms of transport, there is the suburban railway, Multi-Modal Transport System (MMTS), and a new light rail transit system LRTS. The new Hyderabad metro, which opened in 2017, has three lines and 57 stations. It was developed on a concession basis, on an elevated track that was opened in phases and that connects with the new Rajiv Gandhi International Airport (HYD) in the south at Shamshabad.

There is an Inner and Outer Ring Road, each carrying large volumes of road traffic and the main Hyderabad Elevated Expressway to the airport. With its inter-connecting national highways, the city is dominated by a large number of buses that link with Mahatma Gandhi Bus Station.

Transactional Forces and Typology

As a strategic location, Hyderabad evolved with resourceful people since its founding and early dynasties to the time of the Mughals. This was recognised by the British East India Company who came to establish themselves at early Secunderabad following an alliance with the Nizams. With the development of the railway and supporting infrastructure, manufacturing industries were established which were to be its beginning as a major regional centre, as a place of attraction for migrants and workers. The historic Old City grew around the southern bank of the River Musi and the New City grew as an urbanised area around the northern bank, more as a twin city.

As a riverine city that has been developed in stages from the original fort area and cantonment, with its artificial lake, roads and railway, its area has now extended and surpasses its original perimeter. The original area of Hyderabad has now grown along with Sikandrabad to comprise a wholly enlarged and agglomerated metropolitan region. With its outer orbital service road, the city has been shaped by radio-centric development along with poly-nodal centres of concentrated development, interspersed with formal residential areas and many interstitial areas of unplanned informal settlements. Such contrasts reflect the duality of other major mega cities of India, on the one hand richly developed in certain zones, on the other hand inadequately provisioned and under-serviced in other areas without having essential basic infrastructure.

5.5.3 KARACHI, PAKISTAN

Background and Origins

Karachi is located in Pakistan's Sindh Province on the Arabian Sea, between the Hub River and the delta of the great Indus River. There is evidence of early settlements and a port in Assyrian times, and it was developed under the Mughals and fortified by Baluchi settlers. The capital of Sindh Province was transferred from Hyderabad to Karachi in 1840. Due to its location, it rose to prominence during the British East India Company's trading activities and with its desirable cotton exports from the region. This demand increased with the strategic opening of the Suez Canal in 1869, and the city grew and expanded with the construction of the Indian continent's railway system under British rule.

After two demanding world wars, with the Pakistan Movement for the creation of an independent Pakistan, partition of the continent took place in 1947–8, which resulted in a huge movement of peoples with an additional half-million Muslims seeking resettlement in Karachi and Sindh Province from India. Karachi initially became the federal capital of Pakistan, before it was transferred later to Rawalpindi in 1958 and then Islamabad, and grew significantly in many developing areas thereafter.

With its hot climate, the scale, width and depth of the two rivers of Malir and Lyari change dramatically depending on the extent of the summer rains. The 'Old Town' is a largely unplanned area with substantial informal settlements, whereas the New Town was established as the British developed the Karachi Cantonment as a military garrison. The Orangi district is another now sprawling informal district, with separate middle-class areas in Nazimabad, and wealthier areas in Defence and Clifton near the beach, where large commercial corporations have recently gravitated.

Modern City – Buildings and Arteries

The spine of the main downtown area is along I.I. Chundrigar Road, which has become a major financial and commercial district with buildings such as Habib Bank and the MCB Tower. It is a busy area with many multinational corporations and with an active manufacturing sector. Large industrial estates have been developed across Karachi such as Korangi Town, Landhi, Bin Qasim and SITE, along the super-highway north-east towards Hyderabad.

In terms of infrastructure, the city has the old Karachi Circular Railway which requires rehabilitation, the mainline railway which is being upgraded under Chinese economic assistance, and bus rapid transit (BRT) Metrobus which is being developed. Public transport is generally poor across the city and is mainly provided by informal minibuses and private tuk-tuks. New highways and by-passes have been built, but traffic flows, safety and congestion remain of particular concern. Jinnah International Airport (KHI) is the main airport in the east.

The city is administered by the Karachi Metropolitan Corporation (KMC), and the Karachi Development Authority (KDA), divided into seven districts and separate cantonments. Clean water, drainage, sewerage and waste management remain a considerable challenge for a constantly growing metropolitan area. Over 50% are Urdu speaking, and in recent years there have been several waves of migrants, with Pushtans arriving from the north and Afghanistan, and peoples from the Punjab and eastern Bengal.

Economically, after the growth in population in the 1960s, a policy of industrialisation was pursued, which led to the development of new industrial areas with heavier industries such as steel mills. This required the expansion of facilities for the import and export of bulk materials such that a new harbour was built to the east of Karachi at Port Qasim. This has now been considerably expanded with coal and cement, fertiliser, oil-handling facilities and containers. Around these facilities a new town of Bin Qasim has also been developed. The main airport at Karachi is Jinnah International Airport (KHI), located to the east of the city at the original site at Drigh Road.

Karachi is home to a number of private universities, such as Aga Khan University, and also private hospitals, such as the Aga Khan Hospital, and has a large number of mosques, including the Grand Mosque. Some interesting Mughal revival buildings remain in the city, including the Hindu Gymkhana and Mohatta Palace. The modern memorial to the country's founder Mohammed Ali Jinnah, Mazar-e-Quaid, the National Mausoleum, was built in the centre of the city in 1971.

The population of the city is approximately 15 million, with 20 million in the metropolitan area. The density of population of the whole city region is approximately 48,800 pers./sq. km., one of the densest world cities.

There are several new modern commercial malls across the city and the Bahria Icon Tower in Clifton District. With increasing areas of informal settlements, however, it is an enormously sprawling and challenging city to inhabit with its high density, mix of different religious groups and deteriorating infrastructure where congestion, climate and security are challenging.

Climatically, the coastal plain is subject to extreme temperatures and the low-lying and swampy area is subject to exceptional flooding, with dry riverbeds converted to major flood courses across peripheral urbanised areas in wider Sindh Province at times of heavy inundations.

Transactional Forces and Typology

Located on the Arabian Sea in a coastal area of mangroves and swamps between the Hub and Malir rivers, the harbour was developed behind the sand spit, which opened up the region to maritime trade and transit. Whilst bordering the Indus River Delta and neighbouring mountainous countries of Afghanistan and Iran, Karachi was sited on ancient trading routes across Asia. With the realisation of the Suez Canal in the late 19th century, this changed the pattern of maritime trade and was to play a pivotal role in shipping commodities to and from India and beyond with exports of cotton and wheat. The navigation channel was dredged, and a causeway was built along with new wharves at the Port of Karachi.

The city extended radially inland with the realisation of the railways, and new areas of settlement were populated from the central Karachi Cantonment and Railway Colony, Clifton and Defense District northwards to Nazimabad, Orangi and North Karachi. More recently, with more industrial development, the population of the city has expanded eastwards to Korangi and beyond on available land where significant new settlement has occurred. With the main influx of population at the time of independence, further waves of migration have ensued, as for example from Afghanistan and North West Pakistan, which has led to the development of many areas of informal settlement and an intensive informal economy.

As a coastal settlement, the city has grown inwards from the central harbour area. Although initially formally planned, development has extended along all highway axes with built up areas on each side. From the south-east peninsula, with its regular street grid and lanes, the early cantonment and civil lines and other quarters have developed on either side of the Lyari River and Expressway, out past service roads and new sectors as, for example, to North Karachi.

In between are Cooperative Housing Society areas, planned blocks and colonies, but interspersed with many informal settlements as large as around Orangi. Beyond are more industrial areas extending to the east. With its highly mixed ethnic population and continued in-migration to the city, Karachi is a sprawling metropolitan area with a pattern of infill and informal settlements developing on any available areas of unbuilt land. With its demographic challenges, its population density is high, as new areas of settlement continue to be developed in outer marginal areas.

5.5.4 DHAKA (DACCA), BANGLADESH

Background and Origins

Dhaka is located in the lower reaches of the Ganges Delta in Bangladesh in the low-lying floodplain in the Bay of Bengal. During the Mughal period, it grew on the eastern banks of the Buriganga River, and **Dacca** was proclaimed capital of Mughal Bengal in 1608 CE, named Jahangirabad in honour of Emperor Jahangir. It became a prosperous centre on the regional route of the Great Trunk Road connecting North India with Central Asia and the Port of Chittagong.

As a hub for weavers and traders, the city became a centre for muslin, cotton and jute industries. Settlements and buildings developed along the riverbanks, and the city became quite affluent under the Mughals with buildings, gardens and the Bara Katra caravanserai, with its original quadrangular courtyard. The city developed around rivers and lakes and was criss-crossed by man-made canals rather than roads. At Buckland Bund, Sadarghat Wharf served as the terminal for water-borne transport.

In the mid-18th century, the British East India Company took control of the Principality of Bengal and created large plantations of sugar, tea and jute – the muslin industry was heavily taxed. It lost its power as the influence of neighbouring Calcutta grew and a British base was developed in the Dacca Cantonment. During the 19th century, basic infrastructure was developed with the introduction of a metre-gauge railway, water supply and riverside port at Narayangang. New administrative and cultural institutions were developed across the city.

After the abortive attempted partition of Bengal in 1905, in order to separate the larger Muslim Eastern areas from the Hindu Western areas of Bengal, the region was reunited in 1911. However, this was the beginning of a longer-term process that led after World War II to eventual separation in 1947, and as part of the partition of India and Pakistan, with the creation of East Pakistan from East Bengal in 1955.

Modern City – Buildings and Arteries

As capital of East Bengal, the developing city was subject to huge migration and urbanisation during the 1950s. But after a struggle for self-autonomy and a period of civil war, the eventual formalisation of the People's Republic of Bangladesh occurred in 1971 after the breakup of Pakistan with the renamed Dhaka as capital of the country.

A master plan for the city's new capital at Sher-e-Bangla-Nagar was drawn up, with a new Capitol complex for the government. This was inaugurated in the 1980s as a new National Parliament building was realised (designed by Louis Kahn and Muzharud Islam), with the Main (Bhaban) Parliament Chamber, South and Presidential plazas and lake around the Jatiya Sangsad complex. The Bangabhaban remains the official residence of the president. After a period of interruptions by several military dictatorships, parliamentary democracy was eventually restored in 1991.

More recently the low-lying city has continued to expand, with high rates of population growth and in-migration by numerous Bihari refugees. With its geographical and topographical location, the waterways have become choked and subject to frequent flooding during monsoons, major storms and cyclones.

The informal housing sector is under great stress, as the metropolitan area now accommodates a great density of persons, many of these climate refugees from the water, making it one of the most densely populated cities in the world. As a result, the city is very congested, over-populated, very polluted in many areas and at constant risk of flooding from rain, tidal or monsoon events, with firm ground at a premium.

Urban transport is difficult with many cycle rickshaws on the road. In the Motijheel and Mohakhali areas there has been substantial commercial development and building. Different areas are interspersed with temples and mosques, having differing distinctive styles, with some older colonial buildings remaining. These include the original Mughal Lalbagh Fort, the Nawab's old residence and the Rose Garden Palace. It is a vibrant and colourful city with its bazaars, mix of cultures and religious festivals.

In terms of infrastructure, besides the numerous waterways, there is the striking modern Kamalapur Railway Station (designed by Robert Boughey), which is one of the busiest railway stations. It provides suburban services and connects with other major cities, such as Chittagong and Kolkata. With its estuarine Port of Dhaka, a new Port of Pangaon has been built to handle container traffic downstream.

The elevated Dhaka Metro mass rapid transit system is under development with Japanese finance. The national highways radiate out from the city and interconnect with the Asian highway network beyond between neighbouring India and Myanmar. The Dhaka Elevated Expressway connects with Hazrat Shahjalal International Airport (DAC) at the north of the city. South of Dhaka the Dhaka to Mawa Expressway N8 leads to the major new combined road and rail Padma Bridge crossing 6.15 km. long across the River Ganges. This is the lowest crossing of the river after the Jamuna Bridge crossing upstream.

With its large and continuously growing population and demanding topography and waterways, it is one of the most difficult city regions to get around in. Yet, it is an intoxicating place with particular regional Bengali character. The population of the city is approximately 9 million, with 21 million in the metropolitan area. The population density of the whole city region is approximately 47,900 pers./sq. km., one of the densest world cities. An area of densest population is the Karail Bustee in Mahakali, a low-lying area. There are many other similar areas of informal settlements also, such as at Duaripara.

Transactional Forces and Typology

With its location in the lower delta of the River Ganges, Dhaka has been shaped by the surrounding water courses. In a tropical region, with monsoon seasons and cyclones, it is frequently flooded, as many areas lie close to sea level. Dependent mainly on water-borne transport, the realisation and maintenance of urban infrastructure, streets and drainage and utilities is constantly challenging.

With its agricultural perimeter, businesses developed based essentially around local materials of cotton and leather, jute and hessian, with a large jute mill constructed outside the city. Since its formalisation as the capital of Bangladesh, significant numbers of migrants have continued to move into the city from rural areas. With this continuation, the number and size of informal settlements has grown, especially on the perimeter where large populations reside, as at Kamrangirchar Thana. The central core area, however, has been consolidated with an area of commercial modern buildings realised in Motjiheel Thana where financial services and commerce have developed.

In its tropical location, the riverine city of Dhaka has developed to the north and east of the Buriganga River within the Outer Circular Road and adjacent railway. Besides the new capital complex at Sher-e-Bangla Nagar, and administrative buildings adjacent to local Tejgaon Airport, during the 1960s new residential areas were planned with new towns or satellite centres that included Gulshan, Banani and Bushundhara.

Due to the nature of the very flat underlying topography, the city has little shape and comprises dense interconnecting areas of informal settlements, with some contrasting formal sectors and residential blocks. With soft sedimented deposits to the south around the delta, which constrains any proper building, feasible permanent development only remains in marginally higher land to the north. Development of vacant land is being undertaken in East Dhaka and in the emerging Savar District. With recurrent tropical monsoons and potential flooding from cyclones, further climate change will inevitably exacerbate already inherent risks from its low-lying settlement, which seems unavoidable.

5.5.5 BANGKOK, THAILAND

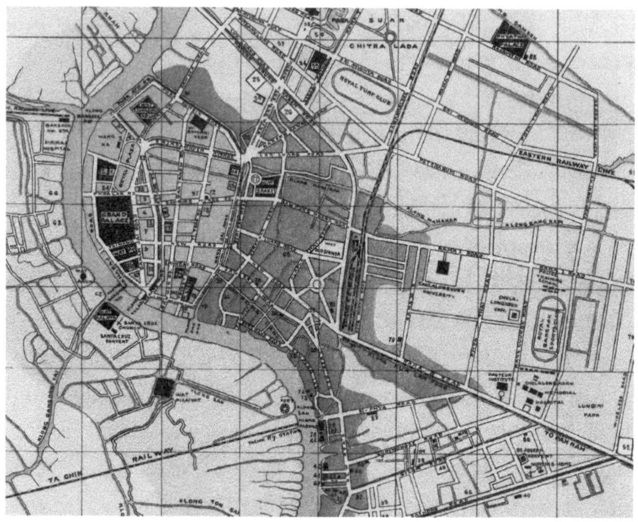

FIGURE 5.5.2 Bangkok – early configuration, 1932 (Siam Rail). *Source*: Maphouse.

Background and Origins

The city of **Bangkok**, or Krung Thep, lies on the Chao Phraya River with its old city centred around Rattanakosin Island, its royal heart, and with Thonburi on the other bank. With the former ancient capital Ayutthaya with its kings and 'wats' to the north, it is not far from the sea to the south and the Bight of Bangkok.

It was in 1781 CE that the reconquest of Siam from the Burmese was accomplished and the area of Taksin's Thonburi was regained as King Rama founded the Chakkri dynasty. In order to rebuild the capital, his court was moved across the river to Rattanakosin where a canal had been cut through from the bed of the Chao Phraya River to create an island. As further klongs were dug, the City Pillar was raised the next year and construction began of the Wat Phra Kaeo Temple with the Emerald Buddha.

The Grand Palace was created on the left bank and with the Inner Palace, which was inhabited solely by women, and the Chakkri Maha Prasat completing the Royal Enclave. The entire Royal Plaza, its enclosure and surrounding buildings is a highly ornamental place.

Wat Pho was built adjacently and has the oldest and largest reclining Buddha, symbolising arrival at Nirvana, covered in gold leaf. Surrounded by 'chedis', it contains the ashes and remains of sacred Buddha images. Nearby now is the National Museum that contains the Buddhaisawan Chapel with its rich murals and gilt images.

Several other imposing wats, such as Wat Arun, were built on the other bank at Thonburi with prominent stairs and towers ('prangs') and three symbolic levels. In addition, Wat Traimit is an ancient Buddha made of gold, and Wat Benjamabophit is made of Carrara marble.

Bangkok developed as a trading centre and received many Chinese immigrants during the 19th century and expanded as trade developed. Organised around the river and its network of canals, as the city was built on soft marine clay, it was sensitive to subsidence. Sanitation also remained a difficulty in the city, and outbreaks of cholera were frequent. Water supply was improved, some canals were filled in and overbuilt as road construction commenced. Markets existed, such as the Damnoen Saduak floating market, where vendors circulate with commodities in sampans, with other markets such as Chatuchak market by the clock tower.

Modern City – Buildings and Arteries

In 1960, a formal master plan for the city was developed by British planner Lichfield, but the city continued to grow organically with ribbon development rather than in accordance with the land-use plan. The commercial area of the city contains numerous markets, shop-houses and Chinatown, with alleys, temples and former tobacco warehouses.

Lumphini Park is a large park near the university, and the main governmental institutions are also located here. Many new major commercial shopping malls have also been developed, with commercial areas and business districts along the busy Si Lom and Sathon roads, where major buildings have been erected, and also along Sukhumvit Road.

The National Assembly of Thailand is located in a new building, Sappaya-Sapasathan, located on the banks of the river at Kiak Kai to replace the former Parliament Building. Government House contains the offices of the prime minister.

In terms of transport, in recent years a high level of investment has been made in new expressways and public transport in order to aim to overcome the chronic road congestion of cars, motorcycles and tuk-tuks. New public transport links are being built elevated above ground, both highways and urban metro, seeking to overcome difficult ground conditions.

With the two MRT lines, Blue and Purple, and two elevated BTS Skytrain lines on the Silom and Sukhumvit axes, these carry a large ridership across the city and to the suburbs and also the rail-air link to the new Suvarnabhumi Airport (BKK) (designed by Jahn). With the sprawling size and scale of the city, it is a major challenge to manage this demand.

The Don Muang Tollway was also built as an elevated expressway across the city to the other airport and other road bridges across the Chao Phraya River, such as the Krungthep bascule bridge. The Port of Bangkok is located at Khlong Toei in the bow of the river with a new deep-sea container port built beyond at Laem Chabang in the Bight of Bangkok.

With its location in a flood plain surrounded by rice fields, originally there were no restrictions on peripheral expansion in the outer areas of Bangkok. Through the previous participation of local village headmen in negotiating road access to outline plots of land during the 1970s–80s, this initially enabled some custom-built housing and residential resources to be improved. However, by the mid-1980s, approximately 1 million people had housed themselves in slum areas or 'salaam' of wooden buildings on stilts in low-lying areas. Connected by cat-walks with no sanitation and very high human densities, many were rented from public or private landlords.

The city is managed from city hall by the Bangkok Metropolitan Administration (BMA) as a special administrative area with a governor and divided into 50 districts ('khet'). The population of the city is approximately 9 million, with about one-third of Chinese origin, and 15 million in the expanding Greater Bangkok Metropolitan

Region. The population density of the whole city region is approximately 11,600 pers./sq. km.

A new programme of renewal was attempted along with infrastructure upgrading by the National Housing Authority that created new formal 'land and house' packages as row houses on sites farther out from the city centre. Affordable plots were also released as informal residential sub-divisions on the periphery but were often lacking services. Hence the challenges of upgrading the existing housing stock and enabling new settlements to be built in a planned manner continued.

With its siting on the river and a series of waterways and canals ('klongs'), Bangkok is lacking a proper arterial road network and basic infrastructure, located as it is on a flat, low-lying plain on the Chao Phraya delta. There is a limited ducted drainage and sewerage system, which is especially challenging for the growing city, and an incomplete system of piped potable water. Deep wells need to be bored, canals have been filled in, land subsidence has occurred and septic tanks are inadequate, resulting in chronic sanitary issues.

Bangkok is a congested and dense business centre as the Thai capital, but it has many attractive cultural areas. It remains a spiritual place, with each family home or dwelling typically having its own 'spirit house' in the garden as its own form of mini-temple.

Transactional Forces and Typology

Originally the city's primary means of transport was the waterway system with the Chao Phraya as the lifeblood of the area. With underlying soft clays, considerable subsidence has occurred due to groundwater pumping. Canals have been filled in, and many areas of the city are below sea level. With coastal erosion and rising sea levels, the city is at great risk of flooding. In the absence of a comprehensive road network, new public transport systems have been built elevated above the city, which has transformed the original fabric of the city.

With its traditional fishing and agricultural activities, handcrafts and gold jewellery, it is gradually being transformed from its crafts background. As a newly industrialising area, its focus has moved to automotive and electrical manufacturing and expansion of service industries, which is changing the nature of the city. With a high rate of migration from rural areas, the city has a large informal sector, with many seeking active employment and lacking permanent accommodation.

As an historic riverine city, Bangkok was developed around its canals with floating markets and with its old temples and wats. With its core at a large bow in the river, its three main klongs are arranged in opposing concave arcs to the Chao Phraya. Over time the predominant characteristics of the waterways have receded, being overbuilt with modern infrastructure, structures and buildings.

Now with a densely developed centre, the city of Bangkok is extending radially on either side of the river, with an emphasis on a near north-south axis. Developments are also emerging in outer areas at poly-nodal centres with infill of informal settlements at generally low-lying areas. Its challenge remains as a result of its geographic territory, with many restricted swampy areas unsuitable for built development. Beyond, the greater metropolitan area continues to be expanded north towards Don Muang and historic Ayutthaya and east around the new road and rail axes past Sukhumvit and towards the new airport.

5.5.6 MANILA, PHILIPPINES

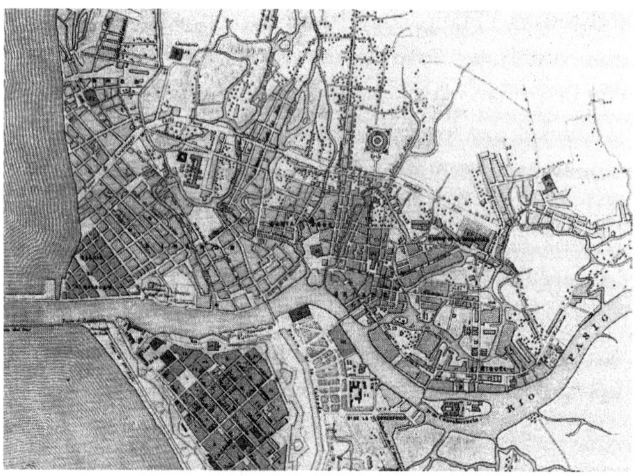

FIGURE 5.5.3 Manila – early configuration, 1848.

Source: Historic Urban Plans.

Background and Origins

Manila, on the shores of Manila Bay on the large island of Luzon, is the capital of the Philippines archipelago. The initial inhabitants were Aboriginal before the migration of Malay Polynesians who assimilated them. As the settlement of Tondo, it was developed as a fortified area on either side of the Pasig River and further settled by the Hindu Majapahit. In the 16th century CE Manila was overrun by the Spanish conquistadors who declared it as a territory of New Spain. With the small district of Intramuros it was made the capital of the Spanish East Indies and the Philippines, developed as a walled city besides the port, following the guidelines of the Spanish Laws of the Indies.

Its role expanded as it became a centre of triangular 'galleon' trade between the Philippines, Spain and Mexico. With spices also from Indonesia, silk from China and silver from Peru, this led to exchanges along the Spice Route, Silk Road and Silver trade, which led to making it the first truly global city with connections between different continents, from Asia to the Pacific.

Within the city, separate neighbourhoods of mainly Chinese settlers and Mexicans developed in Binondo and Ermita respectively. The Spanish continued their domination until the Philippines in seeking independence, after the Battle of Manila, had Manila ceded by Spain to the USA in 1901. A new charter was constituted then for the City of Manila, which comprised Intramuros and 11 other municipal districts into a consolidated capital.

American Daniel Burnham was commissioned in 1905 to prepare a city master plan with new roads, parkways and buildings that included a planned government centre in Wallace Field, from Rizal Park to Taft Avenue with a quadrangle and lagoon around Bagumbayan. Its realisation, however, was interrupted by World War II when Manila was overrun by the Japanese and severely destroyed, becoming as heavily devastated a city in Asia, as Warsaw was in Europe.

Modern City – Buildings and Arteries

A planned area for a new Quezon City, named after a former president, was developed based on a new master plan by planners Parsons and Frost, which was intended to act as national capital on existing barrios. After World War II and independence at the Treaty of Manila in 1946, it was proclaimed the national capital of the Philippines in 1948 to relieve the old overcrowded city of Manila.

This new city was to be developed around a park, with grand Quezon and Panang avenues, with a new Capitol building for the legislature and separate Supreme Court. This was not fully realised except around the geometric Diliman quadrangle and its four triangular areas, and the grand ellipse (Memorial Circle). With several districts, it includes Novaliches and La Loma, Cubao and Diliman, the original San Francisco del Monte and University of the Philippines campus.

Overall, the city has grown enormously through both population growth and urbanisation and is now divided into six congressional districts made up of 150 barangays as follows:

- 1st District – Tondo 1, harbour and container port densely populated on a landfill.
- 2nd District – Tondo east with main railway station and Tutuban Centre.
- 3rd District – Binondo, Quaipo, San Nicolas and downtown Manila along the Pasig River.
- 4th District – Sampaloc and the University Belt, including University of Santo Tomas.
- 5th District – Intramuros with Manila Cathedral, Ermita, Malate and Paco.
- 6th District – Pandacan, Santa Ana and Santa Mesa.

Most areas of the rebuilt city grew post-war and with increased urbanisation developed into a rapidly industrialising city. Accordingly, in 1975 the Metro Manila region was created that originally included Quezon City, San Juan, Makati, Caloocan and Pasay, designated by President Marcos to include further 13 neighbouring urbanised areas, with his wife as governor. The following year, however, the National Capital Region was created and the capital was to remain at Manila City.

Now an even larger region is recognised by NEDA as 'Greater Manila', which comprises Metro Manila and the adjoining provinces of Bulacan (north), Cavite and Laguna (south) and Rizal (east). In each of these areas, and also in Batangas, there is a very large volume of informal settlements. Makati centre has grown into the main financial area of Manila, around the Ayala Triangle with its tower blocks, corporate headquarters of PB Com, Gramercy Residences, central cluster and outgrown Binondo.

There are also individual new centres. Bonifacio Global City is located between the main Epifanio de los Santos Avenue (EDSA) and the C-5 Circumferential Road. As a commercial centre, it comprises the PSE Tower with the Philippine Stock Exchange, Metro Bank Centre and numerous residual condominiums. This adjoins Ortigas Centre, another real estate complex developed by the Pasig River, and another developing area is Filinvest City alongside Bonifacio City.

Whilst previously having an eclectic mix of architecture, with the original San Augustin church in Intramuros and other characteristic styled buildings, little remains except a few reconstructed colonial buildings, such as the Marble Palace, Ayuntamiento de Manila and Art Deco buildings such as the Metropolitan Theatre. Roxas Boulevard acts as a corniche across the central bay.

The whole urban area is in an unstable equatorial climatic region which is sensitive to periodic seismic activity and tsunamis, with annual monsoons, sinking land and regular flooding. Recently a series of reclamation projects has been proposed, such as the new planned Manila Bay – City of Pearl and Waterfront City to be developed in Manila Bay. To date only Horizon Manila is under development on separate areas of reclaimed land.

The Port of Manila is the largest seaport in the Philippines and with its International Container Terminal is the principal centre of economic activity of imports and exports. Besides the heavy petroleum industry and chemicals, with modern electronics, increased retail and commercial activities, the city has become a major commercial, finance and media centre. Metro Manila's Ninoy Aquino International Airport (MNL) is located south of Makati off EDSA.

The Central Bank and other banks, including the Asian Development Bank (ADB), the national administration and government agencies are located in and around the centre of Manila. Congress now sits at the Batasang Pambansa complex in Quezon City with the Supreme Court seated at Bonifacio Global City.

In terms of infrastructure, the city has a dense pattern of street networks and is extremely congested with cars, light vehicles and jeepneys at most times of the day with auto-rickshaws being replaced by e-trikes.

In order to transport high volumes of passengers across the city, there are two lines of light rail transit (LRT) and peripheral mass rapid transit (MRT) and in addition the elevated Metro Manila Skyway. The Philippines National Railways (PNR) operates commuter rail services to the provinces north and south of the centre.

The main infrastructure includes the South Luzon Expressway (SLX) and elevated Skyway that connects with North Luzon Expressway (NLX), the major Magallanes Interchange and Circumferential Road C-5. Bonifacio City comprises the Core B High Street, Forbes Town Centre, Uptown Bonifacio and Grand Central Park.

The housing sector is challenging with a very extensive informal sector and constant pressure to build more intensively with higher-rise accommodation. Most formal developments have been supplemented by informal settlements that generally gravitate around new infrastructure.

The former US Airfield at Fort McKinley was handed back to the Philippines after independence to become Fort Bonifacio. Ultimately, in 1995, under the Bases Conversion Act, this area has been developed by real estate consortia into Bonifacio Global City.

Subic Bay Freeport Zone (SBFZ) and Clark Freeport and Special Economic Zone (CFEZ) are also newly developed areas. With the closure of the US Naval Base to the northwest of Manila, Subic Bay, along with Olongapo City, has been developed as a new trade zone. Similarly, Clark Base has also been redeveloped, along with Los Angeles City, and has evolved as a free trade zone with Clark Global City after the effects of the eruption of Mount Pinatubo in 1991 and the spread of lahar were overcome.

The current total population is approximately 13 million people in Metropolitan Manila, with 2 million in the centre and 3 million in Makati with several very dense areas. The population density of the whole city region is very high at approximately 27,100 pers./sq. km.

The expansion of Quezon City and Bonifacio City continues to form part of the largest growing areas within overall Metro Manila. There has been extensive population growth and intensive business processing outsourcing activities. It is a colourful region with an eclectic and multi-cultural young working population, predominantly Catholic. Overall, this results in a hugely developed megalopolitan agglomeration of Manila, comprising both Metro Manila and most of the provinces of Central Luzon, now elaborated as Calabarzon (Cavite, Laguna, Batangas, Rizal and Quezon), an area of almost 40 million people.

Transactional Forces and Typology

From its earliest activities at the settlement of Tondo as a trade hub in the Pasig River Delta, Manila's strategic location on Luzon Island promoted its development in trade with China. Taken over by the Spanish conquistadors, the walled city of the present Intramuros District was developed as it became the basis for the Manila-Acapulco galleon trade with Mexico and Spain, exchanging silver with silk, and special gems with spices.

In more recent times, as a result of the Philippine-American War, the city was ruled by a US-appointed governor general under a new charter from the beginning of the 20th century. The subsequent makings of a new capital were planned around the Diliman quadrangle with North, East and West avenues based on the Burnham Plan. But it was not until after World War II and the end of Japanese occupation and with independence that the foundations of the contemporary city were really laid.

With urbanisation and industrialisation, despite the setback of a period of martial law, under new mayors the city developed with its port and basic trade in textiles, tobacco and sugar, food processing, rope and timber. With the development of the petroleum industry and oil-processing depots, and with containerisation, the Philippine Port Authority developed to manage the North and South harbours and the major extended landfill International Container Terminal connected by the North Luzon Expressway Harbour Link.

With its continuing industrialisation and enlargement and expansion of commercial areas of Quezon City and Makati, pressure has remained on demand for new residential areas and housing where people have been encouraged to relocate outside the main centres. Despite this, significant areas of informal settlements still remain across the urbanised region.

Located in Manila Bay, on the west coast of the island of Luzon, the city of Manila has developed as a radial city, now with poly-nodal centres both within and around the circumference from the historic walled Intramuros to the new Quezon City, Makati City and Bonifacio Global City and with other centres in Caloocan, Cubao, Ortigas and San Juan. Yet farther outwards, Greater Manila has developed not just as a metropolitan agglomeration but with the expanding adjoining provinces from Batangas and Laguna and the Rizal and Bulacan lakes towards Pampanga and Bataan, creating an enveloping megalopolitan region. Here the natural environment has been shaped by lakes and the features of Mount Makiling and volcanically active Mount Pinatubo just to the north, which in itself has been reshaping parts of the region.

5.5.7 JAKARTA, INDONESIA

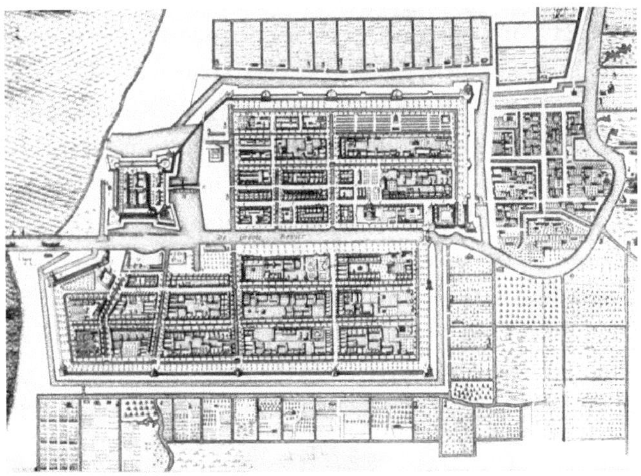

FIGURE 5.5.4 Batavia – early configuration, 1669.

Source: Historic Urban Plans.

Background and Origins

Jakarta, at the mouth of the Ciliwung River on the north coast of the island of Java in the Sunda Strait is the modern capital of Indonesia. It is low lying in a generally flat alluvial plain and with high seasonal rain and poor drainage and subject to frequent flooding. It was formerly known as 'Sunda Kelapa' as a trading port of the Hindu Sunda kingdom. With the arrival of the Portuguese the city was renamed 'Jayakarta'.

Taken over by the Dutch and renamed 'Batavia' who settled there in the 17th century CE and established as the de facto capital of the Dutch East Indies under the influential Dutch United East India Company (VOC). With its fort and ring of canals and system of 'grachten' (drainage), early growth occurred around Taman Fatahillah (Stadhuis Plein) as a result of its trading in cotton and batiks, mainly led by Chinese merchants.

During the 1730s, it was swept by various epidemics of malaria, trade declined and Chinese people were moved outside the walls of the old city, Kota south to Glodok. As development shifted inland farther to the south from the coast, the Dutch built up an area known as Weltevreden for the colonial administration around main squares of Buffelsveld and Paradeplaats, that evolved into Waterlooplein and Koningsplein with its palace. The latter was subsequently redesigned as Merdeka (Independence) Square with the National Monument.

After occupation during World War II and after the War of Independence, with the Republicans centred in Yogyakarta, full independence was secured eventually under Sukarno from the Netherlands in 1949 of the United States of Indonesia with Jakarta as capital, which included Sumatra, Java, parts of Borneo, and the outlying Thousand Islands that are a chain of small natural islands. Western New Guinea was not integrated until 1969, being subsequently known as Irian Jaya, or West Papua.

Modern City – Buildings and Arteries

In terms of heritage, the city revolves around Merdeka (Independence) Square, known as Lapangan Merdeka, with the great National Monument (Monas) completed in 1976, the old district of Kota Tua with Batavia Stadhuis, Jakarta Cathedral and various Dutch colonial buildings in Taman Fatahillah and Taman Suropati. The other open space is Lapangan Banteng accompanied by great modern Istiqlal Mosque, and recent Irian Jaya monument. More stylish areas remain in Cikini and Menteng with the Welcome Monument fountain, Salamat Datang.

In the Golden Triangle of central Jakarta, between Medan Merdeka and Thamion, Sudirman and Gatot Subroto, there are major buildings in the CBD extending southwards. These include several new major shopping malls and commercial developments, as the Autograph and Gama towers, 'Wisma 46', and other tall buildings, such as Menara Astra and Thamion Nine.

With its rich culture, handicrafts and cuisine, Jakarta has an eclectic mix of Betawi, Javanese and Sundanese cultures. With both trading and service sectors, the capital has active textile, chemical and electrical manufacturing sectors, provides banking and financial services, and houses the headquarters of many large Indonesian conglomerates.

The Ciliwung River was channelled by the Dutch into canals, with the two main branches that run broadly north-south. Flood canals (Banjir Canal) were planned to be built following a major flood in 1918 with the West Flood Canal, and the East Flood Canal was still undergoing development in 2015, crossing heavily populated areas. New measures are required for further dikes or relief tunnels. With low-lying land having the propensity to flood, better drainage and flood protection is required, since the land is also subsiding rapidly due to overuse and the extraction of groundwater.

Due to the gradual physical sinking, a Giant Sea Wall Project planned around the coastal perimeter of Jakarta under the National Capital Integrated Coastal Development (NCICD) plan, managed by Kuiper Compagnons Rotterdam. This is to constitute a series of dikes and lagoons and artificial islands shaped like a 'garuda' to protect the city. It will also incorporate urban development schemes, new neighbourhoods and a peripheral toll road and infrastructure. Of enormous cost, inspired by Dubai's offshore islands, with extensive environmental challenges, it remains at the planning stage.

There are several main avenues, such as Jalan Sudiriman and Jalan Thamrin, which are very congested most hours. The Inner and Outer Ring Roads surround the city with five radiating toll roads, with an odd and even daily number plate system to restrict the number of cars driving in and out of the city, and car-free Sundays. New industrial estates have been established in the outer areas of Cikerang in Bekasi for new manufacturing activities with the Cikerang dry port and container facilities. South Jakarta is very developed and comprises areas of very densely occupied and different-styled kampungs. In terms of rail, there are both the Greater Jakarta Commuter Rail and LRT and the Jakarta MRT and LRT, as well as Trans-Jakarta bus rapid transit and numerous bus companies, private minibuses and auto-rickshaws and tuk-tuks.

In 1966, Jakarta was declared a Special Capital Region, with the status of province, and considerable new developments were undertaken with the transition to the New Order, which continued until slowed by the Asian financial crisis. With substantial growth, Jakarta was divided into five administrative districts (kota) with its own governor and provincial parliament for the whole metropolitan region. The capital region includes the Province of Jakarta, with five districts, and the satellite cities and regencies of Bogor, Depok, Tangerang, South Tangerang, and Bekasi – formalised in 1976 and known collectively as JaBoDeTaBek with the creation of the Jabodetabek Corporation Body for the region of Greater Jakarta.

Hence, Jakarta is a huge and highly congested capital and ever-expanding city that places increasing demands on transport, housing and sanitation services. The population of the city is approximately 11 million, with 33 million in the greater metropolitan area. The population density of the whole city region is approximately 13,400 pers./sq. km.

Due to extreme congestion and the high rate of urbanisation across Java, at the beginning of the 21st century, the government decided to move the capital out of Java. After research by the National Development Planning Agency (Bappenas) in 2021, a new Capital City Authority was created for the development of a new capital to be located in East Kalimantan on the island of Borneo, named Nusantara. The first phase and basic infrastructure are currently under development there. On a coastal site with hilly landscape, it will comprise a whole new capital region with administrative and residential areas not far from the neighbouring city of Samarinda. (See also Chapter 4.5.8.)

The busy Port of Jakarta is at Tanjung Priok with container terminal and planned New Priok extension. Soekarno-Hatta International Airport (CGK) is located to the west of the city with the other Halim Perdana Kusuma International Airport (HLP) to the east.

'Desa-kota' Areas

In many urbanised areas in the Global South, there are urban hinterlands that essentially comprise agricultural or small-scale rural industrial activities, of a significant density that co-exist together and are intermingled. These have been labelled 'desa-kota' (village-city) by Terry McGee.

They are especially evident in South-East Asia in Indonesia and Java, in particular in Greater Jakarta, beyond and around JaBoDeTaBek, as described. Outside peri-urban zones, they tend to sprawl linearly along corridors and arterial routes, having intensive, agricultural practices with mainly female participants, on unregulated land areas of rice or other cultivation with dense informal settlements.

Transactional Forces and Typology

With its location in the Sunda Strait on the northern coast of the island of Java, the people of the historic Hindu Sunda Kingdom produced black pepper that was the source for local trade. With the introduction of coffee by the Dutch, this increased trade through the port of Sunda Kelapa. Here on the Malaccan spice route, this reinforced its role as a trading post on the Indonesian archipelago around which the Dutch East India Company established Batavia as its base around the Ciliwung River.

Whilst independence was long-awaited after World War II, as capital of Indonesia, Jakarta was to grow enormously as a result of in-migration, urbanisation and industrialisation. With its agriculture, crafts and textile origins, the development of the city had to be accommodated around low-lying wetlands, with channelled drainage and canals. Given these constraints and the challenging monsoon climate, the mix of Javanese, Betawi, Sundanese and Chinese peoples established themselves in different communities around and across the city.

As a coastal port and with draining rivers, the city of Jakarta had taken shape beyond Jakarta Bay around its channelled paths and canals and historic core, extending yet farther south as suitable land has been taken up with new settlements. Radiating outwards from Merdeka Square, with increasing migration and urbanisation, Jakarta now encompasses the satellite cities of Bogor, Depok, Tangerang and Bekasi into the awkward nomenclature of the megalopolitan agglomeration of JaBoDeTaBek.

Of increasingly poly-nodal structure with ribbon development along axes of often unsuitable land overbuilt with informal settlements, this complex mixed city strives to provide adequate services to its residents. In the meantime, it aims eventually to refocus settlers to the less developed new inner lands of Nusantara in Kalimantan as new capital, in order to relieve congested Java with its low-lying coastal areas with increasing risk from rising sea levels, far from the heart of Kota Jakarta.

5.5.8 KUALA LUMPUR, MALAYSIA

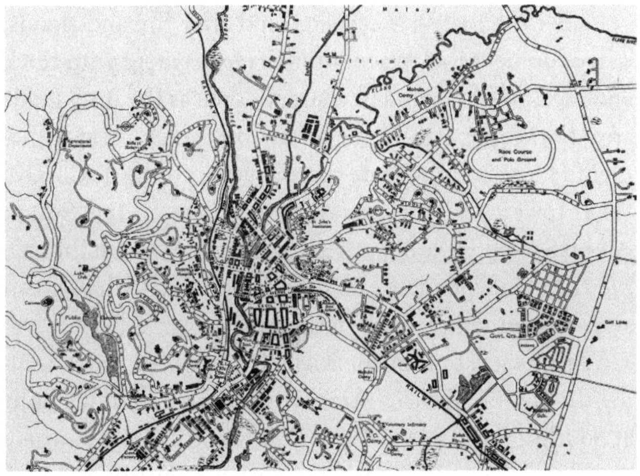

FIGURE 5.5.5 Kuala Lumpur – early configuration, 1928.
Source: Maphouse.

Background and Origins

Kuala Lumpur ('muddy confluence') with its original wooden and 'atap' palm frond buildings, grew from a trading port where the Klang and Gombak rivers join. On the western coast of a long peninsula with coastal areas of mangrove swamps and beaches, Malaysia is separated from Indonesia by the Strait of Malacca, with the island state of Singapore to the south beyond Johor. Originally populated by the indigenous Orang Ashi people, as a rich tropical rainforest, the region was controlled by Malay sultans. With its mainland states that include the Federated Malay States of Selangor, Pahang, Perak and Negeri Sembilan, and the separate Sabah and Sarawak states, it is a tropical region with seasonal monsoons and rains.

Given its strategic location on the streets of Malacca, traders and settlers from both India and China established settlements on the western coast. With the Portuguese and Dutch jostling for presence, the British East India Company took control of Penang, Malacca and also Singapore through the Straits Settlement in 1824. Becoming a Crown colony and with the formation of the Federated Malay Straits, its status evolved as a British protectorate.

It developed quite rapidly from the old Market Square as a centre in the region under the Sultanate of Selangor from the coast and active Port Klang to the beaches of Port Dickson. Inland the geography climbs to the Genting Highlands of Malaya and the rainforests beyond. In Selangor the development of local industries of rubber and tobacco grew, encouraged by demand and Chinese developers were attracted by the local tin mines in Ampang.

Under the British 'Residents' and after fire and floods, redevelopment took place with the replacement with brick and tiled buildings and new streets, such as the Brickfields area. In the city, significant landmarks were built at Parliament House and the State Secretariat, with the distinctive clock tower, and the government offices in the Sultan Abdul Samad Building in 1897 that overlooks the Padang, Merdeka (Independence), and Selangor Club with its green and adjacent temples and mosques.

With its location in the Klang River Valley and adjoining the coastal Malacca Strait, Port Swettenham was built at Southpoint, now known as Port Klang. From a swampy area, berths were built and the harbour constructed that was to become a strategic location on the maritime South-East Asian Spice Route. With the construction of a railway in 1890 between the port and Kuala Lumpur, this was to benefit access to and from the capital. With further modern development of Port Klang at West Port and North Port, the latter has now grown to become the major container terminal in the region.

In terms of infrastructure, the centre is very developed with its old Moorish-style main railway station and railway connecting Kuala Lumpur and Klang that encouraged development, and the recent network of highways and expressways. Its original market areas traditionally had narrow shops with 'five-foot ways' developed by the Straits Chinese, and were populated by cosmopolitan residents, an eclectic mix of Malay (Bumiputra), Chinese and Indian people.

Along with Singapore, the country was overrun by the Japanese in World War II. Post-war with the creation of the Federation of Malaya in 1948, in the following period, the Malayan Emergency developed until eventual ending and attainment of independence in 1957 and creation of the Federation of Malaysia in 1963, with Singapore separating in 1965. New Houses of Parliament were constructed by the Public Works Department (designed by Ivor Shipley), as the legislative assembly of the federal government. In 1974 the Federal Territory of Kuala Lumpur was created with KL as the federal capital ceded from Selangor where Shah Alam was made state capital.

Modern City – Buildings and Arteries

The old Historic Quarter is the heart of the city with the commercial Golden Triangle area and Bukit Bintang, Merdeka Square and the modern tall skyscrapers in KL City Centre and with individual kampungs between the rivers and the distinctive Chinatown district. Whilst most of the old shop-houses have gone, except around the old Market Square, new buildings have been built aiming to incorporate traditional Malay features and Islamic geometric motifs.

More recently the executive and judicial capital was moved to nearby Putrajaya in 1999 as Kuala Lumpur was restructured as a federal territory and administrative capital of Malaysia. Adjoining this now is a new area having aspirations to be the 'Silicon Valley' of the emerging IT sector which has grown enormously as the Kuala Lumpur metropolitan area stretches along the Klang River valley to Port Klang supported by the Kuala Lumpur Smart City Plan.

Kuala Lumpur has become a major financial and commercial hub in the region with new significant buildings such as Merdeka 118 (designed by Fender Katsalidis), Exchange 106 in the Tun Razak Exchange area, with the Malaysian oil and gas company Petronas Twin Towers (designed by Pelli), the tall KL Telecommunications Tower, Ilham Tower and the Telekom Tower. KL Sentral is a new commercial area at Brickfields. With many new commercial malls, it has a retail hub with very active textile and fabric sectors. A number of recreational parks remain, with Lake Gardens, botanical gardens and individual forest reserves that act beneficially as a small urban lung along with new initiatives as 'Greening KL'.

With the recent trend of extensive urban migration, the built-up area has grown extensively with the federal territory having approximately 2 million people and urban population exceeding 15 million in the metropolitan area. The Klang Valley Metropolitan Area, or Greater Kuala Lumpur, now includes the federal territories of both Kuala Lumpur and Putrajaya and the Selangor districts of Petaling, Klang, Gombak, Sepang, Hulu Langat and Kuala Langat, that effectively stretches between Port Klang, Rawang and Semenyih and the Titiwangaa Mountains in the north-east.

In terms of infrastructure there has been extensive development of public transport networks with the commuter rail system, LRT lines and MRT lines, Rapid KL and central KL monorail mainly elevated, each part of the Klang Valley Integrated Transit System (RPDKL). Separate lines connect with Putrajaya, Port Klang, Gombak, Ampang and the main airports of KLIA, radiating out from KL Sentral Station to Kuala Lumpur International Airport (KUL) in the south at Sepang beyond Putrajaya, opened in 1998 (designed by Kurokawa). A direct high-speed railway to Singapore is under planning and development. The city has an Inner, Middle and Outer Ring Road and a complex network of highways, with the North-South Expressway Central Link and the South Klang Valley Expressway.

In 1996, the Multi-Media Super Corridor (MSC Malaysia) was established by the government as a special economic zone managed by Multi-Media Development Corporation, later Malaysian Digital Economy Corp. as a high technology district known as 'Cyberjaya', stretching between the Petronas Twin towers and KL International Airport.

Located mainly on a former oil palm plantation, with new infrastructure many IT companies were established in the business park along with the campus of the new Multi-Media University and the recreational Lake Gardens area was created. Located adjacent to Putrajaya, this is connected by the tolled KL–Putrajaya Highway, now the Maju Expressway MEX), that acts as the backbone of this MSC Corridor.

Transactional Forces and Typology

From its early settlement, the tropical rainforest area around Kuala Lumpur has been the source of oil palm, tin and rubber, each resource being eagerly exploited in the region. With emphasis on trading, as the port was developed in the Malacca Strait on the Maritime Spice Route, whilst the river connection along the Klang Valley to Kuala Lumpur was essential, it was the railway that enhanced this.

In trading these commodities, Kuala Lumpur subsequently expanded as the federal capital of peninsular Malaysia and its island states of Sarawak and Sabah. With its central location on the peninsula, the creation of Putrajaya on the adjoining site as the new administrative capital is acting as a relief to the fast-growing capital since its expansion as one of the fast-growing Asian 'tiger' economies in the metropolitanised Greater KL region. (See also Chapter 4.5.8. Putrajaya.)

As a riverine city located just inland between tropical rainforests, highlands and the coastal area, Kuala Lumpur has been developed around the Klang River Valley. With its historic core, it comprises a complex network of streets and highways and individual kampungs on undulating terrain. With continued development and increased urbanisation, the separate new capital area of Putrajaya to the south and modern area of Cyberjaya provide some relief to the central urbanising forces. However, the interaction between the city regions remains with increased poly-nodal development and densification on remaining available land areas with mixed Malay, Indian and Chinese enclaves.

5.5.9 HO CHI MINH CITY (SAIGON), VIETNAM

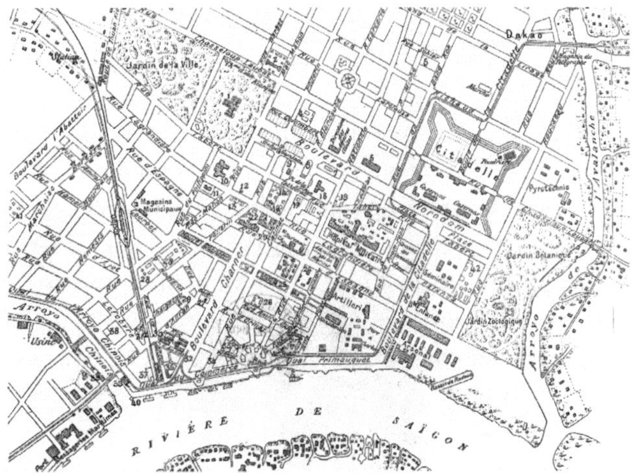

FIGURE 5.5.6 Saigon – early configuration, 1902.
Source: British Library.

Background and Origins

Ho Chi Minh City, formerly Saigon, is located in the southeast region of Vietnam, on the coast in tropical savannah by the delta of the Mekong River on a low-lying flood plain of the Saigon River. The first settlement in the area was a temple around the 4th century CE and a Champa settlement at Baigaur in the 11th century. Through many dynasties as a trading centre as part of the Khmer Empire, it was known as 'Prey Nokor' then as 'Gia Dinh', with the settlement of the Vietnamese people here. In the 17th century, after the Trinh-Nguyen civil war, this cut off access from Cambodia to the East China Sea and the city expanded with increased Chinese settlement.

In 1787, under the Treaty of Versailles, the French government had agreed to support Nguyen Phuc Anh and oversaw the construction of the citadel of renamed Saigon, according to the design of Vauban's defensive principles of 'Eight Diagrams'. Although subsequently destroyed, lower Cochinchina was eventually ceded to France under the Treaty of Saigon in 1862 with Saigon as its capital.

There followed the French colonial era at the end of the 19th century during which the city was developed significantly with colonial-style buildings that included the Governor's Palace, Notre Dame Cathedral and Ben Thanh Market, in the spirit of the original 'Les Halles' in Paris. At the time of the French conquest from the 1860s, as 'Sai Gon' it became an active trading centre and port on the Saigon River, rich in fishing and with agricultural cultivation.

Later, during the 1930s, a new region of Saigon-Cholon was developed around the city. During World War II, Vichy France had accepted Indo-China's occupation by Japan. The French had attempted to extend their influence in the north under the 'Pacification of Tonkin' and entrench a French protectorate around the Hong (Red) River Delta that included Hanoi and Haiphong and also the central area around Annam. But with Japan's surrender in the region, at the end of the world war in 1945, Ho Chi Minh was to lead the Viet Minh in an uprising in the August Revolution and with the capture of Hanoi, the Democratic Republic of Vietnam (North Vietnam) was established.

Vietnamese strongly resisted others seeking to retain control of the country, and a long campaign of insurgency ensued. Whilst the State of Vietnam had been proclaimed in 1949 and with the French Indochina Federation that encompassed both Laos and Cambodia, the First Indochina War was pursued up to 1954, until the French's final defeat at Dien Bien Phu.

With leader Ho Chi Minh in the North backed by the People's Republic of China (PRC), and the French in the South by the USA, much of the country then was under the control of the Viet Minh. In 1954, under the Geneva Conference, the country was formally divided along the 17th Parallel and the Republic of Vietnam was created (South Vietnam) leaving the Communists in the North and around Hanoi. This partition was the beginning of the prolonged battle between North and South, and the ensuing Second Indochina War, with the Viet Cong acting as the National Liberation Front of South Vietnam (NLF).

Vietnam War

From the beginnings of the war in 1955, this escalated significantly with American intervention in 1965 with the Army and Marines to support the South. With a major US air offensive and bombing campaign over the next two to three years, this peaked with the Tet Offensive from Hanoi in February 1968 with the Viet Cong acting across the whole country.

Saigon and the Capital Military District (CMD) were the centre of much fighting. With the Viet Cong fighting guerrilla warfare in the countryside and using the Cu Chi Tunnels outside Saigon, they proclaimed themselves as the Provisional Revolutionary Government (PRG) in the South. Further offensives continued and it was not until President Johnson ordered in March 1968 a reduction of the aerial bombardment, ending in October that year, that the beginning of peace negotiations in Paris ensued. These Peace Accords were prolonged and lasted for five years, until under President Nixon they were signed in January 1973, which removed the USA from the conflict in Vietnam.

But as agreement provisions were flouted between North and South, fighting broke out again. With the Ho Chi Minh Campaign through Hué and Da Nang, without support from the USA, Saigon eventually fell and surrendered to the North Vietnamese in April 1975.

As a result, the Provisional Revolutionary Government of the Republic of South Vietnam merged with the Democratic Republic of Vietnam to create the integrated Socialist Republic of Vietnam with Hanoi as its capital. A programme of forced collectivisation of agriculture was introduced and a series of 're-education camps' introduced. Finally, in 1976, the city was renamed Thanh pho Ho Chi Minh after the unification of the country at the end of the lengthy Vietnamese War. In consequence, a huge programme of rebuilding was required that needed enormous resources and commitment. Although supported by the Soviets, there was only gradual progress, and many Chinese left the country.

Modern City – Buildings and Arteries

Modern Saigon, having been renamed Ho Chi Minh City, now comprises the old City of Saigon, Cholon, Gia Dinh and 16 urban districts and five rural districts. As an administrative executive, a 13-member People's Committee was established. The city is very low lying in an area sensitive to flooding of both the Saigon and Dong Nai rivers, characterised by its neighbouring rivers and connecting canals to the Mekong Delta.

The city has some French colonial buildings in the former French Quarter that remain, such as Notre Dame Cathedral and the Municipal Opera House off Le Duan, Hôtel de Ville, the general post office and a number of historic hotels. The renamed and rebuilt Reunification Palace is on the site of the Norodom Palace of the former French governor of Indochina. With its China Town (Cholon), markets and classic Khmer temples and pagodas, it has been rebuilt as an attractive modern Asian city.

During the 1980s the whole country had suffered as a result of the earlier years of intensive destruction of urban places and defoliation of rural areas. A new programme of 'doi noi' (economic restructuring) was introduced as the country was gradually opened up to foreign direct investment, which contributed to growth in the economy. As it entered the 21st century, the city had the makings of a new economic hub with significant developments downtown and in the Central District, and despite the Asian financial crisis, it managed to rebound.

Ho Chi Minh City has become a centre of manufacturing and a dynamic area. After years of conflict and a long period of rebuilding, it gradually re-established itself with a strong public sector. With the High Tech Park and Software Park, it has promoted high-tech businesses and start-up companies. There are a number of new commercial areas, and several new urban centres have been developed, as at Thu Tiem and Phu My Hung. Significant new buildings are the super-tall, stepped Landmark 81 (designed by Atkins) built in 2018, Bitexco Financial Tower – Saigon Skydeck (designed by Carlos Zapata) and the Saigon Centre.

A new, elevated, urban metro system is under realisation in phases by MAUR. There are many public bus routes, and private and motorcycle taxis which are popular. It is planned to connect the entire country via the North-South Expressway, which is being built progressively on a concession basis in sections from Hanoi to Can Tho. The Ho Chi Minh City to Trung Luong section and to Long Thanh section have been completed.

A new road tunnel built under the river crosses the city east-west, Vuot Song Sai Gon Tunnel, opened in 2011. There are three other central bridges that cross the meandering Sai Gon River, namely Thu Threm, cable-stayed Phu My and Ba Son bridges.

In terms of rail, the two main stations in Ho Chi Minh City are Song Than and Sai Gon. The main North-South Railway is only an old single-track, metre-gauge railway from Hanoi to Ho Chi Minh City, built originally by the French during 1900–36. With its 1,725 km. length, its structures suffered considerable destruction during the years of the Indochina Wars and is in very poor condition.

Sai Gon Port has port areas to the south of the city and is very active along with export processing zones and supporting industrial parks. The city is linked to neighbouring Cambodia by inland waterways. Tan San Nhat International Airport (SGN) is centrally located with the new Long Thanh International Airport under completion. The current population is approximately 10 million in the city and 21 million in the metropolitan area.

Transactional Forces and Typology

With its location just to the north-east of the Mekong River Delta, from the early dynasties, the city grew up around the Sai Gon River in a heavily forested, swampy region, where travel was dependent on waterways. With coastal trade with the south-eastern seaboard of China, its recent history was shaped as capital of the French Protectorate of Cochinchina. With its central French Quarter, this was the focus of the city as the old canals were drained and filled and parks and avenues were built. After the devastating Vietnam War, and with significant rebuilding, the city has developed into an important business and manufacturing hub in South-East Asia, from its dedicated export processing zones and industrial parks.

As a riverine city in very swampy land, as the colonial occupiers drained the marshlands and reclaimed land, they laid out the central grid of the city with boulevards and built monuments in the French manner, such as the Hôtel de Ville, Notre Dame and Les Halles Centrales. West of the centre is Cholon, which is a distinctive Chinese district whose market has created a bustling trading area. Subsequently, new commercial areas were developed, mainly on the Right Bank of the river, with new neighbourhoods stretching from downtown. Despite the formal planned centre, there are many low-lying areas where informal settlements are prevalent. With new river crossings urban connectivity has improved, but the city remains at risk of flood events in the flat coastal delta region. These figures illustrate the modern configuration of the Mega Asian cities as described in the text and the evolution of their modern population.

FIGURE 5.5.7A Mumbai, modern configuration.

FIGURE 5.5.7B Greater Mumbai – modern configuration.

FIGURE 5.5.8 Hyderabad – modern configuration.

FIGURE 5.5.9 Karachi – modern configuration.

FIGURE 5.5.10 Dhaka – modern configuration.

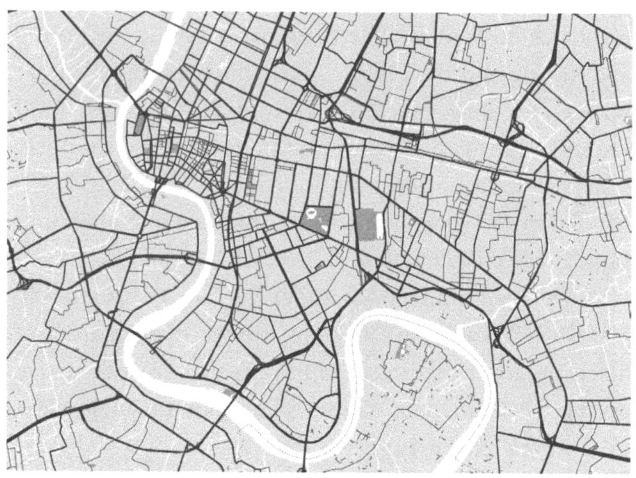

FIGURE 5.5.11 Bangkok – modern configuration.

FIGURE 5.5.12 Manila – modern configuration.

FIGURE 5.5.13 Kuala Lumpur – modern configuration.

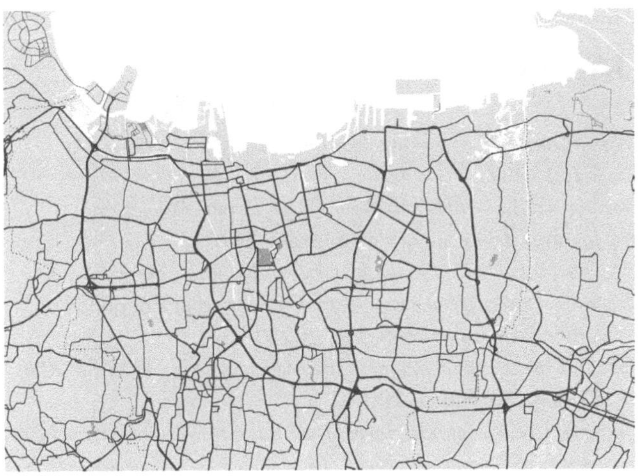

FIGURE 5.5.14 Jakarta – modern configuration.

FIGURE 5.5.15 Ho Chi Minh City – modern configuration.

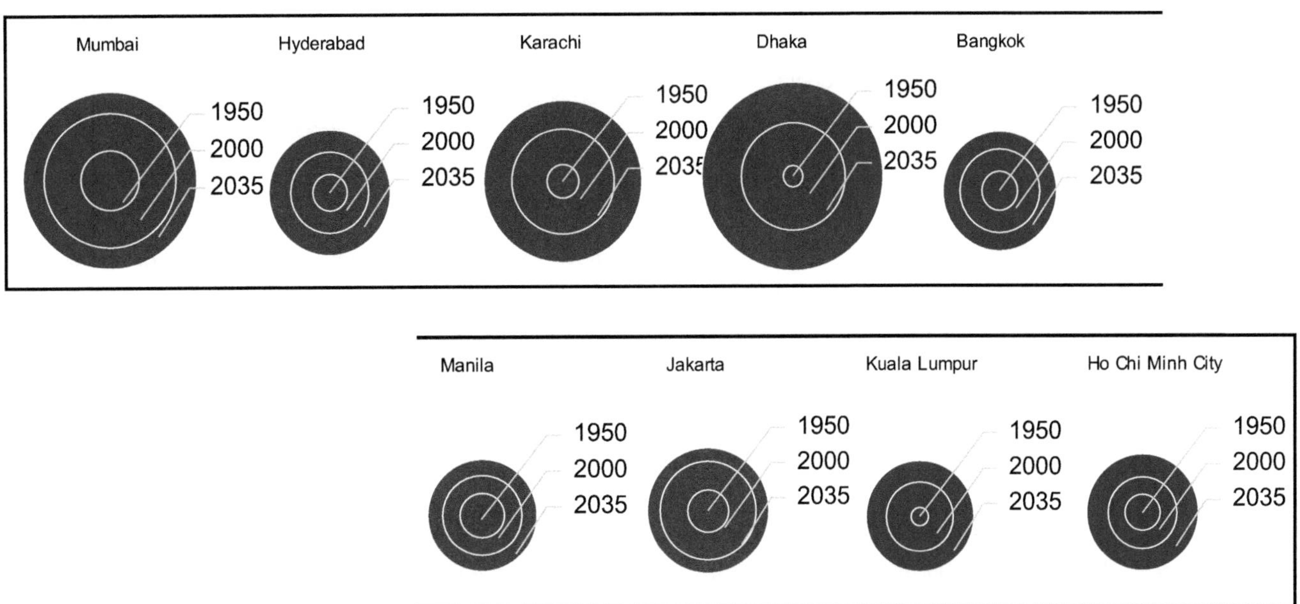

FIGURE 5.5.16 Mega Asian city populations.

Data source: Global Database of Metropolises 2020 – UN-Habitat (2020), UN DESA Population Division.

Further Reading

MUMBAI (BOMBAY)

Heitzman, James – *The City in South Asia*, Routledge, 2008.
Mehta, Suketu – *Maximum City – Bombay Lost and Found*, Headline Review, 2005.
Michell, George & Davies, Philip, eds. – *The Penguin Guide to the Monuments of India*, Penguin, 1990.

HYDERABAD

Michell, George, ed. – *Islamic Architecture of Deccan India*, ACC Art Books, 2018.
Vottery, Madhu – *A Guide to the Heritage of Hyderabad*, Rapa Publications, 2010.

DHAKA (DACCA)

Abusaleh K, Islam R & Islam N – *Globalization and Urban Culture in Dhaka*, Routledge, 2024.
Ahmed, Sharif – *Dacca, a Study in Urban History and Development*, Routledge, 2020.
Ahmed, Sharifuddin – *400 Years of Capital Dhaka and Beyond, Economy and Culture*, Asiatic Society of Bangladesh, 2010.
Bertuzzo, Eliza – *Fragmented Dhaka, Analysing Everyday Life with H Lefebvre's Theory of Production of Space*, Franz Steiner Verlag, 2009.
Bird J, Li Y, Rahman H & Rama M – *Toward Great Dhaka, a New Urban Development Paradigm Eastward*, World Bank Publisher, 2018.
Rabbani, Golam – *From Mughal Outpost to Metropolis*, University Press Ltd., 1997.

KARACHI

Gayet, Laurent – *Karachi, Ordered Disorder and the Struggle for the City*, Oxford University Press, 2014.
Khuhro, Hamida & Mooraj, Anwar – *Megacity of Our Times*, Oxford University Press, 2003.

BANGKOK

Chua, Lawrence – *Bangkok Utopia*, University of Hawaii Press, 2021.
O'Neill, Maryvelma – *Bangkok, a Cultural History*, Oxford University Press, 2008.
Waugh, Alec – *Bangkok, the Story of a City*, W H Allen, 1971.

MANILA

Garrido, Marco – *The Patchwork City, Class, Space & Politics in Metro Manila*, University of Chicago Press, 2019.
Hsieh, Ellen – *The Archaeology of Early Colonial Manila, a Hybrid City in Global History*, University Press of Florida, 2025.
Zaragoza, Ramon – *Old Manila*, Oxford University Press, 1990.

JAKARTA

Abeyasekere, Susan – *Jakarta, a History*, Oxford University Press, 1987.
Kusno, Abidin – *After the New Order, Space, Politics and Jakarta*, University of Hawaii Press, 2013.
McGee, Terry – *The South East Asian City*, Bell, 1967.
Roitman, Sonia & Rukmann, Deden – *Routledge Handbook of Urban Indonesia*, Routledge, 2023.
Silver, Christopher – *Planning the Megacity, Jakarta in the 20th Century*, Routledge, 2007.
Van der Linde, Herald – *Jakarta, History of a Misunderstood City*, Marshall Cavendish, 2022.

KUALA LUMPUR

Gullick JM – *A History of Kuala Lumpur, 1856–1939, Malaysian Branch*, Royal Asiatic Society, 2017.
Kozlowski M, Mehan A & Nawratek K – *Kuala Lumpur, Community, Infrastructure and Urban Inclusivity*, Routledge, 2020.
Loo, Yat – *Architecture and Urban Form in Kuala Lumpur*, Routledge, 2016.
Turnbull, Mary – *A Short History of Malaysia, Singapore & Brunei*, Cassell, 1979.

HO CHI MINH CITY (SAIGON)

Cham, Tran Minh – *Le Viet-nam, Histoire et Civilisation*, Alliance Vietnamienne, 1984.
Kano, Stanley – *Vietnam, a History*, Penguin, 1997.
Sheehan, Neil – *Two Cities, Hanoi and Saigon*, Jonathan Cape, 1992.

Part VI

Megalopolitan Agglomerations

6.0 Introduction – Megalopolitan Agglomerations[1]

Historically, the centre of a city (urbs) has typically grown around its core, with extended development reaching out into the 'sub-urbs'. As the reach has broadened and further centres start to develop, other areas have been labelled, as for example 'ex-urbs'. However, as multiple nodes have developed with improved links and connectivity, this has led increasingly to agglomeration of urban areas as they coalesce and grow through accretion. This model of urbanisation has led not just to enlarged metropolitan areas but now to even larger 'megalopolitan' regions having still larger populations – such has been the trend from 'polis' to 'metropolis' and on to 'megalopolis'.

With mega cities, particularly in Asia and South America, these are gaining yet larger extended urbanised areas on the periphery. In different continents, these metropolitan areas are now coalescing with other growing metropolitan areas to form megalopolitan agglomerations. This is particularly evident in certain areas of Western Europe, North America and South-East Asia.

In continental **Europe**, combined metropolitan areas with multiple cities are urbanised regions of:

- Rhine-Ruhr: Cologne, Düsseldorf, Dortmund and Essen, Germany,
- Randstad: Amsterdam, Rotterdam, Den Haag and Utrecht, Netherlands, and
- Upper Silesia: Katowice, Poland and Ostrava, Czech Republic.

Other very large but more nucleated metropolitan areas are:

- Greater Moscow, Greater London and Greater Paris, each having populations of approximately 15 million people.

In the **USA**, data which is statistically collated in terms of standard metropolitan statistical areas (MSA) indicate metropolitan regions that have a core of high population density and surrounding urbanised area. Where there are adjoining urbanised areas, these are assembled into combined statistical areas (CSA) that have integrated connections forming a yet larger urban region. These constitute, in America, such areas of agglomeration that exceed 15 million people, such as:

North-Eastern Seaboard:

- Boston, Cambridge and Worcester, Massachusetts,
- Newark and Jersey City, New Jersey and New York,
- Philadelphia and Reading, Pennsylvania and Camden, New Jersey and
- Washington DC, Baltimore, Maryland and Columbia and Arlington Virginia.

West Coast:

- San Francisco, San Jose and Oakland, California,
- Seattle and Tacoma, Washington and
- Los Angeles, Long Beach and Anaheim, California.

1 The Texas-Mexico Automotive Supercluster region is where the Mexican car industry in Matamoros under the Border Industrialisation Program has 27 vehicle assembly plants, 'maquiladoras' (duty-free factories), that work cross-border in new vehicle assembly and production with Brownsville, Texas.

DOI: 10.4324/9781003353348-32

South:

- Dallas, Fort Worth and Arlington, Texas.

In **Mexico** there are the large urbanised areas of Mexico City, Matamoros and Tijuana.[1] In **South America** the largest urbanised and continuously urbanising areas are in Brazil at São Paulo and Rio de Janeiro, Buenos Aires in Argentina, Santiago in Chile, and Bogotá in Colombia.

Across **Africa**, major urban areas extend from the enormous urbanised area of Cairo in the north to the Gauteng municipalities around Johannesburg in South Africa. Ever-growing megalopolitan-type agglomerations across tropical African regions, with significant areas of informal settlements are:

- Lagos, Kinshasa and Brazzaville.

In **South and South-East Asia**, the region having the largest size of urban populations is the Indian sub-continent with the countries of **India**, Pakistan and Bangladesh that have cities of great density that carry enormous volumes of people in excess of 20 million and are continuously growing with many areas of informal settlements, especially in the cities of:

- Mumbai, Delhi and Kolkata, and
- Karachi and Dhaka.

In **Japan**, the dense, urban Tokaido Corridor holds a large volume of people and the megalopolitan agglomerations which contain over 20 million people comprise:

- Tokyo–Yokohama and Osaka–Kobe.

Other continuously urbanising areas are Jakarta in Indonesia and Manila in the Philippines.

In **China**, metropolitan areas are growing ever larger where government policy is to knit together clusters of cities to form new megalopolitan regions under the direction of the Chinese National Development and Reform Commission. This is reaching yet another order of magnitude for regions where the aggregate population is around 100 million people. Here the largest megalopolitan regions include:

- Jing-Jin-Ji Megalopolis: Beijing, Tianjin and Ji (Hebei),
- Yangtze River Delta Megalopolis: Shanghai and Suzhou, Nanjing, Hangzhou and Ningbo (Jiangsu/Zhejiang),
- Pearl River Delta Megalopolis: Greater Bay Area: Guangzhou, Dongguan, Shenzhen, Zhuhai, Hong Kong and Macau (Guangdong),
- Yangtze Mid-stream Megalopolis: Wuhan, Changsa and Nanchang (Hubei/Hunan/Jiangxi),
- Guanzhong Plain Megalopolis: Xi'an, Shaanxi and Gansu, and
- Chengyu Megalopolis: Chengdu and Chongqing (Sichuan/Chongqing).

Other clusters that are also growing significantly are the areas around Qingdao (Shandong), Zhengzhou (Henan), Shenyang and Dalian (Liaoning), and Beibu Gulf (Nanning, Zhanjiang and Haikou). Administratively China comprises four major municipalities (Beijing, Shanghai, Chongqing and Tianjin) that are each very heavily urbanised.

This **Part VI** accordingly addresses the different types of megalopolitan agglomerations that are emerging across continents, and especially in South-East Asia, where the forces of aggregation have led to the agglomeration of urbanised areas into built-up areas having characteristics of extended poly-nodal cities.

These types and scale are highlighted in the following tables: Table 6.01 – Megalopolitan Agglomeration Types and Table 6.02 – Agglomeration Population Growth.

Table 6.03 – Population Densities of City Regions, is generated from 'Population Weighted Densities of City Regions' as derived from the Global Human Settlement Layer (GHSL) data furnished by EC-JRC Copernicus and the Center for International Earth Sciences Information Network (CIESIN) at Columbia University, USA that integrates both satellite data and national census data.

Table 6.0.1
Megalopolitan Agglomeration Types

		City Region	Topography	Plan	Trans. Forces	Type
6.01	**MEGALOPOLITAN AGGLOMERATIONS**					
6.1	***Megalopolitan Europe***					
		Randstad	Lowlands	Ring cities	Services	Modern
		Rhine-Ruhr	Riverine	Poly-centric	Industrial-coal	Modern
6.2	***Megalopolitan Americas***					
		US North-Eastern Seaboard	Seaboard	Poly-centric	Services	Megalopolitan
		Dallas-Fort Worth	Plain	Dumb-bell	Hub	Twinned
		Mexico City	Plateau	Axial-organic	Capital	Mexican
6.3	***Megalopolitan Japan***					
		Tokyo-Yokohama	Coastal/bay	Radio-centric	Capital	Megalopolitan
		Osaka-Kobe	Coastal/bay	Linear-organic	Industrial	Megalopolitan
6.4	***Megalopolitan India***					
		Greater Delhi	River plain	Radio-centric	Capital	Megalopolitan
		Greater Kolkata	Riverine	Organic	Sta. Capital	Hybrid
6.5	***Megalopolitan China***					
		Jing-Jin-Ji	Plain/river	Poly-centric	Technology	Megalopolitan
		Qingdao	Coastal/bay	Organic	Industrial	Modern
		Guangzhou	Delta	Poly-centric	Technology	Global-megalopolitan
		Chengdu	Plain	Concentric	Industrial	Sichuan

Table 6.0.2
Agglomerations Population Growth

6.02	POPULATION & DENSITIES		2020	
	Country	**City Region**	**Population million**	**Pop. Density Pers./sq. km.**
	India/Pakistan/Bangladesh	**Mumbai**	22.9	52,400
		Karachi	18.7	48,800
		Dhaka	26.8	47,900
		Kolkata	26.7	36,400
		Delhi	30.3	29,000
	Near East	**Cairo**	24.5	44,900
		Istanbul	14.3	25,200
	China	**Shanghai**	27.8	27,900
		Beijing	19.0	20,200
		Guangzhou	43.8	20,000
	SE Asia	**Manila**	24.8	27,100
		Seoul	22.7	19,800
		Jakarta	38.7	13,400
		Bangkok	17.4	11,600
	Japan	**Tokyo**	34.1	10,200
		Osaka	15.6	8,100
	Americas	**São Paulo**	19.7	14,500
		Mexico City	17.8	13,200

Table 6.0.3
Population Densities of City Regions

6.03	POPULATION & DENSITIES		2020	
	Country	**City Region**	**Population million**	**Pop. Weighted Density 2020 pers./sq. km.**
	India/Pakistan/Bangladesh	**Mumbai**	22.9	52,400
		Karachi	18.7	48,800
		Dhaka	26.8	47,900
		Kolkata	26.7	36,400
		Delhi	30.3	29,000
	Near East	**Cairo**	24.5	44,900
		Istanbul	14.3	25,200
	China	**Shanghai**	27.8	27,900
		Beijing	19.0	20,200
		Guangzhou	43.8	20,000
	SE Asia	**Manila**	24.8	27,100
		Seoul	22.7	19,800
		Jakarta	38.7	13,400
		Bangkok	17.4	11,600
	Japan	**Tokyo**	34.1	10,200
		Osaka	15.6	8,100
	Americas	**São Paulo**	19.7	14,500
		Mexico City	17.8	13,200

Source: Global Human Settlements Layer (CIESIN & EC-JRC Copernicus 2023).

6.1 Megalopolitan Europe

Randstad Region and Rhine-Ruhr Region

6.1.1 MEGALOPOLITAN EUROPE: RANDSTAD REGION

Megalopolitan Region

The **Randstad Region**, or 'ring city', is the name given to the conglomeration around the Rotterdam, Dordrecht and Den Hague region and the Amsterdam, Haarlem and Utrecht region within the central Netherlands. It contains major infrastructure, such as the ports of Rotterdam and Amsterdam and Schiphol Airport (AMS), and it has a very dense population. The northern ring comprises Haarlem, IJmuiden, Amsterdam and reaches towards Utrecht, which although being at the geographical centre of the country, may be considered as being outside this ring. Whilst Amsterdam is the financial and commercial centre, Den Hague is the administrative capital and centre of government. The southern ring comprises Leiden, Den Hague, Rotterdam and Dordrecht. Whilst Rotterdam is the most industrialised area, it also adjoins the university cities of Leiden and Delft.

With a dense network of highways and railways, the country of the Netherlands benefits from a highly integrated transport network for both passengers and goods. Due to its location, much of the freight transit traffic that flows across Europe passes through the country and has to be accommodated, for example on the new Betuwe Rail freight line. With transport and logistics being pivotal for freight and with 'just-in-time' manufacturing, logistics centres have been developed alongside the main highway axes, particularly the A1, A2 and A3 highways. Such development pressures remain on limited available land areas. Due to its low-lying geography and lack of physical space for building, since World War II many polders have been reclaimed and newly built areas developed in the IJsselmer, such as the new towns of Almere and Lelystad. With population growth pressures continuing, this reclamation is continuing as the area is a highly populated urban region of north-west Europe.

For workers and commuters, there are intensive light rail and bus services, and Randstad Rail provides rapid inter-urban transit between the individual cities across the Randstad. As an inherently flat region, there is also a dense network of well-planned intra-urban cycle ways within each city. The Randstad conurbation as a whole has a population of approximately 8.5 million with approximately 1 million and 0.65 million in the largest cities of Amsterdam and Rotterdam respectively.

DOI: 10.4324/9781003353348-33

Rotterdam

From its early founding and construction of dikes to regulate the rivers Schie and Rotte, the city of **Rotterdam** was formed in the 13th century CE. Originally a swampy area, there were four 'Schie' waterways that connected inland towards Delft. With the river dammed at Schie Dam, a network of polders and canals was subsequently developed and tolls charged on passing ships. As the canal was widened between Delft and Overschie, this was connected to Delfthaven that was to become the main shipping route in and out of the country.

Subsequently, in 1892, the Nieuw Maas ('new waterway') canal leading directly to the Rhine-Meuse-Scheldt delta in the North Sea was constructed, which has led to it becoming the main port at the mouth of these rivers. Due to its strategic location, Rotterdam has grown considerably to become now the major European international shipping port.

Despite total destruction during World War II, the whole city has been completely rebuilt and redeveloped. From its initial triangular area, the docks are located at the north bank at Oudenhaven, Leuvenhaven and Delfshaven. The city is divided with the larger Rijnhaven, Maashaven and Waalhaven on the south shore of the Nieuwe Maas as the docks have been extended westwards towards to the river mouth. Here Europoort has also grown extensively as containerisation developed and with Petroleum Haven also. The inner city is centred around Hofplein at the end of Coolsingel and Stadhuisplein. In terms of infrastructure, the Central Railway Station was opened in 1957 at Kruisplein at the end of Westersingel, with the railway bridged across the Maas over Noordereiland. The main railway then runs in a tunnel, Maastunnel, past Euromast under the Havenroute.

The city is enclosed by a rhomboid of highways, the A20 to the north, the A15 to the south and the A16 to the east.

More road tunnels were built, such as the Benelux and Botlek tunnels to replace the old lifting bridge, the iconic cable-stayed Erasmus Bridge, called the 'Swan', and the Wilhelmsbrug (Wilhelms Bridge). North of the city is the combined 'Rotterdam–The Hague' Airport (RTM).

In terms of architecture, it is a thoroughly modern city rebuilt with striking modern developments. The Lijnbaan shopping centre was rebuilt (designed by Van den Brock and Bakema) in 1953. More recent are the Euromast Tower (1960), the Markthal building (designed by MVRDV), the Cube Houses, New Orleans Tower, Maastoren, Zalmhaventoren and De Rotterdam. In the 1990s, a new business centre was developed at Kop van Zuid on the south bank. It is home to the regional headquarters of major industrial companies, such as Unilever, Shell, Maersk and Eneco, and Erasmus University and several other educational institutions are located here. It is a significant business centre with a city population of approximately 0.65 million, contained within the overall conurbation of the Randstad.

Regional Transactional Forces and Typology

As a maritime nation, during the 17th century, when the Dutch East India Company (VOC) was in the ascendancy, the Netherlands was an active trading nation through the ports of Antwerp, Amsterdam, Delft and Dordrecht. Rotterdam was one of the seats of the six chambers of the VOC. Its main growth came from the completion of the Nieuw Maas waterway that secured its direct access to the North Sea.

With its access to the Rhine, Meuse and Scheldt rivers, the waterway provides access into Western Europe and the Ruhr. This has led post–World War II to its development as a significant gateway for freight and with its logistics-handling facilities becoming Europe's largest seaport. Shipping is the main industry with companies such as Royal Nedlloyd and Moller Maersk, as Rotterdam has become a major port for containers and petro-chemicals reinforcing its role as a centre for major industrial conglomerates.

The city is shaped by its waterways and by man-made actions in harnessing them for maritime purposes. With the realisation of the Nieuw Maas, the construction of original wharves and docks, annihilation in World War II and subsequent reconstruction, the city's economy revolves entirely around the port. But it remains dependent on flood protection from the huge storm surge barrier, the Maeslantkering, part of the Delta Works completed in 1997. Road and rail infrastructure has been constructed to interconnect with the main waterways, making bridges and tunnels vital components in the workings of the region. The Port of Rotterdam acts as the main dynamo in the Randstad megalopolitan region around which business services have grown. With its built-up areas both north and south of the waterway, these are being extended linearly in their riverine context.

6.1.2 RHINE-RUHR REGION, GERMANY

Megalopolitan Region

The **Rhine-Ruhr Region** is a major metropolitan conurbation in the most populous state of North Rhine-Westphalia in western Germany that encompasses Dortmund, Bochum, Essen and Duisburg in the Ruhr area, and the cities of Mönchengladbach, Düsseldorf, Wuppertal, Leverkusen, Cologne and Bonn. Once part of the Holy Roman Empire, Rhineland and then Prussia, it is economically a very dynamic area.

The Ruhr has always been a centre of heavy industry with the larger cities of Düsseldorf, Dortmund, Essen and Cologne and also specialises in the services sector. There are a large number of major international companies that are headquartered here and it constitutes approximately 1/7th of the annual German GDP, the powerhouse of modern northern Germany.

In terms of infrastructure, it contains a dense network of highways and autobahn, and railways served by both VRR and DB with rapid transit suburban and urban 'S'- and 'U'-bahn systems. There are navigable waterways with the Duisburg inner harbour connecting with the River Rhine, and Dortmund Port and Ems Canal connecting with the North Sea and the Ruhr River that carries bulk freight traffic.

With a total population of around 10 million people, the whole urbanised area that constitutes the Rhine-Ruhr Region comprises the largest principal cities as follows:

Dortmund, Bochum Essen, Duisberg – 5.2 million; Düsseldorf, Neuss and Ratingen – 1.5 million; Mönnchengladbach – 0.3 million; Wuppertal – 0.3 million; Cologne – 1.9 million and Bonn – 0.9 million.

Dortmund is located by the Emscher and Ruhr rivers as tributaries of the River Rhine in the modern state of North Rhine Westphalia and surrounded by other large cities such as Bochum, Hagen and Recklinghausen. Under the Holy Roman Empire, it became a free imperial city and chief city in the Netherlands Circle of the Hanseatic League, known as Dortmunde. Whilst neighbouring Cologne exerted greater influence in the following centuries, in Napoleonic times, with the dissolution of the Holy Roman Empire, Dortmund came under the Confederation of the Rhine within the

Grand Duchy of Berg until its collapse and was incorporated into the Kingdom of Prussia at the Congress of Vienna. In a geographical area rich with coal seams, the Ruhr State Mining Authority was established, which was the beginning of the development of intensive coal-mining activities.

During the 19th century, the steel-making industry developed around this district, and with increased population immigrating from Eastern Europe and growing industrialisation, Dortmund became a significant centre of economic activity with steel factories such as Westfalenhutte of Hoesch AG.

After World War I, the Ruhr was occupied by France and Belgium, and after a period of rearmament, these factories became targeted by Allied bombers during World War II, when the majority of the city was totally destroyed. Post-war, initially under a British zone of military occupation, the city had to be totally rebuilt. The inner-city area and individual quarters were redeveloped, following the original mediaeval layout that was bisected by the main pedestrianised street Westenhellveg.

The northern part of the city, Nordstadt, was more industrialised and to the south Horde has been incorporated into the city. The area around Lake Phoenix, the former industrial site of Thyssen Krupp with Phoenix See and Port Promenade, has been redeveloped.

Administratively, the city was subdivided with 12 administrative boroughs, including Inner City North, West and East, as an independent city, which is not part of a Landkreis. The ring road now runs along the line of the old city wall. With the modern high-rise RWE and HCC Buildings, and Florianturm TV Tower, the skyline is that of a modern city with a large park, Westfalenpark.

New Autobahns (Ruhr-schnellweg) interconnect along the old Hanseatic trade routes with high-speed inter-city DB trains from the Hauptbahnof. There is a dense network of public transport, regional and stadt-bahn, light rail and bus, run by Verkehrsbund Rhein-Ruhr (VRR), and H-Bahn Monorail connects the TU Dortmund University campuses (part of UA Ruhr with Bochum and Duisberg-Essen).

Dortmund's port, which is at the end of the Dortmund-Ems Canal connecting with the North Sea, was opened in 1895. With its docks and piers, it is one of the largest canal ports and remains very active as a trans-shipment centre. Dortmund Airport (DTM) contains the Halls of Westphalia, which are used as a major exhibition centre each year.

With its former emphasis on heavy industrial activities and breweries, it is now the centre for many SME companies and has refocused on high technology industries. It has become a leading supplier of bio-medical and robotic services, IT and communication systems and with the Technologie Park Dortmund acting as a commercial research centre linked to TU Dortmund University. This region of the Ruhr is now a densely populated area merging with the adjoining cities of Essen and Duisberg.

Essen, as another of the largest cities on the Ruhr and Emscher rivers in the Rhine-Ruhr metropolitan region, was the site of very early settlement of Germanic peoples when the city was walled and declared a free imperial city under the Holy Roman Empire. Its resources were first recognised with the finding of coal many centuries ago, but it was not until 1811 that Krupp founded the first cast-steel factory, which was the beginning of the extensive steel industry and the fabrication of armaments that was eventually to follow.

Following destruction in World War II, major industries re-established themselves after the war under Krupp and with its amalgamation with Thyssen, and the utility companies RWE and EON and other large construction companies developed. As a centre of major trade fairs at the Messe in Essen, with Mulheim Airport (ESS), it has become a major industrial centre along with the triangle of cities of Duisberg, Dortmund and Düsseldorf, each having their own airports.

The Zollverein Industrial Complex is a notable landmark of the former coal-mining activities in North Rhine Westphalia. The region is now wholly inter-connected through a network of modern Autobahn across the Ruhr with the A40 Ruhr-Schnellweg running east-west through the city. Rail transport is excellent in the region with the Inter-city DB, Regional, VRR and S-bahn, and with the U-bahn, tramways and bus lines, it has a highly integrated transport system.

To the south of the city, the boroughs of Kettwig and Werden have been annexed and retain older preserved buildings and parks and the Moltkeviertel quarter. Other notable buildings are the Aalto Theatre and Villa Hengel, the Krupp family residence and Gartenstadt Margarethenhohe as a suburban garden city was developed in 1906 by Margarethe Krupp (designed by Metzendorf) on garden city principles in the forest for factory workers.

With its natural mineral resources, this has contributed to its development as a significant industrial region in the west of Germany as one of the key centres in the metropolitanised area of the Ruhr.

Regional Transactional Forces and Typology

With its geographic location, the Rhine-Ruhr Region has the geophysical feature of two major rivers with land rich in coal seams. With waterways acting as the main routes of transport, these served to support the old trading routes of the Hellweg to the ports of the Hanseatic League. During the Middle Ages settlements grew along the rivers and at crossing points. But it was the discovery of coal and its exploitation, along with the construction of canals to facilitate transport that led to the development of the iron and steel industry in the region.

After the Industrial Revolution and as railways were built, industrialisation expanded, and with the growing demand for labour, populations were attracted to the region that gravitated around the mines and factories. Urbanisation developed such that the region now comprises the three densely urbanised areas centred around Dortmund-Essen-Duisberg, Düsseldorf and Bonn-Cologne.

From steam engines, coke and iron foundries and steel works, major manufacturers such as Thyssen and Krupp emerged. With the realisation of the Dortmund-Ems Canal and the Rhine-Herne Canal, this enhanced connections with the ports. By the early 20th century, the Ruhr Region had become the centre of industrialisation, which led to increased urbanisation.

Beyond the upheavals of both world wars, its significance was recognised in the creation of the European Coal and Steel Community in 1952. As reconstruction took hold, this secured steel production for a period until the mid-1970s when demand declined and manufacturing plants and steel mills were closed. This was in turn to lead to the transformation from older manufacturing industries to new technologies and service industries in the region around which modern businesses were to develop.

From their original settlement, individual towns and cities developed by the rivers along which trade was conducted. As population gravitated to larger centres, the region developed new nodes connected by lines of communication. This has resulted in poly-nodal urban development with higher density levels, resulting in today's megalopolitan area of the urbanised Rhine-Ruhr Region that is approaching 10 million inhabitants. The following figures illustrate the modern configuration of the agglomerations as described and show the evolution of their modern population.

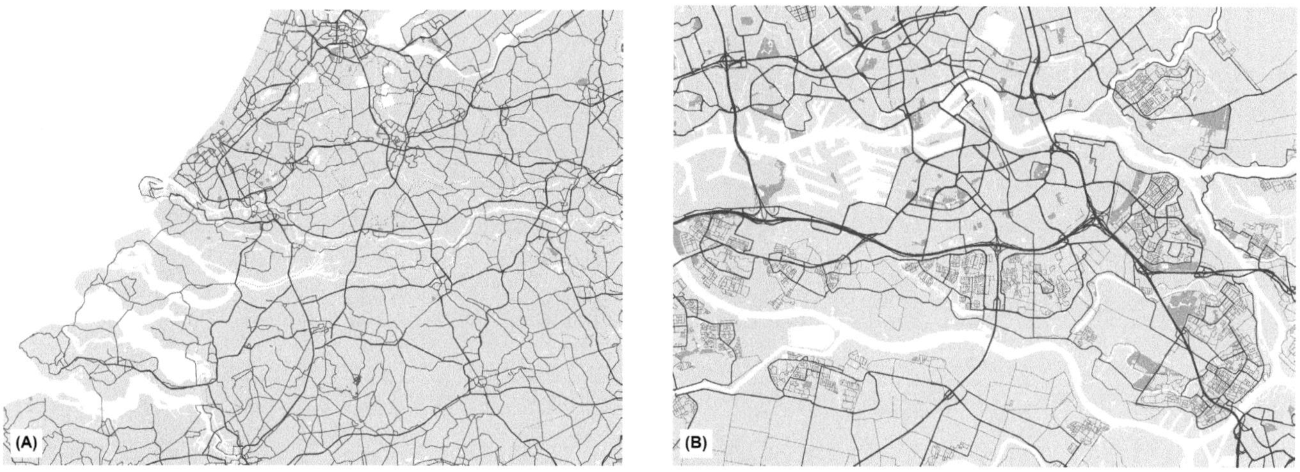

FIGURE 6.1.1A Randstad Region – modern configuration.

FIGURE 6.1.1B Rotterdam – modern configuration.

FIGURE 6.1.2 Rhine-Ruhr – modern configuration.

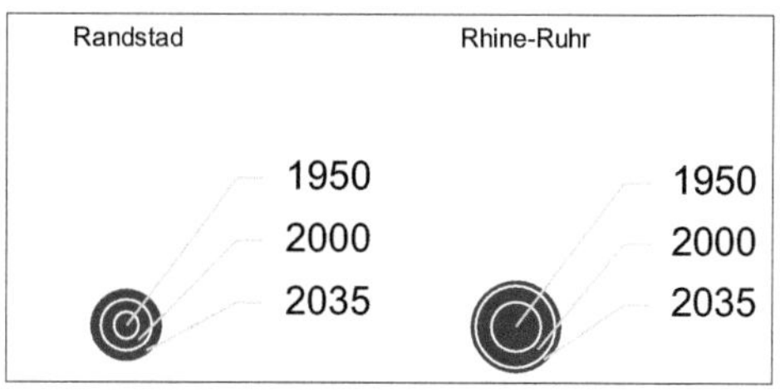

FIGURE 6.1.3 Megalopolitan Europe populations.

Data source: Global Database of Metropolises 2020 – UN-Habitat (2020), UN DESA Population Division.

Further Reading

RANDSTAD

Zonneveld, Wil & Nadin, Vincent, eds. – *The Randstad, a Polycentric Metropolis*, Routledge, 2022.

ROTTERDAM

Groenendijk, Paul & Vollaard, Piet – *Rotterdam Architecture City*, NAI Publisher, 2015.

Linders, Jannes & Swart, Siebe – *The Port of Rotterdam*, NAI Publisher, 2015.

Van der Houwen, Arnout – *The Port of Rotterdam and the Use of Inter-Modal Transportation*, Lambert Academic Publisher, 2010.

RHINE-RUHR

Mellor, REH – *The Rhine, a Study in the Geography of Water Transport*, Routledge, 2023.

Rauhaut, Christoph & Lehmann, Niels – *Fragments of Metropolis, Rhein & Ruhr, Expressionist Heritage*, Hirmer, 2017.

6.2 Megalopolitan Americas

North-Eastern Seaboard, Dallas–Fort Worth, Mexico City

6.2.1 MEGALOPOLITAN AMERICAS: NORTH-EASTERN SEABOARD, DALLAS–FORT WORTH, MEXICO CITY

Megalopolitan Region

On the east coast of the USA, along the **North-Eastern Seaboard**, is a territory that has evolved around developed corridors and has become 'megalopolitan', transcending state boundaries. This constitutes four substantially built-up areas containing principal cities, suburbs and their connecting infrastructure, which comprise essentially:

- Boston and areas along the Massachusetts Turnpike,
- New York City and its boroughs and Newark, Jersey City, Trenton and areas on the New Jersey Turnpike,
- Philadelphia and Camden/Reading, Wilmington, and
- Baltimore, Maryland; Arlington, Virginia; and Washington DC and the surrounding Beltway.

Each of these areas is defined and determined by its infrastructure, of national interstate and circumferential highways that have enabled interconnectivity between different poles of development and zones of attraction. Without a singular defined centre, there are a number of metropolitan trading places that have coalesced, with cities and emerging built-up zones that in their entirety have become 'megalopolitan', in the original language of Jean Gottman. This North-Eastern Seaboard agglomeration includes:

- Boston, Massachusetts, which expands radially outwards from the CBD along the Massachusetts Turnpike (I-90) where it intersects with the Route 128 inner beltway and then with the outer beltway of I-495 **(Greater Boston)**.
- New York City and Manhattan and its adjoining boroughs as the centre of a densely populated area along with the adjacent industrialised areas of New Jersey. Coalescing with this is the Newark, New Jersey, area containing a number of built-up areas from Jersey City, Newark and Hoboken on the Hudson River, and westwards along the I-78 to the intersection with I-287 and south-west beyond along Route 1 and the New Jersey Turnpike towards Trenton **(Greater New York)**.
- Philadelphia–Camden straddles the Delaware Valley and the industrialised areas and counties on either side of the river and around the I-95 in both Pennsylvania and New Jersey **(Greater Philadelphia)**.
- Baltimore–Washington in Maryland and Virginia stretches from Columbia, down I-95 to the I-495 Washington Beltway, and westwards from Rosslyn and Bethesda, past Tyson's Corner to Reston and Dulles. This includes established built-up areas, new towns and districts and major infrastructure such as Dulles International Airport and the Dulles Tollway that act as generators of activity, trading places and zones of attraction around the District of Columbia **(Greater Washington)**.

This megalopolitan region accordingly comprises essentially four major cities and surrounding urbanised areas in adjoining states. With high levels of population and core areas of high population density, the amalgamation of built form and transport arteries has created an extended urbanised agglomeration over many square kilometres. With its coastal ambiance and Atlantic-facing sea routes, maritime trade traditionally was inter-continental and with cabotage up and down the coast. Now with the decline of manufacturing and conversion to a more service-based economy, the benefits of inter-connectivity and proximity have re-invigorated urbanised areas as a new poly-nodal structure of work and home communities has emerged.

The Boston CSA contains an overall population of approximately 8.5 million and is overseen by the Metropolitan Area Planning Council (MAPCo). The New York Metropolitan Area, or Tri-State CSA, with New Jersey and Connecticut contains approximately 24 million people. The Philadelphia CSA contains around 7.5 million people overseen by the Delaware Valley Regional Planning Commission (DVRPC). The Washington-Baltimore CSA contains approximately 10 million people overseen by the Metropolitan Washington Council of Government (MWCoG) and the Baltimore Metropolitan Council (BMC).

DOI: 10.4324/9781003353348-34

Regional Transactional Forces and Typology

Accordingly, as an amalgamation of cities and urbanised counties, this North-Eastern Seaboard area contains approximately 50 million people who reside and work in the region. With its networks of railroads and highways, and inter-connectivity between homes and workplaces, it has evolved into a poly-nuclear region having major cities and peri-urban areas. With declining industries, post-war each conurbation has had to rebuild, transform and reinvigorate its businesses, in addition to renewing older housing areas in the inner cities.

With their Atlantic-facing seaports, New York, Baltimore and Boston have been able to reconfigure themselves so as to maintain their trading activities around modernised ports and airports. There are major airports at Boston Logan (BOS), New York (JFK and LGA), Newark (EWR), Philadelphia (PHL), Baltimore (BWI) and Washington DC Dulles (IAD).

With the decline of inner-city areas, however, migration of residents to the suburbs has continued, with economic activity focusing rather around peripheral beltways and 'edge cities' where increased interactions occur. Around Washington, for example, to this has been added the federal capital District of Columbia and all of its accompanying administrative offices.

As a megalopolitan region, this area of the American North-Eastern Coast differs from its counterpart on the West Coast. With an older housing stock, land development has been generally radial outwards from the older urban cores and beyond brownfield sites. In California, in contrast, where new land platting and development has been much more recent, this has generally been undertaken on greenfield sites on extended grid layouts. There, office parks and commercial retail malls are the most prevalent newer types of development with virgin infrastructure. As old inner-cities strive to reinvent and regenerate themselves, new developments are taking place around outer growth poles by leap-frogging to newly urbanising areas, which is leading to a contrasting and evolving poly-nodal structure.

6.2.2 DALLAS–FORT WORTH–ARLINGTON, TEXAS, USA

Megalopolitan Region

In the centre of the state of Texas, a new megalopolitan region has been developing around the twin cities of Dallas and Fort Worth with Arlington in between. Known as the Metroplex, this has emerged post–World War II, driven by its decision to jointly develop the Dallas Fort Worth International Airport.

Dallas, Fort Worth and Arlington

Dallas, which is located inland in the centre of Texas, was originally populated by the Caddo indigenous peoples. It was colonised by the Spanish south of the Red River and formed the northern boundary of the New Spain Viceroyalty, now the border between Texas and Oklahoma. After Mexico declared independence from Spain, Texas sought its own independence from Mexico, becoming the Texas Republic in 1836, and was subsequently annexed by the United States and admitted to the Union in 1846, with Dallas being incorporated as a city in 1856.

The city was originally founded on a wagon crossing of the Trinity River and White Rock Crossing. The surrounding Texan region is a major producer of cotton and with its prairies is well suited for cattle ranching, around which activities and the Fort Worth stock-yards the city developed. With the subsequent construction of railroads and the Houston and Texas Central and Texas and Pacific railroads, it grew as a significant trading centre.

As the city industrialised, it grew as a manufacturing hub and during World War II, it manufactured Jeep vehicles and aeroplanes as part of the war effort. After the war, this was extended to focus on communications equipment with new manufacturers being established, such as Texas Instruments. With the river running NW-SE to the south of the city, Dallas has developed in neighbourhoods with the central downtown and historic districts, South Side, South and East Dallas, with University Park and White Rock Lake to the north. Real estate developed and with neighbouring

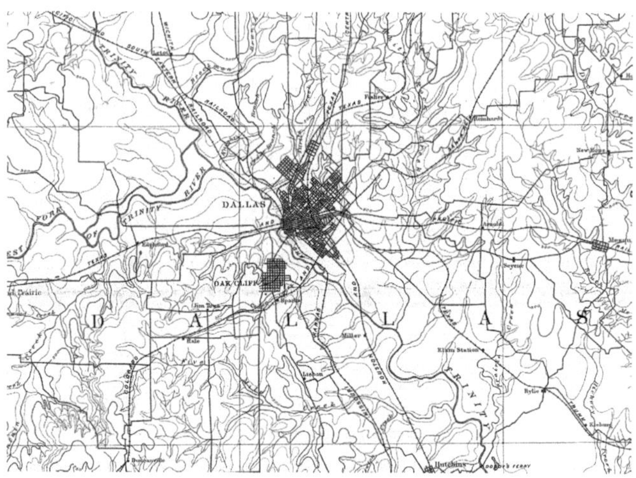

FIGURE 6.2.1 Dallas – early configuration, 1909.
Source: UC Texas.

Dallas County and Fort Worth, both cities developed as a highly attractive growth pole in the region with a growing workforce, mainly Hispanic, and the city of Arlington wedged in between.

Central Downtown is compact, with a group of commercial buildings around Bank of America Plaza, Fountains Place, Thanksgiving and JPMorgan Chase towers and the new city hall and Reunion Tower. With its Uptown area and park built partially over the freeway, North Dallas has commercial malls such as the Galleria and residential areas farther east with Fair Park, the site of the 1936 Texas Centennial Exposition, and the Bishop Arts district to the south-west. The area around the river has been enhanced with the Trinity River and Levee Project.

The city is well served with its inner loop and the hub-and-spoke network of major highways, the Central Expressway having the High Five Interchange between outer I-635 and I-75 on the northern perimeter and connections with Arlington and Fort Worth to the west. The city has its own airport at Dallas Love Field. With the new DART transit system, this at-grade light rail system serves the central area and interconnects out to the Carrollton and Richardson suburbs in the north, and the Trinity Railway Express also runs to the Dallas Fort Worth Airport (DFW) that has become a pivotal hub in the region.

Fort Worth is situated in the Texas Blackland Prairies region, with fertile black soil, with much of the rural land being previously ranch land. It was considered as the place in America 'where the West began'. At the time of the Mexican–American War, a series of forts was designed along the western frontier by Major General Worth between Eagle Pass and the confluence of West Fork and Clear Fork of the Trinity River, with one of the camps on the riverbank being named Fort Worth in 1849.

Great herds of long-horn cattle were traditionally driven north along the Chisholm trail with a stop at Fort Worth to railheads in Kansas before the arrival of the railway at Fort Worth. Stock-yards were subsequently established in the city and the meat-packing business began, led by the Armour and Swift companies.

Besides the original farming and ranching industry, many new businesses have since gravitated and become established in the region, beyond manufacturing to defence since World War II. An inshore natural gas field was then discovered under much of the land, with large gas wells and shale fields. With this, growth in the oil industry began and the establishment of corporate headquarters, such as Exxon-Mobil in Dallas, along with Fluor and other major industrial petro-chemical companies. The Downtown area encompasses Houston and Commerce streets, leading up to Paddock Park, adjoining the West Fork Trinity River. The Cultural District comprises a group of art museums just to the west of the river, an area which adjoins Trinity Park and the botanic gardens.

At the turn of the 20th century, historically the city and the north were covered by a network of electric street cars and an inter-urban railway that connected with Dallas and the intervening suburbs. Now superseded by a dense network of highways, Fort Worth is right at the centre of the intersection of the I-35 and I-30 freeways. There was an original airport at Meacham Park and a master plan for the development of Fort Worth by Victor Gruen, but this was never implemented.

It was the opening of the new major Dallas Fort Worth International Airport (DFW) in 1974 which was transformational, so large as to have its own zip code, midway between Fort Worth and Dallas in Tarrant and Dallas counties. It has significantly reshaped the infrastructure arrangements with Fort Worth and Dallas, now combined within a major dumb-bell layout of highways centred on the airport, as the movement centre around which all vehicles gravitate. With its four major runways exceeding 4,000 m. in length and six terminals connected by Sky Link, the airport now transports approximately 75 passengers per annum.

The interstate highway I-35 splits east and west, and each section runs north-south, dividing the Metroplex in two with Dallas to the east and Fort Worth to the west. The I-20 runs east-west with central **Arlington** in the Mid-Cities area. These cities are surrounded by a network of highways with a dual hub-and-spoke system, with an inner loop (I-30, I-35 and I-40), and outer I-635 and peripheral tolled highway. The North Texas Tollway Authority (NTTA) manages the toll roads in the region with loops around both Dallas and Fort Worth.

Megalopolitan Region

Accordingly, the Dallas–Fort Worth–Arlington conurbation has developed as a metropolitan area in the state of Texas, comprising 11 counties now known as the Metroplex. The population of the city of Dallas itself is approximately 1.3 million and of Fort Worth approximately 1 million, with Arlington having 0.4 million. Combined, the metropolitan area accommodates approximately 7.6 million people.

The CSA of Dallas–Fort Worth (TX-OK) is approaching 10 million people, which comprises 20 counties in North Central Texas and one county in South Central Oklahoma, and is continuing to grow.

The City of **Arlington** is located midway between Fort Worth and Dallas as a trading centre on Johnson Creek. General Motors established a major automobile assembly plant in 1954, around which significant development followed with the Union Pacific Railway, formerly the Texas and Pacific Railroad, running through the centre of the city.

Regional Transactional Forces and Typology

From its prairies, trails and the stock-yards of Fort Worth, cattle ranching was the original driving force, along with the cultivation of cotton. But it was exploitation of the industrial oil and gas sector that was to drive subsequent development in the region.

Following its early railroad lines, the highway network was developed next, followed by the modern aviation sector. From the earlier independent airports at Fort Worth and Dallas respectively at Meacham Field and Love Field, it was the decision with FAA assistance to develop a new major joint airport that led to the development of the combined Dallas Fort Worth International Airport (DFW) that has become such a major activity generator and airline hub in the region.

As a modern Texan prairie city, Dallas as a twin city with Fort Worth is typified by the major international airport midway between each of them. Along with the city of Arlington, the region accordingly has become a significant megalopolitan area – the Metroplex.

With its network of interconnecting highways and in a dumb-bell pattern, with the airport at the geographical centre, it is a significant economic force in the area. As a modern American automobile-orientated region that has been built upon the strength of the oil and gas sector in southern Texas and the Gulf of Mexico, it has become a major airport hub around which further development is gravitating.

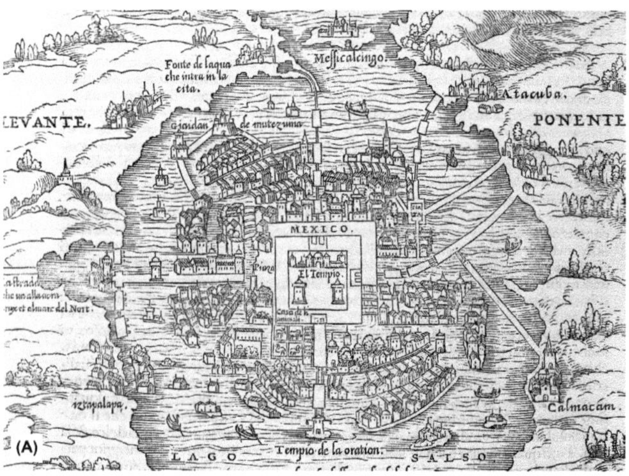

FIGURE 6.2.2A Tenochtitlan – early configuration, 1556 (Janvier).

Source: British Library.

6.2.3 MEXICO CITY, MEXICO

Background and Origins

Mexico City is located in the high plateau, at an elevation of approximately 2,240 m., in the centre of Mexico on the site of the ancient Aztec city Tenochtitlan, on an island in former Lake Texcoco in the Basin of Mexico Valley surrounded by volcanoes. The majority of this lake subsequently dried out as the surrounding area was built up and as the ground has continued to shrink and sink.

The original indigenous inhabitants in this region of Meso-America were Olmecs, who built ceremonial centres along with earthen pyramids. Teotihuacan emerged, a city-state just north-east of the present Mexico City, and flourished from 0 to 400 CE around its citadel (Ciudadela) with temples overlooked by the Cerro Gordo Mountain. However, as Teotihuacan declined it was superseded by the Toltec culture who had settled around Tula in 950 CE under their Aztec god Quetzalcoatl and with Totonecs, Zapotecs and Miztecs at Oaxaca and Monte Alban surrounded by the Mayan culture beyond in the Yucatan in Uxmaal and cities such as Tikal and Chichen Itza. The 'Mexica' people imagined the world as a kind of flat, layered disk between the underworld and heaven, with the points of the compass having symbolic meaning and the temple as the cosmic centre.

The succeeding culture of the Nahuatl-speaking Aztecs became the dominant power in the Central Valley of Mexico, and they were inspired to settle at a site shared with the Tlatelolco where they found an 'eagle perched on a cactus devouring a rattlesnake' and founded **Tenochtitlan** on the island lake of Texcoco around 1325. This city was carefully planned in quadrants with an orthogonal layout, sub-divided and connected by causeways, and was developed further under Moctezuma. The city-states of Tenochtitlan, Tlacopan and Texcoco were to form a triple alliance under new ruler Itzcoatl.

In general, their cities had street plans aligned with the equinox aimed so as to channel divine powers of the universe from the stars. Here they built dams and causeways across the lake and aqueducts for fresh water with Templo Mayor, a steeply inclined, tiered, stepped pyramid with twin temples where human sacrifices were made to the gods, in the centre.

This place was to prevail as an active trading centre via Tlaxcala in the Sierra Madre until the arrival of the Spanish conquistadors and hidalgos under Hernán Cortés. All was to change when, after some treachery and misunderstandings with Moctezuma, the Spaniards overcame the Aztecs in the Siege of Tenochtitlan at the Battle of the Toltec Canal, La Noche Triste, the 'sad night', in 1520. This conquest was to be the beginning of future establishment of Nueva España and of their centre and dominant city, renamed 'Mexico'.

As the surrounding Lake Texcoco shrank and receded, leaving lakes Xochimilco and Chalco, the city began to grow and expand. With new laws also with respect to the indigenous population, the city was redeveloped in accordance with the Spanish standardised model of colonial designs under the Laws of the Indies around an area named Coyoacán, the 'place of the Coyotes', and San Ángel with its squares, just south of Templo Mayor where new plazas, churches and monasteries were built in Spanish colonial architectural style. The other area developed was around Plaza de la Constitucion, known as the 'Zócalo', where an enormous open public square was surrounded by the Palacio Nacional de Moctezuma II, with the Catedral Metropolitana and adjoining Sagravio Metropolitano church which remain, although they are gradually sinking into the former lake clay bed.

With the loss of the lake, fresh water supply became more difficult; wells were bored, but in the longer run water had to be imported via long aqueducts from the hills. Around the central Zócalo, the surrounding city was developed on a grid pattern, with 'traza' according to a prescriptive design, and infrastructure was developed and drained to prevent flooding and to reduce the prevalence of mosquitoes. The Spanish built several churches and palaces and established the seat of government.

The land of Mexico came to be managed as a viceroyalty by the Spanish with the introduction of colonial estates, 'latifundias' and 'haciendas'. However, the local Indian and Creole population resented this, and after a revolt led by Hidalgo, in 1820 the army seized power and declared independence. The city was chosen as the capital of Mexico and an area of 'two leagues square' centred on the square of Zócalo with its 8,800 acres, was defined as federal territory, but excluding Coyoacán.

Much of the remainder of the 19th century featured a period of mixed political disputes with interventions of both French and American troops and different 'caudillos'. With the Mexican–American War and the storming of Chapultepec Castle, the Americans took control of the city, as well as the lands of northern Mexico and Texas. There were periods of reform, in particular under President Porfirio Díaz, from 1880 when the city was extensively developed and transformed, with new infrastructure, social facilities, schools and hospitals built and which came to be known as the Porfirian period.

This was to change under the Mexican Revolution between 1910 and 1920 with the rural resistance of Pancho Villa and Emiliano Zapata, despite the constitutionalist faction ultimately prevailing. With the formation of the Partido Nacional Revolucionario (PNR), this became the dominant party in the country under various nomenclatures such as PRM and PRI, for the remainder of the 20th century.

Modern City – Buildings and Arteries

The modern Mexico City is divided into a number of individual districts. To the west of the centre along Avenue Miguel Hidalgo, Alameda Central Park was developed with its poplar trees and statues next to the Palacio de Bellas Artes Theatre. Running south-west to north-east beyond, the wide boulevard Paseo de la Reforma links the centre with Bosque de Chapultepec with its castle and monuments, and the domed Monumento a la Revolución originally intended as new Senate at the renamed Plaza de la República. With its numerous traffic circles, 'glorietas' and fine houses, it is a long and distinguished boulevard giving name to the district as 'Reforma'.

Beyond the old historic centre, there are a number of distinctive neighbourhoods – Condesa, Colonia Rosa, Santa Maria la Ribera and San Rafael – and individual residential districts such as Benito Juarez and Miguel Hidalgo. Many new private residential areas now have gated communities in more affluent areas, as for example around Coyoacán and San Ángel, formerly Tenanitla.

In the 1950s, the university campus of the Universidad Nacional Autonoma de Mexico (UNAM) was built in the south with its modernist architecture, and the Torre Latinoamericana, the first tall skyscraper in the region, was also built in 1956. With the Olympic Stadium nearby (designed by Frank L Wright), during 1968 the Summer Olympics were held in the city. The population continued to grow, and the city expanded considerably supported by the construction of a new underground metro system.

However, with increased in-migration the informal sector has also grown, traffic has developed, air pollution has increased and problems ensue with groundwater.

Water supply is a problem and has to be secured from the Lerma and Cutzamala watersheds many kilometres away and stored in reservoirs. With its proximity to the mountains, and increased traffic densities, air quality also remains a problem.

The commercial districts are Vallejo and Santa Fe, with outer industrial districts and Linda Vista and Ciudad Universitaria. The Avenue de los Insurgentes houses many commercial offices with the Mexican Stock Exchange, Bolsa de Valores, in the Paseo de Valores. New planned neighbourhoods were also developed, such as the Cuidad Satelite and Jardines dei Padregal.

Many previous peripheral rural areas have become subsumed into urban areas, with inward migration and outward encroachment that have put huge pressure on the supporting infrastructure and social facilities. With a major refuse fire at Iztapalapa, a major earthquake in 1985 and social unrest and armed rebellion in Chiapas, there were changes in government and continued struggles against drug cartels in 1997.

Across the city are the markets of Tepeto, the main fish market La Nueva Viga, La Merced Market, multiple mobile markets (tianguis) and ubiquitous street vendors. There are the large urban parks of Bosque de Chapultepec, Parque Mexico, España, Hundido and outer Desierto de los Leones National Park in the Sierra de la Cruces mountain range.

The metro system (operated by SDC) that opened in 1969 was developed in stages and has 12 metro lines, some rubber tyred using traction technology of the RATP Paris Métro, that extend out to the suburbs with very high ridership. The suburban railway line is the Tren Suburbano, but the main mode of public transport is buses with a number of large, mid-sized and small buses; trolley buses; the central dedicated BRT system, Metrobus, extended along the main avenues and 'Mexbus' reaching out to the suburbs.

In terms of highways, the Circuito Interior (inner ring), and the grand project of Anillo Periférico, with its tolled upper deck (segundo piso), and Circuito Exterior Mexiquenese comprises a series of outer toll roads. These were built in stages during 2006–12 in order to avoid the urbanised area, some forming part of Federal Highway 57D and Arco Norte Federal Highway 40D. The central Viaducto Miguel Alemain crosses the city directly east-west in order to carry urban traffic and has a concrete duct for river-borne water. Mexico City International Airport (Benito Juarez) (CDM) is located to the east of the city.

Greater Mexico City

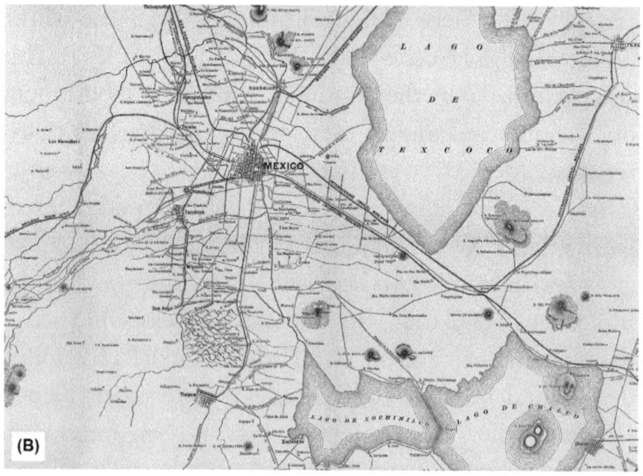

FIGURE 6.2.2B Mexico City – early configuration 1889.
Source: Maphouse.

In 1928 the Federal District of Mexico was created with 16 boroughs, one of which includes Xochimilco which is centred on the southern shore of Lake Xochimilco. As the city grew around farther urban centres, new areas were annexed and the federal district prospered. From the 1950s onwards, the conurbation of Mexico City grew into a much-extended Greater Mexico area. This evolved in 1993 when the constitution was further amended to combine the federal district and Mexico City into one entity.

The population of the core city of Mexico has grown extensively between 2000 and 2020, and the envelope of the city has evolved significantly from its original districts. From the 16 boroughs of Mexico, the city of Tizayuca in Hidalgo State has been added, and also areas from formerly neighbouring states of Mexico and Hidalgo in 1980. In 2009, Greater Mexico City was defined to include the 16 boroughs, having the two largest Itzapalapa and Gustavo Madero, and 59 municipalities.

Then, in 2016, the federal district became effectively a new state known as the Ciudad de Mexico, or CDMX, effectively an additional federal state, having its own government and legislative assembly with the city divided into 16 'alcadias', formally 'delegaciones'. The modern metropolitan area of Greater Mexico City has now grown enormously with very populous areas such as Ciudad Nezahualcóyotl (Neza), Ecatepec de Morelos, Naucalpan de Juárez and Chimalhuacán. These also comprise huge areas of 'favelas' that have been extended on any available land and up hillsides. With the provision of new cable-car connections, these informal settlements are becoming quasi-formalised as the informal economy is being driven by creative small industries.

The federal district has a population of approximately 9 million and the metropolitan area has a population of approximately 21 million – the population density in the whole Mexico City region is approximately 13,200 pers./sq. km. The wider megalopolitan area can be considered as extending farther out to comprise Puebla, Toluca, Cuernavaca, Tlaxcala and Tula, a much larger area of approximately 19,500 sq. km., having a total population of over 32 million. This is continuing to grow as net migration to the city is continuing from Latin and Central American states as well as from Europe and Asia.

All of this growth is extending over the valley of Mexico or Anahuac, formerly a system of inter-connected lakes. Surrounded by mountains on all sides, this heavily urbanised area is located in a valley that now traps exhaust emissions and industrial pollution. With continuous monitoring, a policy of 'Hoy no circula' is being implemented, or 'one day without a car', for each household. When the Metropolitan Index of Air Quality (IMECA) exceeds certain thresholds, this is extended to two days per week, policed according to the numeration of individual car registration plates.

Regional Transactional Forces and Typology

In the modern period, the Mexican economy has grown with new industries and manufacturing developing, after earlier emphasis on import substitution, before businesses have evolved in this century to concentrate on service industries. As a result, the metropolitan area has grown with interspersed nodes and poly-nodal centres as the enlarging urbanised area has extended outwards.

With increased congestion, this has led to the trend for outward development as the suburbs are growing with the continued outer expansion of Mexico City. Whilst dense inner municipal areas remain, the outlying peripheral marginal areas are extending, with substantial dichotomies of social separation between rich and poor areas, for example, between Tecamachalco and El Molinito, and Chemapa and Bosque.

The typology of the city region is most unusual, from its establishment on the basis of three separate elevated city areas of Tenochtitlan, Tlacopan and Tlateloco in Lake Texcoco. As the lake was drained and water subsided, these areas were developed on grid plans, with just sections of canals remaining as evidence of the former lake, as seen at southern Xochimilco and Chalco. With former raised 'chinampas' where local agricultural production flourished, early marketplaces had developed around which the Aztec population gravitated.

Today, these different centres have lost their individual distinguishing identities with amorphous development across the flat former basin area. With dense inter-connecting networks, limited individual areas of character remain, such as the historic centre around Zócalo, Reforma and Chapultepec, Coyoacán and Xochimilco's that have distinctive local characteristics.

All of these are knitted together with a dense network of streets and highways now overlain with a comprehensive metro network and circumscribed by the Outer Periférico Highway and inner ring. As the built extensions continue to be realised outward, these lower-density areas are interspersed with denser poly-nodal centres across and around the areas of Greater Mexico City. The following figures illustrate the modern configuration of the megalopolitan regions as described and the evolution of their modern population.

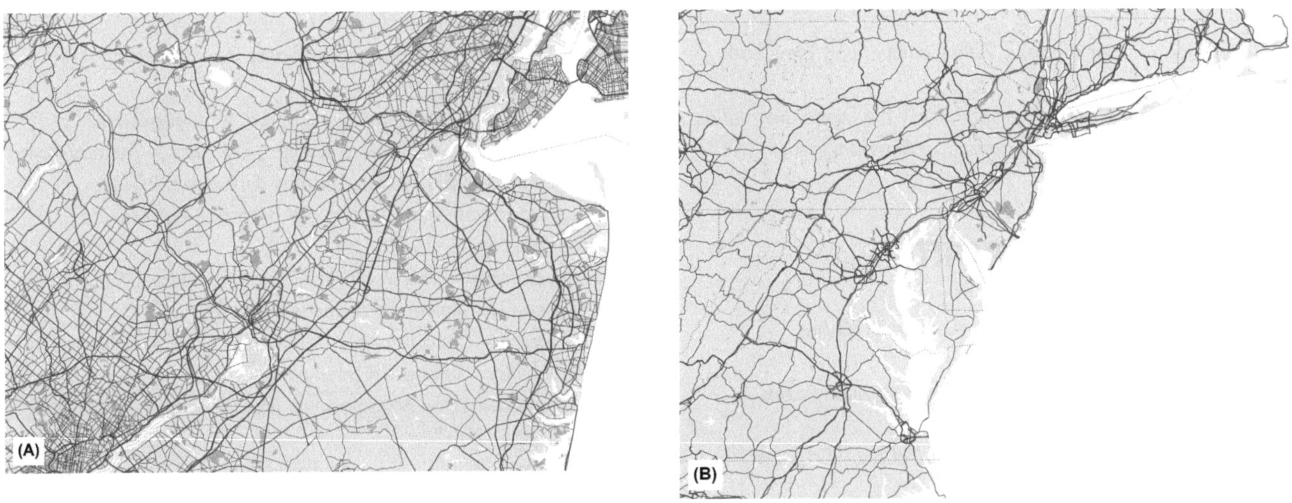

(A) (B)

FIGURE 6.2.3A AND B US North-Eastern Seaboard – modern configuration.

FIGURE 6.2.4A AND B Mexico City and Greater Mexico – modern configuration.

FIGURE 6.2.4A AND B (*Continued*)

FIGURE 6.2.5 Dallas–Fort Worth – modern configuration.

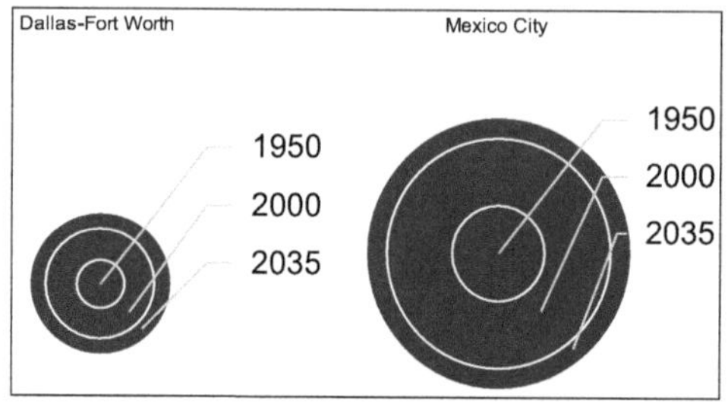

FIGURE 6.2.6 Megalopolitan American populations.

Data source: Global Database of Metropolises 2020 – UN-Habitat (2020), UN DESA Population Division.

Further Reading

NORTH-EASTERN SEABOARD, USA

Gottman, Jean – *Megalopolis*, 20th Century Fund New York, 1961.

Kotkin, Joel – *The New Geography, How the Digital Revolution is Reshaping the American Landscape*, Random House, 2000.

DALLAS – FORT WORTH

Bleakley, Bruce – *Dallas Aviation*, Arcadia Publisher, 2011.

Fitzgerald, Ken – *Dallas, Then and Now*, Pavilion Books, 2013.

Roark, Carol – *Fort Worth and Tarrant County*, TCU Press, 2003.

MEXICO CITY

Adams, Richard – *Prehistoric Mesoamerica*, University of Oklahoma Press, 1991.

Bross, Benjamin – *Mexico City's Zócalo*, Routledge, 2021.

Coe, Michael, Urcid, Javier & Koontz, Rex, eds. – *Mexico from the Olmecs to the Aztecs*, Thames & Hudson, 2019.

Kandell, Jonathan – *La Capital, the Biography of Mexico City*, Random House, 1988.

Mundy, Barbara – *The Death of Aztec Tenochtitlan, the Life of Mexico City*, University of Texas Press, 2018.

Rinke, Stefan – *Conquistadors and Aztecs, a History of the Fall of Tenochtitlan*, Oxford University Press, 2023.

Sa, Lucia – *Life in the Megalopolis, Mexico City and São Paulo*, Routledge, 2007.

Ward, Peter – *Mexico City, the Production and Reproduction of an Urban Environment*, John Wiley, 1990.

6.3 Megalopolitan Japan

Tokyo–Yokohama and Osaka–Kobe

6.3.1 TOKYO-YOKOHAMA REGION, JAPAN - KEIHIN

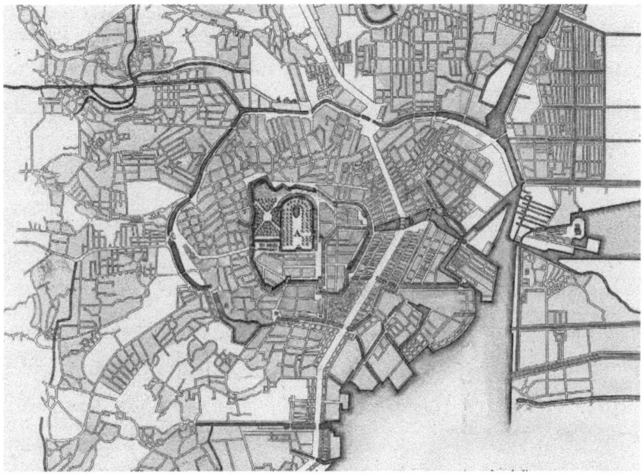

FIGURE 6.3.1 Edo-Tokyo – early configuration, 1752 (Bellin).
Source: Maphouse.

Background and Origins

Tokyo, the present capital of Japan, is in the Kanto region on Honshu Island behind the Buso Peninsula where the Sumida River enters Tokyo Bay. It was formerly populated by Jomon people and was previously the administrative capital under the Tokugawa shogunate, then known as **Edo**.

In earliest times, the bay was a coastal area of tidal marshes ('Umite') where shellfish were abundant and harvested and where several rivers, the Sumida, Tama, Tone and Edo, deposited mud into the bay. Following generations developed agriculture around Yayoi Hill, with the beginnings of rice farming and using irrigation from the rivers, creating market gardens and orchards.

In earlier times, around 710 CE, the capital was located during the Yamato period at **Heijo-kyo** (Nara). Here the city was planned on a gridiron basis of 8 by 9 'bo' based upon the Chinese model of Chang'an. It developed as a religious centre, and one 'cho' (2.45 acres) of land was granted to noblemen according to their rank.

The capital was subsequently moved to **Heian-kyo** (now Kyoto) in the 9th century and was developed under a similar grid plan. It remained there for 1,000 years until the Meiji restoration, when having gradually lost power, the capital was transferred to Edo.

Edo itself was originally a small, fortified town in the 12th century until the castle was built in 1456, with strengthened lines of defence and moats, located at the mouth of the Hirakawa River. With its irregular spiral of canals that surrounded the citadel, this labyrinthian structure acted to strengthen its fortification for the shoguns. The castle and the moat were expanded in the area of Marunouchi ('within the moat') and Yotsuya ('four valleys') beyond. In 1603, the Nishanbashi Bridge was built at the centre of the city, the point from which five important road corridors emanate. Various riverbed routes also were re-aligned and sections channelled into the main Okawa River.

During the 17th century, under the Tokugawa Ieyasu, Edo grew in size and stature with the residence of the 'daimyo' and with the castle acting as its nucleus. The Tokugawa rulers introduced the practice of 'sankin kotai', requiring feudal samurai to commit to alternate attendance in Edo every other year. This led to shoguns having to build their own individual mansions in the city just for their own periodic occupation, which led to considerable development of the city and being populated with merchants and craftsmen. During these times the city was crossed by canals and water courses, some of which were subsequently re-routed and eventually filled up, resulting in narrow streets and wooden buildings across the Low City ('Shitamachi').

With much of the city built of wood, there was extensive destruction by fire in the Meireki earthquake of 1657, after which the city was rebuilt with fire watch towers and grew subsequently with increasing coastal trade. With Edo Castle and its canals and embankments commanding the bay, this resulted in a stable period in Japan for a long period of 250 years, becoming the world's largest city in 1720.

But in 1853, the American commodore Matthew Perry and his Black Ships entered the bay, demanding the opening up of the cities of Shimoda and Hakodate

DOI: 10.4324/9781003353348-35

to foreign trade, which was difficult to resist. Consequently, in 1868, the shogunate was terminated and imperial power was restored under Emperor Meiji. With the imperial court transferred back from Kyoto to Edo, the city was renamed as **To-kyoto** (the eastern capital), shortened to Tokyo, and Kyoto was renamed Sai-kyo (western capital).

In 1872, a fire destroyed the Ginza area of Tokyo, where the Mint was located, and was rebuilt as Bricktown. The generally higher ground of the Yamanote area to the west of the castle was appropriated by the wealthiest daimyo, with artisans and others condemned to the lower ground to the east, Shitamachi, which has shaped its subsequent development. In 1888, the Urban Improvement Ordinance was passed that aimed to enhance basic city infrastructure.

Then, in 1923, much of the city was destroyed by the Great Kanto earthquake, following which a master plan for redevelopment was prepared by Mayor Shinpei with a focus on modernisation and industrialisation. During World War II, which was catastrophic for Japan, major destruction of the city occurred in 1945 as a result of significant Allied bombing and fires. With the surrender of Japan at the end of the war and after a period of subsequent American occupation under General MacArthur, a more stable period of expansion developed after rebuilding with reversion to prosperity.

A large part of central Tokyo remains occupied by the Imperial Palace surrounded by its extensive moats. With its ramparts from the old castle, the citadel and watch towers such as Fujimi-yagura, it is enhanced by fine surrounding gardens.

Just north outside, the Yasukuni-jinga monumental torii shrine war memorial rests controversially. In Kanda there are two famous shrines, Yushima Seido (Sacred Hall) to Confucius, and Kanda Myojin to Daikoku and Ebisu (farming and fishing) that previously stood in front of Edo Castle. Odaiba is an outer area where gun batteries were originally located. The city of Tokyo comprises the municipalities of the Tama districts and Nishi Tama to the west, the central wards of Chiyoda, Chuo and Minato, Shinjuku and Shibuya, with other outlying wards.

In the late 20th century, land around the Marunouchi was redeveloped with construction of the Mitsubishi headquarters, known as the 'London Block'. The city consolidated as industrialisation commenced, with new railways, electric trams, telegraph lines and Hibiya Park. The Ryounkaku ('clouds surpassing') octagonal tower was built in 1890 as the first tall tower in Tokyo, but it was damaged by an earthquake.

Modern City – Tokyo – Buildings and Arteries

The modern city of Tokyo has grown enormously and constitutes several distinctive districts. To the east is the main railway station and the commercial centre of the city, lined with dense office buildings, and the Marunouchi district and Yurakucho district with the major Tokyo International Forum (designed by Vinoly) and site of the former Imperial Hotel (designed by Frank L Wright) in the Hibya district.

East of the railway and expressway is the Ginza shopping area in former Bricktown, with Shiodama centre to the south, as most of the former timber buildings have now disappeared. To the north are the Nihombashi and Kanda neighbourhoods, as well as the former Shitamachi area and Nihombashi Bridge, the location from which all distances in Japan are measured, along with the beginning of the Tokaido road running between Tokyo and Kyoto, and the Chuo and Yamanote railway lines.

South-west of the Imperial Palace are the districts of Akasaka and Ropponji with Akasaka Tower, the midtown tower and farther east the Tokyo Tower that emulates the Eiffel Tower in Paris. The garden 21_21 Design Site (designed by Andao Tadao and Issey Miyake) also emphasises the midtown area as a design hub. In northern Tokyo, there are more traditional districts such as Asakusa with the Senso-ji Buddhist Temple, Hozomon gate, Shinto shrines and the kabuki theatre district. Other areas include Ueno Park with its museums, Yanaka with its temples and the Rikugi-en gardens named after the six principles of traditional Japanese poetry.

In western Tokyo, Shinjuku is a modern area of the city, with the main railway terminal stations of Japan Railways, Keio, Odakyu and Seibu where the bullet train connects Tokyo and Osaka, and other main cities that link up the four main islands of Japan. In 1964, the new Shinkansen high-speed rail line was opened to connect directly on the old Tokaido Corridor from Tokyo to Nagoya and Osaka.

Next to the Tokyo Metropolitan Government (TMG) towers (designed by Tange) are the Shinjuku Gyoen Gardens and nearby is the Meiji-Jinju Shinto shrine to Emperor Meiji, the centre of annual festivals with its inner and outer gardens. The National Stadium was built for the 1964 Summer Olympics with Yoyogi Stadium (designed by Tange) by the former Washington Heights.

The most contemporary part of Tokyo is Shibuya, with its main station, road intersections and infamous 'scramble' street crossing; it also has stores such as Prada (designed by Hertzog de Meuron) and many others along the Ometesando Hill.

The area of Ikebukuro is central around the railway station, and the south-eastern area of Tokyo revolves around Tokyo Bay with its original fish and seafood wholesale markets. Now the substantial waterfront area of Odaibo has been developed on reclaimed land, linked by the double-decked Rainbow Bridge and the Yurikamome Monorail and with the brutalist Fuji TV Headquarters (designed by Kenzo Tange).

Post–World War II, the country has overseen rapid reconstruction and developed a highly industrialised modern economy. Besides heavy industries, national Japanese companies with governmental support through MITI have expanded into new technological and electronics sectors. With automobile manufacturers, electronic giants and information technologies have grown, supported by their own financial infrastructure.

With the unique Japanese national style and codes of business, the focus was on broadening and revitalising the financial sector through wider participation and leadership in this sector across the South-East Asia region into a global presence. Pre-war, businesses had been run by family-controlled networks, or 'zaibatsu', that were effectively cartels. Post-war, successor companies evolved under the oversight of MITI as 'keiretsu' alliances, or groupings from which the conglomerates of Mitsui, Mitsubishi, Sumitomo and Yasuda have grown into substantial corporations.

The new Tokyo Narita International Airport (NRT) was finally opened in 1978 after environmental protests. Concomitantly a huge amount of commercial building also occurred and on the 400th anniversary of the government's move to Edo symbolically in 2003 new skyscrapers were completed in the Roppongi Hills and Shiodomi City. The super-tall Tokyo Skytree was built in 2010, the world's tallest tower with surrounding Skytown on former Tobu Railway land, and the Summer Olympics were held again in Tokyo in 2021, delayed one year by Covid-19.

The infrastructure especially of urban rail and metro has been very well developed, along with the extended rail and high-speed railway lines across the country and with a dense network of expressways built through, across and around the capital.

The first railway lines developed were the suburban railways to which were added the circular Yamanote line and the Kaisei electric railway, each with 3′6″ gauge track. Later, railways were developed with standard-gauge track. The Tokyo metro has been developed with several underground lines and also the Toei subway, managed respectively by the Teito Rapid Transit Authority (TRTA) and the Tokyo Metropolitan Bureau of Transport (TMG).

The population of the city of Tokyo is approximately 14 million, with 37 million in the metropolitan area. The population density of the whole Tokyo city region is approximately 10,200 pers./sq. km. The Tokyo prefecture is now administered by the Tokyo Metropolitan Government (TMG) for the Greater Tokyo Area, which comprises 23 wards. The City of Tokyo had been merged with the prefecture in 1943 to form the Metropolitan Prefecture of Tokyo with an integrated TMG. During the 1960s, Kenzo Tange developed a master plan for major built development across Tokyo Bay, but this was never pursued.

Tokyo Bay is now crossed by the major tolled Tokyo Bay Aqua-Line bridge and tunnel combination that was completed in 1997 to link Kawasaki and central Tokyo with Kisarazu in Chiba to the east. With the construction of an artificial island, this combined bridge and tunnel combination shortens communications between the central city and the Boso peninsula.

With its location at the boundary of three different Asian and Pacific tectonic plates, Tokyo has a history of, and is at risk of, earthquakes. It is located also in a tropical typhoon zone and around different river systems which create another risk of flooding. Considerable development has been done in this respect in the construction of levees and dams, and with the Metropolitan Area Outer Underground Discharge Tunnel as a flood water diversion channel to the Edo River and Saitama.

The administrative boundary of the urban Tokyo metropolis has an anomaly in that it actually contains two external linear Pacific Island chains, the Izu Islands and the volcanic Ogasawara Islands including Iwo Jima, which stretches 1,000 km. south of the capital city into the Pacific, also administered by TMG. With its traditional tea houses, No and Kabuki theatre, sumo wrestling, kimono clothing, and woodblock art and design, Tokyo is a colourful city with deeply developed culture and traditions. Interestingly, since the average residential building lifespan is only about 25 years, this contributes to its evolution as a metabolising city.

Transactional Forces and Typology

With Tokyo's location in a bay of tidal marshes, earliest human activities were marine based between the outlying islands, with basic agriculture until irrigation enabled rice farming to be developed around the rivers. With its wooden buildings, it was not until it was fortified that Edo grew in status as a centre and prospered under the Tokugawa shogunate across the whole Kanto region. Despite periodic setbacks of earthquakes and fires, the Pax Tokugawa held until coming to an end in 1868 with the Meiji converting Edo to the imperial eastern capital of Tokyo.

As industrialisation gathered pace at the end of the 19th century, population growth and urbanisation increased intensively at the turn of the 20th century. With the setback of the 1923 Kanto earthquake, and the catastrophic destructive impact of World War II, and devastation from firebombing, the city had to totally rebuild itself. It did this by miraculous transformation and through the transistor-led electronic post-war industry. Beyond this, annexation and land reclamation has led to extensive growth in the geographical size of the city and adjoining the built-up cities of Yokohama and Kawasaki, creating an urbanised agglomeration of megalopolitan proportions.

From its origins as Edo, Tokyo became a compact, coastal, monocentric city developed around Tokyo Bay, with the river as the eastern limit. Post-war, after devastation and with reconstruction, the city expanded and developed as further insular areas were annexed. Its extensive outer expansion along with the neighbouring prefectures has resulted in the urbanised area being enlarged, with more land being reclaimed and developed into a poly-nodal structure. Whilst the city has extended around the bay, administratively this has reached a long way eastwards to comprise separate wards. As the junction of the cities of Yokohama, Kawasaki and Chiba around the bay, Tokyo is the centre of an expanding major megalopolitan agglomeration in the south-east of Honshu Island.

Modern City – Yokohama

One of the largest coastal cities adjoining Tokyo is **Yokohama** on the Ooka River, originally founded around a sandbar. It was one of the first treaty ports to be opened under the Meiji in 1860, which developed rapidly around Kannai and its canals into a port city. This became one of the earliest areas to be industrialised around the international Osanbashi pier. Subsequently, factories were built along the waterfront and with the development of power plants, new industries, shipyards and marine side developments followed. The first suburban railway had been built to link Tokyo from Shimbashi with Yokohama at the end of the 19th century.

After major destruction in the 1923 Great Kanto earthquake and as a result of intensive World War II firebombing, the city had to be totally rebuilt. Substantial industrialisation of heavy industries then developed here with the establishment of, for example, Kawasaki Heavy Industries.

Within the Kanagawa prefecture, Yokohama has been developed extensively now with the Minato Mirai 21 zone (Port of Future 21st Century) the massive commercial port complex on reclaimed land with dedicated connecting rail link. Its main built developments include Landmark Tower, Marine Tower, piers for cruise ships and a new stadium. There is also the older Chinatown district.

The municipal Yokohama subway was opened in 1972 and connected with the new Shinkansen Station and existing Central and East Japan Railway company lines. The cable-stayed Yokohama Bay Bridge across the bay was opened in 1989.

With the districts of Yamanouchi, Shinko and Kitinakadori, the port area has been completely redeveloped from former industrial to commercial activities. Separate areas of new land reclamation have been undertaken in the bay, starting from Haneda Airport (HND) in Tokyo, past Kawasaki, linked with the new Metropolitan Bayshore Expressway. These areas act as handling areas for all of the industrial activities and are continuing to be extended outwards farther into the bay.

The population of Yokohama is approximately 3 million and it is the capital of the Kanagawa prefecture.

Megalopolitan Region – Greater Tokyo – Keihin

Greater Tokyo is now considered as the area that comprises Tokyo, Kawasaki and Yokohama, known as the **Keihin** industrial zone. As a virtually continuous built-up area, this contains not only a significant resident population, but also intensive industrial and commercial workplaces. With such a high-level density of development, this is wholly dependent on the efficient provision of communication infrastructure and supporting public services. In this respect the metropolitan government has managed to keep pace, in particular in developing its rail and metro infrastructure, which continues to offer a high level of service.

As a result of intensive urban extension centred around the former port areas, the forces of development are shaping poly-nodal centres around the coastal bay, with linear extensions along the coastline but also infill and creation of new land areas in existing bays. These, connected by supporting road and rail infrastructure, have led to the formation of new land areas, mainly exploited for commercial uses with exhibition halls and increasingly recreational also with parks and theme parks. As an outlet, these serve as some relief for the congested and more over-built central areas.

As a group of major coastal port areas in Tokyo Bay, due to their geographic characteristics, the form of development is creating a new genre of Japanese urban development, with the formation of reclaimed land areas offshore in the bay. This is substantially increasing the available land area for industrial, commercial and logistics purposes. Such development as is evident individually at Tokyo, Kawasaki and Yokohama, typifies the solutions in response to growth pressures of industrialisation in this coastal region.

Whilst fortunate to have the coastal zones available for development, the environmental implications are significant in terms of hydrology and sea water levels, with the needs of environmental protection being paramount in this climate-affected region. As rivers flow and lose their deposits, surge tides and tsunamis act in a contra direction, increasing potential events of conflict. This then tends to lead to much greater required project development, in terms of physical infrastructure and tidal defences.

As radial development inland is constrained by the natural geography of inclines and hillsides, the opposing coastal development is equally challenged by the more expensive ground reclamation and consolidation in silty bay areas, exposed to severe oceanic and marine forces.

6.3.2 OSAKA-KOBE REGION, JAPAN – KEIHANSHIN

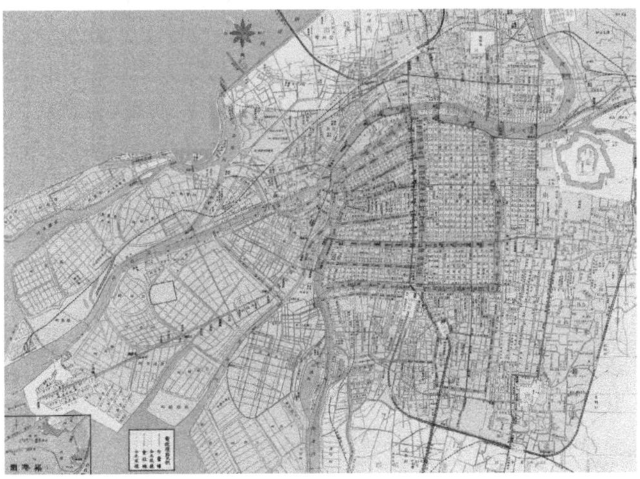

FIGURE 6.3.2 Osaka – early configuration, 1911 (Kinoshita).
Source: Maphouse.

Background and Origins

Osaka lies in the central rice-growing plain of Kansai in the bay on the confluence of the Ogawa and Neya-gawa rivers on Honshu Island. With sediments deposited from the rivers, the original settlement region of Yamato, formerly known as Naniwa, is adjacent to the old cities of Kyoto and Nara. During the Kofun period around the 5th century CE, it had developed as a coastal port, trade extended with the other islands of Japan and subsequently Naniwa Palace was built. In the 16th century, Osaka Castle was built and was fought over by Toyotomi Hideyori and Tokugawa Ieyasu in the siege of Osaka approximately 30 years later.

With its central castle and moats, the rebuilt Osaka-jo was a substantial castle. With its religious temples, Shitenno-ji and Ishiyama Hongan-ji, and trading activities, it developed as a mercantile city and became the 'merchant's capital' during the Tokugawa era. Whilst many changes developed under the Meiji period, it next evolved into an industrialising centre. Despite the setback of the Great Kanto earthquake in 1923, and following significant destruction in World War II, it was totally rebuilt to become a cosmopolitan city after rapid industrialisation occurred during the post-war period.

Modern City – Osaka – Buildings and Arteries

Modern Osaka has a densely developed centre around its rivers and coastal bay surrounded by the built-up areas of satellite cities. Central Osaka is divided into wards between Kita (north) with the commercial Umeda and Shin-Osaka areas, and Minami (south), having the Chuo-dori railway as the dividing line, and with the Osaka-jo castle area to the east and the business districts of Semba and Tennoji to the south.

Nakanoshima is a thin island on the Ogawa River where city hall was built in 1923, along with other administrative and institutional buildings that act as a focus of the city in the central area. Several skyscrapers are sited densely around the main train stations, with the Umeda Sky Building near Osaka-Umeda Station. Abeno Harukas is a major building (designed by César Pelli) at Osaka-Abenobashi Station and the retro Tsutenkaku Tower at Tennoji nearby.

In the Bay Area, main developments around Osaka Bay include the (WTC) Cosmo Tower, Osaka Amenity Park Tower and Universal Studios on reclaimed areas of land that are being developed, and Sakai City.

Osaka has an extensive suburban rail and urban metro network, an overground circular loop line railway and major modern highways and expressways. The city is connected by Shinkansen to other main cities, and in addition to Osaka International Airport (ITM) at Itami is the new Kansai International Airport (KIX) (designed by Piano), built on reclaimed land out in Osaka Bay.

In 1970 Osaka Expo was held at Expocity with another Expo in 2025 on a dedicated site to the north of the city connected by monorail, now the basis of a new city at Suita. The government introduced in 1974 a policy of promoting the development of underground shopping malls, which led to the extensive number of underground complexes in central Osaka.

As a centre of commerce, the main Osaka Securities Exchange is based at Osaka and is now one of Japan's main media hubs. Several major industrial companies have their headquarters in Osaka, including Sharp, Panasonic and Sanyo.

The population of the city of Osaka is approximately 2.7 million, and it is capital of the Osaka prefecture. The CSA has a population of approximately 20 million in the so-called **Keihanshin** (Kyoto–Osaka–Kobe) metropolitan area which, as the region around Osaka Bay, is considered as Greater Osaka. The density of population of the whole Osaka City Region is approximately 8,100 pers./sq. km.

Transactional Forces and Typology

The broad, flat coastal plain served as an area of intensive agriculture and fishing in earlier times. Osaka's location on the coastal bay, with its long wide river, made it a most suitable port location. As urbanisation and industrialisation occurred, this central area around the port developed around marine and maritime activities. With its main grid extending linearly southwards, it also spread around the bay westwards to link with Kobe where its own harbour facilities were also being developed. Here, beyond agriculture of rice and beef for which it is best known, heavy industries and steel production were developed. With industrial sites at a premium, having a large open bay, land reclamation was feasible with the creation of new built zones for specialist activities.

The original settlement gravitated around the river and its floodplain into the estuary and the bay. Due to the topography, the built environment has extended to the south and also circumferentially around the bay westwards towards Kobe. With industrialisation, the harbour has been expanded significantly, with land reclamation taking place in several areas. This has served both to increase developable land area and to provide new locations for sites of industrial and urban development.

Taking advantage of this, the major infrastructure of Kansai International Airport (KIX) has been created on filled land reclamation, creating a significant new facility linked by bridge, with road and rail access out into the sea in Osaka Bay. With continued urbanisation and industrialisation, it is likely that this type of development will continue around the bay where Osaka joins with Kobe and where similar types of reclaimed land development are also occurring, such as the new Kobe International Airport (UKB). This, however, does create particular environmental challenges of coastal development, not least given its location in an area of geophysical instability and subject to periodic typhoons and tsunamis.

Modern City – Kobe

Kobe is an historical port city in its own right, as one of the original treaty ports in Japan, which in 1868 allowed foreign settlement at the Bund, lying to the west of Osaka across the bay. Beyond its original settlement, much land has been reclaimed to build the new Port Island and Rokko Island in the bay.

However, heavily destroyed in World War II, a major setback also occurred with the destructive Hanshin Earthquake in 1995. There was considerable devastation, subsequent to which a protective sea wall was constructed around the whole main harbour. Disaster reduction measures have also been introduced that are aimed at building resistance against earthquakes and tsunamis.

The city has been fully rebuilt, comprising several areas of the commercial centre, Nankon-machi, Chinatown, the Harbour and Island developments and Kitano to the north with its wooden houses. The Harbour Island with its distinctive, red twisted-cable Port Tower is newly developed, along with Rokko Island. Sannomiya is a developed commercial centre of the city around the central station and the new developed area of Meriken Park ('American').

Since its original formation as a treaty port, marine activities developed around the harbour focusing on trade between the islands and beyond, across South-East Asia and China. Ship repair facilities were developed in the port along with other marine support activities. The city has developed westwards from the centre but also an extensive programme of land reclamation has been undertaken out into the bay. This has provided additional areas for specialist activities in addition to the dedicated reclaimed area of the new Kobe Airport (UKB), opened in 2006.

Economically, Kobe is a vibrant part of the Kansai region with its container port, and major industries such as Kawasaki Heavy Industries, Mitsubishi, Kobe Steel and shipbuilding activities.

In terms of transport arteries, the principal highways that cross the city are the Meishin and Hanshin

Expressways (tolled). Railways include the JR Kobe line, the Hankyu line and the Hanshin line to Osaka and the Sango Electric railway, and there is a subway and funicular lines and cable cars also out of the city.

West of the city is Akashi Kaikyo Ohasi, the world's longest suspension bridge, opened in 1998, that crosses the straits to Awaji-shima Island and Shikoku Island that connects these islands and which survived the last major earthquake.

The population of Kobe itself is approximately 1.5 million and acts as the capital of the Hyogo prefecture. The city has nine wards.

Megalopolitan Region – Greater Osaka – Keihanshin

With increased levels of urbanisation and industrialisation, both of the centres of Osaka and Kobe have developed as key poles that have attracted significant development. Due to the geographical nature of Osaka Bay, the coastal crescent is now wholly developed from the western side of Kobe around to the Yodo River and Yodogawa around Osaka and south beyond Sakai to Izumisano. The urbanised area extends outwards towards Kyoto, encompassing these cities in the prefectures of Osaka, Kyoto and Hyogo.

South of Osaka by Rinku New Town, the new Kansai International Airport Island has been constructed and created as a dedicated airport zone, another major area of land reclamation in the bay. This constitutes the current model in the region, given scarce remaining developable land, of reclaiming new areas around the coastal bay from the sea. Whilst profitable new zones of real estate, the full environmental and hydrological impacts in a sensitive seismic and oceanic zone remain challenging.

Typologically, this constitutes the modern form of coastal bay development here, constructed around a marine crescent with poly-nodal centres, and also stretching inland, tightly connected by road and rail transport links that knit these areas closely together. Rail has been particularly beneficial with both its elevated high-speed and sub-surface urban connections, in acting as provider of effective public transport services across the megalopolitan region.

In a similar manner to the Tokyo–Yokohama Keihin region, the Osaka–Kobe Keihanshin region has developed as another megalopolitan region. With the benefit of large areas of coastal land reclamation, this has afforded additional land mass for built development. Despite its high cost and challenging environmental constraints, this method has been used to enable ongoing industrialisation and in particular to provide manufacturing and logistics hubs and modern maritime trading zones to be expanded in this growing region of South-East Asia. The following figures illustrate the modern configuration of the megalopolitan agglomerations of Japan as described and the evolution of their population.

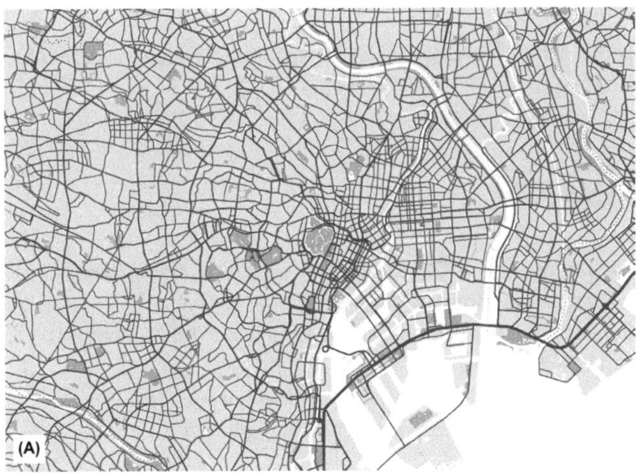

FIGURE 6.3.3A AND B Tokyo and Tokyo–Yokohama – modern configuration.

FIGURE 6.3.3A AND B (*Continued*)

FIGURE 6.3.4A AND B Osaka and Osaka–Kobe – modern configuration.

FIGURE 6.3.4A AND B (*Continued*)

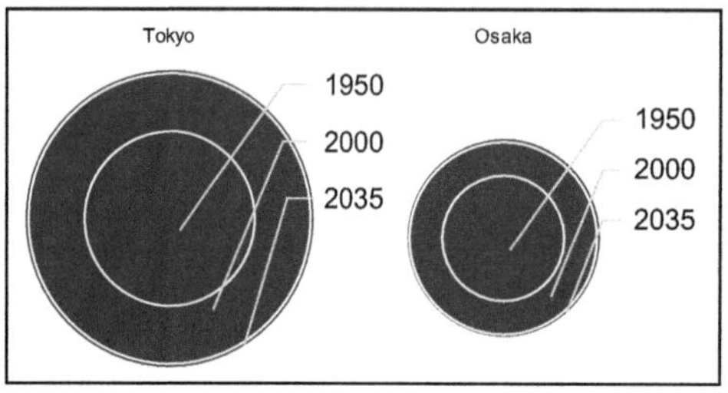

FIGURE 6.3.5 Megalopolitan Japan populations.
Data source: Global Database of Metropolises 2020 – UN-Habitat (2020), UN DESA Population Division.

Further Reading

TOKYO

Almazan J, McReynolds J & Saito N, eds. – *Emergent Tokyo, Designing the Spontaneous City*, ORO Editions, 2024.
Clements, Jonathan – *Short History of Tokyo*, Haus Publishing, 2018.
Cybriwsky, Roman – *Tokyo, the Shogun's City at the 21st Century*, John Wiley, 1998.
Harding, Christopher – *Japan Story*, Penguin Books, 2019.
Richie, Donald – *Tokyo, a View of the City*, Reaktion Books, 1999.
Seidensticker, Edward – *A History of Tokyo, 1867–1989, from Edo to Showa, the Emergence of the World's Greatest City*, Tuttle Publishing, 2024.
Shelton, Barrie – *Learning from the Japanese City, Looking East in Urban Design*, Routledge, 2012.
Sherman, Anna – *The Bells of Old Tokyo*, Picador, 2019.
Song, Jiewon – *Global Tokyo, Heritage, Urban Development and the Transformation of Authenticity*, Palgrave Macmillan, 2020.
Sorensen, André – *The Making of Urban Japan*, Routledge, 2002.
Ward, Philip – *Japanese Capitals*, Oleander, 1985.

YOKOHAMA

Makimura, Yashiro – *Yokohama and the Silk Trade*, Lexington Books, 2017.
Poole, Otis – *The Death of Old Yokohama in the Great Japanese Earthquake of 1923*, Routledge, 2010.

OSAKA

Cronin, Michael – *Osaka Modern, the City in the Japanese Imaginary*, Harvard University Press, 2017.
Sansom, George – *Japan, a Short Cultural History*, Stanford University Press, 1931.
Whelan, Christal – *Kansai Cool, a Journey into the Cultural Heartland of Japan*, Tuttle Publisher, 2018.

KOBE

Edgington, David – *Reconstructing Kobe, the Geography of Crisis and Opportunity*, UBC Press, 2011.

6.4 Megalopolitan India

Greater Delhi and Greater Kolkata

6.4.1 GREATER DELHI, INDIA

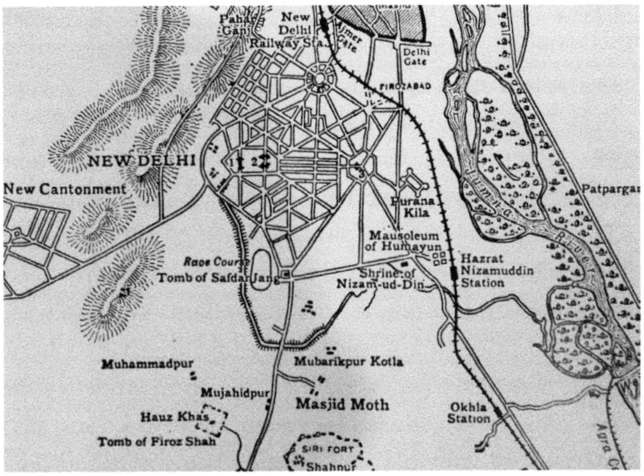

FIGURE 6.4.1 Delhi – early configuration, 1926 (Trott).

Source: Maphouse.

Background and Origins – Old Delhi

Delhi, as the capital of India in the north of the country, consists of both Old Delhi and New Delhi. Now within the National Capital Territory that adjoins the states of Haryana and Punjab, 'Old Delhi' is characterised by sites of former older city places and Mughal monuments, whereas 'New Delhi' was planned and built by the British Raj as an imperial capital in 1930 on a dedicated site on Raisina Hill.

Evidence of earliest civilisation from 1450 BCE was around the site of Indraprastha when it was a heavily forested region. The area of Delhi as now recognised was developed near the Yamuna River over many years from seven individual sites that at different times had acted as capitals, notably:

- Qila Rai Pithora	c. 1180 CE
- Siri fort	c. 1300
- Tughlaqabad, Aravallis	c. 1320
- Jahanpanah, nr Chiragh	c. 1340
- Ferozabad	c. 1350
- Purana Qila, Shergarh	c. 1540
- Shahjahanabad and Lal Qila	c. 1640

The area came under Turkic Muslim dominance around 1200 CE, followed by periods of administration under the Delhi Sultanate and the Mughal Empire. The last city, Shahjahanabad, served as the capital of the Mughal Empire and constituted what is generally known as the walled 'Old City' today, before losing power to the Marathas from the Deccan Plateau.

In the Old City Lal Qila, the Red Fort, with its walled battlements was the seat of the former Mughal emperor, completed around 1650 CE with striking arched gateways. The Grand Mosque Jama Masjid nearby was completed at about the same time, with its courtyard and twin minarets and marble domes. The oldest mosque in Delhi is at the Qutb Minar with its old sandstone minaret in the south of the city. There are a number of tombs of former Mughal emperors located around the city, notably Humayun, Isa Khan and Safdarjung, with other tombs found in the Mehrauli archaeological park.

The expanding British East India Company was granted fiscal administration of eastern India in 1765, and after the traumatic Indian Mutiny and War of Independence in 1857, the British Crown replaced the East India Company, as Delhi was incorporated into the region of the Punjab and set about developing the city during the 19th century.

The Old City was densely populated, and under the Military Cantonment Act of 1864, the British implemented public health legislation aimed at improving sanitation. Military reservations were established, with residential properties for their occupying staff assigned according to rank, the physical separation of the European and Indian populations determined along the 'civil lines' for government employees. Much of the early peripheral accommodation in Delhi was in 'bustees', flimsy informal huts of 'kachcha' construction without services for the local population or proper sanitation.

Under the British Raj there was a tradition of holding grand ceremonials as, for example, the Durbar held in Delhi in 1903 to celebrate King Edward VII's succession in Coronation Park, and again in 1911 for King George V's accession. These were impressive events with many princes

DOI: 10.4324/9781003353348-36

of the princely states represented in attendance in the full celebration of the British Raj under control of the viceroy.

The capital of India was moved from Calcutta to the inland area of Delhi in 1912, which in having its former associations with Mughal and sultanate history, was nearer to Shimla in the north, the British Raj's summer capital.

In 1919 a diarchy system of governance was set up with the governor of a province having direct control over law and order and finance, but with other areas such as education and health being transferred to provincial councils.

New Delhi

With its former city sites and historic monuments, Delhi was considered most suitable by the British as the new capital of India. A site was sought between the Delhi Ridge and old Purana Qila that was considered appropriate for the location of Government House from where the principal ancient monuments and the River Yamuna could be seen. Besides the densely packed old 'native city', the new civil lines and military cantonment were laid out and ceremonial areas with main axes, radiating avenues and monuments (see Chapter 4.5.2).

As a whole, New Delhi, designed by Lutyens, was realised in an integrated manner that had a sense of scale and grandeur with main axes and triangles that were not only different from the Old City but had a stratified spatial order and strict ranking of housing for officers and employees by sector. It was not inaugurated until 1930 when under changing circumstances, such imperial posing was not to be required much longer. After World War II these buildings were transferred to India's own government once independence was achieved in 1947, with Parliament House and a new Central Secretariat and administration.

Delhi also had become under the British the centre of the railway system in India with eight lines converging around the Old City. A new railway station was planned in the centre facing an urban square between the existing and the new city.

Modern City and National Capital

Huge transformations took place in Delhi after its independence, with a significant increase in population due to inward migration and with ongoing growth which has continued to stretch and create further urbanised and suburban areas of high density, an extending conurbation in all directions.

In 1956, the Union Territory of Delhi was created under the States Reorganisation Act, and in 1991 the Union Territory was further modified and renamed the **National Capital Territory** (NCT) of Delhi. This emerging city has continued to grow extensively and became further consolidated with continuing in-migration to the national capital of India. It is managed by the Municipal Corporation of Delhi and divided into three district areas.

As prevailing capital, a modern plan by PM Modi has been initiated to entirely rebuild the central area under the Central Vista Redevelopment Project alongside Kastavya Path, formerly Raj Path, with a new complex of governmental buildings that has changed the former character of the Government Centre. The new triangular Parliament Building seats both Lok Sabha and Rajya Sabha, horseshoe-shaped Lower and Upper Houses, along with a Public Constitution Hall at its centre.

In order to handle the continuing increase in population, the transportation network has been significantly enhanced with the construction of an underground metro system. The Delhi metro was developed in phases to reach a wide coverage and is able now to carry large volumes of passengers daily across the city supported by Delhi Suburban Railway.

The new metro run by DMRC was commenced in 1998 and comprises an integrated system of 10 lines and over 250 stations with both elevated and underground lines and standard- and broad-gauge track. Phase I was built on broad gauge within Delhi State, the Red, Blue and Yellow Lines, with Phase II extended into the neighbouring modern states of Haryana and Uttar Pradesh. Phase III added both radial and ring lines, Pink, Violet and Magenta. Still new routes are planned yet farther out under Phase IV. There is also the rapid transit metro around Gurgaon.

In terms of road transport, Delhi has both an Inner and Outer Ring Road and a dense network of interconnecting streets and roads. Externally the new tolled Delhi–Noida Direct Flyway (DND) connects with the developing Noida District in Uttar Pradesh, and the Delhi–Gurgaon Expressway into Haryana and the tolled Delhi–Faridabad Skyway that runs south to Faridabad in Haryana with a number of other highways. The network of national highways around the city now connects the Golden Quadrilateral of the major Indian metropolitan cities of Delhi, Kolkata, Chennai and Mumbai.

With significant demographic changes, the capital of the Federal Republic of India has had its challenges – religious, political and economic – especially under the prevailing caste system. With 29 states and seven union territories now including Delhi, and with its linguistic and cultural variations, there remain significant challenges in the development of infrastructure, social, health and education facilities across the capital and adjoining states. The current population of the city is approximately 11 million, with 26 million in the metropolitan area. The population density of the whole Delhi city region is approximately 29,000 pers./sq. km.

In the wider National Capital Region (NCR) of Delhi, with approximately 40 million inhabitants, the focus has been on the development of a series of satellite cities in order to take the pressure off the central area of Delhi. These include **Faridabad**, an area that developed with industry in the 1950s and which is now a major industrial hub in Haryana State south-east of Delhi and has a large number of major industrial companies. Greater **Noida**, the New Okhla Industrial Development Authority, is a new planned satellite city in adjoining Uttar Pradesh and has become an employment hub for migrant workers, set up in 1976 by Rajiv Gandhi as a special economic zone as an initiative during the state of emergency.

Ghaziabad to the east is at the junction of the East India Railway and the Sind, Punjab and Delhi Railway. Here there are new industries that have been established since 1967 under the Development Authority, now part of Uttar Pradesh (UP). **Gurgaon** (Gurugram) is an entirely new satellite area developed by DLF, a major private real estate company just to the south-west of Delhi in Haryana State. The Indira Gandhi International Airport (DEL) is to the south-west of Delhi with other new airports at the emerging cities of Ghaziabad and NOIDA (DXN) at Jewar.

With its rapid growth and influx of population, there have been several plans to manage the growth of Delhi. Subsequent to the creation of the National Capital Region in 1985, there was a strategy aimed at orchestrating the growth of the capital and neighbouring areas and decentralising the capital through the creation of centres within a range of 100–400 km. to act as 'counter-magnets'. Whilst these satellite cities were given economic incentives, the aspirations were not fully realised. With a master plan in 2021 followed by Plan for 2041, the current plan is more incremental, with staged milestones for improving transport and affordable housing for this mega city region with the focus on a cluster of developed nodes. Areas such as Gurgaon and NOIDA are already virtually major cities themselves. Complemented by the Smart Cities Mission, originating in 2015 across India, focused on infrastructure, energy management and mobility, this needs to be further targeted around developing nodes and clusters of individual city challenges.

Transactional Forces and Typology

With its ancient origins from Indraprastha on the banks of the River Yamuna and its evolution through Mughal times to the present day, the area of influence of Delhi has grown as a trading centre. In different periods, historically differing cities have been sited around the location of the present Delhi with their own individual fortifications and landmarks as it grew as a centre of power under the Islamic Delhi Sultanate and then the Mughals.

Its earlier purposes that comprised defence, administration and religious centre through the ages were extended through becoming the centre of the East India Company and the headquarters of the British Raj. With construction of the Grand Trunk Road and the railways, this further increased the pivotal significance of Delhi's trading role across the region.

'Old Delhi' evolved organically from its location by the river and with its own individual fortifications, whereas 'New Delhi' was planned as a distinct imperial city with two main axes suited for ceremonial occasions, radiating boulevards, hexagonal districts and cross streets. Since independence, its role as the federal capital has been consolidated, and the attraction of the city has grown yet more significant with further waves of migration into the city and changes being made to city institutions to reflect this growth.

Besides its governmental role, its commercial and trading role has continued to grow, leading to the development of satellite cities around the centre. As National Capital Territory its population has grown greatly and its areas of influence are extending into neighbouring modern Haryana and Punjab states. This extended metropolitan area of Greater Delhi is now of megalopolitan proportions, having in excess of 40 million inhabitants, that feature as a series of poly-nodal settlements across the whole urbanised region.

6.4.2 GREATER KOLKATA (CALCUTTA), INDIA

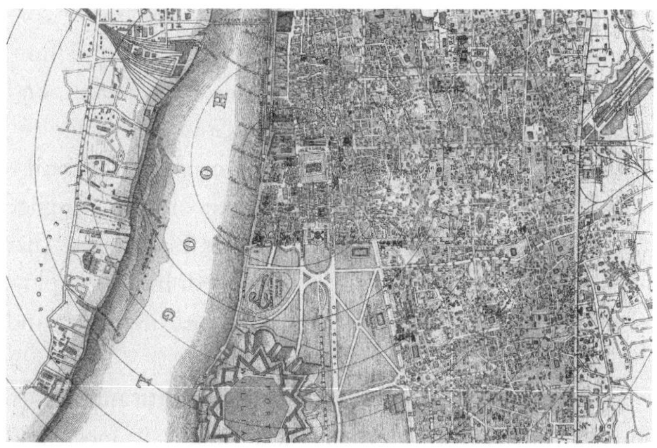

FIGURE 6.4.2 Calcutta – early configuration, 1892 (Letts).
Source: Maphouse.

Background and Origins

The city of Calcutta has developed essentially on the east bank of the Hooghly River, opposite Howrah, which flows into the Bay of Bengal. It was renamed **Kolkata** in 2001 and is now the capital of the state of West Bengal, adjoining the Ganges River Delta and not far from the Brahmaputra River Delta in neighbouring Bangladesh.

The three riverside villages that predated this city were ruled originally by the Nawab of Bengal under Mughal suzerainty. Located at Chattanuty on marshy ground near Saltwater Lake, and preceded by Portuguese and Dutch occupiers, Fort William was built in 1696 by the British East India Company on the east bank as Kalikshetra was developed with its Kalighat Temple and a port was expanded for increasing trade across South-East Asia.

There were, however, skirmishes with the occupiers and humiliation by the Nawab of Siraj-ud-Daula at Plassey, after the infamous Black Hole incident in 1756. After the Treaty of Allahabad in 1765, the British established and entrenched themselves in the territory with the development of barracks and residences; the original fort was strengthened and moated, and trading halls and warehouses were built, as well as institutions to support the cotton-growing industry around the city.

The first governor, Robert Clive, originally annexed 900 sq. miles of land south of Calcutta, known as the Bengal 24 Parganas before being succeeded as governor general by Warren Hastings in 1772. The next year, under Lord North's Regulating Act, the East India Company, with tax-gathering rights, was reined in and subordinated to more authority and regulation from the British Parliament.

With increasing prosperity, the banks of the River Hooghly and along the Strand gradually developed and became studded with elegant mansions such as the Marble Palace, such that Calcutta came to be known as the 'City of Palaces'. With the Maidan adjoining the fort and newly constructed roads, the 'White Town' developed around Tank Square with its mansions and colonnaded verandas, and administrative buildings such as the Government House, Supreme Court, town hall and the red-brick Writers Building of the East India Company, north of Chowringhee Road. This was in distinct contrast to the 'Black Town' that comprised the northern part of the city for Indian employees. These neighbourhoods were not healthy areas, and malaria and other diseases were prevalent.

The increase in trade in cotton, indigo, jute and opium brought increased migration and mixes of races and multi-culturality into the city, similar to other international ports, with Chinese, Africans, Europeans and Malays mixing with Bengalis. The Asiatick Society was established and aimed to better understand and promote Hinduism and Sanskrit, and the College of Fort William was set up to train the civil servants (clerks or 'writers') of Bengal, as distinct from the troops, in local languages and traditions.

Bazaars and markets became a feature across different areas of town as it developed and became increasingly populated. Sanitation was a persistent problem, and in 1803 a Town Improvements Committee was established to improve drainage, clear the riverbanks, build new 'ghats', and improve the urban infrastructure. However, this did not percolate into the worst districts, which steadily grew worse generating the 'Big Calcutta Stink', as described by Kipling in *The City of Dreadful Night*, where cholera and other diseases prevailed.

Railways were introduced, with the Howrah Station built on the west bank and the Bengal–Nagpur Railway and Sealdah Station on the east bank for the South Eastern Railway. The East India Railway was extended to Delhi in 1865 with the construction of a broad-gauge rail line. There was also a circular railway broad gauge built around the city, known as the Chakra Line, that was built initially to service the port, Kiddipore Docks and Cossipore Gun and Shell Factory, and the barracks at Dum Dum.

After its development, the imperial city was at its peak in the second half of the 19th century. Industry developed in the region with coal discovered in Raniganj. Railways were extended to transport this, rail workshops were built for support and factories were developed with steam-driven power. With abundant local natural resources, power looms produced gunny sacks from jute for packaging commodities, tea was being shipped from up-country Darjeeling and cotton products were being woven. Agency houses acted as traders and factories were used for the different commodities that were exported. Jute mills were built to create 'bags' for the significant trades in grain, rice and sugar.

However, with the division of Bengal in 1905, West Bengal was enlarged with people migrating from newly created Bihar and Orissa states that were predominantly Hindu, and also into East Bengal that became mainly Muslim. In 1911, the status of the city of Calcutta as the centre of India was diminished as the British transferred the capital to New Delhi, and it began to lose some of its former influence. The annual festival is called 'Durga Puja', an occasion for great celebrations.

The buildings of Government House (now Raj Bhavan), the general post office in Dalhousie Square, the High Court, Royal Exchange and Great Eastern Hotel and BBD Bag Square with its adjoining tank (lake) framed the centre of the colonial part of Calcutta around the 'Writers Building' of the East India Company, which formed the heart of the city.

The most celebrated landmark constructed was the Victoria Memorial that was completed in 1921 by the parade ground and with the white St Paul's Cathedral nearby. Presidency College formed an integral part of the educational sector of the University of Calcutta. The former parade ground of the Maidan still remains a large open area around Fort William, with its famous Eden Gardens park and cricket ground.

Housing for workers in the jute mills was built in rows of red-brick terraces, and adjoining areas were becoming very urbanised. Despite this, there was much poverty and with the caste issues, meandering sacred cows, open drains, densely populated areas and migrants, sanitary conditions were very poor. Burning bodies on the ghats on the Hooghly River was a common sight.

Densely populated 'bustees' spread from Budge Budge to Kanchrapara, which were essentially collections of huts where the 'thika' tenant or hut owner built dwellings on municipal land that they then rented out to human bustee dwellers. Also 'khatals' were areas in bustees where land was rented out just for animals to be stabled for a small fee.

The population of Calcutta had already reached 1 million by around 1850. The municipal Calcutta Corporation was founded in 1876, and the improved Entally Drainage system was commissioned in 1896. In 1911, the Calcutta Improvement Trust was established as a result of a report by E Richards in 1914 in identifying means to improve and expand the city. His findings identified huge areas of slums, with narrow lanes and no real streets, with inadequate drainage and sanitation that required improvement.

The streets were inadequate for vehicles and housing was essentially classed as a 'rank slum'. Land parcels were often further subdivided under 'partition suits' where, if inheriting sons disagreed, habitations became subdivided. Better transport and communications were required, along with habitation and sanitation improvements.

The orporation finances were in a bad state. With the great tank, Lal Dig, water supply was available and a pumping plant at Chandpal Ghat was built to lift water from the river. The water supply became better managed once a pumping station was constructed at Palta. Subsequently the administration of the Calcutta City Corporation was reformed in 1923, leading to corresponding municipal reforms.

Transport

In a predominantly low-lying area having intensive monsoon seasons, the rivers across the region were wide but shallow and needed constant dredging in order to maintain access to riverside quays. The wide Hooghly River, whilst an asset, remains a continuing challenge with its constantly changing sandbanks, shoals and tidal bores. The most suitable sites for wharves and docks were sought as far upstream as possible on either bank and near the creeks and Tolly's Nullah canal. Large ships often could not reach the city, and heavy freight loads had to be transferred into smaller lighters.

More recent man-made developments have had a significant impact on modern-day Kolkata. In the north, the great Farakka Barrage Dam on the Ganges River was completed by the Indian government in 1975, which served to exert some control on the flows of the Hooghly River and in managing the displacement of silt towards the Bay of Bengal. By controlling the flow of the Ganges, this also impacts neighbouring Bangladesh, with its own Brahmaputra River that combines with the Ganges to form the Padma River. This has become the subject of several disputes, in seeking to balance the water flows for each country.

Whilst the depth of navigable water has increased between the sea and Kidderpore Docks of the Port of Calcutta, numerous island mud banks remain. The general estuarine problem is of the outflow silting up and the inflow becoming more saline. As a result, a new dock complex has been built downstream on the peninsula at Haldia, which has been developed on a wider site that has now become the main industrialised port for the city.

The first Howrah Bridge was a floating pontoon bridge built in 1874. This was replaced with a modernised large truss cantilever bridge in 1943 (Rabindra Setu) that had particular military significance as a strategic crossing. Another Hooghly bridge was built upstream at Bally.

A multi-span road and railway bridge, the Vivekananda Setu Bridge, with steel trusses on caissons, was commenced in 1926 and opened in 1931. As a result of its aging, the Second Vivekananda (Nivedita) Setu modern extra-dosal highway bridge was built and opened more recently in 2007, developed as a toll concession by NHAI and adjoining the original one that forms part of the ring road to the city. The other Hooghly bridge that has been built to the south is the Vidyasagar Setu Bridge that was opened in 1992 as a modern cable-stayed bridge that is also a tolled road bridge. This acts as the southern river crossing for the city.

Tramways introduced across the city before 1900 and managed by a limited liability company were subsequently removed, but there has been a recent plan to aim to reintroduce them. The Lower Circular Road is a perpetually congested road with local traffic, and Chowringhee Road runs north-south from the Maidan to Kalighat, similarly very congested.

In the 1970s a Metropolitan Transport Project was planned with the assistance of Russian and East German engineers, and a deep trench was excavated for an underground cut and cover rail line to run from Tollygunge via the centre of the city out to Dum Dum Airport. This is now the location of Netanji Subhas Chandra Bose International Airport (CCU).

The metro line was eventually opened in 1984 and extended in stages to New Garia, known as the Blue Line 1 (broad gauge) running north-south. More recently another line, the Green Line (standard gauge) running east-west, has been planned to run underneath the river from Howrah and then elevated out to Salt Lake City. Part of this is operational despite significant construction, land and financing issues and delays. Four further lines by the Metro Rail Corporation are also planned.

After the discovery of coal, a whole iron and steel industry came to be developed in the adjoining western region from Kolkata with blast furnaces and rolling mills at Durgapur, Jamshedpur and Asansol. With new steelworks at Rourkala and Kulti, engineering at Midnapore and Ranchi, factories at Patna and Siliguri, and oil refinery and locomotive works, Bengal became a most industrialised area with its own mineral and oil resources.

The Tata family founded Jamshedpur Works and the Birla Brothers, Hindustan Motors at Uttarpara, both of which have grown into major conglomerations. In neighbouring Jharkhand, a new Bokaro Steel City was built in 1965 as a Global Active City that has become home to Bharat Petroleum and the Steel Authority of India.

Modern City – Buildings and Arteries

The region of Bengal has had a long unfortunate history of both famines and divisions. In 1943 in particular there was a very severe Bengal famine during World War II and after the war with independence and partition, with its separation of Hindus and Muslims, it suffered considerable tensions with the creation of neighbouring East Pakistan from Eastern Bengal and the migration of peoples in and out of West Bengal.

In the late 1960s, as the city government, the United Front with the Communist CPI(M) took over power from the Congress Party, which was a significant change in direction. Budgets of the municipal corporation were reviewed, large landholdings were reassessed and a revolutionary mood ensued with 'gheraos' of state businesses, strikes, protests and disruption, some quite violent. Statues were pulled down, streets were renamed and a general municipal overhaul was undertaken. This was to change in 1969, however, with the closing down of the Calcutta Assembly by the government in New Delhi and the imposition of the president's direct rule. During this period, the economy had stagnated and started to de-industrialise.

In 1961, the Calcutta Metropolitan Planning Organisation (CMPO) was set up to manage the development of the city. Later, in 1967, the British planner Buchanan and others reported on Calcutta as being a 'city in crisis'. Goals for the city were laid out in the Base Development Plan for the Calcutta Metropolitan District 1966–86, as a strategic development plan for the city. This has since been superseded by the subsequent Kolkata Metropolitan Development Authority (KMDA).

Assisted by the World Bank and Ford Foundation, the plan was instrumental in trying to shape the development of the city from the Basic Development Plan with the reclamation of salt marshes in the east, new housing and new highways and communications across the city. The Twin Cities of Calcutta to the east and Howrah opposite to the west, with the townships of Bansberia and Kalyani in the north were to be developed with industry, highways and rail connections to western Durgapur and Asansol. The southern access was to be extended to connect with the new port of Haldia. This plan was to prevent the situation worsening with bustee lands compulsorily repurchased, sites upgraded and people to be rehoused in 'sites and services' projects. However, little real progress was made.

In 1971, after the Indo-Pakistan War and the creation of independent Bangladesh, this resulted in huge movement of populations, with an exodus of Muslims and immigration of Hindus into the central urban area of Calcutta. With the large transfer and movement of people, the bustees grew in the north around Cossipore and in the 24 Parganas.

The city was eventually renamed **Kolkata** in 2001, and currently has approximately 5 million in the city whilst the metropolitan area of Kolkata has grown to approximately 15 million people. Now stabilised and growing again, there has been an increase in employees in the services sector and focusing on IT. The population density of the whole Kolkata city region is approximately 36,400 pers./sq. km.

Externally, Greater Kolkata has grown enormously with its 35 municipalities and sprawling informal settlements. There are a very large number of dwellers in 'bustees' with unauthorised informal areas that are virtual slums devoid of basic services along many of the main roads and thoroughfares. As a response, Mother Teresa had developed the 'Missionaries of Charity', which spread beyond the convents of the city and Nirmal Hriday to assist the very ill and impoverished women and children across the region during the second half of the 20th century.

Transactional Forces and Typology

The beginnings of the city developed around the Hooghly River that was an access route on the western side of the mouth of the River Ganges. Here at a bend in the river, a trading post at Sutanuti was developed around existing villages that were to grow into the future Calcutta with Howrah on the opposite bank. As large ships were able to gain access using lighters, for trade in materials and jute, subsequent silting forced a new port to be developed and used downstream at Haldia.

With an abundance of raw materials in the region and sources of coal, other industries were also developed that caused the city to grow into a significant processing and manufacturing hub with its own iron and steel works. Whilst acting primarily as the capital of India, it was a vital administrative centre that at the time of the British occupation had good communications that were developed revolving around the introduction of the railways that complemented the waterways.

As a riverine city, settlements grew on each side of the River Hooghly, and linearly around Fort William on the east bank. With only one large, original, fixed pontoon bridge crossing, each side developed quite distinctly with numerous small ferry links. The eastern area is low lying with extensive wetlands which have been canalised in some areas.

The new areas of Salt Lake City and New Town were planned and developed to the north-east in the 1960s, having a distinctive difference from the many informal, unplanned and poorer areas that have grown up around the city. A new extensive Kolkata metro is under development and construction, part underground and part elevated, that aims to alleviate the heavily congested trafficked radial highway routes in and out of the city.

With its continuing urbanisation and in-migration, this is leading to the ever-expanding area of Greater Kolkata and its surrounding metropolitan region with development increasingly focused on poly-nodal centres around and on the periphery of the original core by the Hooghly River. The following figures illustrate the modern configuration of the megalopolitan agglomerations of India and the evolution of their modern population.

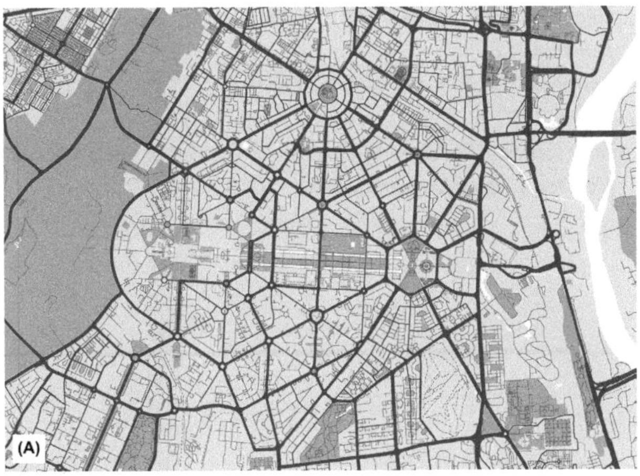

FIGURE 6.4.3A AND B Delhi and Greater Delhi – modern configuration.

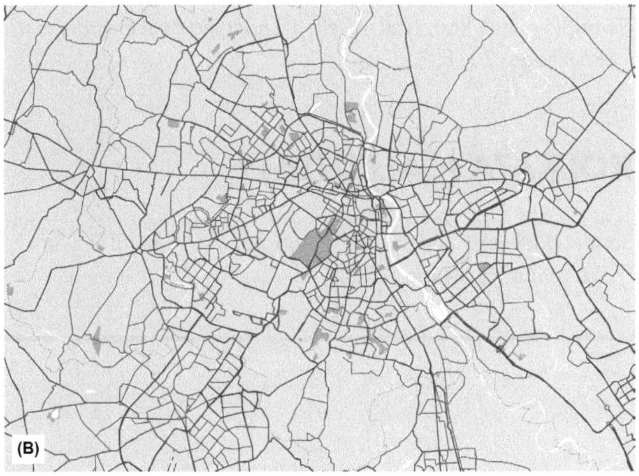

FIGURE 6.4.3A AND B *(Continued)*

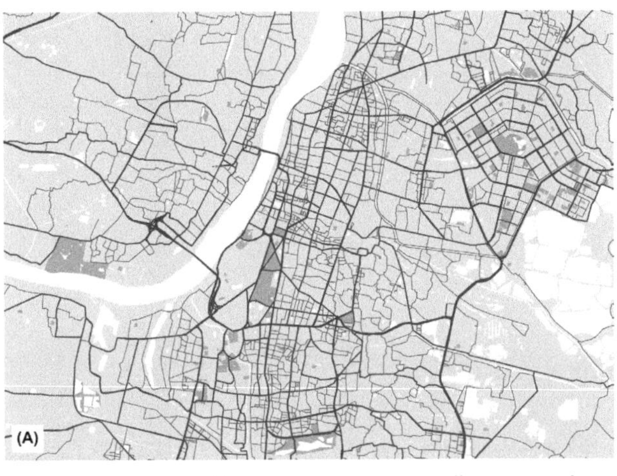

FIGURE 6.4.4A AND B Kolkata and Greater Kolkata – modern configuration.

FIGURE 6.4.4 AAND B *(Continued)*

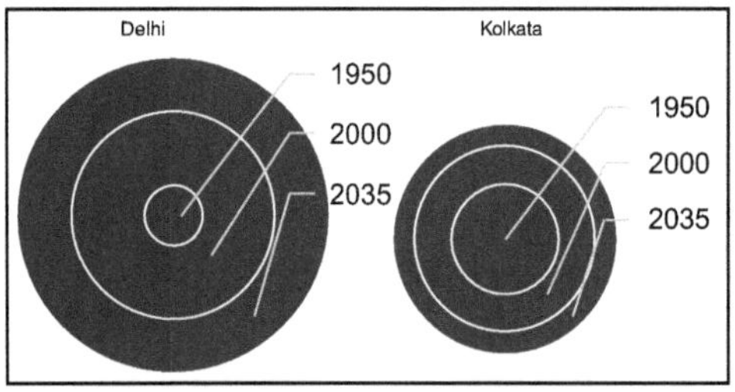

FIGURE 6.4.5 Megalopolitan India populations.

Data source: Global Database of Metropolises 2020 – UN-Habitat (2020), UN DESA Population Division.

Further Reading

DELHI

Collins, Larry & Lapierre, Dominique – *Freedom at Midnight*, Vikas Publishing House, 1997.

Nehru, Jawaharlal – *The Discovery of India*, Penguin, 2004.

Sadani, Rashmi – The Moving City, Scenes from the Delhi Metro and the Social Life of Infrastructure, University of California Press, 2022.

Tiwari, Piyush & Rao, Jyoti – *Delhi's Changing Built Environment*, Routledge, 2018.

KOLKATA (CALCUTTA)

Bandyopadhyay, Sumana, ed. – *Kolkata, the Colonial City in Transition*, Routledge, 2024.

Banerjee, Sumanta – *Memoirs of Roads, Calcutta from Colonial Urbanization to Global Modernization*, Oxford University Press, 2016.

Bhattacharyya, Debjani – *Empire and Ecology in the Bengal Delta, the Making of Calcutta*, Cambridge University Press, 2019.

Campbell, Mark – *Architecture and Urbanism in a Contact Zone, Histories of Difference, Migrancy and Dwelling in Kolkata*, Routledge, 2023.

Kipling, Rudyard – *The City of Dreadful Night*, Sampson Low Marston, 1891.

Moorhouse, Geoffrey – *Calcutta*, Weidenfeld & Nicholson, 1971.

Raymer, Steve – *Redeeming Calcutta, a Portrait of India's Imperial Capital*, Oxford University Press, 2013.

6.5 Megalopolitan China

Beijing, Tianjin/Hebei, Qingdao/Shandong, Guangzhou/Guangdong, Chengdu/Chongqing

6.5.1 BEIJING-TIANJIN-HEBEI MEGALOPOLITAN REGION, PEOPLE'S REPUBLIC OF CHINA

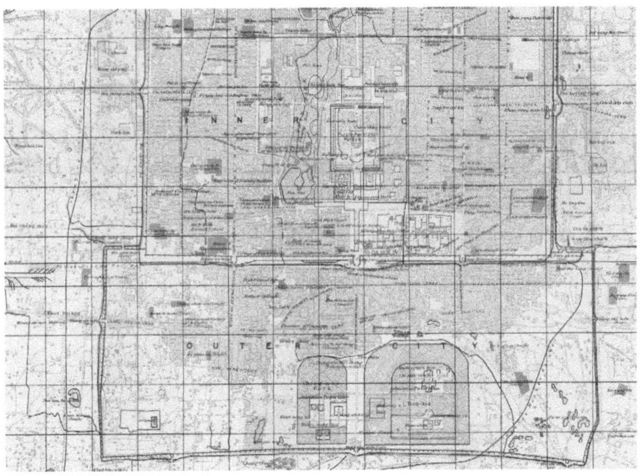

FIGURE 6.5.1 Beijing – early configuration, 1923.

Source: Maphouse.

Modern City – Beijing

Modern Beijing is an enormous city that now anchors the **Jing-Jin-Ji** metropolitan region of the **Beijing-Tianjin-Hebei** capital economic zone. Its infrastructure has been developed into a major hub of highways and railways with high-speed inter-urban trains and a comprehensive urban metro and subway network. The Beijing municipality is huge and comprises 16 large administrative divisions and subsequent townships.

With its coal-burning power plants, the air quality in the metropolitan region is often poor, especially in winter, and dust storms from the desert plains are also frequent. However, conversion to electric vehicle fleets is contributing significantly to reducing pollution from car emissions. Many large state-owned enterprises have their headquarters in Beijing, which has a sophisticated industrial base along with an increasingly service-based economy. The Central Military Command is also based in Beijing.

In former times, under Mao, most of the urban population belonged to a 'danwei'. This was a work-unit model that included both workplace and residential facilities such that population being constrained within their own unit, there was little need to travel or interconnect with other independent realms as they were relatively self-sustaining.

However, as economic reforms were introduced, with separation between workplace and residence, new affordable housing was required, which tended to be developed in peripheral urban areas. Such 'land conversion' led to the intensive development of cheap, formerly agricultural land that was sacrificed, leading to significant peri-urban development around the city. This was also dependent on individuals securing urban 'hukou' permits to enable permanent residence in the city. Over time this has led to an ongoing redistribution of population from the outlying peripheral areas.

The most evident change has been the development of highways around the capital. Initially this led to the destruction of old city walls in order for land to be opened up for the first ring of highways, but this has been increased with the addition of the second to the sixth new rings.

In the 1990s demand grew as individual car ownership took off, departing from the former dependence on bicycles for local transport that had been prevalent for many years. At the beginning of urbanisation, bicycles had been the mode of transport for the masses, when there were around 540 million bicycles across China.

Progressing to mopeds and motorcycles, subsequently cars with internal combustion engines took over. With the development of its own manufacturing capacity, vehicle sales rapidly increased for the local population and in 2020 across the PRC, there were approximately 240 million privately owned vehicles on the roads (or approximately 140 cars/1,000 people). In 2002, Beijing had 1 million vehicles, rising in 2020 to approximately 5.5 million registered passenger vehicles in Beijing.

However, quota controls are now being introduced for the acquisition of new vehicles through auction or lottery processes. The central area of the city is heavily congested with traffic at most times, despite the introduction of a number-plate rationing system (with two selected end

DOI: 10.4324/9781003353348-37

numbers per week being disallowed) and a number of designated rush-hour bus lanes. There has, however, been a distinct progressive evolution in the use of electric vehicles and scooters that is assisting in limiting rising pollution levels across the city.

Major expressways have been built around the city, including the outer orbital Capital Area Loop Expressway (G 95), and additional new ring roads have been built.

In terms of rail, there are now 16 metro lines, the first circular underground line being built in 1969 and crossed by the linear east-west Line 2. These have been developed and extended significantly during the last 20 years. New high-speed rail lines radiate out to the major cities of Tianjin, Shanghai, Guangzhou, Xianjian and Zhangjiakou from the Central, North, South and West stations, which are linked and inter-connected with the main metro lines. The two major international airports of the original Beijing Capital (PEK) to the north-east and the new Beijing Daxing (PKX) (designed by Hadid/AdeP) in the south serve significant volumes of passenger and freight traffic and are connected by airport express subway lines.

As the city has developed, commemorative events have been the opportunity to remodel different areas of the city. For the Asian Games in 1990 and especially for the 50th Anniversary of the PRC in 1999, buildings were refurbished, polluting taxis and microvans were banned and new plantings in several parts of the city were created. Tiananmen Square was transformed and Pong-An Avenue widened. It was the 2008 Summer Olympiad that focused energies with construction of the Olympic Green, Village and Stadium to the north of the Forbidden City. However, many courtyard houses ('siheyuan') and 'hutong' neighbourhoods were demolished and Zhejiang Village was razed, losing some of the former character of the city.

In terms of built development, the downtown CBD of modern Beijing today contains several iconic buildings, such as the CCTV Headquarters (designed by OMA), the Linked Hybrid Building (designed by Holl) and a number of towers such as the China World Trade Centre Tower III (designed by SOM) and the CITIC Tower (designed by KPF) over 500 m. high. With the iconic 'Birds Nest' Olympic Stadium (designed by Hertzog and de Meuron), other featured landmarks were realised. Further significant built development has been the promotion and realisation of large urban commercial malls (see Support Material: Annexe 6.5A – Beijing Significant Modern Buildings).

Besides the Olympic district, new areas were also developed on the edge of the city, such as Jiangguomen District to the east between the 3rd and 4th Outer Ring Roads and the Financial District to the west. These large developments have served as new growth poles at either side of the city, around new landmark buildings.

New economic zones within Beijing include Zhongguancun 'silicon village', Beijing Financial District, Beijing CBD, Yirzhuang Economic and Technological Development area and Beijing Airport Economic Zone. For residential areas, the model of expansion has been the development of major districts such as Huilongguan north of Beijing, that are organised in extensive super-blocks around transit stations.

Both the Asian Infrastructure Investment Bank and the Silk Road Fund are based in Beijing. These encourage the Chinese government's international development strategy of the Belt and Road Initiative that is driving the development of continuous marine routes and highway and railroad land routes across countries of the Asian continent towards Europe, emulating the paths once used by former Silk Route travellers (see Support Material: Annexe 6.5B – China – Belt and Road Initiative).

An enormous urbanised place under constant change, a new modern Chinese dynasty of manufacturing and production has comprehensively established itself following the Cultural Revolution, driven by high rates of urbanisation with in-migration from rural areas. The current population is approximately 22 million in the city. The density of the population in the whole Beijing city region is approximately 20,200 pers./sq. km.

Tianjin (Tientsin)

Background and Origins

Tianjin is a coastal city just south-east of Beijing by the Bohai Gulf of the Yellow Sea. Tianjin municipality, which borders Beijing municipality, has now become the fourth-largest city in China, within the Bohai Economic Rim and forming part of the Jing-Jin-Ji metropolis.

On land created by sedimentation from the Yellow River, it developed in the estuary as a trading port, named Tientsin, enhanced by the creation of the interior man-made Grand Canal that was commenced around 600 CE during the Sui dynasty, and was eventually to connect Beijing with Hangzhou. During the Song and Yuan dynasties, this became a pivotal route in the transfer of grain across the country to the north.

As part of the Hai River system delta, the Taku Forts were located in the estuary. However, with the Opium Wars in the 19th century and after the Arrow Incident, the Treaty of Tientsin, signed in 1860, ended with Tientsin being opened to trade through a series of treaties between the Xianfeng emperor and Britain, France, the USA and Russia. With these concessions and with other countries, trade developed as Tientsin became a significant treaty port along with Shanghai and Canton.

Under the Boxer Uprising in 1900, an Eight-Nation Alliance was required to recapture Tientsin and then retain garrisons so as to maintain access to Beijing. Later, during the 1930s, under the Sino-Japanese Wars, the control of the municipality fell to Japan, which it retained until the end of World War II. Subsequently under the Chinese Civil War, the city was only eventually brought under control by the CCP, following the Pingjin Campaign at the beginning of 1949 with the formation of the PRC.

Modern City – Tianjin

Post-war the city and port developed significantly. In 1976 there was a major earthquake in Tangshan that caused severe damage. Major reservoirs had to be rebuilt for flood control and water supply purposes, such as under the South-North Water Transfer Project developed under China's national series of Five-Year Plans. After the Central Route, the Eastern Route was completed in 2014, enabling the transfer of 1 billion cu. m./annum from the Yangtze River.

The central area of the city around the Hai River has been renovated, as well as the area around the former Italian Concession. With Liberation Bridge built in 1927, the area is now dominated by the tall World Financial Centre (designed by SOM).

The first underground metro was opened in Tianjin in 1976, with two more lines opened in 2012 and two more under construction as of 2025. The main railway station at Laolongtou, originally opened in 1892, was rebuilt in 1988. The high-speed train line was opened in 2008 between Beijing and Tianjin to coincide with the Summer Olympic Games, for which Tianjin was co-host city, with stations at Tianjin North and West as well as Tianjin Main station.

In 1984 the Tianjin Economic Technological Development Area (TEDA) and the Tianjin Binhai New Area were created with the coastal port as their core, with the Tianjin CTF Finance Centre (designed by SOM) and the adjoining Circum Binhai City Belt. Here there are significant industries of automobile and aircraft manufacturers, petrochemicals and textiles. The Tianjin Tanggu Port was redeveloped as part of neighbouring Hebei Province, and industry developed especially around the port area. The Tianjin Port Free Trade Zone was created in 1991 with a Logistics Zone at Tianjin Binhai International Airport (TSN) and the Nangang Industrial Zone.

Concerning commercial activity, Tianjin has a number of significant research institutes and universities and key buildings, such as the Jin Tower-World Financial Centre (designed by SOM), the TV Tower, the China 117 Tower, the Tianjin 'Eye' and the Sino-Singapore Tianjin Eco-City.

There are three main ring roads, the Inner, Middle and Outer Ring with the main expressways of the National Trunk system that connect with the national highways. The Jingjintang Expressway was the first to be built to connect Beijing with Tianjin under World Bank finance, and the Jingji Expressway was opened in 2003.

All of the major development now is essentially gravitating around the Binhai New Area, a pilot free trade zone to the east of Tianjin in Bohai Bay. The population of Tianjin is approximately 15 million.

Hebei (Ji) Province

Beijing and Tianjing are circumscribed by Hebei Province, which comprises 11 prefecture-level divisions. As a resource-rich region, especially with coal and the North China Oilfield, it has become a highly industrialised region with extensive iron and steel production and manufacturing industries. With a growing population, it has developed a number of economic and technological development zones at Shijiazhueng. With its complementary universities of science and technology and research institutes, it is a resourceful economic region in terms of human, physical and technical resources.

Megalopolitan Region

The extensive urbanised and metropolitan areas of Beijing and Tianjin together now constitute the backbone of a megalopolitan region in Hebei province, from the highlands down to the coastal Bo Hai. With their intermediate and adjoining industrial areas and natural mineral resources, these poly-nodal settlements have been populated to form larger agglomerations. This region, rather than accommodating growth around a single corridor, is developing across the whole Hebei province, anchored by the pull of each of the two major municipalities of Beijing and Tianjin. As a result, it has become one of the most significant megalopolitan areas in modern China.

Regional Transactional Forces and Typology

With its long, dynastic history through the ages, Beijing has been of prime significance variously as northern capital and now as primate national capital today. With the adjoining Tianjin municipality, this has grown in economic significance from its role as an original treaty port to become the vital port in supporting the capital region today. With industrialisation right across the adjoining Hebei region (Ji province), and with its intensive coal and steel industry, this urbanised zone has come to be known as the Jing-Jin-Ji region.

From its earlier settlements around the Hai River, this developed area has been complemented by a dense network of railway and highway infrastructure. With increased migration and urbanisation, this infrastructure has been focused on modern urban metro systems and high-speed inter-urban railways. As combined adjoining metropolitan areas, this region can be characterised as an interconnected poly-nodal system having multiple centres of growth and development.

The Xiong'an New Area has also been created recently to the south-west of Beijing as a state-led development, in order to serve as a development pole for this economically vibrant triangle. Overall, with the cities of Beijing and Tianjin each having individual populations of approximately 22 million and 15 million respectively, the combined metropolitanised areas now have a population in aggregate exceeding 90 million, creating an inherently megalopolitan region.

6.5.2 QINGDAO (TSINGTAO), SHANDONG, PEOPLE'S REPUBLIC OF CHINA

Background and Origins

Qingdao is located in Shandong Province on the Yellow Sea of China. With its wide, protected natural bay, it has served as a significant harbour over the ages. Originally inhabited under the Eastern Zhou dynasty, it has been continuously settled and occupied. Under the later Qing Empire, when it was known as Tsingtao, its fortifications were developed and it became a strategic coastal city.

More recently, at the beginning of the 20th century, it was taken over and occupied by Germany, who had imperial designs on developing a colony and naval base in the region. Accordingly, a concession was negotiated focused on Kiaochau (Jiaozhou) Bay in 1898, with its centre around Market Street (Marktstrasse) in Tsingtao.

It was developed as a strategic naval base with several wharves, and an urbanised area was developed in the centre and a railway was constructed (Jiaoji Railway). The city centre retains historical buildings from this former German colonial area in the Shinan District and Badaguan area with the former Governor's Office building and Guest House near Qingdao Bay and old St Michael's Cathedral.

This period of occupation was, however, not to last for long, and during World War I, the city was invaded by Japan, who retained control for several years until its reversion to the Republic of China after the Treaty of Versailles. The city was again invaded and occupied by Japan in 1938 in World War II, but with their defeat it was incorporated as the People's Republic of China (PRC) at the end of the war.

Modern City – Buildings and Arteries

Under the PRC, due to its strategic location and with its port, the main wharves were developed in protected Jiaozhou Bay, and areas of manufacturing and shipbuilding were developed with significant economic activity ensuing around this.

More recently, the port has been modernised with new communication links and expressways under the National Trunk Highway System, and Qingdao has grown into one of the largest fully automated container ports. In 1984, it was designated as a special economic and technology development zone around which manufacturing has gravitated.

In 2011, the long Jiaozhou–Haiwan Bridge to connect Qingdao with Huangdao at the opposite side of the bay was opened with three navigable bridge sections. At the same time, a modern tunnel between Qingdao and Huangdao was opened across the shorter bay opening that carries the modern metro. The new metro system was constructed to extend across the city along with this line that crosses under the bay.

The eastward-facing coast of the city retains a number of attractive natural rocky bays of Tuandao, Qingdao, Huiquan, Taiping and Fushan. In 2008, with the Summer Olympics held in China, Qingdao developed the Sailing Centre for the sailing competitions that were held around Marina City in Fushan Bay. With the tall TV Tower, there are a number of skyscrapers built around the new central financial centre.

The city has developed a number of major industries including, for example, Haier white goods and the Tsingtao Brewery that originated during the German period. The population is approximately 7.5 million in the city and 10 million in the metropolitan area. The main airport is

Qingdao International Airport (TAO), located at Jiaodong to the north-west of the city.

Megalopolitan Region

With its unique geographical location on the Shandong Peninsula, the city has urbanised around its core, located on the eastern side of the bay. It also has attractive bays and wide beaches and a protected natural harbour. With major new infrastructure developments, in particular the long bay highway crossing and the cross-harbour tunnel, urbanisation has taken place on both sides of the bay and around the coastal perimeter.

With its favourable natural harbour and good communications, this has attracted development, bringing more businesses into the region. With many buildings and built-up areas regenerated, increasingly the region is developing as a dense metropolitan area with separate poles of development along the coast and inland, making it an attractive location for businesses. As these take hold, it is developing into a greater megalopolitan region.

Regional Transactional Forces and Typology

As a coastal city with agricultural hinterland, the configuration of Qingdao has acted as a place of attraction with its natural harbour and broad coastal bay area. Besides its activity as a regional trading centre, the natural harbour has been of strategic significance. With the modern infrastructure links built across and under the bay, this has led to further development also on the western side in the district of Huangdao, which has new dedicated poles of development.

From its early development on the eastern peninsula, the city has been enveloped and surrounded around the coastal bay with circumferential development occurring beyond. From a small, localised centre and its harbour, Qingdao has evolved into a highly urbanised and growing coastal metropolitan region. As further discrete nodes of planned development continue, this is resulting in the region, with its increased internal migration, beginning to demonstrate megalopolitan characteristics. All of this has been predicated by the two key infrastructure links, the long Cross Bay Bridge and the short Bay Crossing Tunnel that have opened up direct connections to southern districts.

6.5.3 GUANGZHOU AND GUANGDONG, PEOPLE'S REPUBLIC OF CHINA

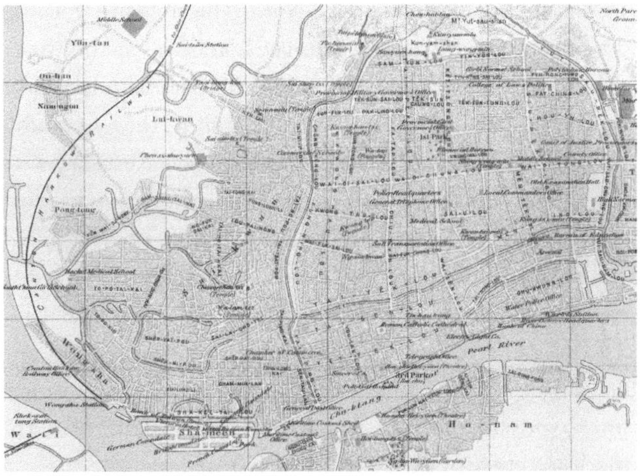

FIGURE 6.5.2 Canton – early configuration, 1924.

Source: Maphouse.

Delta Region

The **Pearl River Delta** is an alluvial region where three river branches – the Xi, Bei and Dong Jiang – converge to flow into the South China Sea. In a flat, wide region, crisscrossed by water channels and patterned with islands, settlements developed around the inland estuary where traders exchanged local commodities with travellers on the historic maritime Silk Routes. This dendritic system of rivers and tributaries creates a branching pattern of waterways across the whole region.

Here was the only location during the Qing government where foreign traders were allowed, at the place called Canton, named from the Portuguese 'cantao', around which settlement grew at a small location called Shameen Island at a branch in the Pearl River.

Subsequently with the outer port of Macau settled by the Portuguese and the British-occupied Hong Kong Island, where Aberdeen in the south afforded a protected port and location for replenishment, trade began to develop from inland China, regulated by Chinese customs under the 'canton' system. Within the extensive delta of the Pearl River and its numerous water arteries, the first settlements of foreigners developed at areas where navigation was feasible.

It was after the Opium Wars in 1840 that a number of treaty ports were agreed to be opened along the coast, which provided the focus for individual centres of trade to be developed. Upstream anchorages at Canton had enabled foreign trade and transactions to be undertaken at the earliest established Factories, or hongs.

Such places were forms of joint stock companies that were leased to foreign firms for trading, which were managed as a Cohong merchants' guild. The first site that was developed linearly across Shamian Island was at the waterside on Thirteen Factory Street opposite Honam Island on the Pearl River. From here and Whampoa, they were able to connect downstream at the entrance of the estuary with the separate enclaves of Macau and Hong Kong and the subsequent foreign concessions under the treaty port regime.

Canton on the Pearl River, with its products of rice, tea, porcelain and silk, was the origin of the early maritime Silk Route. With the development of modern China and opening up of the Chinese economy, this has grown almost unrecognisably into the sprawling city of **Guangzhou**, the modern centre of the province of Guangdong.

With its associated districts of Dongguan, Zongshan and Foshan, and with the fast-growing centres of Shenzhen and Zhuhai, having been designated as special enterprise zones (SEZ) in 1980, under economic inducements this has resulted in the great urban agglomeration of such a highly productive area within Guangdong province in the South-East Asian region that it has become.

Guangzhou (Canton)

Background and Origins

Guangzhou, formerly known as Canton, is the capital of the modern province of Guangdong on the Pearl River Delta. As a significant port upstream, it acted as origin of the maritime Silk Road, from the time of the Han dynasty during which tea, porcelain and silk were traded. During the later Yuan and Ming dynasties private foreign trade was restricted and it remained inward looking, but under the Ming dynasty it developed as a provincial capital as a walled city laid out with a grid structure.

After the Manchu invasions and the beginning of the Qing dynasty in 1644, whilst the old walled city was protected, a new area was developed, the Tatar Quarter. The Xiguan area was developed to the south-west outside the confines of the wall, also with narrow streets and waterways, in a more independent manner. The basic city of Canton was a brick-walled city with eight main gates and divided in two between the old northern and newer southern part. Along the riverbank and older Xiguan were fishermen and Tanka boat people living on junks.

By the 19th century, the city had become established between the Old and New City, with wealthier quarters to the west of the city and military camp and parade ground in the east, with areas in the province administered by the mandarinate.

The approach to Canton was through the Pearl River estuary, past old forts and batteries on either side of the Bocca Tigris (mouth of the tiger) up-river to the island of Whampoa, the Yellow Anchorage, Huangpu where large foreign ships were moored and cargoes were transferred into smaller Chinese sampans. They were met by hong merchants, who managed dues and customs as trade expanded.

Under the canton system, all trade was focused on the single port of Canton with its Thirteen Factories on Shamian Island located along the eastern bank of the Pearl River operating under the Cohong monopoly with its main trade of tea, porcelain and silk. Here it developed as an economic centre, supported by its guilds as a place of trade.

During the 1840s the Opium Wars developed between the Chinese and the British. This was ended with the Treaty of Nanking in 1842 under which Canton lost its privileged trade status. Five treaty ports were opened up at Canton, Shanghai, Ningbo, Fuzhou and Xiamen, and Hong Kong Island was ceded to Britain. These became new gateways for trade and concession areas developed under a supplementary treaty that benefitted from the principle of extra-territoriality offering 'most favoured nation' status. After the treaty there was extensive increase in trade and the city grew substantially outward from around central Shamian Island in Liwan District.

The result of these treaty concession arrangements and the introduction of newly developed ocean-going steamships was to significantly impact the development of the region. With Canton effectively as the 'head' and Shanghai as the 'tail' of the treaty port arrangements, these were to become privileged enclaves.

In 1907, a city expansion plan led to the extension of the Bund District and the Honam District on the southern riverbank. With a new Public Works Bureau and the introduction of a system of zoning, sections of the city wall were dismantled and new roads and the Changdi District were built. The Kowloon to Canton Railway had been realised and was operational with a new station in Canton, and the Honam District was developed farther to the south with a new bridge across the Pearl River.

In 1932, a master plan supported development around a new civic centre, with new districts and blocks, improvement of the waterfront and a process of land aggregation and restructuring whilst retaining former shop-houses and colonnades. The city retains the two large open areas of Yuexiu Park and Liuhuahu Park.

As a process of building and modernisation began, infrastructure improvements were required, along with the multiple waterways, riverbanks and crossings. Canton developed in different sectors, an administrative area to the north, Haizhu industrial district to the south, low-cost housing in the east and Dongshan services farther east, including a number of public parks. The Lingnan style of design or vernacular architecture was retained in places with its characteristic motifs, colours and gardens. Compact housing was built comprising Li-long houses and T'ang-lau terraces. Several parks and gardens across the city still encapsulate the renowned Eight Sights of Guangzhou, of rivers and mountains as originally described at the time of the Song dynasty.

But significant events that affected the city were to occur in the 1920s when Sun Yat-sen made it his base and briefly the first capital, now named Guangzhou, before moving to Nanjing. With the Kuomintang Nationalists under Chiang Kai-shek in civil war with the Communist Party, this was a short-lived period of the Republic of China.

Much was to change under the PRC with the following Great Leap Forward and Great Proletarian Cultural Revolution, but it was subsequent economic reforms and the socialist market economy that enabled take-off of the industrial manufacturing sector and mixed economy in the region. After the death of Mao, Deng Xiaopeng promoted work-unit enterprises under a 'responsibility system' that would reward productivity. These reforms were well suited to the economic environment of coastal Guangdong province, that as a compact Cantonese-speaking unit energetically responded from its favoured trading position in the Pearl River Delta.

Modern City – Buildings and Arteries

In line with its trading history, the first International Trade Fair was held in Guangzhou in 1957, which continues to be held there annually. Traditional local manufacture had been focused on porcelain and carved jade, wood and embroidery. With the opening up of the economy, significant changes ensued that emanated in increased industrialisation and economic development across the region.

In 1975, Zhou Enlai had called for a drive to implement the four 'modernisations'. With a gradual re-opening of the country Deng also pushed for further reforms and in 1979 announced the creation of special economic zones in Guangzhou and Guangdong province, and five years later Guangzhou was designated as an 'open coastal city'.

Development in and around Guangzhou focused on either side of the Pearl River and its port. Central Haizhu District is bordered on all sides by branches of the river, and the modern city has grown with the annexation of the new districts of Panyu and Huadu, and now comprises 11 urban districts that include central Haizhu and Liwan, Tianhe and Baiyun districts.

Major modern buildings which have been developed in Guangzhou are the CITIC Plaza, Guangzhou Circle, Pearl River Tower (designed by SOM), the IFC Twin Towers (CTF and IFC) and the Canton Tower (designed by Arup/IBA), built specially for the 2010 Asian Games with its twisted lattice and elliptical cross-section. With the development of the central downtown area, the city now has many commercial malls that have changed the streetscape, for example Teem Plaza, Victoria Plaza, Flower City Square and Shangxiajin Street.

At the core a new CBD was created in Tianhe District, and New Zhujiang City was built with its iconic opera house (designed by Hadid) and the Guangdong Museum (designed by Yim). New Pearl City and areas around the new high-speed train station have also been developed. A whole new university zone has been created in Changzhou that includes South China University of Technology.

The population of Guangzhou is estimated at approximately 13 million people, and in the Pearl River Delta Bay Area Economic Zone it is much greater. The population density of the whole Guangzhou city region is approximately 20,000 persons/sq. km. There are nine separate districts, including Shenzhen, Dongguan, Zhongshan, Foshan, Jiangmen and Zhuhai, which in total form an enormous urban agglomeration (Ref – National Central Cities). Most of the population are Han Chinese that speak Cantonese, whilst in-migrants generally speak Mandarin.

Each year there is a large seasonal volume of temporary migrants for up to six months of approximately 10 million migrant workers that gravitate to the area, and generally without permanent urban 'hukou' household registration, which limits their access to decent urban housing and public services (see Support Material: Annexe 6.5C – Chinese Housing, Villages within the City).

In terms of transport, the city of Guangzhou has a large number of long bridges that cross the Pearl River when entering from the south that comprise the Humen Great, Nansha, Huangpu, Guangzhou (ring expressway), Pazhou, Huanan (expressway), Liede, Guangzhou, Haiyong, Jiangwan, Haizhu, Jiefang and Renmi (inner ring) bridges.

The first three Guangzhou Metro (GMTR) lines that run north-south and east-west were built crossing the river, and then a further four lines were realised. With connections to Foshan, which has its own MRT and also separate light rail lines, there is a comprehensive network of public transport in and around the urbanised areas.

In terms of rail, an extensive investment in new high-speed rail lines has taken place with fast inter-city connections between Guangzhou and the major cities of Beijing, Wuhan and Shanghai. With the new stations of Guangzhou, Dongguan, Shenzhen North and Hong Kong West Kowloon Station, this has grown in scale to become the connected inter-city system that comprises the Greater Bay Area Transit Network.

With the realisation of the special economic zones (SEZ), improvement in interconnection of the regional infrastructure was required, which led to the development of major highway and bridge links. This includes the Guangzhou–Shenzhen Expressway via Dongguan, around which development has gravitated, and on to Hong Kong on the eastern side of the Pearl River.

With improved eastern access, another link on the western side of the Pearl River has been developed south from Guangzhou to Zhuhai and Macau where previously several river crossings were required by ferry. These two routes were to be capped to the north with a Guangzhou circumferential road. The very long causeway which has been built across the estuary bay to connect Hong Kong with Zhuhai and Macau is the long, combined bridge and tunnel crossing, the HKZM Bridge, 55 km. in length.

An even newer connection under development between Shenzhen and Zhongshan across the Pearl River estuary is a combined 24 km. bridge and tunnel crossing Lingdingyang, constructed as another fixed link across the river to connect these two major growing conurbations. These constitute direct routes that form new permanent links across the entire megalopolitan region.

Since the 1990s, a regional master plan has been pursued for the Pearl River Delta, and in 2000 the city boundaries of Guangzhou were extended to include Panyu District in the south and Huadu District in the north. New infrastructure was also promoted and a deep-water port at the tip of the new Nansha District was realised.

With the expansion of the major individual ports and airports at Hong Kong, Shenzhen, Zhuhai, Foshan and Guangzhou, each of these are acting as significant new poles of development in the region. Guangzhou Baiyun International Airport (CAN) was opened in 2004 on the outskirts of Guangzhou to replace the original Canton Airport. A further Pearl River airport is under construction as Foshan Gaoming International Airport.

In addition, new facilities were built for the Canton International Trade Fair in Haizhu District, and the whole island University City, with ten university campuses, was realised. In 2005 the Greater Guanzhou Metropolis was created that comprises the Central City Area, Panyu District, Huadu District, Zengchen City and Conghua City.

With continuing extensive development of economic zones around this Pearl Inner River Delta, it has become a major manufacturing hub. This includes areas such as the industrial Guangzhou Economic and Technological Development Zone, the Guangzhou Nansha Export Processing Zone and Guangzhou Free Trade Zone and Science City.

Guangdong Province

The wider Guangdong province constitutes each of the historic cities of Foshan, Huizhou and Zhaoqing, the port cities of Canton and Huangpu, Jiangmen, Shantou and Zhanjiang and the resource-rich cities of Maoming and Shaoguan with coal and ore. During the 1980s, the province of Guangdong formerly had prefecture-level capitals in the Inner Delta of Canton, Foshan, Jiangmen, Shenzhen, Zhuhai and Haikou. However, as growth developed, in 1988, these became superseded by metropolitan regions headed by the outer port cities of Yangjiang and Shanwei (west and east) and inner Qingyuan and Heyuan due to their strategic locations on road and rail corridors.

During the 1990s, as the growth continued, the Inner Delta area developed around the denominated first-tier cities of Dongguan, Baoan, Doumen and Panyu and the second-tier cities of Foshan, Nanhai, Shunde, Xinghui and Zhongshan. With growth developing on either side of the Bay Area, this was reinforced by the creation of SEZ's on either side of the Pearl River at Zhuhai and Shenzhen respectively.

Zhuhai and Macau

Zhuhai is on the south-west bank of the Pearl River in Guangdong province, adjoining the special administrative region of Macau. It was formerly essentially an agricultural and fishing district comprising several offshore islands. It was designated as a SEZ in 1980 just after the similar creation of the Shenzhen SEZ.

With a deep-water port, it has attracted a significant number of large manufacturing industries, includes electronics, biotechnology and petrochemical. With a high-tech industrial development zone and free trade zone, and with its high-speed rail connections, it forms a part of the Pearl

River Delta Ring Expressway now linked with Guangzhou. Its population is approximately 2.5 million.

The very long, tolled highway Hong Kong–Zhuhai–Macau Bridge (HKZM), with its two artificial islands, undersea immersed tube tunnel section and central bridge sections across the Pearl River Delta, was realised and opened in 2018. Promoted by Gordon Wu from Hong Kong, it was supported by the Zhuhai administration and provides a direct link for road freight traffic across the bay to connect these three major cities in the Bay Area.

The growth of **Macau**, with its historic Portuguese centre, continues as a special administrative region (SAR) of China with its established casinos and gaming industry anchoring it as a centre of tourism and recreation in the region. Macau International Airport (MFM) was built on reclaimed land off Taipa in the estuary in 1995 just before handback and with a new container port. Since return to the PRC, despite a permanent population of under 1 million people, economic activity has continued to be maintained.

Shenzhen

Shenzhen is located on the east bank of the Pearl River estuary in the province of Guangdong just at the northern border of Hong Kong. Originally, it featured traditional Hakka walled villages that were typical features in the region. Previously within Xin'an County (then Bao'an) in South China, by the historic town of Nantou, under the Tang dynasty its seat was moved to Dongguan. Local trading in the area originally was in salt, spices, rice and pearls.

With the Opium Wars during the 19th century, and as a result of the Treaty of Nanking, the areas of Hong Kong Island and Kowloon Peninsula were separated from Xin'an and ceded to Britain. Subsequently, the New Territories were also leased for 99 years to Britain under the Second Convention of Peking. Close links were developed with the construction of the Kowloon–Canton Railway with connections of the British and Chinese sections and final stop at Shenzhen Station, known as 'Shum Chun', cut through Beacon Hill Tunnel, which opened in 1910.

However, this line was not fully utilised until the 1970s when, with the intention to build a series of new towns in the New Territories, it was enhanced and dualled. The railway was corporatised and expanded as the Western Corridor Railway, which was to become KCR East Rail and eventually taken over by Hong Kong MTR Corporation and integrated with their metro system. At this period there was a significant increase of in-migration of people to the region.

With its strategic coastal location, the Guangdong Provincial Government, recognising the strength of potential port activities, agreed to the establishment of an initial new industrial zone at Shekou. In order to increase foreign trade, a new coastal SEZ at Shenzhen was then nominated. Formalised in 1978, Shenzhen SEZ was established under a preferential regime that would promote foreign investment.

Whilst initially requiring special border arrangements, this zone became highly successful. At the time of initial development a large number of informal 'villages-in-the-city' were realised by new workers. These were very dense ad-hoc developments, generally located nearby new factories. As the area rapidly urbanised, they were taken over, rebuilt and upgraded under major programmes of urban renovation around 2005. This, however, had the effect of making many areas of housing unaffordable for workers, creating a deficit in affordable housing.

In 2010, the SEZ was substantially increased again with the creation of the Qianhai New District around the semi-circular bay and the Qianhai–Shenzhen–Hong Kong Modern Service Industry Cooperation Zone. Along with the Shekou industrial zone this was integrated into a new Guangdong free trade zone in 2015 for the promotion of modern service industries having favourable tax rates. The whole area developed into a major hub for high-tech industries, consumer electronics and IT components. With a major new container terminal developed, the Port of Shenzhen, this has grown to comprise the largest trading port in Guangdong on the Pearl River.

From a population of less than 0.1 million in 1980, the population of the city of Shenzhen has reached approximately 15 million and 23 million in the metropolitan area (although only 5.7 million have registered 'hukou').

The jurisdiction of the city has been divided into individual districts which are contained within the SEZ. From the first international foreign trade centre, the Guomao Building, the city has risen vertically with major new landmark developments such as the Ping An Finance Centre (designed by KPF), the China Resources Tower (designed by KPF), Shun Hing Square (designed by Cheung) and Kingkey KK100 and the new civic centre. Shenzhen has its own stock exchange and also hosts ChiNext, along with many high-tech companies such as Huawei, ZTE and TenCent with its Seafront Towers headquarters. New universities and institutes of technology also have been established that are enhancing the skills base in the area. Foxconn has a major Apple iPhone manufacturing facility there.

With the new Shenzhen Bao'an International Airport (SZX) on the coast in the bay and new rapid transit connections, Shenzhen is fully integrated with high-speed rail line connections between Guangzhou and Hong Kong. A new bridge crossing the Pearl River west from Shenzhen to Zhongshan is also under realisation mid-estuary, which will further consolidate transport links between each side of the river.

Dongguan

The other major urbanised area is **Dongguan** in Guangdong province on the Dong River flowing into the Pearl River Inner Delta just to the north-west of Shenzhen. From being formerly predominantly agricultural, created a city in 1987, it has been transformed into a significant industrial and manufacturing district where several major foreign manufacturers are located. It has six main districts and incorporates a number of individual towns with a population of approximately 10 million, of which approximately 50% are locally registered.

In terms of communications, Dongguan lies on the Guangzhou to West Kowloon high-speed railway line and on the north-south axis G-4 Expressway between Beijing and Hong Kong and the Humen Bridge that spans the Pearl River to connect with Nansha and Zhongshan. It is also home to the South China Mall, the largest retail mall in the world.

Megalopolitan Region

It was mainly as a result of the creation of SEZs at Shenzhen and Zhuhai that major industrialisation took off in this region. Around Hong Kong new areas of land reclamation are continuing to be undertaken around West Kowloon, which has enabled further development to be undertaken. In Guangzhou and Dongguan also, from their agricultural beginnings, more land has been consolidated and rivers across the delta have been bridged, facilitating communications and intensive zones of industrial production.

All of this has led to the poly-nodal realisation of metropolitan areas across and around the Pearl River Delta region. Accordingly, the expansion of this South China metropolis has continued to such an extent that it has resulted now in the creation of the heavily urbanised and industrial agglomerated region of the Greater Bay Area of the Pearl River in southern China, a true megalopolitan region.

Regional Transactional Forces and Typology

From the multiple entrances to the dendritic Pearl River Delta with its individual islands and water courses, settlements were to develop on the most advantageous locations on the waterway. From the beginnings of shared trade on Shamian Island with its wharves and warehouses, the centres of Guangzhou and Macau and Hong Kong materialised, with the whole area becoming one of the most productive trading zones.

From the Portuguese settlement in Macau, the British settlement in Hong Kong and the Kowloon Peninsula and the New Territories, each promoted external trade and services, gradually extending their restricted geography with new areas of reclaimed land. Each early city centre has followed its own growth pattern, from the original Thirteen Factories of Canton to the modern manufacturing facilities of Guangzhou. From Xin'an came the formation of Bao'an and Dongguan and the creation of Shenzhen on the eastern side, promoted as a special industrial zone now with its extensive modern container port, and with Zhuhai on the western side of the Pearl River.

With the reversion of the leased areas of Hong Kong and Macau back to China and the creation of two special administrative regions, this has enabled the Greater Bay Area to develop more as an integrated whole. With its new inter-connected integrated infrastructure of road and rail routes, this has led to its development as an agglomerated megalopolitan region in southern China with around 100 million people, having significant productive and manufacturing capacity across this whole area.

From its coastal location with marine waterways shaped by the Pearl River and its delta, this region has developed from its beginnings of essentially separate individual port locations into the poly-nodal development of accreting land areas, each transformed into manufacturing areas and commercial districts interlinked with complex transport networks to form major hubs.

Within Guangdong province there is now a constellation of cities from the former triangle of Canton, Macau and Hong Kong to the new conurbations south of Guangzhou of Zhuhai, Zhongshan and Foshan on the western side of the Pearl River and Dongguan and Shenzhen on the eastern side of the estuary, having Guangzhou at the apex of the triangle. With the two new lengthy Bay Area fixed crossings of the HKMZ and Lingdinyang, these links across the bay greatly shorten transport routes and are redistributing regional patterns of freight traffic.

With the western conurbation of Guangzhou and Foshan, 'Guang-Fo', and eastern Dongguan, there are competing commercial and industrial areas across the region, which is covered by a dendritic pattern of waterways. Without permanent accommodation, however, adequate facilities are required for the 'floating population' where unskilled labour migrates in response to evolving employment opportunities. These seasonal shifting patterns of population need to be accommodated under the challenge of dualism of workers either with or without urban 'hukous'.

The urbanised centres of Guangzhou, Zhuhai and Dongguan, Shenzhen and Hong Kong all combined in and around the Pearl River Delta, with their extensive rail, road, sea and air connections, now constitute one of the most dynamic economic megalopolitan agglomerations in South-East Asia.

6.5.4 CHENGDU, SICHUAN, PEOPLE'S REPUBLIC OF CHINA

Background and Origins

Chengdu is capital of the province of Sichuan located on the western edge of the Sichuan Basin, located in the Chengdu Plain. To the west and north-west are the Qionglai and Longmen mountains and forests. The Jin and Fu rivers flow through the city and connect with the Min River, a tributary of the Yangtze River. Chengdu was inhabited since the earliest times under the ancient Shu Kingdom, bronze civilisation and Baodun culture.

In 311 BCE the Dacheng (big city) and Shaocheng (small city) were laid out with walls and structure modelled on Xianyang, the Qin capital. As capital during the Three Kingdoms era under the Qin dynasty, it prospered during the Han dynasty. Besides agriculture, with its favourable land and irrigation system, the Dujiangyan Scheme, used for flood control with its levees and weirs, brocade was produced and other commodities, such as tobacco, were grown in the region.

During the 17th century, it was briefly known as 'Xijing' (Western Capital). Later, during World War II, when the Kuomintang government retreated to Sichuan under Chiang Kai-shek, it resisted Japanese advances. It was the last city in the civil war to be taken by the PLA in 1949.

Modern City – Buildings and Arteries

The city of Chengdu has its old core and temples, Jinli Street and Kuanzhai, wide and narrow alleys, and within the city there are distinct residential and commercial areas. The Old Anshun Covered Bridge crosses the Jinjiang River. It has been developed outwards with districts surrounded by a series of outer circumferential roads. Post-war the city has steadily industrialised and great progress has been made in modernisation.

The main central districts are Qingjang, Jinniu, Chenghua, Jinjiang and Wuhou with a central district of Xihuamen with new railway connections to the north and east. Growing as a technological hub with distinct specialised areas, there is a plan for areas to be pedestrianised under the concept of a 'park city', and a new satellite city, 'Great City', is to be created.

With the Funan River and Shahe River renovations, there are several areas that have been upgraded. Renmin Park, formerly Shaocheng Park, is an attractive recreational park with other outer parks. The Chunxi Road is a pedestrianised retail area and with Jinxiu Garden (developed by Singapore Capital).

Under the Five Main Roads and One Bridge Project, significant infrastructure with new highways and expressways has been developed that includes the new Chengdu metro and BRT lines, high-speed rail lines, major new East, South and West stations and the Cross Europe freight rail line that runs through to Poland, the China Europe Railway Express.

The city has a developed Finance Centre for its banking institutions and new business districts of Jiaozi Park and Xinnan World Business District. Sichuan University has integrated Chengdu and West China faculties, and there are a number of new universities and scientific research institutes.

These comprise a high-tech industrial development zone' for IT enterprises that includes many of the major international electronics companies along with a global delivery centre in Chengdu and a high-tech comprehensive free trade zone, export processing zone and National Cross Straits Technological & Industrial Development Park.

Chengdu Tianfu Software Park has significant outsourcing businesses for software services. In addition, a large logistics centre comprises very large logistics industries and significant automobile manufacturing businesses along with major car assembly and aircraft manufacturing plants.

In 2011, Tianfu New Area was created, an area planned for new high-tech innovative industries and service facilities. The New Century Global Centre was constructed, 400 by 500 m. in overall dimensions, an enormous commercial complex and exhibition centre to the south of the city.

In addition, the new Tianfu International Airport (TFU) (designed by Aéroports de Paris) has been built to the south-east, as a second airport in Chengdu after Shuangliu Airport (CTU) with an aviation development zone and aerotropolis all connected by major expressways. The Meishan California Smart City is also a new planned district.

The population of the city is approximately 5 million with approximately 20 million in the greater urbanised area.

Regional Transactional Forces and Typology

From its role as the capital of the ancient Shu Kingdom on the Chengdu plain, and city of brocade and hibiscus, it has retained its significance under subsequent Chinese dynasties. As the capital of Sichuan province, it has evolved from its early agricultural background to highly active modern manufacturing sectors. Economically it has attracted significant inward investment and has become a national electronics and IT hub in China with major chip manufacturers (Intel and AMD), computing services suppliers and software developers. It is now designated as one of China's 'national central cities' around which urban development will be focused.

With its central location, industrially it has developed as a logistics centre and outsourcing centre with automobile and defence manufacturing industries and its technological, high-tech industrial zones, dedicated trade and export processing zones and free trade zone. Commercially, it has major convention and exhibition facilities and retail centres. New residential areas have also been built, extending outwards to the south-east of the city.

As a city type, the city was built and has been developed around its historic central core. Modern developments have led to the realisation of ongoing circumferential development with a series of ring roads that encompass the city. With integrated neighbourhoods and districts, it has continued to grow and expand radially outwards and towards the periphery. It now has the characteristics of a very large centralised Chinese city having a classic radio-centric development pattern enhanced with local and traditional Sichuan characteristics in its much larger metropolitan area. The following figures illustrate the modern configuration of the megalopolitan agglomerations in China as described and the evolution of their modern population.

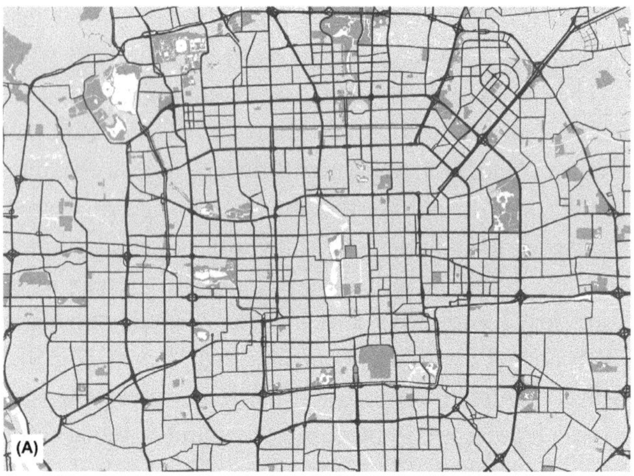

FIGURE 6.5.3A Greater Beijing – modern configuration.

FIGURE 6.5.3B Beijing - Tianjin – modern configuration.

FIGURE 6.5.4 Qingdao – modern configuration.

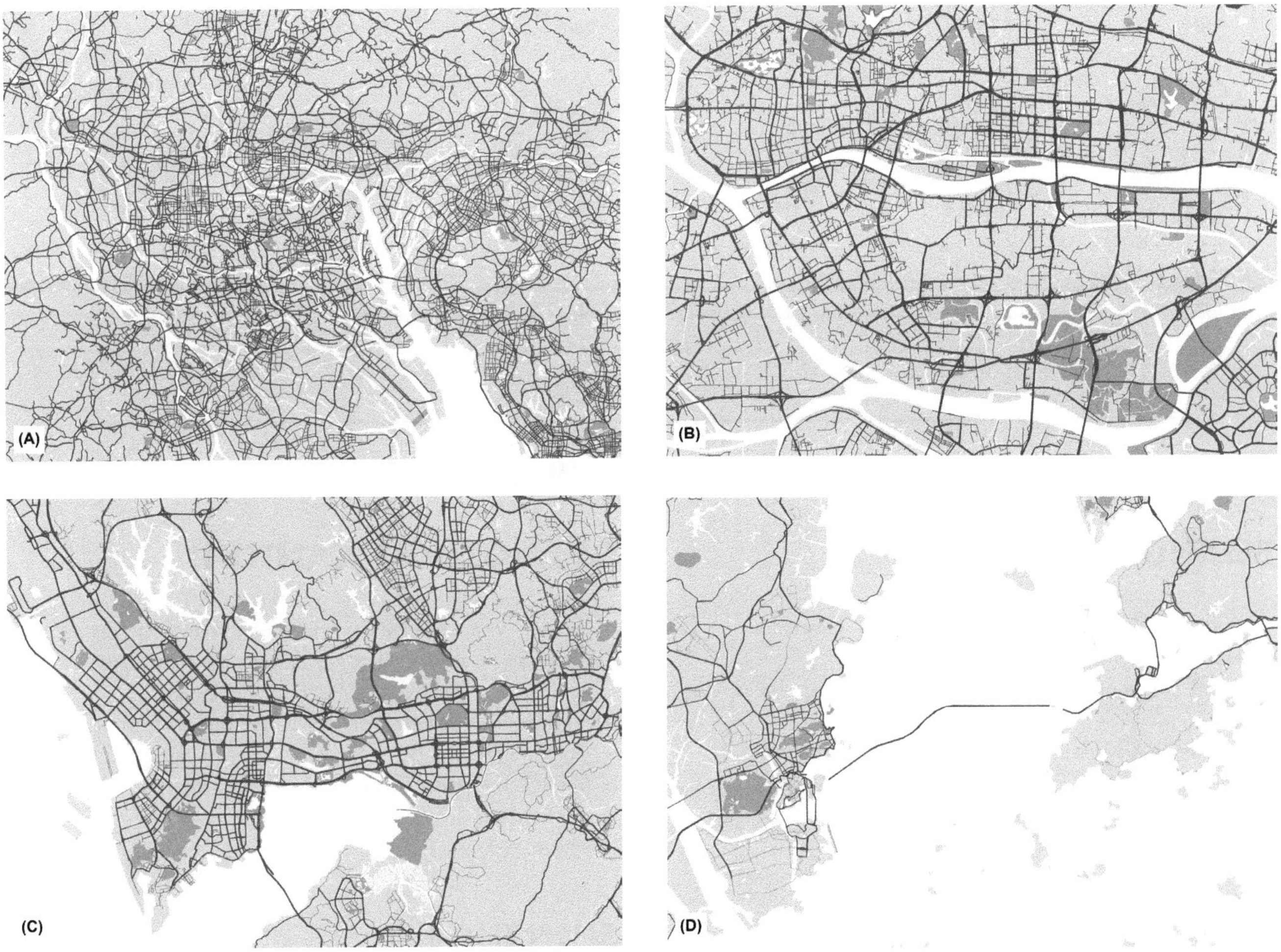

FIGURE 6.5.5A AND B, C AND D Guangzhou and Canton, Shenzhen and Macau/Hong Kong – modern configurations.

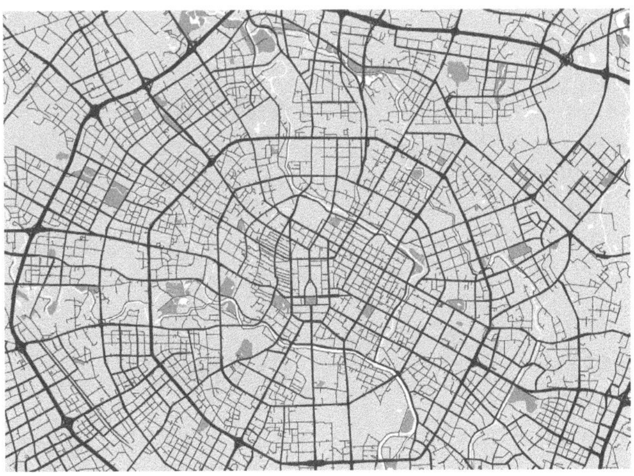

FIGURE 6.5.6 Chengdu – modern configuration.

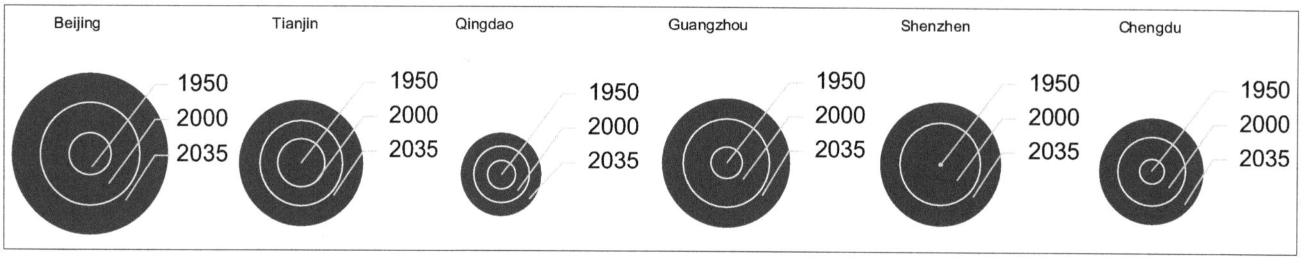

FIGURE 6.5.7 Megalopolitan China populations.
Data source: Global Database of Metropolises 2020 – UN-Habitat (2020), UN DESA Population Division.

Further Reading

BEIJING

Campanella, Thomas – *The Concrete Dragon*, Princeton Architectural Press, 2008.
Clements, Jonathan – *Beijing, a Short History*, Haus Publishing, 2022.
Curtis, Simon & Klaus, Ian – *The Belt and Road City*, Yale University Press, 2014.
Friedmann, John – *China's Urban Transition*, University of Minnesota Press, 2005.
Hillman, Jonathan – *The Digital Silk Road*, Harper Collins, 2021.
Lorge, Peter – *Imperial China*, One World Publications, 2021.
Miller, Tom – *China's Urban Billion*, Zed Books, 2012.
Miller, Tom – *China's Asian Dream*, Zed Books, 2019.

TIANJIN (TIENTSIN)

Paltemaa, Lauri – *Managing Famine, Flood and Earthquake in China, Tianjin 1958–85*, Routledge, 2015.
Spence, Johnathan – *The Search for Modern China*, WW Norton, 2013.

QINGDAO (TSINGTAO)

Fenby, Jonathan – *The Siege of Tsingtao*, Viking China, 2015.
So, Fion WL – *Germany's Colony in China, Colonialism, Protection and Economic Development in Qingdao and Shandong, 1898 – 1914*, Routledge, 2019.

GUANGZHOU (CANTON)

Al, Stefan – *Villages in the City*, Hong Kong University Press, 2014.
Cookson Smith, Peter – *Shaping Canton, the Mountains Are High and Emperor far Away*, ORO Editions, 2023.
Fay, Peter – *The Opium War, 1840–42*, University of North Carolina Press, 1997.
Li, Dun – *The Ageless Chinese*, Macmillan, 1978.
Van Dyke, Paul – *Whampoa and the Canton Trade*, Hong Kong University Press, 2020.

SHENZHEN

Du, Juan – *The Shenzhen Experiment, the Story of China's Instant City*, Harvard University Press, 2020.
Hu, Richard – *The Shenzhen Phenomenon, from Fishing Village to Global Knowledge City*, Routledge, 2020.

MACAU

Porter, Johnathan – *Macau, the Imaginary City*, Taylor & Francis, 1999.
Wei, George – *Macao, the Formation of a Global City*, Routledge, 2014.

CHENGDU

Stapleton, Kristin – *Civilizing Chengdu, Chinese Urban Reform, 1895–1937*, Harvard University Press, 2000.
Yumin, Ye & Le Gates, Richard – *Coordinating Urban and Rural Development in China, Learning from Chengdu*, Ed. Elgar Publisher, 2013.

Part VII

Summary
7.0 Synopsis

The goal has been to follow the development and urbanisation of *'global city typologies'* in terms of their 'transactional forces in urbanised development', from their origins and functions via their physical places and arteries. This has addressed the evolution of such transactional forces in the development of built places and urban networks for inhabitants in their everyday needs of work, accommodation, recreation, welfare and transportation.

The overall scope has been to broadly assess those historical forces and evolving factors that have impacted the creation and realisation of newly developing urbanised places and cities.

From consideration of their original natural settings and human settlements, the analysis has been aimed at determining different typologies of established and emerging cities through review of the development of their differing urbanised forms and communication networks, their organisation, function and comparative layouts.

The separate Parts with sets of individual chapters have been grouped around analysing 125 different *established, planned and emerging global cities* according to different typologies and thematic categories that comprise:

- Established historic cities – imperial, theistic, academic and ecclesiastic;
- Early trading cities – mercantile, maritime, island and mountain;
- Planned cities – new capitals, former Soviet, divided, industrial and modern American;
- Emerging global cities – innovative and transnational;
- Global South – mega African, mega South American and mega Asian cities' and
- Megalopolitan agglomerations.

so as to observe the morphology, evolution and shape of both their *private* built developments and their *public* infrastructure.

Based upon prevailing forces that encompass globalisation and urbanisation, urban forms have incurred substantial growth pressures which are only being accommodated to different degrees. Such transactional forces have evolved in various manners in established, planned and emerging global cities across different continents.

The aim has been to address many of these transactional factors from differing historic, cultural, social, technical and organisational perspectives in the built environment, through to consideration of their impact on the economy of modern global cities.

The goal for most cities is to provide a harmonious place to live and work in, safely with proximity for market transactions, having purposeful fit, essential services, good urban form and experience of a sense of place, in emergent and sustainable environments that enhance our physical, mental and economic well-being.

These objectives of carefully handling the transactional challenges in the evolution and morphology of cities and managing growth pressures in the integration of places and networks for general mobility remain demanding and particularly in terms of resilience under prevailing forces of climate change.

Beyond traditional cities with their cores and surrounding suburbs, the analysis has been broadened to cover emerging cities, mega cities and extending cities and urbanised areas especially in the **Global South**, many having extensive informal areas, to whole areas of polynodal development that are generating new megalopolitan agglomerations with continuously extending peripheries

DOI: 10.4324/9781003353348-39

(see Support Material: Annexe 7.0 – Evolution of Largest Cities and Urbanised Areas).

Such different typologies are now increasingly evident with the duality of new formally planned cities versus those much more informal places that are evolving generally in peripheral areas around existing settlements and on available land that often has only marginal physical characteristics.

The functionality and ambience of these areas, with decent services and communications for different mixes of inhabitants, is the real aspiration for future urbanised areas in different regions around the world. This remains the global challenge for their provision in a sustainable, renewable and equitable manner in order to provide protection, purpose and comfort for dense groupings of people in an increasingly urbanised world environment.

Bibliography and Further Reading

GENERAL

International Dictionary of Historic Places – *Ring, Watson, and Schellinger*, Routledge, 1996.
Volume 1 – The Americas
Volume 2 – Northern Europe
Volume 3 – Southern Europe
Volume 4 – Middle East and Africa
Volume 5 – Asia and Oceania

Almandoz, Arturo, ed. – *Planning Latin America's Capital Cities, 1850 – 1950*, Routledge, 2002.

Bakirtzis, Nikolas & Zavagno, Luca – *The Routledge Handbook of the Byzantine City*, Routledge, 2024.

Bates, Crispin & Mio, Minoru – *Cities in South Asia*, Routledge, 2015.

Bharne, Vinayak – *The Emerging Asian City*, Routledge, 2013.

Carter, Donald K. – *Remaking Post-Industrial Cities: Lessons from North America and Europe*, Routledge, 2016.

Elsheshtawy, Yasser – *Planning Middle Eastern Cities: An Urban Kaleidoscope*, Routledge, 2004.

Gonzalez-Perez, J M, Irazabal, C & Lois-Gonzalez, R, eds. – *The Routledge Handbook of Urban Studies in Latin America and the Caribbean*, Routledge, 2024.

Gordon, David – *Planning Twentieth Century Capital Cities*, Routledge, 2006.

Gunzburger Makas, Emily & Damljanovic Conley, Tanja – *Capital Cities in the Aftermath of Empires: Planning in Central and Southeastern Europe*, Routledge, 2010.

Hall, Thomas – *Planning Europe's Capital Cities: Aspects of Nineteenth-Century Urban Development*, Routledge, 1997.

Hamnett, Stephen & Forbes, Dean, eds. – *Planning Asian Cities, Risks and Resilience*, Routledge, 2012.

Heathcott, Joseph, ed. – *The Routledge Handbook of Infrastructure Design*, Routledge, 2023.

Hein, Carola – *Port Cities: Dynamic Landscapes and Global Networks*, Routledge, 2011.

Heitzman, James – *The City in South Asia*, Routledge, 2008.

Hu, Richard, ed. – *Routledge Handbook of Asian Cities*, Routledge, 2023.

Lee, Robert & McNamara, Paul – *Port Cities and Their Hinterlands*, Routledge, 2023.

Le Galès, Patrick & Robinson, Jennifer, eds. – *The Routledge Handbook of Comparative Global Urban Studies*, Routledge, 2023.

Nel-lo, Oriol & Mele, Renata – *Cities in the 21st Century*, Routledge, 2016.

Nelson, Arthur & Lang, Robert – *Megalopolitan America*, Routledge, 2019.

Padawangi, Rita – *Routledge Handbook of Urbanization in Southeast Asia*, Routledge, 2018.

Parnell, Susan & Oldfield, Sophie – *The Routledge Handbook on Cities of the Global South*, Routledge, 2014.

Rowe, Peter – *East Asia Modern, Shaping the Contemporary City*, Reaktion Books, 2005.

Rukmana, Deden – *The Routledge Handbook of Planning Megacities in the Global South*, Routledge, 2020.

Segbers, Klaus – *The Making of Global City Regions: Johannesburg, Mumbai, Sao Paulo & Shanghai*, Johns Hopkins University Press, 2007.

Shirley, Ian & Neill, Carol – *Asian and Pacific Cities: Development Patterns*, Routledge, 2013.

Silver, Christopher – *Planning the Megacity: Jakarta in the Twentieth Century*, Routledge, 2008.

Sridar, Kala & Mavrotas, George – *Urbanization in the Global South*, Routledge, 2021.

Tian, Li & Guo, Yan – *Peri-Urban China, Land Use, Growth and Integrated Urban-Rural Development*, Routledge, 2019.

Vormann, Borris – *Global Port Cities in North America: Urbanization Processes and Global Production Networks*, Routledge, 2015.

Waddell, Gene – *Asian Building Types*, Ind., 2022.

Werthmann, Christian – *Informal Urbanization in Latin America*, Routledge, 2021.

Wu, Fulong – *Globalization and the Chinese City*, Routledge, 2005.

Yacobi, H & Nasasra, M, eds. – *Routledge Handbook on Middle East Cities*, Routledge, 2019.

PARTICULAR

Abrams, Charles – *The City is the Frontier*, Harper & Row, 1967.
Angel, Shlomo – *Planet of Cities*, Lincoln Institute of Land Policy, 2012.
Barnett, Jonathan – *City Design, Modernist, Traditional, Green and Systems Properties*, Routledge, 2016.
Batty, Michael – *Inventing Future Cities*, MIT Press, 2018.
Benevolo, Leonardo – *The European City*, Blackwell, 1993.
Berger, Alan & Kotkin, Joel, eds. – *Infinite Suburbia*, Princeton Architectural Press, 2017.
Bertaud, Alain – *Order Without Design, How Markets Shape Cities*, MIT Press, 2018.
Blumenfeld, Hans – *The Modern Metropolis*, MIT Press, 1967.
Breese, Gerald – *Urbanization in Newly Developing Countries*, Prentice-Hall, 1966.
Briggs, Asa – *Victorian Cities*, Penguin, 1968.
Bronin, Sara – *Key to the City, How Zoning Shapes our World*, WW Norton, 2024.
Brunn, Stanley & Williams, Jack – *Cities of the World, World Regional Urban Development*, Harper & Row, 1983.
Burdett, Ricky & Rode, Philipp, eds. – *Shaping Cities in an Urban Age* (London School of Economics & Alfred Herrhausen Ges.), Phaidon, 2016.
Calder, Kent – *Global Political Cities*, The Brookings Institution, 2021.
Campanella, Thomas – *The Corporate Dragon, China's Urban Revolution*, Princeton Architectural Press, 2008.
Chandler, Tertius & Fox, Gerald – *3000 Years of Urban Growth*, Academic Press, 1974.
Churchill, Henry – *The City Is the People*, Reynal & Hitchcock, 1945.
Clark, Greg – *Global Cities*, The Brookings Institution, 2016.
Clark, Greg & Moonen, Tim – *World Cities and Nation States*, Wiley Blackwell, 2017.
Clark, Peter, ed. – *The Oxford Handbook of Cities in World History*, Oxford University Press, 2016.
Curtis, Simon & Klaus, Ian – *The Belt and Road City*, Yale University Press, 2024.
Darwin, John – *Unlocking the World, Port Cities and Globalization*, Allen Lane, 2020.
Davis, Mike – *Planet of Slums*, Verso Books, 2006.
Davis, Mike – *Magical Urbanism, Latinos Reinvent the US City*, Verso Books, 2000.
Fernandez-Armesto, Felipe – *The Americas, the History of a Hemisphere*, Weidenfeld & Nicolson, 2003.
Fitzsimmons, A., fwd. – *World's Best Cities*, National Geographic, 2014.
Frankopan, Peter – *The Silk Roads*, Bloomsbury, 2015.
Furtado, Peter, ed. – *Great Cities Through Travellers' Eyes*, Thames & Hudson, 2019.
Garreau, Joel – *Edge City*, Doubleday, 1991.
Geddes, Patrick – *Cities in Evolution*, Williams & Norgate, 1915.
Gehl, Jan – *Cities for People*, Island Press, 2010.
Girouard, Mark – *Cities & People*, Yale University Press, 1985.
Glaeser, Edward – *Triumph of the City*, Penguin Press, 2011.
Glaeser, Edward, ed. – *Urban Empires*, Routledge, 2021.
Glaeser, Edward & Cutler, David – *Survival of the City*, Basic Books, 2021.
Goldin, Ian & Lee-Devlin, Tom – *Age of the City*, Bloomsbury Press, 2023.
Gottmann, Jean – *Megalopolis*, 20th Century Fund New York, 1961.
Hakim, Besim – *Arabic – Islamic Cities*, Keegan Paul International Ltd., 2008.
Hall, Peter – *The World Cities*, McGraw Hill, 1966.
Hall, Peter – *Cities of Tomorrow*, Blackwell, 1988 (3rd edition 2002).
Hall, Peter – *Cities in Civilization*, Weidenfeld & Nicolson, 1998.
Howard, Ebenezer – *Garden Cities of Tomorrow*, Faber & Faber, 1946.
Hunt, Tristram – *Ten Cities That Made an Empire*, Allen Lane, 2014.
Jacobs, Jane – *The Death and Life of Great American Cities*, Random House, 1961.
Jacobs, Jane – *The Economy of Cities*, Penguin, 1969.
Jacobs, Jane – *Cities and the Wealth of Nations*, Random House, 1984.
Kasarda, John & Lindsay, Greg – *Aerotropolis*, Allen Lane, 2011.
Knox, Paul – *Biographies of 50 City Districts, Palimpsests*, Birkhauser, 2012.
Knox, Paul – *Atlas of Cities*, Princeton University Press, 2014.
Knox, Paul – *Better by Design?* Virginia Tech Publishing, 2020.
Kostof, Spiro – *The City Shaped*, Thames & Hudson, 1991.
Kostof, Spiro – *The City Assembled*, Thames & Hudson, 1992.
Kotkin, Joel – *The City*, Modern Library, 2005.
Lang, Jon – *Routledge Companion to 20th & Early 21st Century Urban Design*, Taylor & Francis, 2021.
Lewis, Nigel C – *Design & Order – Principles in the Planning & Making of Place*, Wiley Blackwell, 2020.
Lynch, Kevin – *The Image of the City*, MIT Press, 1960.
Lynch, Kevin, ed. – *Growing Up in Cities*, MIT Press – UNESCO, 1977.
Lynch, Kevin – *Theory of Good City Form*, MIT Press, 1981.
Marshall, Stephen – *Cities, Design and Evolution*, Routledge, 2009.
Marshall, Tim – *Divided*, Elliot & Thompson, 2018.
McEvedy, Colin – *Cities of the Classical World*, Allen Lane, 2011.
McGee, Terry – *The South East Asian City*, Bell, 1967.
McGuirk, Justin – *Radical Cities*, Verso, 2014.
Montgomery, Charles – *Happy City*, Penguin Random House, 2013.
Morris, A E J – *History of Urban Form*, Prentice Hall, 1994.
Morris, Jan – *Among the Cities*, Viking, 1985.
Mumford, Eric – *Designing the Modern City*, Yale University Press, 2018.
Mumford, Lewis – *The Culture of Cities*, Secker & Warburg, 1938.
Mumford, Lewis – *The City in History*, Penguin, 1961.
Norwich, J Julius, ed. – *The Great Cities in History*, Thames & Hudson, 2009.
Olsen, Donald – *The City as a Work of Art*, Yale University Press, 1985.

Ovenden, Mark, ed. – *Metro Maps of the World*, Capital Transport Publishing, 2003.
Ovenden, Mark – *Underground Cities*, Frances Lincoln, 2020.
Park, Robert & Burgess, Ernest – *The City*, University of Chicago Press, 1925.
Portes, Alejandro & Armony, Ariel – *Emerging Global Cities*, Columbia University Press, 2023.
Reader, John – *Cities*, Vintage, 2004.
Reps, John – *The Making of Urban America*, Princeton University Press, 1965.
Richards, Brian – *Transport in Cities*, Architecture Design & Technology Press, 1990.
Rose, Jonathan – *The Well-Tempered City*, Harper Collins, 2016.
Sassen, Saskia – *Cities in a World Economy*, Sage Publisher, 2019.
Schellinger P & Salkin R – *International Dictionary of Historic Places*, Vol 5, Routledge, 1996.
Schenk, Leonhard – *Designing Cities*, Birkhauser, 2013.
Seabrook, Jeremy – *Cities*, Pluto Press, 2007.
Sitte, Camillo – *The Art of Designing Cities*, Reinhold, 1945.
Sjoberg, Gideon – *The Pre-Industrial City*, The Free Press, 1960.
Smith, Monica – *Cities – The First 6,000 Years*, Simon & Schuster, 2019.
Solesburg, William – *World Cities, City Worlds*, Matador, 2013.
Sudjic, Deyan – *The Language of Cities*, Allen Lane, 2016.
Thompson, J Michael – *Great Cities and Their Traffic*, Victor Gallancz, 1977.
Thunberg, Greta, ed. – *The Climate Book*, Allen Lane, 2022.
Tunnard, Christopher – *The City of Man*, Charles Scribner's Sons, 1953.
Warner, Sam B – *The Urban Wilderness*, University of California Press, 1995.
Weber, Max – *The City*, Free Press, 1968.
Whyte, William H. – *City*, Doubleday, 1988.
Wilson, Ben – *Metropolis*, Jonathan Cape, 2020.

Index

For Product Safety Concerns and Information please contact our EU representative GPSR@taylorandfrancis.com Taylor & Francis Verlag GmbH, Kaufingerstraße 24, 80331 München, Germany

* 9 7 8 1 0 3 2 4 0 4 9 6 7 *

T - #0126 - 110725 - C0 - 280/210/19 - PB - 9781032404967 - Gloss Lamination